Microplastic Pollution

This reference book reviews various aspects of microplastics, from their sources and manifestation in terrestrial, aquatic, and air environments to their fate in wastewater treatment systems. It also covers sampling, analysis, and detection methods for microplastics, along with advanced instrumentation for quantification. Further, the book presents health risk analysis and the toxicity of microplastic contamination, including their ecotoxicological impact on the environment and health risks associated with their accumulation in the tropical food chain and food web. The chapters also present studies exploring the health risks associated with microplastic additives and their interactions with other pollutants. Towards the end, the book focuses on plastics and microplastics management, exploring advanced technologies for bioplastics production, the biodegradation of plastics and bioplastics, and the role of nanotechnology in plastic management. This book serves as an important source for researchers, policymakers, and environmentalists concerned about the impact of microplastics on ecosystems and human health.

Sustainable Industrial and Environmental Bioprocesses

This book series aims to provide a comprehensive collection of books focusing on bioprocesses in industrial and environmental biotechnology. The multidisciplinary content encompasses the chemical and biochemical engineering, industrial microbiology, and energy biosciences, all with a central theme of sustainable development and circular economy principles. The books align with the Sustainable Development Goals (SDGs) and offer state-of-the-art information and in-depth knowledge on the subject matter. The book in these series also emphasizes on the application of emerging tools, such as machine learning and artificial intelligence, for the advancement of bioprocesses. While primarily targeting academicians and researchers, the series is valuable for policy planners and industry professionals, with carefully tailored contents to cater to their specific needs and interests.

Series Editor: Dr Ashok Pandey

Professor Ashok Pandey is currently Executive Director, Centre for Energy and Environmental Sustainability-India, Lucknow. Formerly, he was Distinguished Scientist at the Centre for Innovation and Translational Research, CSIR-Indian Institute of Toxicology Research, Lucknow, India; Eminent Scientist at the Center of Innovative and Applied Bioprocessing, Mohali; and Chief Scientist and Head, Centre for Biofuels, CSIR-National Institute for Interdisciplinary Science and Technology, Trivandrum. He is HSBS National Innovation Chair (Biotechnology) and is/has been a visiting/distinguished professor in many countries. His major research and technological development interests are industrial and environmental biotechnology and energy biosciences, focusing on biomass to biofuels and chemicals, waste to wealth and energy, etc. Professor Pandey is the recipient of many national and international awards and honors, which include Fellow of five international and four national academies such as the World Academy of Sciences; Indian National Science Academy, National Academy of Sciences, etc; Distinguished Fellow, the Biotech Research Society, India; Academician of European Academy of Sciences and Arts, etc.

Decentralized Sanitation and Water Treatment: Treatment in Cold Environments and Techno-Economic Aspects
R.D. Tyagi, Ashok Pandey, Patrick Drogui, Bhoomika Yadav, Sridhar Pilli and Jonathan W.C. Wong

Biodegradation of Toxic and Hazardous Chemicals: Detection and Mineralization
Kashyap K. Dubey, Kamal. K. Pant, Ashok Pandey and Maria Angeles Sanromán

Biodegradation of Toxic and Hazardous Chemicals: Remediation and Resource Recovery
Kashyap K. Dubey, Kamal K. Pant, Ashok Pandey and Maria Angeles Sanromán

Biomass Hydrolysing Enzymes: Basics, Advancements, and Applications
Reeta Rani Singhania, Anil Kumar Patel, Héctor A. Ruiz and Ashok Pandey

Waste Management in Climate Change and Sustainability- Lignocellulosic Waste
Sunita Varjani, Izharul Haq, Ashok Pandey, Vijai Kumar Gupta, and Xuan-Thanh Bui

Ashwagandha for Quality of Life: Scientific Evidence
Sunil C Kaul and Renu Wadhwa

Microfluidics in Food Processing: Technologies and Applications
Ayon Tarafdar, Ranjna Sirohi, Barjinder Pal Kaur, Ashok Pandey, and Claude-Gilles Dussap

Sustainable Technologies in Food Waste Management
Ranjna Sirohi, Ayon Tarafdar, Luciana Porto de Souza Vandenberghe, Mohammad J. Taherzadeh, and Ashok Pandey

Microplastic Pollution: Occurrence, Health Risk and Challenges
Navish Kataria, Vinod Kumar Garg, Changseok Han and Eldon R. Rene

For more information on this series, please visit www.routledge.com/Sustainable-Industrial-and- Environmental-Bioprocesses/book-series/SIEB

Microplastic Pollution
Occurrence, Health Risk and Challenges

Edited by
Navish Kataria, Vinod Kumar Garg,
Changseok Han and Eldon R. Rene

CRC Press is an imprint of the
Taylor & Francis Group, an **informa** business

Designed cover image: Shutterstock

First edition published 2025
by CRC Press
2385 NW Executive Center Drive, Suite 320, Boca Raton FL 33431

and by CRC Press
4 Park Square, Milton Park, Abingdon, Oxon, OX14 4RN

CRC Press is an imprint of Taylor & Francis Group, LLC

© 2025 selection and editorial matter Navish Kataria, Vinod Kumar Garg, Changseok Han and Eldon R. Rene, individual chapters, the contributors

Reasonable efforts have been made to publish reliable data and information, but the author and publisher cannot assume responsibility for the validity of all materials or the consequences of their use. The authors and publishers have attempted to trace the copyright holders of all material reproduced in this publication and apologize to copyright holders if permission to publish in this form has not been obtained. If any copyright material has not been acknowledged please write and let us know so we may rectify in any future reprint.

Except as permitted under U.S. Copyright Law, no part of this book may be reprinted, reproduced, transmitted, or utilized in any form by any electronic, mechanical, or other means, now known or hereafter invented, including photocopying, microfilming, and recording, or in any information storage or retrieval system, without written permission from the publishers.

For permission to photocopy or use material electronically from this work, access www.copyright.com or contact the Copyright Clearance Center, Inc. (CCC), 222 Rosewood Drive, Danvers, MA 01923, 978-750-8400. For works that are not available on CCC please contact mpkbookspermissions@tandf.co.uk

Trademark notice: Product or corporate names may be trademarks or registered trademarks and are used only for identification and explanation without intent to infringe.

ISBN: 9781032706535 (hbk)
ISBN: 9781032706566 (pbk)
ISBN: 9781032706573 (ebk)

DOI: 10.1201/9781032706573

Typeset in Times
by Newgen Publishing UK

Contents

List of Figures ix
List of Tables xi
Preface xiii
About the Editors xv
Contributors xvii

Chapter 1   Microplastics in the Environment: An Introduction 1

Sangita Yadav and Navish Kataria

Chapter 2   Sources, Effects, and Fate of Microplastics in Aquatic Environment 12

Biswanath Mahanty, Mohd. Zafar, Shishir Kumar Behera, Tejas R. Atri, and Eldon R. Rene

Chapter 3   Microplastic in the Air: Sources, Distribution, and Environmental Implications 28

Somvir Bajar, Neha Yadav, and Kavita Yadav

Chapter 4   Fate and Control of Microplastics in Municipal Wastewater Treatment Systems 54

Linhua Fan, Li Gao, and Arash Mohseni

Chapter 5   Sources and Occurrence of Microplastics in the Terrestrial Environment 68

Sarva Mangala Praveena

Chapter 6   Analytical Techniques of Microplastic in an Aquatic Environment: Sampling, Extraction, and Identification 85

Abel Egbemhenghe, Chika J. Okorie, Toluwalase Ojeyemi, Hussein K. Okoro, Ebuka Chizitere Emenike, Bridget Dunoi Ayoku, Kingsley O. Iwuozor, and Adewale George Adeniyi

Chapter 7   Advanced Instrumentation for Quantification of Microplastics 105

Jatinder Singh Randhawa

Chapter 8   Microplastics in the Aquatic Environment: Analytical and Mitigation Methods 124

Md Abdullah Al Noman, S. Shanthakumar, Shishir Kumar Behera, S. Srinivasan, Eldon R. Rene, and Tejas R. Atri vi Contents

Chapter 9 Analytical and Detection Techniques for Microplastics 147

Anu Kumari, Meenu Yadav, and Rachna Bhateria

Chapter 10 Microplastic Contamination in Water, Sediment, and Biota in
Mangrove Forests 167

Rungpilin Jittalerk and Sandhya Babel

Chapter 11 Microplastics Pollution in the Aquatic Environments: Occurrences,
Ecological Impacts, and Remediation Technologies 189

*Sumarlin Shangdiar, Kassian T.T. Amesho, Timoteus Kadhila,
Chingakham Chinglenthoiba, Sioni Iikela, Nastassia Thandiwe Sithole,
and Mohd Nizam Lani*

Chapter 12 Health and Environmental Impact of Microplastics: A Closer View 206

*Niranjan Koirala, Arjun Sharma, Saru Gautam, Nabin Chaulagain,
Proestos Charalampos, and Jian Bo Xiao*

Chapter 13 Uptake, Accumulation, and Ecotoxicological Impacts of Microplastic on Plant
Production and Soil Ecosystem 228

*Sangita Yadav, Navish Kataria, Jun Wei Roy Chong, Pawan Kumar Rose,
Seema Joshi, Pau Loke Show, and Kuan Shiong Khoo*

Chapter 14 Ecotoxicological Impact of Microplastics in the Environment 245

*Ajay Valiyaveettil Salimkumar, Mary Carolin Kurisingal Cleetus,
Judith Osaretin Ehigie, Cyril Oziegbe Onogbosele, Dorcas Akua Essel,
Ransford Parry, Bindhi S. Kumar, M.P. Prabhakaran, and V.J. Rejish Kumar*

Chapter 15 Microbial Degradation of Plastic Polymers: Organism Diversity, Mechanism,
and Influencing Factors Perspective 266

*Pawan Kumar Rose, Nishita Narwal, Rakesh Kumar, Navish Kataria,
Sangita Yadav, and Kuan Shiong Khoo*

Chapter 16 Advanced Technologies for the Production of Bioplastics 286

*Kassian T.T. Amesho, E.I. Edoun, Sumarlin Shangdiar, Abner Kukeyinge Shopati,
Timoteus Kadhila, Nastassia Thandiwe Sithole, Chandra Mohan, Sioni Iikela,
and Manoj Chandra Garg*

Chapter 17 Role of Nanotechnology in Plastic and Microplastic Management 303

*Manisha Gulati, Khushboo Singhal, Malya, Piyush Kumar Gupta,
Deepansh Sharma, and Sunny Dholpuria*

Contents

Chapter 18 Life-cycle Assessment of Microplastics in the Environment 321

Kassian T.T. Amesho, Sumarlin Shangdiar, E.I. Edoun,
Abner Kukeyinge Shopati, Timoteus Kadhila, Sioni Iikela,
Nastassia Thandiwe Sithole, Chandra Mohan, and Manoj Chandra Garg

Index 333

Figures

1.1	Classification of microplastics based on source, size, and shape	2
1.2	Sources, transportation, uptake, and trophic transfer of MPs/NPs in the food web	3
1.3	MPs and NPs bioaccumulation and translocation in various trophic levels of the food chain	6
2.1	Sources, fate, and transport of MPs from different environmental compartments to the aquatic environment	17
3.1	Classification of microplastics based on origin	30
3.2	Sources and transport of microplastics	33
3.3	Sampling and analysis procedure for airborne microplastics	35
3.4	Sampling devices for airborne microplastics	36
3.5	Fate of airborne microplastics	41
3.6	Impact of airborne microplastics on humans	43
4.1	Schematic diagram of the sources of microplastics in WWTPs	55
4.2	A comparison of microplastics removal efficiency (mean values) for the two AS-based tertiary treatment systems	62
4.3	A comparison of the average microplastics removal efficiency for Plants B and C, and the additional plant investigated	63
5.1	Primary and secondary sources of microplastics in the terrestrial environment	70
5.2	Fragmentation process of larger plastic items in a soil environment	71
5.3	Fragmentation process of various plastics waste into various microplastic particle shapes	72
5.4	Microplastic particle shapes found in surface soil from selected agricultural farms	77
5.5	Plastics usage in selected farms which contributed to microplastic particles in the surface soil sample	78
5.6	Plastic usage practices in selected farms and their contribution to microplastics contamination in surface soil samples	79
6.1	A visual representation depicting the presence and fate of MPs in aquatic ecosystems	89
6.2	Tools utilized for collecting microplastics in seawater at various depths: (a) neuston net, (b) manta trawl, (c) catamaran, for sampling MPs on the surface, and (d) bongo nets for capturing MPs in mid-water	91
7.1	Numerous advanced analytical methods for microplastics	106
7.2	Physicochemical characterization methods for microplastic detection	107
7.3	Commonly used analytical methods for microplastics analysis on the basis of microplastic size	110
8.1	Methods of microplastics analysis in water, sediment, and biota samples	129
8.2	Removal, degradation, and mitigation methods for MPs	134
9.1	Different microplastic analysis techniques	148
10.1	Microplastics detected in the Bang Pu mangrove forest, Thailand	168
10.2	Plastics and microplastics trapped by pneumatophores in the Bang Pu mangrove forest, Thailand	171
10.3	Microplastic analysis in mangrove biota	176
10.4	Pathways for microplastic contamination in mangrove forests	179
11.1	Keyword network analysis on "Microplastics" related articles from 2018 to 2024 found in Web of Science (WoS) database	190
11.2	Sources of microplastics in aquatic environments	191
11.3	Schematic view of MPs' distribution and fate	193
11.4	Processes related to ecological impacts of microplastics in aquatic ecosystems	194

ix

11.5	Diverse potential long-term global impacts of accumulating and poorly reversible plastic pollution 197
12.1	Sources and types of microplastic 207
12.2	Types of microplastics according to their chemical composition 208
12.3	Weathering or aging process of microplastics...209
12.4	Formation of microplastic from different sources..210
12.5	Sources of microplastics 211
12.6	Percentages of sources of microplastic impacts..212
12.7	Impacts of different sources of microplastics on human health...215
12.8	Process of microplastics impacting the environment in different ways.................................218
13.1	Uptake, translocation and toxicity of microplastics to the soil–plant ecosystem 235
14.1	Distribution of MPs in different environments 250
14.2	Oxidative stress due to micro-/nanoplastics (MNPs). CAT, catalase; GPx, glutathione peroxidase; GR, glutathione reductase; GSH, glutathione; GSSH, glutathione disulfide; ROS, reactive oxygen species; MDA, malonic dialdehyde; SOD, superoxide dismutases...251
14.3	Sources and transfer of MPs to humans and the potential health risks..................................253
15.1	Organisms and mechanisms associated with plastic degradation 273
15.2	The enzymes associated with plastic degradation..276
15.3	Factor affecting plastic degradation 277
16.1	Keyword network analysis on "Bioplastic"-related articles from 2018 to 2024 found in Web of Science (WoS) database..287
16.2	Production process of bioplastics, which incorporates mechanical, biological, and thermochemical methods. These methods have the potential to replace petroleum-based plastics in high-demand applications. The numbers in brackets indicate the energy content associated with each step of the process..288
16.3	The non-renewable energy consumption and global warming potential for (a) production alone (cradle-to-gate) and (b) production throughout its lifespan (cradle-to-grave) for different types of bioplastics and petroleum-based plastics..........................296
16.4	(a) Essential elements for standardizing TEA/LCA in sustainable plastic management. (b) Comparative analysis of ^{3}E indicator parameters between petroleum-based and bioplastic materials..297
16.5	The study's scope, focusing on the comparison between bioplastics and petroleum-based plastics. This comparison spans from raw material acquisition through plastic production to end-of-life treatment. Additionally, the figure outlines the methodology employed for standardizing independent LCA and TEA studies, facilitating the comparison of ^{3}E (Environmental, Economic, and Energy) indicators................................298
17.1	Impact of microplastics on various organs of a developing fetus 304
17.2	Degradation of plastic compound from the enviornment by using different NPs..................305
17.3	Methods of plastic recycling: enzymatic polymerization and catalytic polymerization.....305
17.4	Enzymatic degradation of plastics 306
17.5	Estimation of various sources causing microplastic pollution in the world's ocean..........310
17.6	NT-based approaches and their mechanism employed for mitigating microplastic pollution 311
17.7	Microplastic degradation by magnetic coil 312
18.1	Keyword network analysis on "Microplastics, Life cycle assessment"-related articles from 2018 to 2024 found in Web of Science (WoS) database 322

Tables

1.1	Major strategies for the removal of MPs	7
2.1	Referred sources of microplastics in the environment—representative examples	14
2.2	Occurrence of microplastics in aquatic environments, and surface and groundwater	20
3.1	Different analytical techniques used for identifying microplastics	38
4.1	Additional information on the WWTPs under this study	60
4.2	Abundance of microplastics (25 μm–5 mm) in the sludge/biosolids samples of Plant A	64
4.3	Abundance of microplastics in different types of sludge reported in some previous studies	64
7.1	Advantages and disadvantages of the currently used microplastics identification techniques	116
8.1	Typical microplastic pretreatment methods	128
8.2	Advantages and disadvantages of different analytical methods for detecting MPs	130
8.3	Efficiency of the different methods used for MPs removal and degradation	135
9.1	Comparison of different microplastics samples	149
9.2	Comparison of various treatment methods used for microplastics analysis	151
9.3	Comparison of various physical characterization techniques for microplastic analysis	153
9.4	Comparison of different analytical techniques used for analysis of microplastics	154
9.5	Comparison of different microscopy techniques used for detection of microplastics	161
10.1	Microplastic contamination in mangrove sediment	169
10.2	Microplastic contamination in mangrove water	172
10.3	Microplastic contamination in mangrove biota	173
11.1	Sources of microplastics in aquatic environments	192
11.2	Distribution and fate of microplastics in different aquatic environments	192
11.3	Effects of microplastics on different organisms	196
11.4	Comparison of remediation technologies for microplastic pollution	200
13.1	Effects of microplastics on soil functions	231
13.2	Uptake and impacts of microplastics on terrestrial plants	236
14.1	Distribution of MPs in different environmental compartments from selected recent research studies (2022–2023)	249
14.2	Summary of the effects of different types of MPs on different species	254
14.3	MPs in food, drinking water, and air	255
16.1	Different types of biological wastes used for the production of biodegradable plastics, their source feedstock, and applications	291
16.2	Overview of the distinct characteristics and objectives of life cycle assessment (LCA) and techno-economic analysis (TEA) in the context of bioplastic production	298
17.1	Effect of microplastics on various organs and glands of the human body	304
17.2	Global aquatic ecosystems and microplastic contents in them.	310
18.1	Types of LCA and their applications	325
18.2	Challenges associated with LCA studies on microplastics	327
18.3	Overview of current policies and regulations on microplastics	329

Preface

In the current scenario, microplastics is considered as an emerging contaminant and ubiquitous in the environment. Recent research highlights the existence and distribution of microplastic traces in every component of the environment and their related adverse impact on wildlife and human health. It will draw the attention of the audience to microplastic concerns worldwide. This book also discusses the emerging microplastic detection techniques and advanced instrumental approach for the quantification of microplastics in air, soil, freshwater, marine, sediments, food chain, and even living organisms. Microplastics carries its additives and several other pollutants in different components of the environment. It also elaborates the distribution of microplastic-associated contaminants and their ecotoxicological impact on living organisms.

This book *Microplastic Pollution: Occurrence, Health Risk and Challenges* elaborates the research advances on microplastic pollution, including sources, transportation behavior, health risk analysis, advanced analytical techniques, modeling approaches, and associated pollutant interactions with wildlife and ecology. It provides an overview of a sustainable approach to plastic management by society and government policy in sustainable plastic waste management. This book also explores the future scope and challenges in next-generation bioplastic development. The chapters provide readers with rich reference information sources on important topics in this field. There are several illustrations so that scholars can grasp the flow of the content and also the transition to the next topic.

This book contains 18 chapters majorly focusing on microplastic pollution and its occurrence, health risk potential, and future challenges in microplastic management. Each chapter of the book contains valuable information for the worldwide audience including students, researchers, scientists and engineers, marine ecologists, environmental managers, policymakers, NGOs, and activists who are working in a holistic approach toward plastic waste management and environment conservation. Chapter 1 is the introduction chapter of microplastics in the environment. Chapters 2–5 are related to the sources and occurrences of microplastics in aquatic, atmospheric, wastewater, and terrestrial environments; they not only provide an overview of the properties and effects of microplastics but also outline the details on their sources, fate, and control of microplastics in wastewater treatment systems. Chapters 6 and 7 discuss the analytical methods and advanced instrumentation for quantification of microplastics, and Chapter 8 provides explicit information related to analytical techniques related to microplastics and their mitigation methods. Chapter 9 includes detection with analysis of microplastics. Chapters 10–14 elaborate on the health risk analysis and the toxicity of microplastic contamination. They also include the ecotoxicological impacts of microplastics, additives in microplastics, associated contaminants on the environment, and their transportation, accumulation, and health risks in the food chain and food web. Chapters 15–18 focus on future scope and challenges in microplastic management. Chapter 15 includes the various methods, strategies, and approaches used in the biodegradation of microplastics. Chapter 16 highlights the future approach in bioplastic production and mainly focuses on the advanced technologies used in the production of bioplastics to overcome microplastic pollution. Chapter 17 represents the role of nanotechnology in the management of plastic and microplastics. Chapter 18 elaborates on the life-cycle assessment of microplastics in the environment.

Finally, we are thankful to all the chapter contributors, learned reviewers, production team, and editorial team of Taylor & Francis Group/CRC Press, who supported us at every stage of conceptualization, proposal revision, initiation, operation, proofreading, and completion of the book, including the production process.

Navish Kataria, Vinod Kumar Garg, Changseok Han and Eldon R. Rene
Book editors

xiii

About the Editors

Navish Kataria is currently working as an Assistant Professor in the Department of Environmental Sciences, J.C. Bose University of Science and Technology, YMCA, Faridabad, India, since January 2020. He completed his Ph.D. in Environmental Science and Engineering in 2018. He was awarded *Dr. D.S Kothari Post-Doctorate Fellowship* in 2019 by UGC, New Delhi. He has recently received a UGC start-up Research Grant for research work. He has published >40 research articles in high-impact international journals and 11 book chapters. Dr. Kataria has published three books (two with CRC Press and one with Springer). He has also served as managing guest editor in the reputed journal *ESPR* (Springer journal) with an Impact Factor of 5.2 and published an editorial. He is serving as Review Editor of *Frontiers in Environmental Chemistry* (Section: Sorption Technologies) and reviews other reputed journals, including some by Elsevier, Springer, MPDI, and De Gruyter. His major research interests are environmental pollution management, water, and wastewater treatment technologies, green/sustainable technology pollution, and biowaste management.

Vinod Kumar Garg is presently working as a Professor at the Department of Environmental Science and Technology, Central University of Punjab, Bathinda, Punjab, India. Previously, he worked at Guru Jambheshwar University of Science and Technology and CCS Haryana Agricultural University, Hisar, India, in different capacities. He has published more than 200 research and review articles and two books. Prof. Garg was awarded the Thomson Reuters Research Excellence—India Citation Awards 2012. He is an active member of various scientific societies and organizations, including the Biotech Research Society of India, Indian Nuclear Society, etc. He was elected a Fellow of the Biotech Research Society of India in 2011.

Changseok Han is currently working as an Associate Professor at the Department of Environmental Engineering and Program in Environmental & Polymer Engineering in INHA University, Incheon (Korea). He has published more than 100 research and review papers with an h-index of 43 in the field of environmental engineering and nanotechnology. Recently, he has focused on the development of standard protocols for sampling and pretreatment of soil and sediment samples to determine microplastics in the environment. Also, Dr. Han has studied to develop reliable advanced oxidation technologies for decomposing micropollutants in the environment. He is currently an Editor-in-Chief of *Journal of Environmental Analysis, Health, and Toxicology* in Korea, an Editor of Separation and Purification Technology, a Subject Editor of Process Safety and Environmental Protection, an Associate Editor of *Environmental Analysis Health and Toxicology*, and editorial board member of different journals, including *Chemical Engineering Journal Advances*, *Frontiers in Catalysis*, *Applied Nano and Advanced Materials Science* and *Technology*.

Eldon R. Rene is currently working as an Assistant Professor (Senior) in resource recovery technology at the IHE Delft Institute for Water Education, The Netherlands. He obtained his University Teaching Qualification (UTQ) diploma from IHE Delft and a Ph.D. in Chemical Engineering from the Indian Institute of Technology Madras (India). The focus of Dr. Rene's multidisciplinary research interests, with his students and collaborators, is related to the design and testing of resilient bioreactor configurations for the treatment of waste gases and industrial wastewater, industrial resource management, and the application of artificial intelligence tools to model and predict the performance of bioreactors. A chemical engineer by profession with a specialization in industrial pollution control, Dr. Rene has conducted research to address a range of issues related to the environment and energy.

Contributors

Adewale George Adeniyi
Department of Chemical Engineering
University of Ilorin
Ilorin, Nigeria
Chemical Engineering Department
Landmark University
Omu-Aran, Nigeria

Kassian T.T. Amesho
Centre for Environmental Studies
The International University of Management
(IUM)
Main Campus, Dorado Park, Windhoek,
Namibia

Tejas R. Atri
Industrial Ecology Research Group
School of Chemical Engineering
Vellore Institute of Technology
Vellore, Tamil Nadu, India

Bridget Dunoi Ayoku
Department of Pure and Industrial Chemistry
University of Port Harcourt
Rivers State, Nigeria

Sandhya Babel
School of Bio-Chemical Engineering &
Technology
Sirindhorn International Institute of Technology
Thammasat University
Rangsit Center, Pathum Thani, Thailand

Somvir Bajar
Department of Environmental Sciences
J.C. Bose University of Science and
Technology
YMCA
Faridabad, Haryana, India

Shishir Kumar Behera
Industrial Ecology Research Group
School of Chemical Engineering
Vellore Institute of Technology
Vellore, Tamil Nadu, India

Rachna Bhateria
Department of Environmental Sciences
Maharshi Dayanand University
Rohtak, Haryana, India

Proestos Charalampos
Laboratory of Food Chemistry
Department of Chemistry
National and Kapodistrian University of Athens
Zografou
Athens, Greece

Nabin Chaulagain
Department of Biotechnology
SANN International College
Purbanchal University
Kathmandu, Bagmati Province, Nepal

Chingakham Chinglenthoiba
Centre for Research Impact and Outcome
(CRIO)
Chitkara University
Rajpura, Punjab, India

Jun Wei Roy Chong
Department of Chemical and Environmental
Engineering
Faculty of Science and Engineering
University of Nottingham Malaysia
Jalan Broga
Semenyih, Selangor Darul Ehsan, Malaysia

Mary Carolin Kurisingal Cleetus
Faculty of Science and Technology
Research Centre for Experimental Marine
Biology and Biotechnology (PIE-UPV/EHU)
Areatza Pasealekua
Plentzia-Biskaia
Basque Country, Spain
Faculty of Science and Technology
University of Liege
Liege, Belgium

Sunny Dholpuria
Department of Life Sciences
J.C. Bose University of Science and
   Technology
YMCA
Faridabad, Haryana, India

E.I. Edoun
Tshwane School for Business and Society
Faculty of Management of Sciences,
Tshwane University of Technology (TUT)
Pretoria, South Africa

Abel Egbemhenghe
Department of Chemistry and Biochemistry
College of Art and Science
Texas Tech University, Lubbock, Texas, USA
Department of Chemistry
Lagos State University
Ojo, Lagos, Nigeria

Judith Osaretin Ehigie
Faculty of Science and Technology
Research Centre for Experimental Marine
   Biology and Biotechnology (PIE-UPV/EHU)
Areatza Pasealekua
Plentzia-Biskaia
Basque Country, Spain
Helmholtz Centre for Environmental Research
   – UFZ
Permoserstraße
Leipzig, Germany

Ebuka Chizitere Emenike
Department of Pure and Industrial Chemistry
Nnamdi Azikiwe University
Awka, Nigeria
Environmental-Analytical & Material
   Research Group
Department of Industrial Chemistry
University of Ilorin
Ilorin, Nigeria

Dorcas Akua Essel
Faculty of Science and Technology
Research Centre for Experimental Marine
   Biology and Biotechnology (PIE-UPV/EHU)
Areatza Pasealekua
Plentzia-Biskaia
Basque Country, Spain
Department of Climate Change Impacts in
   Oceans and Coast
AZTI Foundation: Marine Research Center
Spain

Linhua Fan
School of Engineering
RMIT University
Melbourne, VIC, Australia

Li Gao
South East Water Corporation
Frankston, VIC, Australia

Manoj Chandra Garg
Amity Institute of Environmental Science
   (AIES)
Amity University Uttar Pradesh
Gautam Budh Nagar, Noida, India

Saru Gautam
Department of Biotechnology
SANN International College
Purbanchal University
Kathmandu, Bagmati Province, Nepal

Manisha Gulati
Department of Life Sciences
J.C. Bose University of Science and
   Technology
YMCA
Faridabad, Haryana, India

Piyush Kumar Gupta
Department of Life Sciences
School of Basic Sciences and Research
Sharda University
Greater Noida, Uttar Pradesh, India

Contributors

Sioni Iikela
School of Education
Department of Higher Education and Lifelong
    Learning
University of Namibia (UNAM)
Windhoek, Namibia

Kingsley O. Iwuozor
Department of Pure and Industrial Chemistry
Nnamdi Azikiwe University
P.M.B.
Awka, Nigeria
Environmental-Analytical & Material Research
    Group
Department of Industrial Chemistry
University of Ilorin
Ilorin, Nigeria

Rungpilin Jittalerk
School of Bio-Chemical Engineering &
    Technology
Sirindhorn International Institute of Technology
Thammasat University
Rangsit Center, Pathum Thani, Thailand

Seema Joshi
Department of Environmental Studies
Om Sterling Global University
Hisar, India

Timoteus Kadhila
School of Education
Department of Higher Education and Lifelong
    Learning
University of Namibia (UNAM)
Windhoek, Namibia

Navish Kataria
Department of Environmental Science and
    Engineering
J. C. Bose University of Science and
    Technology
YMCA
Faridabad, Haryana, India

Kuan Shiong Khoo
Department of Chemical Engineering and
    Materials Science
Yuan Ze University
Taoyuan, Taiwan

Niranjan Koirala
Nepal Academy of Science and Technology
Gandaki Province Specialized Research Center
Pokhara, Nepal

Bindhi S. Kumar
Research Department of Fisheries and
    Aquaculture
St. Albert's College (Autonomous)
Kochi, Kerala, India

Rakesh Kumar
Department of Biosystems Engineering
Auburn University
Auburn, AL, USA

V.J. Rejish Kumar
Faculty of Ocean Science and Technology
Kerala University of Fisheries and Ocean
    Studies
Panangad, Kochi, Kerala, India
Department of Aquaculture
Kerala University of Fisheries and Ocean
    Studies
Panangad, Kochi, Kerala, India

Anu Kumari
Department of Environmental Sciences
Maharshi Dayanand University
Rohtak, Haryana, India

Mohd Nizam Lani
Microplastic Research Interest Group (MRIG)
Universiti Malaysia Terengganu (UMT)
Kuala Nerus, Terengganu, Malaysia
Faculty of Fisheries and Food Science
Universiti Malaysia Terengganu
Kuala Nerus, Terengganu, Malaysia

Biswanath Mahanty
Division of Biotechnology
Karunya Institute of Technology & Sciences
Karunya Nagar, Coimbatore, Tamil Nadu, India

Malya
Department of Life Sciences
J.C. Bose University of Science and
    Technology
YMCA
Faridabad, Haryana, India

Chandra Mohan
Department of Chemistry
SBAS
K.R. Mangalam University
Gurugram, India

Arash Mohseni
School of Engineering
RMIT University
Melbourne, VIC, Australia

Nishita Narwal
University School of Environment Management
Guru Gobind Singh Indraprastha University
New Delhi, India

Md Abdullah Al Noman
Department of Water Supply
Sanitation and Environmental Engineering
IHE Delft Institute for Water Education
Delft, The Netherlands

Chika J. Okorie
Department of Chemistry and Biochemistry
College of Art and Science
Texas Tech University, Lubbock, Texas, USA
Department of Pure and Industrial Chemistry
Nnamdi Azikiwe University
Awka, Nigeria

Hussein K. Okoro
Environmental-Analytical & Material Research
   Group
Department of Industrial Chemistry
University of Ilorin
Ilorin, Nigeria

Toluwalase Ojeyemi
Department of Environmental Toxicology
Texas Tech University, Lubbock, Texas, USA
Department of Crop Protection and
   Environmental Biology
University of Ibadan
Ibadan, Nigeria

Cyril Oziegbe Onogbosele
Department of Zoology
Faculty of Life Sciences
Ambrose Alli University
Ekpoma, Edo State, Nigeria

Ransford Parry
Faculty of Science and Technology
Research Centre for Experimental Marine
   Biology and Biotechnology (PIE-UPV/EHU)
Areatza Pasealekua
Plentzia, Biskaia
Basque Country, Spain

M.P. Prabhakaran
Department of Aquatic Environment and
   Management
Kerala University of Fisheries and Ocean
   Studies
Panangad, Kochi, Kerala, India

Sarva Mangala Praveena
Department of Environmental and Occupational
   Health
Faculty of Medicine and Health Sciences
Universiti Putra Malaysia
UPM Serdang, Selangor Darul Ehsan, Malaysia

Jatinder Singh Randhawa
Department of Biotechnology
School of Applied and Life Sciences
Uttaranchal University
Dehradun, Uttarakhand, India

Eldon R. Rene
Department of Water Supply
Sanitation and Environmental Engineering
IHE Delft Institute for Water Education
Delft, The Netherlands

Pawan Kumar Rose
Department of Energy and Environmental
   Sciences
Chaudhary Devi Lal University
Sirsa, Haryana, India

Ajay Valiyaveettil Salimkumar
Faculty of Science and Technology
Research Centre for Experimental Marine
   Biology and Biotechnology (PIE-UPV/EHU)
Areatza Pasealekua
Plentzia-Biskaia
Basque Country, Spain

Sumarlin Shangdiar
Institute of Environmental Engineering
National Sun Yat-Sen University (NSYSU),
Kaohsiung, Taiwan

S. Shanthakumar
Centre for Clean Environment
Vellore Institute of Technology
Vellore, Tamil Nadu, India

Arjun Sharma
Department of Biotechnology
SANN International College
Purbanchal University
Kathmandu, Bagmati Province, Nepal

Deepansh Sharma
Department of Life Sciences
J.C. Bose University of Science and
    Technology
YMCA
Faridabad, Haryana, India

Abner Kukeyinge Shopati
Namibia Business School (NBS)
Faculty of Commerce
Management and Law
University of Namibia (UNAM)
Windhoek, Namibia

Pau Loke Show
Department of Chemical and Environmental
    Engineering
University of Nottingham Malaysia
Jalan Broga, Selangor, Malaysia

Khushboo Singhal
Department of Life Sciences
J.C. Bose University of Science and
    Technology
YMCA
Faridabad, Haryana, India

Nastassia Thandiwe Sithole
Department of Chemical Engineering
University of Johannesburg (UJ)
Doornfontein, South Africa

S. Srinivasan
Centre for Clean Environment
Vellore Institute of Technology
Vellore, Tamil Nadu, India

Jian Bo Xiao
Department of Analytical Chemistry and Food
    Science
Faculty of Food Science and Technology
University of Vigo
Vigo, Spain

Kavita Yadav
Department of Environmental Sciences
J.C. Bose University of Science and
    Technology
YMCA
Faridabad, Haryana, India

Meenu Yadav
Department of Environmental Sciences,
    Maharshi Dayanand University
Rohtak, Haryana, India

Neha Yadav
Department of Environmental Sciences
J.C. Bose University of Science and
    Technology
YMCA
Faridabad, Haryana, India

Sangita Yadav
Department of Environmental Science and
    Engineering
Guru Jambheshwer University of Science &
    Technology
Hisar, Haryana, India

Mohd. Zafar
Department of Applied Biotechnology
College of Applied Science and Pharmacy
University of Technology and Applied
    Sciences-Sur
Sultanate of Oman

1 Microplastics in the Environment

An Introduction

Sangita Yadav and Navish Kataria

1.1 INTRODUCTION

The outstanding properties of plastics, such as high durability, resistance to corrosives, low manufacturing cost, and lightweight, have been utilised in a wide range of applications for plastics. In 2022, more than 400.3 million tons of plastics were produced worldwide to meet the increasing demands in packaging, construction, automotive, electronic, and agricultural applications (PlasticsEurope, 2023). Recently, due to the outbreak of COVID-19, the consumption of single-use plastic-based medical wastes has significantly soared. Meanwhile, the extensive production and usage of plastics have caused their massive disposal in the environment. Nearly 80% of all plastics manufactured since 1950 have ended up in the aquatic or terrestrial environment (Kataria et al., 2024). It has been estimated that over 250 million tonnes of plastics would be accumulated in the ocean by 2025. These plastic wastes can be fragmented into debris smaller than 5 mm, termed as microplastics (MPs) (Hu et al., 2022; Yadav et al., 2023). Physical, chemical, and biological causes can decompose plastic litter in the environment, resulting in smaller plastic particles. Microplastics can be divided into primary and secondary microplastics based on their source. Based on size, microplastics can be classified as nano-microplastics, meso-microplastics, and macro-microplastics. On the basis of shape, microplastics can be classified as fragments, pellets, foams, beads, and fibres. Figure 1.1 presents a brief classification of microplastics.

Microplastic wastes in the environment have gradually increased in recent decades, in line with rising worldwide plastic production, and are expected to continue to rise in the future (Priya et al., 2022). Therefore, with the increasing input of microplastics into the environment, microplastic pollution has become a global issue and concern. Microplastics are widely found in different areas of the environment, including air, fresh water, wastewater, seawater, bottled water, tap water, and food (Wang et al., 2021). Microplastics can persist in the environment for a long time and threaten organisms' feeding, growth, reproduction, and physiology, and even may result in death. Due to their adverse environmental effects, discarded plastic wastes and minuscule particles are regarded as possible agents of global change (Chang et al., 2022). Therefore, the accumulation of microplastics in environmental components is gaining attention and becoming a significant concern among global researchers and scientists. The abundance of microplastics in lakes, rivers, estuaries, oceans, and beaches worldwide has been documented in highly populated areas and areas with intensive anthropogenic activities. Because of their small size, microplastics can enter the human food chain through the consumption of seafood as well as other terrestrial food items and subsequently can have an impact on human health (Rose et al., 2023a). Secondly,

DOI: 10.1201/9781032706573-1

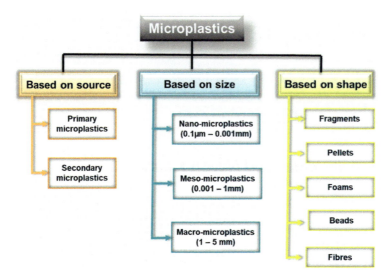

FIGURE 1.1 Classification of microplastics based on source, size, and shape.

plastic waste disposal in municipal waste disposal systems produces poisonous leachate, which can contaminate water and soil (Kataria et al., 2022). It is also responsible for climate change as plastic burning leads to the release of harmful greenhouse gases like carbon dioxide, methane, and nitrous oxide (Khyalia et al., 2022). The unprecedented use of plastic products and improper waste management techniques will continue to increase plastic waste. The irresponsible behaviour of people regarding the use of plastics, dumping plastic products, improper management systems, and associated harmful impacts have turned the planet into a "plastic planet" (Rose et al., 2023a).

Therefore, this chapter provides up-to-date and comprehensive information on the prevalence of microplastics in various environment matrices, including fresh water, atmosphere, soil, marine, and food chain. The primary emphasis is on various sources of microplastics and their routes to humans, along with associated possible health risks. In addition, detailed information on microplastic sampling and quantification techniques, as well as updates on various microplastic recovery or removal techniques, are covered in order to fully understand the impact of microplastics in soil and aquatic ecosystems. This chapter concludes with mitigation strategies, current challenges, and future perspectives of microplastic pollution.

1.2 MICROPLASTIC POLLUTION OCCURRENCE, SOURCES, AND ANALYSIS IN THE ENVIRONMENT

A large amount of plastic waste enters the environment through various pathways due to improper management and disposal practices and causes serious environmental pollution problems. Once in the environment, plastic wastes can slowly break down and generate numerous smaller plastic debris (Zhang et al., 2021). Microplastics can release into the environment via diverse pathways, mainly consisting of human (e.g., synthetic textiles, personal care products), transportation (e.g., erosion of synthetic rubber tyres), and industrial (e.g., plastic pellets) sources, in which most microplastics will accumulate into the ocean through river transport or direct discharge. Figure 1.2 shows the various sources and mechanisms (transfer, transportation, uptake) after discharge into the environment. After being discharged into the environment, these microplastics will undergo environmental processes, such as accumulation, degradation, and migration, under different environmental

Microplastics in the Environment – An Introduction

FIGURE 1.2 Sources, transportation, uptake, and trophic transfer of MPs/NPs in the food web.

conditions, eventually entering the human body through various exposure routes, including inhalation, ingestion, and dermal contact (Prata et al., 2020).

Microplastics are ubiquitous worldwide, and there are two types of sources: primary and secondary. The primary sources of microplastics in the environment refer to microplastics that are intentionally produced either for direct use or as a precursor to other products, and secondary microplastics are formed from the degradation of larger plastic materials, especially marine debris. Examples of primary microplastic sources include plastic pellets from pre-production, industrial abrasives, exfoliants, etc., while the release of secondary microplastics is highly dependent on the properties of the plastic, the extent of weathering, and the energetics of the local environment (Akanyange et al., 2022). When plastics are subjected to UV or solar radiation, they experience weathering degradation and deterioration, losing their tensile properties, discolouring and forming surface fissures, and eventually becoming frail and fragile. Any forces acting on them, such as waves, wind, or human action, can easily break them down into smaller pieces (Kristanti et al., 2023). As regards the origin of MP contamination in water environments, the primary sources are improperly disposed plastic wastes from land and, to a lesser extent, derivatives from marine activities, such as the fishing industry employing plastic equipment (Li et al., 2016). Plastic mulch films, compost and municipal solid waste, biosolids such as anaerobic digestate and sewage sludge, irrigation and flooding of wastewaters, atmospheric deposition, illegal dumping of waste, and plastic-coated fertilisers represent the principal sources of microplastics in soil (Hurley & Nizzetto, 2018). Guo et al. (2020) mentioned that the abundance of microplastics in soil in Australia is up to 4,000 mg/kg and up to 50,000 mg/kg in China. In the industrial area of Sydney, Australia, the concentration of microplastics reaches up to 67,500 mg/kg (Li et al., 2020). Gaston et al. (2020) reported that the abundance of indoor microplastics (15.9 items/m^3) is twice that of outdoor microplastics (7.2 items/m^3).

Due to the increased interest in this field, the analysis of MPs including detection, quantification and impact to human health can be done through several procedures. Large microplastics (1–5 mm) are identified with the naked eye, and an optical microscope can be used for smaller microplastics to obtain images for analysis, which provide the shape and number of the microplastic particles. More advanced techniques, such as SEM-EDS (scanning electron microscopy coupled with energy-dispersive X-ray spectroscopy), were recently reported in the literature to characterise the morphology of ultra-small plastic particles and facilitate the differentiation of microplastics from other plastic-like particles (Goldstein et al., 2017). Moreover, these spectroscopies are complementary vibrational techniques. Generally, signals with vigorous IR intensity have weak Raman intensity and vice versa. For instance, carbonyl groups, being polar functional groups, are well detected by IR, whereas aromatic bonds and double bonds are better identified using Raman spectroscopy (Savoca et al., 2020). FTIR, Py-GC-MS, and Raman spectroscopy are techniques used to analyse the chemical structure and determine the types of MPs. These methods enable accurate chemical identification of the samples, permitting the separation of M/NPs from other impurities and revealing the polymer base and even the presence of additives (Rose et al., 2023b). These are valuable techniques for identifying the polymeric composition of microplastics of different types (with sizes $\leq 2\ \mu m$) since they require small sample amounts and limited sample preparation. Hanun et al. (2023) employed SEM, FTIR, and BET analysis techniques to investigate new and aged PET MPs.

1.3 EFFECT OF MICROPLASTIC POLLUTION

The actual ecological and environmental risks of microplastics are still up for debate despite the fact that numerous ecotoxicological studies have already been conducted as a result of the mismatch between microplastic concentrations and characteristics used in laboratory experiments and those observed in the field (Zhang et al., 2021). Real-world microplastics have a greater variety and complexity in their forms, shapes, sizes, and compositions, which are thought to be connected to their toxicity (Lambert & Wagner, 2016). MPs' exceptional durability, longer stability, compact size, and light weight make them harmful. They can be consumed by birds, mammals, and aquatic species due to their micron size range, and they eventually end up in human food products, where they create problems. A variety of toxic contaminants, including polychlorinated biphenyls (PCBs), polycyclic aromatic hydrocarbons (PAH), polybrominated diphenyl ethers (PBDEs), dichlorodiphenyltrichloroethane (DDT), heavy metals, and endocrine-disrupting substances (EDCs), bind to MPs due to their hydrophobic nature, relatively high surface area/volume ratio, and extraordinary vector capacity (Kurniawan et al., 2021; Sharma et al., 2021). De Souza Machado et al. (2018) investigated the impacts of four forms of microplastics (such as PP and PET) on the bulk density, water-holding capacity, and water stability agglomeration of sandy soil and discovered that only PET had an impact, with the other types of microplastics having no impact. Microplastics may also influence soil porosity by affecting soil permeability, which in turn may affect soil water evaporation.

Microplastics are ingested by or adsorbed on marine species after entering the ocean and accumulate in their gastrointestinal organs and tissues, posing a hazard to their health due to their small density, light volume, and particle size, which are similar to the food of many marine organisms. Consuming low-trophic organisms with microplastic-filled bodies allows high-trophic organisms to accumulate microplastics. Microplastics impact organisms through a variety of processes after ingestion of MPs. When people eat fish that contain microplastics, the microplastics may enter their bodies and pose an unknown hazard to their health and also a threat to biodiversity (Nelms et al., 2018; Akhbarizadeh et al., 2019). Microplastics that invade living organisms endanger biological health in various ways, including oxidative stress, neurological damage, endocrine disruption, immunological damage, and others (Mao et al., 2022). The potential of microplastics to adsorb polycyclic aromatic hydrocarbons has been demonstrated by Avio et al. (2015), underscoring the high bioavailability of these substances following ingestion as well as the toxicological impacts of MPs on numerous molecular and cellular pathways. The effects of polyethylene (PE) microplastic particles on marine organisms were investigated by Oliveira et al. (2013). *Pomatoschistus microps* larvae were exposed to a solution containing 184 µg/L PE microplastics (1–5 µm), and the results showed that the particles inhibited AchE activity, preventing neurotransmitter transfer in fish, and ultimately impairing nerve function.

Recent research using multigenerational exposure has demonstrated that biological impacts may occur at low environmental-relevant concentrations. In fact, the water flea *Daphnia pulex* experienced oxidative damage after three generations of exposure to polystyrene microplastics (1 µg/L) (Liu et al., 2020), while *D. magna*'s survival was decreased after three generations of exposure to polystyrene (2000 particles/mL) (Schür et al., 2020). In high concentrations (1500–1800 mg/L), Yu et al. (2022) discovered that plasticisers in plastics (phthalate esters) can prevent wheat seeds from germinating and lead to programmed cell death in wheat seeds. Additionally, according to Giorgetti et al. (2020), 50-nm polystyrene nanoplastics may be ingested by onion root division cells and result in oxidative stress, cytotoxicity (such as aberrant mitosis), and gene toxicity. Therefore, understanding the types and particle sizes of microplastics is crucial to determining their plant toxicity.

Examining the indirect chemical toxicity brought on by the release of additives and monomers, such as PCBs and propylene, in addition to the particle toxicity of ingested microplastics, is necessary. Microplastics have chemical additives [such as BPA and bis(2-ethylhexyl)phthalate] since these chemicals are integrated into the production of plastic products. Microplastics may transfer and release these harmful compounds into digestive systems after ingestion since these additives are not chemically bonded to the plastic polymer matrix (Wang et al., 2021). Aged microplastics adsorb more PCBs than new microplastics, according to Alimi et al. (2018). Additionally, there are many different adsorption mechanisms involved in the process of organic pollutants being bound to microplastics, and the influence of each mechanism is determined by the characteristics of the plastic, the surrounding environment, and the organic pollutants. Additionally, weathered microplastics in the environment have a rough surface with a lot of specific surface area and a negative charge that can bind to heavy metals and organic contaminants, acting as a transporter of pollutants in the environment. Therefore, more research is needed to understand the combined effects of microplastics and heavy metals or pesticides in agricultural systems (Mao et al., 2022).

MPs can potentially harm human health in ways like secondary genotoxicity and inflammation, and their accumulation might trigger or strengthen an immunological response. The amount of microplastics that have been collected in the human body is obviously challenging to estimate accurately. Only a small number of studies have evaluated human exposure to MPs, taking into account total intake from various routes (Pironti et al., 2021). A probabilistic approach was recently utilised in a study to assess the lifetime accumulation of MPs in children and adults (Mohamed Nor et al., 2021). To assess the intake of microplastics, the model took into account the ingestion of food (fish, salt, molluscs, and crustaceans), liquids (tap water, bottled water, beer, and milk), and atmospheric inhalation. In addition, it measured the amount of MP accumulated through intestinal absorption and biliary excretion. The findings showed that microplastics contributed significantly less to total chemical intake than other contaminants with similar exposure pathways that were more dangerous, such as benzo(a) pyrene, di(2-ethylhexyl)phthalate, 3,3′,4,4′,5-pentachlorobiphenyl, and lead. Despite the aforementioned, human exposure to MPs is still not negligible, necessitating research into how they affect people. Microplastic has the potential for transport from lower trophic to high trophic levels via primary producers and consumers in the food chain, as depicted in Figure 1.3. Producers absorb and store them in their various organs, which then pass via various trophic levels and eventually reach humans.

FIGURE 1.3 MPs and NPs bioaccumulation and translocation in various trophic levels of the food chain.

1.4 STRATEGIES FOR MICROPLASTIC POLLUTION MANAGEMENT

Comprehensive and integrated approaches are required for managing MP contamination, with a focus on lowering plastic production and outlawing plastic microbeads that are manufactured deliberately.

1.4.1 Leading Role of Authorities in the Fight against MP Pollution

Legislative, economic, and educational methods are only a few of the integrated measures that authorities can use. With several directives on plastic management that aim to create a "zero-plastic" or "circular plastic economy," the European Commission has set an example for the rest of the world (Ray et al., 2022). The use of primary MPs and microbeads in personal care products is generally prohibited (Rochman et al., 2015). Before the COVID-19 pandemic, numerous nations planned to outlaw single-use plastics (Chen et al., 2021). After the pandemic, world leaders took several steps to address the plastic catastrophe (Vaughan, 2020). Along with additional market mechanisms that act as a "stick and carrot" for other stakeholders, stiffer penalties in terms of both economic and administrative perspectives for violations of plastic waste management legislation. Cleanup efforts and educational initiatives promote scientific research, such as gathering plastic debris from the marine environment and identifying environmentally appropriate substitute materials, as well as serving as tools to increase community awareness (Prata et al., 2019).

1.4.2 Remediation of Microplastic Pollution

In addition to management and proper disposal procedures, advanced remedies must be explored to remove MPs from the environment. Over the last few decades, plastic degradation has been thoroughly studied. However, there are other major strategies for the removal of MPs in the environment: (a) microbial degradation, (b) advanced oxidation processes (AOPs), (c) thermal degradation, (d) adsorption, and (e) membrane treatment. Typically, thermal degradation and AOPs accomplish rapid decomposition and highlight the potential to produce carbon and hydrogen as fuel (Ray et al., 2022). A few techniques to remove MPs from the environment are provided in Table 1.1.

Microplastics in the Environment – An Introduction

TABLE 1.1
Major Strategies for the Removal of MPs

Methods	Description	Advantages	Disadvantages	References
Electrocoagulation	Metal electrodes and electricity are used throughout the procedure, and coagulants are used to break up the colloidal by balancing the charges on the surfaces of the suspended MPs	Simple to operate Cost-effective High efficacy	Electrodes need to be repaired or replaced frequently High electricity use Mechanical damage	Kristanti et al. (2023)
Membrane filtration	MPs are trapped when contaminated water is allowed to pass through the film	Simple to operate Cost-effective Minimal chemical use	Membrane fouling The membrane needs to be serviced or replaced frequently	Shen et al., (2020)
Biological degradation	MPs are degraded into other chemical compounds (CH_4, H_2S, CO_2, H_2O)	Simple to operate environmentally beneficial,	Long time consumption Influenced by environmental factors	Hu et al. (2021)
Magnetic extraction	MPs are separated using magnetic particles, acid, and magnetism	Cost-effective High efficacy	Secondary MP generated that causes pollution	Padervand et al. (2020)
Photocatalytic degradation	Decompose MPs into water and carbon dioxide	Simple to be manufactured economical Does not release hazardous substances	Lower removal efficiency More retention duration	Kristanti et al. (2023)

Nanotechnology-based strategies have been demonstrated to play a crucial role in water and wastewater treatment by combating various challenging MNP pollutants with high efficiency. Adsorption, photocatalysis, and membrane filtration are the main processes in which nanomaterials can help remediate MNP pollution. This is due to the properties of nanomaterials that include a high surface-to-volume ratio, tuneable surface charges, variable functionalities, and lower required dosage (Ouda et al., 2023). One of the most critical advantages of nanomaterials, compared with conventional water technologies, is their ability to integrate unique properties, resulting in multifunctional systems such as nanocomposite membranes that enable particle retention and elimination of contaminants. Further, these multifunctional nanomaterials enable higher process efficiency due to their unique characteristics, such as a high reaction rate (Iravani, 2021).

1.5 BIOPLASTICS: SOLUTION TO PLASTIC POLLUTION

Synthetic polymer disposal methods have not been developed in a way that is environmentally friendly. As a result, their wastes frequently act as ongoing sources of contamination and the discharge of hazardous compounds. Bioplastics refer to innovative bio-based plastic polymers such as PLA (polylactic acid), PHA (polyhydroxyalkanoates), PHB (polyhydroxybutyrate), and starch blends, as well as microbial polymers such as polynucleotides, polypeptides, and polysaccharides (Ibrahim et al., 2021). Biopolymers created from natural sources (plant, animal, or microbial) are thus acknowledged as an alluring and viable replacement for synthetic polymers in a number of ways. Biopolymers are actually any natural polymers created using the cells of living organisms.

They are made of monomeric units that bond together covalently to form larger compounds. The type of biopolymer differs between polynucleotides, polysaccharides, and polypeptides, depending on the monomeric unit used to form the compound. Most biopolymers spontaneously fold into specific structures that decide their biological functions and depend on the primary structures (Chandran et al., 2023). Bioplastics are potential candidates for the sustainable development of ecofriendly products due to their degradability in the natural environment and similarities in physiochemical and mechanical properties with those of synthetic plastics. Biopolymers are not only a guarantee for a greener and pollution-free biosphere but also a safer and healthier environment free from toxic synthetic plastics. To sum up, it can be said for sure that biopolymers are the basis for a sustainable environment (Muneer et al., 2021).

1.6 ORGANISATION OF THIS BOOK

This book aims to summarise the current state of knowledge and the most recent research advancements in emerging contaminants (microplastic) in the environment. The book has been divided into 18 chapters. Chapter 1 is the introductory chapter on microplastics in the environment. After the introductory chapter, the sources and occurrences of microplastics in aquatic, atmospheric, wastewater treatment, and terrestrial environments are presented in Chapters 2, 3, 4, and 5, which not only provide an overview of the properties of different types of microplastics but also outline the details on their sources and also the fate and control of microplastics in municipal wastewater treatment systems. Analytical methods and advanced instrumentation for quantifying MPs are presented in Chapters 6 and 7, respectively, and Chapter 8 provides explicit information related to analytical techniques related to microplastics and their mitigation methods. Chapter 9 includes detection and analysis of microplastics. Chapters 10–14 include health risk analysis and the toxicity of microplastic contamination. These chapters include the ecotoxicological impacts of MPs, additives in MPs, associated contaminants on the environment, and their transportation, accumulation, and health risks in the food chain and web. Chapter 15 includes the biodegradation of MPs. Chapter 16 mainly focuses on the recent technologies used for the production of bioplastics as an emerging solution for the problem of MP pollution. Chapter 17 represents the role of nanotechnology in the management of plastic and MPs. The life-cycle assessment of microplastics in the environment is presented in Chapter 18.

1.7 CONCLUSION

Plastic pollution remains a planetary issue, with plastic waste found in almost every part of the environment, ranging from remote areas of the globe to indoor environments where people spend much of their time. This phenomenon may not see a stop any time soon as the demand and production of plastics, coupled with economic growth, continue to increase worldwide. Plastics are undeniably advantageous for society, but they have turned into an environmental menace, as evidenced by areas with abnormally high concentrations of plastics, especially for MPs. Microplastic concentrations may also continue to surge as fragmentation of plastic debris remains inevitable. MPs can travel a long distance from their point of origin by wind or water, causing the environmental impact to be more significant than expected. Environmental MPs have a negative impact on species, causing the extinction of certain species and impeding plantation growth, as well as contributing to climate change. Humans are mainly exposed to MPs via ingesting or drinking, which accumulates in their bodies and constitutes a human risk due to their fractured form, tiny size, and a wide variety of probable origins, critical issues derived from the concept of pinpointing the specific source of the MPs. Therefore, monitoring processes serve a significant role in mitigating and controlling MP pollution as the fate of MPs can be fully identified. Effective strategies or treatments must be introduced to reduce the number of MPs being discharged to preserve natural systems and sustain quality standards for domestic purposes. As a result, the usage and release of MPs can be severely constrained by enacting comprehensive legislation to regulate plastic use or MP release as part of a worldwide strategy, even before the results of long-term assessments are available.

REFERENCES

Akanyange, S. N., Zhang, Y., Zhao, X., Adom-Asamoah, G., Ature, A. R. A., Anning, C., … & Crittenden, J. C. (2022). A holistic assessment of microplastic ubiquitousness: Pathway for source identification in the environment. *Sustainable Production and Consumption, 33*, 113–145. https://doi.org/10.1016/j.spc.2022.06.020

Akhbarizadeh, R., Moore, F., & Keshavarzi, B. (2019). Investigating microplastics bioaccumulation and biomagnification in seafood from the Persian Gulf: A threat to human health?. *Food Additives & Contaminants: Part A, 36*(11), 1696–1708. https://doi.org/10.1080/19440049.2019.1649473

Alimi, O. S., Farner Budarz, J., Hernandez, L. M., & Tufenkji, N. (2018). Microplastics and nanoplastics in aquatic environments: Aggregation, deposition, and enhanced contaminant transport. *Environmental Science & Technology, 52*(4), 1704–1724. https://doi.org/10.1021/acs.est.7b05559

Avio, C. G., Gorbi, S., Milan, M., Benedetti, M., Fattorini, D., d'Errico, G., … & Regoli, F. (2015). Pollutants bioavailability and toxicological risk from microplastics to marine mussels. *Environmental Pollution, 198*, 211–222. https://doi.org/10.1016/j.envpol.2014.12.021

Chandran, R. R., Thomson, B. I., Natishah, A. J., Mary, J., & Nachiyar, V. (2023). Nanotechnology in plastic degradation. *Biosciences Biotechnology Research Asia, 20*(1), 53–68. http://dx.doi.org/10.13005/bbra/3068

Chang, M., Zhang, C., Li, M., Dong, J., Li, C., Liu, J., … & Stoks, R. (2022). Warming, temperature fluctuations and thermal evolution change the effects of microplastics at an environmentally relevant concentration. *Environmental Pollution, 292*, 118363. https://doi.org/10.1016/j.envpol.2021.118363

Chen, Y., Awasthi, A. K., Wei, F., Tan, Q., & Li, J. (2021). Single-use plastics: Production, usage, disposal, and adverse impacts. *Science of the Total Environment, 752*, 141772. https://doi.org/10.1016/j.scitotenv.2020.141772

de Souza Machado, A. A., Lau, C. W., Till, J., Kloas, W., Lehmann, A., Becker, R., & Rillig, M. C. (2018). Impacts of microplastics on the soil biophysical environment. *Environmental Science & Technology, 52*(17), 9656–9665. https://doi.org/10.1021/acs.est.8b02212

Gaston, E., Woo, M., Steele, C., Sukumaran, S., & Anderson, S. (2020). Microplastics differ between indoor and outdoor air masses: Insights from multiple microscopy methodologies. *Applied Spectroscopy, 74*(9), 1079–1098. https://doi.org/10.1177/0003702820920652

Giorgetti, L., Spanò, C., Muccifora, S., Bottega, S., Barbieri, F., Bellani, L., & Castiglione, M. R. (2020). Exploring the interaction between polystyrene nanoplastics and Allium cepa during germination: Internalization in root cells, induction of toxicity and oxidative stress. *Plant Physiology and Biochemistry, 149*, 170–177. https://doi.org/10.1016/j.plaphy.2020.02.014

Goldstein, J. I., Newbury, D. E., Michael, J. R., Ritchie, N. W., Scott, J. H. J., & Joy, D. C. (2017). *Scanning electron microscopy and X-ray microanalysis*. Springer.

Guo, J. J., Huang, X. P., Xiang, L., Wang, Y. Z., Li, Y. W., Li, H., … & Wong, M. H. (2020). Source, migration and toxicology of microplastics in soil. *Environment International, 137*, 105263. https://doi.org/10.1016/j.envint.2019.105263

Hanun, J. N., Hassan, F., Theresia, L., Chao, H. R., Bu, H. M., Rajendran, S., … & Jiang, J. J. (2023). Weathering effect triggers the sorption enhancement of microplastics against oxybenzone. *Environmental Technology & Innovation, 30*, 103112. https://doi.org/10.1016/j.eti.2023.103112

Hu, K., Tian, W., Yang, Y., Nie, G., Zhou, P., Wang, Y., … & Wang, S. (2021). Microplastics remediation in aqueous systems: Strategies and technologies. *Water Research, 198*, 117144. https://doi.org/10.1016/j.watres.2021.117144

Hu, K., Yang, Y., Zuo, J., Tian, W., Wang, Y., Duan, X., & Wang, S. (2022). Emerging microplastics in the environment: Properties, distributions, and impacts. *Chemosphere, 297*, 134118. https://doi.org/10.1016/j.chemosphere.2022.134118

Hurley, R. R., & Nizzetto, L. (2018). Fate and occurrence of micro (nano) plastics in soils: Knowledge gaps and possible risks. *Current Opinion in Environmental Science & Health, 1*, 6–11. https://doi.org/10.1016/j.coesh.2017.10.006

Ibrahim, N. I., Shahar, F. S., Sultan, M. T. H., Shah, A. U. M., Safri, S. N. A., & Mat Yazik, M. H. (2021). Overview of bioplastic introduction and its applications in product packaging. *Coatings*, *11*(11), 1423. https://doi.org/10.3390/coatings11111423

Iravani, S. (2021). Nanomaterials and nanotechnology for water treatment: recent advances. *Inorganic and Nano-Metal Chemistry*, *51*(12), 1615–1645. https://doi.org/10.1080/24701556.2020.1852253

Kataria, N., Bhushan, D., Gupta, R., Rajendran, S., Teo, M. Y. M., & Khoo, K. S. (2022). Current progress in treatment technologies for plastic waste (bisphenol A) in aquatic environment: Occurrence, toxicity and remediation mechanisms. *Environmental Pollution*, *315*, 120319. https://doi.org/10.1016/j.envpol.2022.120319

Kataria, N., Yadav, S., Garg, V. K., Rene, E. R., Jiang, J. J., Rose, P. K., … & Khoo, K. S. (2024). Occurrence, transport, and toxicity of microplastics in tropical food chains: perspectives view and way forward. *Environmental Geochemistry and Health*, *46*(3), 1–36. https://doi.org/10.1007/s10653-024-01862-2

Khyalia, P., Gahlawat, A., Jugiani, H., Kaur, M., Laura, J. S., & Nandal, M. (2022). Review on the use of microalgae biomass for bioplastics synthesis: A sustainable and green approach to control plastic pollution. *Pollution*, *8*(3), 844–859. 10.22059/POLL.2022.334756.1273

Kristanti, R. A., Hadibarata, T., Wulandari, N. F., Sibero, M. T., Darmayati, Y., & Hatmanti, A. (2023). Overview of microplastics in the environment: Type, source, potential effects and removal strategies. *Bioprocess and Biosystems Engineering*, *46*(3), 429–441. https://doi.org/10.1007/s00449-022-02784-y

Kurniawan, S. B., Said, N. S. M., Imron, M. F., & Abdullah, S. R. S. (2021). Microplastic pollution in the environment: Insights into emerging sources and potential threats. *Environmental Technology & Innovation*, *23*, 101790. https://doi.org/10.1016/j.eti.2021.101790

Lambert, S., & Wagner, M. (2016). Characterisation of nanoplastics during the degradation of polystyrene. *Chemosphere*, *145*, 265–268. https://doi.org/10.1016/j.chemosphere.2015.11.078

Li, J., Song, Y., & Cai, Y. (2020). Focus topics on microplastics in soil: Analytical methods, occurrence, transport, and ecological risks. *Environmental Pollution*, *257*, 113570. https://doi.org/10.1016/j.envpol.2019.113570

Li, W. C., Tse, H. F., & Fok, L. (2016). Plastic waste in the marine environment: A review of sources, occurrence and effects. *Science of the total environment*, *566*, 333–349. https://doi.org/10.1016/j.scitotenv.2016.05.084

Liu, Z., Cai, M., Wu, D., Yu, P., Jiao, Y., Jiang, Q., & Zhao, Y. (2020). Effects of nanoplastics at predicted environmental concentration on Daphnia pulex after exposure through multiple generations. *Environmental Pollution*, *256*, 113506. https://doi.org/10.1016/j.envpol.2019.113506

Mao, X., Xu, Y., Cheng, Z., Yang, Y., Guan, Z., Jiang, L., & Tang, K. (2022). The impact of microplastic pollution on ecological environment: A review. *Frontiers in Bioscience-Landmark*, *27*(2), 46. https://doi.org/10.31083/j.fbl2702046

Mohamed Nor, N. H., Kooi, M., Diepens, N. J., & Koelmans, A. A. (2021). Lifetime accumulation of microplastic in children and adults. *Environmental Science & Technology*, *55*(8), 5084–5096. https://doi.org/10.1021/acs.est.0c07384

Muneer, F., Nadeem, H., Arif, A., & Zaheer, W. (2021). Bioplastics from biopolymers: An eco-friendly and sustainable solution of plastic pollution. *Polymer Science, Series C*, *63*, 47–63. https://doi.org/10.1134/S1811238221010057

Nelms, S. E., Galloway, T. S., Godley, B. J., Jarvis, D. S., & Lindeque, P. K. (2018). Investigating microplastic trophic transfer in marine top predators. *Environmental Pollution*, *238*, 999–1007. https://doi.org/10.1016/j.envpol.2018.02.016

Oliveira, M., Ribeiro, A., Hylland, K., & Guilhermino, L. (2013). Single and combined effects of microplastics and pyrene on juveniles (0+ group) of the common goby Pomatoschistus microps (Teleostei, Gobiidae). *Ecological Indicators*, *34*, 641–647. https://doi.org/10.1016/j.ecolind.2013.06.019

Ouda, M., Banat, F., Hasan, S. W., & Karanikolos, G. N. (2023). Recent advances on nanotechnology-driven strategies for remediation of microplastics and nanoplastics from aqueous environments. *Journal of Water Process Engineering*, *52*, 103543. https://doi.org/10.1016/j.jwpe.2023.103543

Padervand, M., Lichtfouse, E., Robert, D., & Wang, C. (2020). Removal of microplastics from the environment. A review. *Environmental Chemistry Letters*, *18*, 807–828. https://doi.org/10.1007/s10311-020-00983-1

Pironti, C., Ricciardi, M., Motta, O., Miele, Y., Proto, A., & Montano, L. (2021). Microplastics in the environment: intake through the food web, human exposure and toxicological effects. *Toxics*, *9*(9), 224. https://doi.org/10.3390/toxics9090224

PlasticsEurope. (2023). Plastics – the Fast Facts 2023. Assessed on 12 sept.2023. Plastics – the fast Facts 2023 • Plastics Europe.

Prata, J. C., da Costa, J. P., Lopes, I., Duarte, A. C., & Rocha-Santos, T. (2020). Environmental exposure to microplastics: An overview on possible human health effects. *Science of the Total Environment*, *702*, 134455. https://doi.org/10.1016/j.scitotenv.2019.134455

Prata, J. C., Silva, A. L. P., Da Costa, J. P., Mouneyrac, C., Walker, T. R., Duarte, A. C., & Rocha-Santos, T. (2019). Solutions and integrated strategies for the control and mitigation of plastic and microplastic pollution. *International Journal of Environmental Research and Public Health*, *16*(13), 2411. https://doi.org/10.3390/ijerph16132411

Priya, A. K., Jalil, A. A., Dutta, K., Rajendran, S., Vasseghian, Y., Qin, J., & Soto-Moscoso, M. (2022). Microplastics in the environment: Recent developments in characteristic, occurrence, identification and ecological risk. *Chemosphere*, *298*, 134161. https://doi.org/10.1016/j.chemosphere.2022.134161

Ray, S. S., Lee, H. K., Huyen, D. T. T., Chen, S. S., & Kwon, Y. N. (2022). Microplastics waste in environment: A perspective on recycling issues from PPE kits and face masks during the COVID-19 pandemic. *Environmental Technology & Innovation*, *26*, 102290. https://doi.org/10.1016/j.eti.2022.102290

Rochman, C. M., Kross, S. M., Armstrong, J. B., Bogan, M. T., Darling, E. S., Green, S. J., … & Veríssimo, D. (2015). Scientific Evidence Supports a Ban on Microbeads. Environmental Science & Technology, *49*(18), 10759–10761. https://doi.org/10.1021/acs.est.5b05043

Rose, P. K., Jain, M., Kataria, N., Sahoo, P. K., Garg, V. K., & Yadav, A. (2023a). Microplastics in multimedia environment: A systematic review on its fate, transport, quantification, health risk, and remedial measures. *Groundwater for Sustainable Development*, 100889. https://doi.org/10.1016/j.gsd.2022.100889

Rose, P. K., Yadav, S., Kataria, N., & Khoo, K. S. (2023b). Microplastics and nanoplastics in the terrestrial food chain: Uptake, translocation, trophic transfer, ecotoxicology, and human health risk. *TrAC Trends in Analytical Chemistry*, 117249. https://doi.org/10.1016/j.trac.2023.117249

Savoca, S., Bottari, T., Fazio, E., Bonsignore, M., Mancuso, M., Luna, G. M., … & Spanò, N. (2020). Plastics occurrence in juveniles of Engraulis encrasicolus and Sardina pilchardus in the Southern Tyrrhenian Sea. *Science of the Total Environment*, *718*, 137457. https://doi.org/10.1016/j.scitotenv.2020.137457

Schür, C., Zipp, S., Thalau, T., & Wagner, M. (2020). Microplastics but not natural particles induce multigenerational effects in Daphnia magna. *Environmental Pollution*, *260*, 113904. https://doi.org/10.1016/j.envpol.2019.113904

Sharma, S., Basu, S., Shetti, N. P., Nadagouda, M. N., & Aminabhavi, T. M. (2021). Microplastics in the environment: Occurrence, perils, and eradication. *Chemical Engineering Journal*, *408*, 127317. https://doi.org/10.1016/j.cej.2020.127317

Shen, M., Song, B., Zhu, Y., Zeng, G., Zhang, Y., Yang, Y., … & Yi, H. (2020). Removal of microplastics via drinking water treatment: Current knowledge and future directions. *Chemosphere*, *251*, 126612. https://doi.org/10.1016/j.chemosphere.2020.126612

Vaughan, A. (2020, November 11). Social 'bubbles' unlikely to be allowed in UK soon. *New Scientist*. https://doi.org/10.1016/S0262-4079(20)31052-6

Wang, C., Zhao, J., & Xing, B. (2021). Environmental source, fate, and toxicity of microplastics. *Journal of Hazardous Materials*, *407*, 124357. https://doi.org/10.1016/j.jhazmat.2020.124357

Yadav, S., Kataria, N., Khyalia, P., Rose, P. K., Mukherjee, S., Sabherwal, H., … & Khoo, K. S. (2023). Recent analytical techniques, and potential eco-toxicological impacts of textile fibrous microplastics (FMPs) and its associated contaminates: A review. *Chemosphere*, 138495. https://doi.org/10.1016/j.chemosphere.2023.138495

Yu, H., Zhang, Y., Tan, W., & Zhang, Z. (2022). Microplastics as an emerging environmental pollutant in agricultural soils: effects on ecosystems and human health. *Frontiers in Environmental Science*, *10*, 217. https://doi.org/10.3389/fenvs.2022.855292

Zhang, K., Hamidian, A. H., Tubić, A., Zhang, Y., Fang, J. K., Wu, C., & Lam, P. K. (2021). Understanding plastic degradation and microplastic formation in the environment: A review. *Environmental Pollution*, *274*, 116554. https://doi.org/10.1016/j.envpol.2021.116554

2 Sources, Effects, and Fate of Microplastics in Aquatic Environment

Biswanath Mahanty, Mohd. Zafar, Shishir Kumar Behera, Tejas R. Atri, and Eldon R. Rene

2.1 INTRODUCTION

In comparison to materials such as metal or glass, plastics are significantly lighter, rendering them suitable for the production of products such as bottles, and containers, as well as for packaging; additionally, they are typically durable and cost-effective to produce (Narayanan, 2023). Only a small fraction (~9%) of the annual plastic production (i.e., 350 million metric tonnes) is recycled, and the rest ends up in landfills, incinerators, or the environment (Auta et al., 2017). Although the recycling percentage varies from country to country, most of the plastic waste generated ends up in the environment. Tiny plastic particles (< 5 mm) generated through degradation, fragmentation, and weathering of these wastes, referred to as microplastics (MPs), are even more problematic than their larger counterparts (Xu et al., 2020). MPs are becoming a growing concern in aquatic environments due to their far-reaching ecological and environmental implications (Du et al., 2021).

MPs originate from a variety of sources, both direct and indirect, and follow a complex trajectory that results in their entry into the ocean (Chaukura et al., 2021). The direct sources include plastic pollution, where large plastic items degrade into smaller fragments, microbeads added to personal care products, and nurdles used in plastic manufacturing (Auta et al., 2017). Indirect sources include MPs released from tire wear, synthetic textile shedding during washing, and the erosion of paints and coatings. The path to the ocean typically begins with land runoff carrying MPs from urban areas, entering storm water drains, and subsequently flowing into rivers and streams, eventually paving the way to the coastal regions. Coastal erosion is also a contributing factor to MPs pollution along the shoreline. Once in the coastal waters, ocean currents facilitate their dispersion throughout the marine ecosystem, where they can be ingested by marine life, causing bioaccumulation and potential ecological disruptions (Darabi et al., 2021). The abundance of MP in the ecosphere has become a serious concern that MPs are now detected in arrays of commercial food and beverages (Luqman et al., 2021).

MPs exert a range of ecological and health impacts on the aquatic ecosystem, with potential ramifications extending to humans (Gola et al., 2021). They can be ingested by marine organisms, leading to physical harm, reduced feeding efficiency, and, in severe cases, even death (Xu et al., 2020). MPs can also disrupt nutrient cycling and alter sediment composition, affecting the overall health of the ecosystem. Additionally, they can absorb and transport chemical pollutants, potentially concentrating toxic substances as they move up the food chain (Xu et al., 2020). Increased concern arises as they eventually end up in humans, as MPs in seafood and drinking water pose potential ingestion risks (Gola et al., 2021).

MPs engage in a multitude of interactions within the aquatic ecosystems, and their ultimate fate depends on a complex interplay of various factors. They can be ingested by different organisms, potentially leading to bioaccumulation as they move up the food chain. They can also act as sponges

Sources, Effects, and Fate of MPs in Aquatic Environment 13

for chemical pollutants, absorbing and concentrating harmful substances from the surrounding water. They also support microbial growth, altering their characteristics and behavior. Furthermore, they can sorb nutrients and organic matter, potentially affecting nutrient cycling. Despite these diverse interactions, the ultimate fate of MPs remains marked by their remarkable durability and persistence in the environment over extended periods, raising concerns about their long-term cumulative effects on the aquatic ecosystems (Darabi et al., 2021). Water currents and tides facilitate their dispersion, leading to their widespread distribution. Some MPs settle in the sediment, where they accumulate and may impact benthic life.

This chapter aims to provide a comprehensive understanding of MPs in aquatic ecosystems. It will explore the origin, sources, and types of MPs, delve into their pathways in the aquatic ecosystems, discuss their ecological and health impacts, and examine the different interactions they have within aquatic ecosystems. Additionally, it will elucidate their ultimate fate, shedding light on their dispersion throughout the aquatic ecosystem.

2.2 SOURCES, DISPERSAL, AND TRANSPORT OF MPS IN THE AQUATIC ENVIRONMENT

2.2.1 SOURCES OF MPS

The abundance and distribution of plastics in our ecosphere can be of diverse origin guided by natural and anthropogenic activities (Onyedibe et al., 2023). It is estimated that annually about 8 million tonnes of plastic moves into oceans, predominantly from land-based sources and from fishing activity, shipping, and aquaculture (Jambeck et al., 2015). Plastics released to the environment undergo size reduction due to natural forces and can be divided into macro- (>25 mm), meso- (<25 mm–5 mm), micro- (5 mm to 0.1 μm) and nano- (<0.1 μm) plastics. Though polymer composition, particle color, and morphology of MPs can be used to track the potential sources (Fahrenfeld et al., 2019), unequivocal identification may not always be possible. Small plastic particles industrially manufactured through extrusion or grinding for commercial applications (e.g., as abrasives, glitter, microbeads, powders, plastic pellets, and personal care products) are referred to as primary MPs (Laskar & Kumar, 2019). Secondary MPs are generated from the fragmentation of primary MPs or plastic wastes through abrasion and weathering (UV exposure). Depending on the age, abrasion of rope used in maritime hauling activity generates secondary MP fragments (Napper et al., 2022). Unlike the engineered nanoplastics (NPs) produced in the industry, NPs resulting from the degradation of MPs are heterogeneous in different environments. NPs can aggregate, or agglomerate, depending on the charge and hydrophobicity. Like MPs, NPs can also be classified either as primary, i.e., manufactured for use in paintings, drugs, cosmetics, and electronics, or secondary NPs formed from the degradation of MPs. The sources of MPs are heterogeneous, wide, and diverse in their occurrence. A wide variety of industrial activities, industrial waste disposal, transportation (maritime and land-based), and daily activities in our life contributes to MP pollution (Table 2.1).

Though oceans are the ultimate sink, the MPs are primarily generated through different terrestrial activities. The terrestrial activities comprises of urban runoff, inadequate management and disposal practices for plastic waste in landfills, plastic item littering in agricultural practices, industrial operations, road traffic and transportation activities, synthetic textile washing, and natural calamities such as cyclones and floods. Various consumer products or discharged industrial waste are not efficiently removed in wastewater treatment plants (WWTPs). The removal of low-density MPs through settling is not always effective and requires secondary or tertiary treatment (Lares et al., 2018). During the grease removal stage, a significant proportion of MPs (microbeads) can be removed (Murphy et al., 2016). Considering the large volume of effluent, MPs released into water streams even at a low concentration could be an environmental and health concern.

TABLE 2.1

Referred Sources of Microplastics in the Environment–Representative Examples

Source	Nature or Abundance of MPs	References
Use of face mask	Polypropylene fibers	Chen et al. 2021
Microfibers from apparel and textiles	–	Henry et al. 2019
Effluent from WWTPs	$15.70\ (\pm5.23)\ L^{-1}$	Murphy et al. 2016
	$1.2 \times 10^{8}\ d^{-1}$	Naji et al. 2021
Shedding from synthetic (acrylic, nylon, polyester) textiles	$7360\ m^{-2}\ L^{-1}$	Carney et al. 2018
Municipal solid waste landfill (leachate)	$0.42–24.58\ L^{-1}$	He et al. 2019
Municipal sewage sludge[a]	$12.8 \pm 5.2\ g^{-1}$	Rolsky et al. 2020
Agricultural plastic mulching	$0.1–324.5\ kg\ ha^{-1}$	Huang et al. 2020
Wear and tear from tires	$0.81\ kg\ y^{-1}\ person^{-1}$	Kole et al. 2017
Glitter for make-up, craft activities, textile products	–	Yurtsever 2019
Coated fertilizer	90% of all microplastics	Katsumi et al. 2020

[a] Average abundance from 12 countries involved in the study.

Landfills, as solid waste management facilities, are considered to be the major repositories of primary and secondary MPs. The extreme microenvironment in landfills (i.e., pH varying between 4.5 and 9.0, high salinity, shifting temperature, and biogas generation) accelerates the degradation of plastic waste (Sun et al., 2021). The MP load in landfills could be 20,000–91,000 items/kg, which is much higher than in sediments or agricultural soil (Golwala et al., 2021). Data compilation from 12 countries on MP load in sewage sludge reveals a large variability, i.e., $0.45 \pm 0.2\ MP\ g^{-1}$ in the Netherlands to $113 \pm 57\ MP\ g^{-1}$ in Italy (Rolsky et al., 2020). Apart from contaminating the surrounding environment, tiny MPs released through landfill leachate or carried through wind or rain can pollute the surrounding water bodies.

An array of industries (e.g., textile, electronic, automobile) and activities (e.g., construction, urbanization, tourism, improper disposal of plastic packaging) has been linked to MP pollution (An et al., 2020). In freshwater, mainly fibers and pellet forms of MPs dominate. However, synthetic textiles and personal care products are the most dominant source of MPs in wastewater discharge. In the ocean, about 80% of MPs are from land-based discharge, 18% are derived from recreational activities and commercial fishing, and the remaining 2% are from marine industries related to aquaculture (Ali et al., 2021). Textile industries annually produce more than 42 million tons of synthetic fibers and contribute to the accumulation of MP in aquatic environments (Kelly et al., 2019). The abundance of MPs can be correlated with proximity to industrial activities and population density. Microfibers shedding from synthetic textiles, depending on the fabric variant, are a predominant source of MP contamination. It has been estimated that polyester fleece on average sheds 7360 fibers $m^{-2}\ L^{-1}$ in one wash (Carney et al., 2018).

The discharge of MP-laden waste from industries, households, or establishments into sewage and agricultural runoff gets its way into streams and rivers, before reaching the seas and oceans (Chan et al., 2021). Apart from the translocation of MPs from the terrestrial ecosystem, the marine environment gets additional input from waste disposal at sea and stormwater. Fibers are the predominantly occurring physical form, although MPs can be of another form such as fragments, particles, pellets, or spheres. The abundance of MPs in a particular depth of water column, i.e., whether they float or sink, is determined by the polymer composition and density. Thus, understanding the polymer composition and density of MPs is crucial for evaluating their behavior in various environments, their modes of transportation, and their potential effects on the ecosystem. Furthermore, it facilitates the formulation of efficient measures to reduce and eliminate the negative impacts. Shed fibers from

Sources, Effects, and Fate of MPs in Aquatic Environment

clothes and carpets, drying of clothes, fly ash from incineration, and industrial emissions contribute to MPs in the air. Atmospheric fallout of MPs in an urban area could predominantly be of natural and synthetic fibers (Dris et al., 2016). Field-based studies indicate that suspended atmospheric MPs are indeed an important source of MP pollution in the ocean, especially by textile microfibers (Liu et al., 2019). Atmospheric MPs mostly originate from the degradation of larger plastic objects due to processes such as weathering, erosion, and mechanical abrasion. The primary origins of secondary sources are from the emission of MPs resulting from diverse human activities, including industrial operations, vehicular traffic, and the deterioration of plastic wastes.

2.2.2 DISPERSAL AND DEGRADATION OF MPs IN DIFFERENT ENVIRONMENTAL COMPARTMENTS

The proximity of various water systems to pollution sources affects the water quality. Different industrial activities, such as synthetic textile production, urban dust, industrial emissions, fragmented plastic from household furniture, landfills, waste incineration, agriculture soil, and sewage sludge are the predominant sources of MPs. These MPs are continuously transported through air. Annual atmospheric transport of MPs into oceans, though with high uncertainty, could be between 0.013 and 25 million metric tons (Allen et al., 2022). Studies have indicated that atmospheric transport of MP particles in the ocean could be of a similar order of magnitude due to riverine transport to the ocean (Evangeliou et al., 2020). The transport and vertical concentration profile of MPs in the atmosphere are affected by several factors, such as speed, direction, and velocity of wind, temperature, and rainfall. However, meteorological conditions, topology, and thermal circulation also influence their transportation in urban areas. The suspended MPs in the atmosphere travel a long distance or settle through a dry/wet deposition process under the influence of these atmospheric factors. The residence time and final deposition of MPs depend on the wind, precipitation, particle density, and gravity of large-particle sediments. In addition to transport within soil, MPs are transported to nearby air and water environments through natural and anthropogenic activities (O'Connor et al., 2019). The low-density MPs travel a long distance by wind and finally deposit the pollutants in aquatic and terrestrial systems (Ali et al., 2021).

The deposition of high concentrations of MPs in soil can be processed through horizontal and vertical movement, biotic and abiotic transports in soil and groundwater environment, influencing the migration of MPs in soil (Ren et al., 2021). Such MP movement in soil depends on the soil texture, aggregation, porosity, and agronomic activities including plowing and harvesting. Vertical movement of MPs may occur through the soil pores and cracks (Fahrenfeld et al., 2019). The vertical transport of MPs is highly dependent on soil structure, which is influenced by burrows of anecic earthworms, agricultural practices, cracks and crevices in the soil, and plant root penetration. However, long-range horizontal transport is controlled by geophagous earthworms, mosquitoes, and agricultural practices (Hurley & Nizzetto, 2018). The bioturbation process (e.g., plant root growth and uprooting) and occurrence of different plant fauna (e.g., vertebrates, earthworms, and larvae) affect the movement of MPs in soil. Besides, mites and collembolans can actively be involved in the transportation and distribution of MPs in soil (Ali et al., 2021). Fungal mycelia in soil pores improve the translocation of pollutant degrading bacteria, thereby accelerating the degradation of MPs (Cao et al., 2017).

Dynamic climatic forces and coastal transport processes influence the temporal and spatial distribution of MPs in aquatic environments. Stormwater lead to the export of terrestrial MPs and an abundance of MPs in river mouths has been reported during rainy periods (Cheung et al., 2016). The vertical mixing within the water column through strong winds and the associated wave action leads to the turbulence mixing of MPs from the bottom surface. Wave action dominates the low-density debris in the upper profile; however, heavier debris is present in the lower profile under the cross-shore circulation motion. The disintegration of plastic particles is accelerated by the turbulence mixing in beach sediment and it is one of the mechanism responsible for burying MPs inside the sediment. The MP particles can be present available up to 2.0 m depth in the vertical profile of beach sediment, and high-energy oceanographic sea storms are a mechanism responsible for MP burial (Turra et al., 2014).

Large quantities of MPs have been shown to be maintained in the sediment of wetland and beach areas. However, there are very few research works that have been conducted on the physical mechanisms governing their fate and transport (Galgani et al., 2013). Mathematical models have been developed to assess the potential transportation pathways under different climatic conditions (Dichgans et al., 2023; Um et al., 2023). In recent years, various types of models have been documented in the literature, including hybrid models that combine hydrodynamics and statistics, process-based, multimedia, and statistical models (Uzun et al., 2022). Among them, hydrodynamic models are the most popular that have been used to predict the transportation and deposition pathways of MPs in aqueous environments in different geographical regions (Cavalcante et al., 2020).

In a biological environment, aquatic and terrestrial biota play an essential role in transporting and depositing the MPs in soil and marine environments through bioturbation and biofouling processes (Akdogan & Guven, 2019). Different aquatic organisms influence the transportation of MPs, whereas different types of soil can also play an active role. It has been reported that the microarthropods Collembola facilitate the transportation of MP beads in the soil environment (Maaß et al., 2017). It has been reported that earthworms facilitated the vertical transport of MPs, and *Lumbricus terrestris* transported MPs up to 10 cm depth vertically in 21 days (Rillig et al., 2017). Soil microorganisms also play a significant role in the transportation of MPs through ingestion, egestion, burrowing, and adhesion processes (He et al., 2018). MPs are either excreted or accumulated in tissues and transferred to other trophic levels in the aquatic food chain after ingestion. Besides, the biofouling of MPs by marine microorganisms increases their size and density, accelerating the downward transportation in deep marine sediment (Akdogan & Guven, 2019). Similarly, aquatic plant species play a crucial role in the fate and transport of MPs in water systems and are facilitated by cellulosic materials of plant cells through electrostatic forces.

2.2.3 TRANSPORT OF MPS IN THE AQUATIC ENVIRONMENT

The primary and secondary MPs in the aquatic environment arise through various point and nonpoint sources. The effluent of different industries, including WWTPs, is considered as the most critical point source of MPs released into the marine environment. In contrast, runoff, atmospheric deposition, and drainage systems are the most vital nonpoint sources that are difficult to manage when compared to point sources (Tan et al., 2022). Urban stormwater runoff is responsible for carrying most of the MPs from nonpoint sources such as construction activities, artificial turf, landfill leachate, etc. (Bailey et al., 2021). It has been reported that the MPs discharged from nonpoint sources (e.g., surface runoff, domestic sewage, and sewer sediment) are sixfold greater than those released from point sources (Chen et al., 2020). Notably, improper disposal of domestic garbage, including plastic film, organic fertilizer, and sewage sludge is the primary source of MPs in rural areas. About 60–98% of this sludge is commonly used for agricultural purposes, landfilled, composted, and incinerated, thereby releasing a significant portion of MPs into the aquatic environment (Liu et al., 2021). Therefore, MPs arise from different sources and are transported through separate environmental compartments such as the atmosphere, soil, water, and biological systems. Atmospheric transport results in wide dispersal and vertical gradient of MPs depending on the speed, direction of wind, temperature, and precipitation. In addition, meteorological factors, thermal circulation, and topology in urban areas also affect the transportation of MPs. The suspended MPs in the atmosphere travel a long distance and finally settle through dry or wet deposition processes. The atmospheric forces like wind, rain, and MP density influence the residence time of the MPs in the atmosphere. These forces usually transport low-density MPs over long distances, resulting in pollution from aquatic and terrestrial MPs (Horton & Dixon, 2018).

Sources, Effects, and Fate of MPs in Aquatic Environment

Continuously active strong hydrodynamic forces, e.g., wind, tides, waves, and thermal gradient in the aquatic environment dictate the spatiotemporal distribution of MPs. Surface runoff of MP-laden wastes/products from various production and consumption facilities is transported through rivers and released into coastal areas. However, MPs retained in the riparian and hyporheic zone at the river mouth can be significant, particularly during rainy seasons. Also, strong winds and waves lead to the vertical mixing of the water columns, mixing MPs from the bottom.

Several studies have reported that most of the transported MPs are retained in the coastlines' estuaries, beaches, and wetland sediments. Though physical modeling describing those transportation phenomena is limited, numerical models have been developed to conduct the situational analysis of transportation mechanisms and predict the potential transport pathways under different climatic conditions (Hardesty et al., 2017). The advancement and applications of these numerical models can evaluate environmental risks associated with MP pollution and deployment of management policies related to MP pollution control. Thus, from a practical perspective, regulatory actions, increased public knowledge, and behavioral changes on the part of consumers are necessary to effectively address nonpoint sources of MPs.

2.2.4 MECHANISM OF MPs TRANSFER

Weathering, considered as a series of degradation processes including photocatalytic degradation, oxidative degradation, hydrolytic degradation, and biodegradation, deals with the fate and transport of MPs (Figure 2.1). The rate of MP degradation depends on the bulk morphology of MPs and their polymer chemistry along with their exposure to different environmental conditions. Photo-oxidation is one of the most critical MP degradation processes that result in polymer chain cleavage upon exposure and absorbance of solar UV radiation in the presence of oxygen and hydrogen (Ali et al., 2021). In the photo-oxidation process, the MPs present in the soil surface upon exposure to high UV radiation generate free radicals through the photolysis of C–C and C–H bonds present in the polymer structure. The extent of oxidation depends on the temperature, moisture content, UV exposure, soil composition, and crystallinity of MPs (Ng et al., 2018). Upon exposure to mechanical and abrasive forces, the photo- and thermal oxidation process leads to oxidative weathering of MPs and weakening MPs at the macroscale level making them more vulnerable to degradation. At

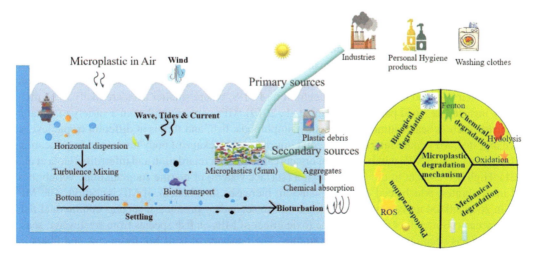

FIGURE 2.1 Sources, fate, and transport of MPs from different environmental compartments to the aquatic environment.

the microscale level, some chemical changes through chain-scission, branching, and cross-linking lead to an increase or decrease in the molecular weight of MPs, and the addition of an oxygenated functional group of esters, ketones have been reported to cause tensile elongation at the break and tensile modulus (Rouillon et al., 2016).

The weathering process is highly affected by wind, waves, physical abrasion, and turbulence, where a higher rate of MP weathering is observed on the beaches than on the water surface. Similarly, MPs collected from the surface water can have a higher rate of cleavage and surface cracking on exposed surfaces, than those in fouled bottom surfaces (Halle et al., 2016). Though biodegradation affects the weathering of MPs, the reaction rate is too slow for substantial degradation of MPs. During MP transformation, the first precursor step that leads to the microbial degradation is converting organic materials to release CO_2 and H_2O (under aerobic process) or CO_2 and CH_4 (under anaerobic process). The degradation of MPs by microorganisms begins with the formation of a biofilm on the surface of MPs, providing a suitable environment for the colonizing microbes. The characteristics of MPs responsible for microbial degradation and biofilm formation include their molecular weight, hydrophobicity, surface roughness, surface electrostatic interaction, water absorption capacity, and chemical and morphological structure (Rummel et al., 2017). Among the various microorganisms, bacteria, fungi, and actinomycetes are the most common groups that have shown their ability to biodegrade MPs. An actinomycetes species, *Rhodococcus ruber* strain C208 has been reported to form a biofilm on the surface of polyethylene (PE) within 30 days and biodegrade almost 8% of the PE (Orr et al., 2004). In a similar study, two bacterial strains, namely, *Rhodococcus* sp. strain 36 and *Bacillus* sp. strain 27 were isolated from mangrove sediment and the polypropylene degradation rates were 6.4% and 4%, respectively (Auta et al., 2018). Besides, many fungal species have been reported for their ability to attach to MP surfaces and degrade them through the formation of chemical bonds, such as ester functional group, carboxyl group, and carbonyl group. Fungi facilitate the circulation and transformation of these MPs in an ecosystem by decreasing the hydrophobicity of MPs.

2.3 EFFECT OF MPS ON AQUATIC ORGANISMS

2.3.1 SINGLE POLLUTANT

Marine organisms are constantly exposed to plastic debris where the ingestion of MPs can result in their physiological damage, including obstruction of the gastrointestinal tract (Mercogliano et al., 2020). MPs can accumulate in phytoplankton or zooplanktonic species and move into small invertebrates in the food chain before being ingested by marine carnivores (sardines, herring, and octopuses). At higher trophic levels, invertebrates (e.g., Crustacea, Echinodermata) and vertebrates (benthic and pelagic fish, marine mammals, and seabirds) ingest MPs while consuming their prey. The MP levels in the gut or tissue of commercial fish can vary depending on the feeding behavior, stages of their life, and the size and concentration of MPs, where smaller MPs are likely to be mistakenly ingested as "food particles" (Giani et al., 2019). MPs can accumulate in a variety of aquatic organisms such as sea cucumbers, copepods, amphipods, fish, and turtles at different tropic levels. Once ingested, MPs cause internal or external injuries and blockage of body tracts, and translocation and accumulation of MPs within different organs cause growth retardation and infertility (Gola et al., 2021). High MP content in the feeding region (surface waters) of marine zooplanktonic vertebrates and invertebrates, which feed on phytoplankton, increases the chances of MP ingestion. The MP load in zooplankton moves through the food web to secondary producers, such as commercial fish and cetaceans (Botterell et al., 2019). The toxic effect of MPs on fish species from freshwater, seas, and oceans has been reported in various studies. The presence of hard MP fragments and fibers in the stomach of fish in areas closer to urban populations has been observed. Oxidative damage and decreased isocitrate dehydrogenase activity have been observed on *Pomatoschistus* sp. in controlled

Sources, Effects, and Fate of MPs in Aquatic Environment

studies spiked with pyrene MP combinations. Reduced survival rate, superoxide dismutase activity, catalase activity, and increased lipid peroxide thiobarbituric acid have been observed in *Litopenaeus vannamei* shrimp juveniles following intramuscular administration (0.1–1.0 µg g^{-1} shrimp) of polyethylene microspheres (Hsieh et al., 2021).

MP polymers of different shapes and sizes are frequently found in the stomachs of whales and dolphins, which can cause blockage of the intestinal tract and lead to digestive impairment (Zhu et al., 2019). Ingested MPs can block/clog the filtering organs of marine organisms (*Balaenoptera physalus* L.) (Fossi et al., 2012). Apart from altered swimming behavior, MP ingestion causes a reduction in stomach capacity, growth rate, and even death in marine turtles. Absorption and accumulation of MPs by algal species such as *Chlorella* sp., *Scenedesmus*, and *Dunaliella* sp. cause oxidative stress and a reduction in photosynthetic activity, thereby impacting other aquatic organisms feeding on these algae (Pencik et al., 2023).

2.3.2 COMBINED POLLUTANTS

The high surface area to volume ratio of MPs make them a good adsorbent or a carrier matrix for dyes, heavy metals, and organic pollutants, such as polycyclic aromatic hydrocarbons, polychlorinated biphenyls (PCBs), polybrominated diphenyl ethers, and dichloro-diphenyl-trichloroethane. The residual monomers, additives, and chemicals released during the partial degradation of plastics and products are potentially toxic to aquatic organisms. However, studies on the quantitative assessment of hydrophobic organic chemicals (HOCs) transfer by MPs are limited. The bioavailability model suggests that the addition of a minute amount of PE or polystyrene to sediment results in an increase in the accumulation of phenanthrene and PCBs. A positive correlation between the amount of ingested plastic particles and the concentrations of PCBs in the tissues of birds also suggests the possible exposure route, though the pollutant transfer pathway (through plastics or contaminated food) may not be always discernable as internal HOC concentration is not linked to their stomach plastic concentrations. However, in marine environments, chemicals sorbed onto plastic can have negative effects on fishes (Bakir et al., 2016). MPs along with heavy metal (Cr) can decrease the acetylcholinesterase (AchE) activity in *Pomatoschistus* sp. The accumulation of MPs in the gills, guts, and liver of *Dainorerio* is known to cause inflammation, increase in superoxide dismutase and catalase activity, and altered metabolic pathways.

2.4 OCCURRENCE AND FATE OF MPS IN THE AQUATIC ENVIRONMENT

2.4.1 OCCURRENCE OF MPS IN THE AQUATIC ENVIRONMENT

MPs have been detected across all the bio-geosphere, from the atmosphere, land, surface waters, sediments, agricultural soils, drinking water, and food products (Mendoza et al., 2021) (Table 2.2). The abundance of MPs in oceans has been a major concern over the recent decades. MPs have even been detected in the most remote Arctic Ocean (Pan et al., 2022). MPs can accumulate in sediment and subsurface layers, and be absorbed by phytoplankton or ingested by crustaceans (Contam, 2016). The presence of MPs has been detected in the stomachs of marine animals, mollusks, crustaceans (Bruzaca et al., 2022; D'Costa, 2022), fish (Huang et al., 2023), and even whales (Moore et al., 2022). Though the ocean is considered as a sink, detection of MPs in the marine environment indicate the possibility of sea-based emissions, reversing its role from sink to source (Goßmann et al., 2023).

In addition to the marine environment, MPs are ubiquitous in surface waters and wastewater effluents. Secondary MPs have been found in water and sediments in lakes across the globe, from North American lakes, Edgbaston Pool in the UK, Vembanad Lake in India (Sruthy & Ramasamy, 2017), to Lake Hovsgol in Mongolia. Primary MPs have been detected in samples of river water,

TABLE 2.2
Occurrence of Microplastics in Aquatic Environments, and Surface and Groundwater

Sl.	Media	Origin/Location	Count/Concentration	Size Range	Polymer Type/Shape	References
1	Sea water	South Chaina Sea	103.4±98.3 m⁻³ [f]	34.2–4996 μm	PET, PS, PE (fibers, granules)	Liu et al. 2023
2	Corals	Trinidad	3.4 ± 1.9 [b]	–	PE, PET	Rani-Borges et al. 2023
3	Ground water	Jiaodong, China	87–6832 [e]	<100 μm	PET, PU	Mu et al. 2022
4	Fresh water fish	Turkey	610 [c]	0.10–4.85 mm	PE, PS, polychloroprene fibers	Boyukalan and Yerli 2023
5	Fish and squid	Madeira	8.75 ± 12.34 [d]		Phthalate esters fibers (majority)	Sambolino et al. 2023
6	Fish	Surabaya River, Indonesia	280.73±162.25 [g]		Fibers	Lestari et al. 2023
7	Commercial fish snack	Mexico				Kutralam-Muniasamy et al. 2023
8	Bottled water	Turkey	2–20 [a]	6–1000 μm		Altunışık 2023
9	Bottled water	Thailand	140 ± 19 [a]	>6.5 μm		Kankanige and Babel 2020
10	Bottled water	Korea	12–58 [a]	40–723 μm	Spherical, rod, fiber	Lee et al. 2021
11	Springs and wells	USA	7 ± 4.3 [h]	<1.5 mm	PE	Panno et al. 2019
12	Groundwater from coastal aquifer	India	0–4.3 [h]	0.12–2.5 mm	PA/PE	Selvam et al. 2021

[a] gL⁻¹; [b] adhered particles based on g⁻¹ soft tissue; [c] total microplastic count from seven varieties of fish (n = 406); [d] the highest observed count in stomach per individual; [e] in <250 μm fraction of road dust; [f] MP particle count; [g] maximum particle counts per gram wet weight; [h] MP count per liter. Abbreviations: PA - Polyamide; PE - Polyethylene; PEP - Poly(ethylene propylene); PET - Polyethylene terephthalate; PS - Polystyrene; PU - Polyurethane

Sources, Effects, and Fate of MPs in Aquatic Environment 21

and sediments across Europe. The Black Sea receives more than 1500 tons of MPs each year via the river Danube alone (Lechner et al., 2014).

A recent study on MPs in urban areas of the Czech Republic has suggested that MP load significantly differs depending on water sources, i.e., river water is much more contaminated than that of reservoirs (Pivokonsky et al., 2018). The sampling period averaged MP count in raw water across the sampling location with 1473 ± 34, 1812 ± 35, and 3605 ± 497 particles L^{-1}, and 40–60% of MP loads were of 1–5 µm (Pivokonsky et al., 2018). However, the chemical identity of submicron particles (<1 µm) in the raw water samples, though not confirmed, could be of MP origin.

2.4.2 FATE OF MPS IN THE AQUATIC ENVIRONMENT

The fate of MPs includes various physical, chemical, and biological degradation processes that lead to ingestion, biofouling, bioaccumulation, and biomagnification in the aquatic environment. The critical threat associated with MPs include their ability to absorb various organic pollutants and serve as a vector in the transportation of these hazardous substances in the aquatic environment (Tourinho et al., 2019). The various mechanisms, including photodegradation, biodegradation, thermos-degradation, and hydrolysis, lead to the aging and weathering of MPs (Figure 2.1). Several modifications on the surface of MPs (e.g., fouling) can be found during the aging and weathering process. Photo-oxidation is one of the prominent aging mechanisms of MPs that results in the formation of a carbonyl group, increased surface polarity, and thereby facilitating the preferential sorption of polar compounds (Figure 2.1) (Tourinho et al., 2019). However, the sorption of both polar and nonpolar compounds can be reduced on the UV-weathered MP surface due to the formation of hydrogen bonds and water clusters (Hüffer et al., 2018). The occurrence of MPs in aquatic ecosystems has been firmly established and has widespread influences on the function of marine ecosystems and food chains. Recent studies have revealed that the fate and bioavailability of MPs are affected by a change in their buoyancy upon integration with aquatic microalgae (Nava & Leoni, 2021).

2.5 CONCLUSIONS

The anthropogenic fallout of MPs in the terrestrial environment either moves through atmospheric transport or discharge through sewage, runoff into freshwater lakes, rivers, and ultimately into the marine environment. In the aquatic environment, horizontal and vertical movement of MPs is greatly influenced by the flow velocity, water depth, underwater topography, wind speed, and the density of the MP particles. The characteristics of MPs, hydrophobicity, weathering, pH and ionic strength of the water influence their ability to adsorb pollutants. The hydrophobic surface of MPs helps in the easy colonization of bacteria or algae to form biofilm. Although the presence of MPs facilitates the accumulation of contaminants in aquatic species, the physical damage (ingestion/obstruction), and behavioral change (feeding habit) in aquatic organisms could be more prominent than the toxicity of the pollutants.

REFERENCES

Akdogan, Z., and Guven, B. 2019. Microplastics in the Environment: A Critical Review of Current Understanding and Identification of Future Research Needs. *Environmental Pollution* 254: 113011. doi:10.1016/j.envpol.2019.113011.

Ali, M.U.U., Lin, S., Yousaf, B., Abbas, Q., Munir, M.A.M., Ali, M.U.U., Rasihd, A., Zheng, C., Kuang, X., and Wong, M.H. 2021. Environmental Emission, Fate and Transformation of Microplastics in Biotic and Abiotic Compartments: Global Status, Recent Advances and Future Perspectives. *Science of The Total Environment* 791: 148422. doi:10.1016/j.scitotenv.2021.148422.

Allen, D., Allen, S., Abbasi, S., Baker, A., Bergmann, M., Brahney, J., Butler, T. et al. 2022. Microplastics and Nanoplastics in the Marine-Atmosphere Environment. *Nature Reviews Earth & Environment* 3 (6): 393–405. doi:10.1038/s43017-022-00292-x.

Altunışık, A. 2023. Microplastic Pollution and Human Risk Assessment in Turkish Bottled Natural and Mineral Waters. *Environmental Science and Pollution Research* 30 (14): 39815–25. doi:10.1007/s11356-022-25054-6.

An, L., Liu, Q., Deng, Y., Wu, W., Gao, Y., and Ling, W. 2020. Sources of Microplastic in the Environment. In: *Microplastics in Terrestrial Environments. The Handbook of Environmental Chemistry* (Vol. 95), ed. He, D., and Luo, Y., Springer, Cham. 143–59. doi:10.1007/698_2020_449.

Auta, H.S., Emenike, C.U., and Fauziah, S.H. 2017. Distribution and Importance of Microplastics in the Marine Environment: A Review of the Sources, Fate, Effects, and Potential Solutions. *Environment International* 102: 165–76. doi:10.1016/j.envint.2017.02.013.

Auta, H.S., Emenike, C.U., Jayanthi, B., and Fauziah, S.H. 2018. Growth Kinetics and Biodeterioration of Polypropylene Microplastics by *Bacillus Sp.* and *Rhodococcus Sp.* Isolated from Mangrove Sediment. *Marine Pollution Bulletin* 127: 15–21. doi:10.1016/j.marpolbul.2017.11.036.

Bailey, K., Sipps, K., Saba, G.K., Arbuckle-Keil, G., Chant, R.J., and Fahrenfeld, N.L. 2021. Quantification and Composition of Microplastics in the Raritan Hudson Estuary: Comparison to Pathways of Entry and Implications for Fate. *Chemosphere* 272: 129886. doi:10.1016/j.chemosphere.2021.129886.

Bakir, A., O'Connor, I.A., Rowland, S.J., Hendriks, A.J., and Thompson, R.C. 2016. Relative Importance of Microplastics as a Pathway for the Transfer of Hydrophobic Organic Chemicals to Marine Life. *Environmental Pollution* 219: 56–65. doi:10.1016/j.envpol.2016.09.046.

Botterell, Z.L.R., Beaumont, N., Dorrington, T., Steinke, M., Thompson, R.C., and Lindeque, P.K. 2019. Bioavailability and Effects of Microplastics on Marine Zooplankton: A Review. *Environmental Pollution* 245: 98–110. doi:10.1016/j.envpol.2018.10.065.

Boyukalan, S., and Yerli, S.V. 2023. Microplastic Pollution at Different Trophic Levels of Freshwater Fish in a Variety of Türkiye`s Lakes and Dams. *Turkish Journal of Fisheries and Aquatic Sciences* 23 (11). doi:10.4194/TRJFAS23747.

Bruzaca, D.N.A., Justino, A.K.S., Mota, G.C.P., Costa, G.A., Lucena-Frédou, F., and Gálvez, A.O. 2022. Occurrence of Microplastics in Bivalve Molluscs Anomalocardia Flexuosa Captured in Pernambuco, Northeast Brazil. *Marine Pollution Bulletin* 179: 113659. doi:10.1016/j.marpolbul.2022.113659.

Cao, D., Wang, X., Luo, X., Liu, G., and Zheng, H. 2017. Effects of Polystyrene Microplastics on the Fitness of Earthworms in an Agricultural Soil. *IOP Conference Series: Earth and Environmental Science* 61: 012148. doi:10.1088/1755-1315/61/1/012148.

Carney, A.B.M., Åström, L., Roslund, S., Petersson, H., Johansson, M., and Persson, N-K. 2018. Quantifying Shedding of Synthetic Fibers from Textiles; a Source of Microplastics Released into the Environment. *Environmental Science and Pollution Research* 25 (2): 1191–99. doi:10.1007/s11356-017-0528-7.

Cavalcante, R.M., Pinheiro, L.S., Teixeira, C.E.P., Paiva, B.P., Fernandes, G.M., Brandão, D.B., Frota, F.F., Filho, F.J.N.S., and Schettini, C.A.F. 2020. Marine Debris on a Tropical Coastline: Abundance, Predominant Sources and Fate in a Region with Multiple Activities (Fortaleza, Ceará, Northeastern Brazil). *Waste Management* 108: 13–20. doi:10.1016/j.wasman.2020.04.026.

Chan, C.K.M., Park, C., Chan, K.M., Mak, D.C.W., Fang, J.K.H., and Mitrano, D.M. 2021. Microplastic Fibre Releases from Industrial Wastewater Effluent: A Textile Wet-Processing Mill in China. *Environmental Chemistry* 18 (3): 93–100. doi:10.1071/EN20143.

Chaukura, N., Kebede, K.K., Innocent, C., Isaac, N., Willis, G., Welldone, M., Thabo, T.I.N., Bhekie, B.M., and Francis, O.A. 2021. Microplastics in the Aquatic Environment—The Occurrence, Sources, Ecological Impacts, Fate, and Remediation Challenges. *Pollutants* 1 (2): 95–118. doi:10.3390/pollutants1020009.

Chen, H., Jia, Q., Zhao, X., Li, L., Nie, Y., Liu, H., and Ye, J. 2020. The Occurrence of Microplastics in Water Bodies in Urban Agglomerations: Impacts of Drainage System Overflow in Wet Weather, Catchment Land-Uses, and Environmental Management Practices. *Water Research* 183: 116073. doi:10.1016/j.watres.2020.116073.

Chen, X., Chen, X., Liu, Q., Zhao, Q., Xiong, X. and Wu, C. 2021. Used Disposable Face Masks Are Significant Sources of Microplastics to Environment. *Environmental Pollution* 285: 117485. doi:10.1016/j.envpol.2021.117485.

Cheung, P.K., Cheung, L.T.O., and Fok, L. 2016. Seasonal Variation in the Abundance of Marine Plastic Debris in the Estuary of a Subtropical Macro-Scale Drainage Basin in South China. *Science of The Total Environment* 562: 658–65. doi:10.1016/j.scitotenv.2016.04.048.

Contam, EFSA Panel on Contaminants in the Food Chain. 2016. Presence of Microplastics and Nanoplastics in Food, with Particular Focus on Seafood. *EFSA Journal* 14 (6). doi:10.2903/j.efsa.2016.4501.

Darabi, M., Majeed, H., Diehl, A., Norton, J., and Zhang, Y. 2021. A Review of Microplastics in Aquatic Sediments: Occurrence, Fate, Transport, and Ecological Impact. *Current Pollution Reports* 7 (1): 40–53. doi:10.1007/s40726-020-00171-3.

D'Costa, A.H. 2022. Microplastics in Decapod Crustaceans: Accumulation, Toxicity and Impacts, a Review. *Science of The Total Environment* 832: 154963. doi:10.1016/j.scitotenv.2022.154963.

Dichgans, F., Boos, J.P., Ahmadi, P., Frei, S., and Fleckenstein, J.H. 2023. Integrated Numerical Modeling to Quantify Transport and Fate of Microplastics in the Hyporheic Zone. *Water Research* 243: 120349. doi:10.1016/j.watres.2023.120349.

Dris, R., Gasperi, J., Saad, M., Mirande, C., and Tassin, B. 2016. Synthetic Fibers in Atmospheric Fallout: A Source of Microplastics in the Environment? *Marine Pollution Bulletin* 104 (1–2): 290–93. doi:10.1016/j.marpolbul.2016.01.006.

Du, S., Zhu, R., Cai. Y., Xu, N., Yap, P.-S., Zhang, Y.Y., He, Y., and Zhang, Y.Y. 2021. Environmental Fate and Impacts of Microplastics in Aquatic Ecosystems: A Review. *RSC Advances* 11 (26): 15762–84. doi:10.1039/D1RA00880C.

Evangeliou, N., Grythe, H., Klimont, Z., Heyes, C., Eckhardt, S., Lopez-Aparicio, S., and Stohl, A. 2020. Atmospheric Transport is a Major Pathway of Microplastics to Remote Regions. *Nature Communications* 11 (1): 3381. doi:10.1038/s41467-020-17201-9.

Fahrenfeld, N.L., Georgia, A-K., Beni, N.N., and Bartelt-Hunt, S.L. 2019. Source Tracking Microplastics in the Freshwater Environment. *TrAC Trends in Analytical Chemistry* 112: 248–54. doi:10.1016/j.trac.2018.11.030.

Fossi, M.C., Panti, C., Guerranti, C., Coppola, D., Giannetti, M., Marsili, L., and Minutoli, R. 2012. Are Baleen Whales Exposed to the Threat of Microplastics? A Case Study of the Mediterranean Fin Whale (Balaenoptera Physalus). *Marine Pollution Bulletin* 64 (11): 2374–79. doi:10.1016/j.marpolbul.2012.08.013.

Galgani, F., Hanke, G., Werner, S., and De Vrees, L. 2013. Marine Litter within the European Marine Strategy Framework Directive. *ICES Journal of Marine Science* 70 (6): 1055–64. doi:10.1093/icesjms/fst122.

Giani, D., Baini, M., Galli, M., Casini, S., and Fossi, M.C. 2019. Microplastics Occurrence in Edible Fish Species (Mullus Barbatus and Merluccius Merluccius) Collected in Three Different Geographical Sub-Areas of the Mediterranean Sea. *Marine Pollution Bulletin* 140: 129–37. doi:10.1016/j.marpolbul.2019.01.005.

Gola, D., Tyagi, P.K., Arya, A., Chauhan, N., Agarwal, M., Singh, S.K., and Gola, S. 2021. The Impact of Microplastics on Marine Environment: A Review. *Environmental Nanotechnology, Monitoring & Management* 16: 100552. doi:10.1016/j.enmm.2021.100552.

Golwala, H., Zhang, X., Iskander, S.M., and Smith, A.L. 2021. Solid Waste: An Overlooked Source of Microplastics to the Environment. *Science of The Total Environment* 769: 144581. doi:10.1016/j.scitotenv.2020.144581.

Goßmann, I., Herzke, D., Held, A., Schulz, J., Nikiforov, V., Georgi, C., Evangeliou, N., Eckhardt, S., Gerdts, G., Wurl, O., Scholz-Böttcher, B.M. 2023. Occurrence and Backtracking of Microplastic Mass Loads Including Tire Wear Particles in Northern Atlantic Air. *Nature Communications* 14 (1): 3707. doi:10.1038/s41467-023-39340-5.

Halle, A., Ladirat, L., Gendre, X., Goudouneche, D., Pusineri, C., Routaboul, C., Tenailleau, C., Duployer, B., and Perez, E. 2016. Understanding the Fragmentation Pattern of Marine Plastic Debris. *Environmental Science & Technology* 50 (11): 5668–75. doi:10.1021/acs.est.6b00594.

Hardesty, B.D., Harari, J., Isobe, A., Lebreton, L., Maximenko, N., Potemra, J., van Sebille, E., Vethaak, A.D., and Wilcox, C. 2017. Using Numerical Model Simulations to Improve the Understanding of Micro-Plastic Distribution and Pathways in the Marine Environment. *Frontiers in Marine Science* 4: 30. doi:10.3389/fmars.2017.00030.

He, D., Luo, Y., Lu, S., Liu, M., Song, Y., and Lei, L. 2018. Microplastics in Soils: Analytical Methods, Pollution Characteristics and Ecological Risks. *TrAC Trends in Analytical Chemistry* 109: 163–72. doi:10.1016/j.trac.2018.10.006.

He, P., Chen, L., Shao, L., Zhang, H., and Lü, F. 2019. Municipal Solid Waste (MSW) Landfill: A Source of Microplastics? Evidence of Microplastics in Landfill Leachate. *Water Research* 159: 38–45. doi:10.1016/j.watres.2019.04.060.

Henry, B., Laitala, K., and Klepp, I.G. 2019. Microfibres from Apparel and Home Textiles: Prospects for Including Microplastics in Environmental Sustainability Assessment. *Science of The Total Environment* 652: 483–94. doi:10.1016/j.scitotenv.2018.10.166.

Horton, A.A., and Dixon, S.J. 2018. Microplastics: An Introduction to Environmental Transport Processes. *WIREs Water* 5(2):e1268. doi:10.1002/wat2.1268.

Hsieh, S.-L., Wu, Y.-C., Xu, R.-Q., Chen, Y.-T., Chen, C.-W., Singhania, R.R., and Dong, C.-D. 2021. Effect of Polyethylene Microplastics on Oxidative Stress and Histopathology Damages in Litopenaeus Vannamei. *Environmental Pollution* 288: 117800. doi:10.1016/j.envpol.2021.117800.

Huang, L., Li, Q.P., Li, H.H., Lin, L., Xu, X., Yuan, X., Koongolla, J.B., and Li, H.H. 2023. Microplastic Contamination in Coral Reef Fishes and Its Potential Risks in the Remote Xisha Areas of the South China Sea. *Marine Pollution Bulletin* 186: 114399. doi:10.1016/j.marpolbul.2022.114399.

Huang, Y., Liu, Q., Jia, W., Yan, C., and Wang, J. 2020. Agricultural Plastic Mulching as a Source of Microplastics in the Terrestrial Environment. *Environmental Pollution* 260: 114096. doi:10.1016/j.envpol.2020.114096.

Hüffer, T., Weniger, A-K., and Hofmann, T. 2018. Sorption of Organic Compounds by Aged Polystyrene Microplastic Particles. *Environmental Pollution* 236: 218–25. doi:10.1016/j.envpol.2018.01.022.

Hurley, R.R., and Nizzetto, L. 2018. Fate and Occurrence of Micro(Nano)Plastics in Soils: Knowledge Gaps and Possible Risks. *Current Opinion in Environmental Science & Health* 1: 6–11. doi:10.1016/j.coesh.2017.10.006.

Jambeck, J.R., Geyer, R., Wilcox, C., Siegler, T.R., Perryman, M., Andrady, A., Narayan, R., and Law, K.L. 2015. Plastic Waste Inputs from Land into the Ocean. *Science* 347 (6223). American Association for the Advancement of Science: 768–71. doi:10.1126/science.1260352.

Kankanige, D., and Babel, S. 2020. Smaller-Sized Micro-Plastics (MPs) Contamination in Single-Use PET-Bottled Water in Thailand. *Science of The Total Environment* 717: 137232. doi:10.1016/j.scitotenv.2020.137232.

Katsumi, N., Kusube, T., Nagao, S. and Okochi, H. 2020. The Role of Coated Fertilizer Used in Paddy Fields as a Source of Microplastics in the Marine Environment. *Marine Pollution Bulletin* 161: 111727. doi:10.1016/j.marpolbul.2020.111727.

Kelly, M.R., Lant, N.J., Kurr, M., and Burgess, J.G. 2019. Importance of Water-Volume on the Release of Microplastic Fibers from Laundry. *Environmental Science & Technology* 53 (20): 11735–44. doi:10.1021/acs.est.9b03022.

Kole, P.J., Löhr, A.J., Belleghem, F.W., and Ragas, A. 2017. Wear and Tear of Tyres: A Stealthy Source of Microplastics in the Environment. *International Journal of Environmental Research and Public Health* 14 (10): 1265. doi:10.3390/ijerph14101265.

Kutralam-Muniasamy, G., Shruti, V.C., Pérez-Guevara, F., Roy, P.D., and Martínez, I.E. 2023. Consumption of Commercially Sold Dried Fish Snack "Charales" Contaminated with Microplastics in Mexico. *Environmental Pollution* 332: 121961. doi:10.1016/j.envpol.2023.121961.

Lares, M., Ncibi, M.C., Sillanpää, M., and Sillanpää, M. 2018. Occurrence, Identification and Removal of Microplastic Particles and Fibers in Conventional Activated Sludge Process and Advanced MBR Technology. *Water Research* 133: 236–46. doi:10.1016/j.watres.2018.01.049.

Laskar, N., and Kumar, U. 2019. Plastics and Microplastics: A Threat to Environment. *Environmental Technology & Innovation* 14 (May): 100352. doi:10.1016/j.eti.2019.100352.

Lechner, A., Keckeis, H., Lumesberger-Loisl, F., Zens, B., Krusch, R., Tritthart, M., Glas, M., and Schludermann, E. 2014. The Danube so Colourful: A Potpourri of Plastic Litter Outnumbers Fish Larvae in Europe's Second Largest River. *Environmental Pollution* 188: 177–81. doi:10.1016/j.envpol.2014.02.006.

Lee, E-H., Lee, S., Chang, Y., and Lee, S-W. 2021. Simple Screening of Microplastics in Bottled Waters and Environmental Freshwaters Using a Novel Fluorophore. *Chemosphere* 285: 131406. doi:10.1016/j.chemosphere.2021.131406.

Lestari, P., Trihadiningrum, Y., and Warmadewanthi, I.D.A.A. 2023. Investigation of Microplastic Ingestion in Commercial Fish from Surabaya River, Indonesia. *Environmental Pollution* 331: 121807. doi:10.1016/j.envpol.2023.121807.

Liu, B., Lu, Y., Deng, H., Huang, H., Wei, N., Jiang, Y.Y., Jiang, Y.Y., Liu, L., Sun, K., and Zheng, H. 2023. Occurrence of Microplastics in the Seawater and Atmosphere of the South China Sea: Pollution Patterns and Interrelationship. *Science of The Total Environment* 889: 164173. doi:10.1016/j.scitotenv.2023.164173.

Liu, K., Wu, T., Wang, X., Song, X., Zong, C., Wei, N., and Li, D. 2019. Consistent Transport of Terrestrial Microplastics to the Ocean through Atmosphere. *Environmental Science & Technology* 53 (18): 10612–19. doi:10.1021/acs.est.9b03427.

Liu, W., Zhang, J., Liu, H., Guo, X., Zhang, X., Yao, X., Cao, Z., and Zhang, T. 2021. A Review of the Removal of Microplastics in Global Wastewater Treatment Plants: Characteristics and Mechanisms. *Environment International* 146: 106277. doi:10.1016/j.envint.2020.106277.

Luqman, A., Nugrahapraja, H., Wahyuono, R.A., Islami, I., Haekal, M.H., Fardiansyah, Y., Putri, B.Q. et al. 2021. Microplastic Contamination in Human Stools, Foods, and Drinking Water Associated with Indonesian Coastal Population. *Environments* 8 (12): 138. doi:10.3390/environments8120138.

Maaß, S., Daphi, D., Lehmann, A., and Rillig, M.C. 2017. Transport of Microplastics by Two Collembolan Species. *Environmental Pollution* 225: 456–59. doi:10.1016/j.envpol.2017.03.009.

Mercogliano, R., Avio, C.G., Regoli, F., Anastasio, A., Colavita, G., and Santonicola, S. 2020. Occurrence of Microplastics in Commercial Seafood under the Perspective of the Human Food Chain. A Review. *Journal of Agricultural and Food Chemistry* 68 (19): 5296–5301. doi:10.1021/acs.jafc.0c01209.

Moore, R.C., Noel, M., Etemadifar, A., Loseto, L., Posacka, A.M., Bendell, L. and Ross, P.S. 2022. Microplastics in Beluga Whale (Delphinapterus Leucas) Prey: An Exploratory Assessment of Trophic Transfer in the Beaufort Sea. *Science of The Total Environment* 806: 150201. doi:10.1016/j.scitotenv.2021.150201.

Mu, H., Wang, Y., Zhang, H., Guo, F., Li, A., Zhang, S., Liu, S. and Liu, T. 2022. High Abundance of Microplastics in Groundwater in Jiaodong Peninsula, China. *Science of The Total Environment* 839: 156318. doi:10.1016/j.scitotenv.2022.156318.

Murphy, F., Ewins, C., Carbonnier, F., and Quinn, B. 2016. Wastewater Treatment Works (WwTW) as a Source of Microplastics in the Aquatic Environment. *Environmental Science & Technology* 50 (11): 5800–5808. doi:10.1021/acs.est.5b05416.

Naji, A., Azadkhah, S., Farahani, H., Uddin, S., and Khan, F.R. 2021. Microplastics in Wastewater Outlets of Bandar Abbas City (Iran): A Potential Point Source of Microplastics into the Persian Gulf. *Chemosphere* 262: 128039. doi:10.1016/j.chemosphere.2020.128039.

Napper, I.E., Wright, L.S., Barrett, A.C., Parker-Jurd, F.N.F., and Thompson, R.C. 2022. Potential Microplastic Release from the Maritime Industry: Abrasion of Rope. *Science of The Total Environment* 804: 150155. doi:10.1016/j.scitotenv.2021.150155.

Narayanan, M. 2023. Origination, Fate, Accumulation, and Impact, of Microplastics in a Marine Ecosystem and Bio/Technological Approach for Remediation: A Review. *Process Safety and Environmental Protection* 177: 472–85.

Nava, V., and Leoni, B. 2021. A Critical Review of Interactions between Microplastics, Microalgae and Aquatic Ecosystem Function. *Water Research* 188: 116476. doi:10.1016/j.watres.2020.116476.

Ng, E-L., Lwanga, E.H., Eldridge, S.M., Johnston, P., Hu, H-W., Geissen, V., and Chen, D. 2018. An Overview of Microplastic and Nanoplastic Pollution in Agroecosystems. *Science of The Total Environment* 627: 1377–88. doi:10.1016/j.scitotenv.2018.01.341.

O'Connor, D., Pan, S., Shen, Z., Song, Y., Jin, Y., Wu, W-M., and Hou, D. 2019. Microplastics Undergo Accelerated Vertical Migration in Sand Soil Due to Small Size and Wet-Dry Cycles. *Environmental Pollution* 249: 527–34. doi:10.1016/j.envpol.2019.03.092.

Onyedibe, V., Kakar, F.L., Okoye, F., Elbeshbishy, E., and Hamza, R. 2023. Sources and Occurrence of Microplastics and Nanoplastics in the Environment. In *Current Developments in Biotechnology and Bioengineering*, ed. R.D. Tyagi, A. Pandey, S. Pilli, A. Pandey, B. Yadav. Elsevier. 33–58. doi:10.1016/B978-0-323-99908-3.00019-1.

Orr, I.G., Hadar, Y., and Sivan, A. 2004. Colonization, Biofilm Formation and Biodegradation of Polyethylene by a Strain of Rhodococcus Ruber. *Applied Microbiology and Biotechnology* 65 (1): 97–104. doi:10.1007/s00253-004-1584-8.

Pan, Z., Liu, Q., Sun, X., Li, W., Zou, Q., Cai, S., and Lin, H. 2022. Widespread Occurrence of Microplastic Pollution in Open Sea Surface Waters: Evidence from the Mid-North Pacific Ocean. *Gondwana Research* 108: 31–40. doi:10.1016/j.gr.2021.10.024.

Panno, S.V., Kelly, W.R., Scott, J., Zheng, W., McNeish, R.E., Holm, N., Hoellein, T.J., and Baranski, E.L. 2019. Microplastic Contamination in Karst Groundwater Systems. *Groundwater* 57 (2): 189–96. doi:10.1111/gwat.12862.

Pencik, O., Molnarova, K., Durdakova, M., Kolackova, M., Klofac, D., Kucsera, A., Capal, P. et al. 2023. Not So Dangerous? PET Microplastics Toxicity on Freshwater Microalgae and Cyanobacteria. *Environmental Pollution* 329: 121628. doi:10.1016/j.envpol.2023.121628.

Pivokonsky, M., Cermakova, L., Novotna, K., Peer, P., Cajthaml, T., and Janda, V. 2018. Occurrence of Microplastics in Raw and Treated Drinking Water. *Science of The Total Environment* 643: 1644–51. doi:10.1016/j.scitotenv.2018.08.102.

Rani-Borges, B., Gomes, E., Maricato, G., Lins, L.H.F.C., de Moraes, B.R., Lima, G.V., Côrtes, L.G.F. et al. 2023. Unveiling the Hidden Threat of Microplastics to Coral Reefs in Remote South Atlantic Islands. *Science of The Total Environment* 897: 165401. doi:10.1016/j.scitotenv.2023.165401.

Ren, Z., Gui, X., Xu, X., Zhao, L., Qiu, H., and Cao, X. 2021. Microplastics in the Soil-Groundwater Environment: Aging, Migration, and Co-Transport of Contaminants – A Critical Review. *Journal of Hazardous Materials* 419: 126455. doi:10.1016/j.jhazmat.2021.126455.

Rillig, M.C., Ingraffia, R., and de Souza Machado, A.A. 2017. Microplastic Incorporation into Soil in Agroecosystems. *Frontiers in Plant Science* 8: 1805. doi:10.3389/fpls.2017.01805.

Mendoza, L.M.R., Vargas, D.L., and Balcer, M. 2021. Microplastics Occurrence and Fate in the Environment. *Current Opinion in Green and Sustainable Chemistry* 32: 100523. doi:10.1016/j.cogsc.2021.100523.

Rolsky, C., Kelkar, V., Driver, E., and Halden, R.U. 2020. Municipal Sewage Sludge as a Source of Microplastics in the Environment. *Current Opinion in Environmental Science & Health* 14: 16–22. doi:10.1016/j.coesh.2019.12.001.

Rouillon, C., Bussiere, P-O., Desnoux, E., Collin, S., Vial, C., Therias, S., and Gardette, J-L. 2016. Is Carbonyl Index a Quantitative Probe to Monitor Polypropylene Photodegradation? *Polymer Degradation and Stability* 128: 200–208. doi:10.1016/j.polymdegradstab.2015.12.011.

Rummel, C.D., Jahnke, A., Gorokhova, E., Kühnel, D., and Schmitt-Jansen, M. 2017. Impacts of Biofilm Formation on the Fate and Potential Effects of Microplastic in the Aquatic Environment. *Environmental Science & Technology Letters* 4 (7): 258–67. doi:10.1021/acs.estlett.7b00164.

Sambolino, A., Iniguez, E., Herrera, I., Kaufmann, M., Dinis, A., and Cordeiro, N. 2023. Microplastic Ingestion and Plastic Additive Detection in Pelagic Squid and Fish: Implications for Bioindicators and Plastic Tracers in Open Oceanic Food Webs. *Science of The Total Environment* 894: 164952. doi:10.1016/j.scitotenv.2023.164952.

Selvam, S., Jesuraja, K., Venkatramanan, S., Roy, P.D., and Kumari, V.J. 2021. Hazardous Microplastic Characteristics and Its Role as a Vector of Heavy Metal in Groundwater and Surface Water of Coastal South India. *Journal of Hazardous Materials* 402: 123786. doi:10.1016/j.jhazmat.2020.123786.

Sruthy, S., and E.V. Ramasamy. 2017. Microplastic Pollution in Vembanad Lake, Kerala, India: The First Report of Microplastics in Lake and Estuarine Sediments in India. *Environmental Pollution* 222: 315–22. doi:10.1016/j.envpol.2016.12.038.

Sun, J., Zhu, Z-R., Li, W-H., Yan, X., Wang, L-K., Zhang, L., Jin, J., Dai, X., and Ni, B-J. 2021. Revisiting Microplastics in Landfill Leachate: Unnoticed Tiny Microplastics and Their Fate in Treatment Works. *Water Research* 190: 116784. doi:10.1016/j.watres.2020.116784.

Tan, Y., Dai, J., Wu, X., Wu, S., and Zhang, J. 2022. Characteristics, Occurrence and Fate of Non-Point Source Microplastic Pollution in Aquatic Environments. *Journal of Cleaner Production* 341: 130766. doi:10.1016/j.jclepro.2022.130766.

Tourinho, P.S., Koči, V., Loureiro, S., and van Gestel, C.A.M. 2019. Partitioning of Chemical Contaminants to Microplastics: Sorption Mechanisms, Environmental Distribution and Effects on Toxicity and Bioaccumulation. *Environmental Pollution* 252: 1246–56. doi:10.1016/j.envpol.2019.06.030.

Turra, A., Manzano, A.B., Dias, R.J.S., Mahiques, M.M., Barbosa, L., Balthazar-Silva, D., and Moreira, F.T. 2014. Three-Dimensional Distribution of Plastic Pellets in Sandy Beaches: Shifting Paradigms. *Scientific Reports* 4 (1): 4435. doi:10.1038/srep04435.

Um, M., Weerackody, D., Gao, L., Mohseni, A., Evans, B., Murdoch, B., Schmidt, J., and Fan, L. 2023. Investigating the Fate and Transport of Microplastics in a Lagoon Wastewater Treatment System Using a Multimedia Model Approach. *Journal of Hazardous Materials* 446: 130694. doi:10.1016/j.jhazmat.2022.130694.

Uzun, P., Farazande, S., and Guven, B. 2022. Mathematical Modeling of Microplastic Abundance, Distribution, and Transport in Water Environments: A Review. *Chemosphere* 288: 132517. doi:10.1016/j.chemosphere.2021.132517.

Xu, S., Ma, J., Ji, R., Pan, K., and Miao, A-J. 2020. Microplastics in Aquatic Environments: Occurrence, Accumulation, and Biological Effects. *Science of The Total Environment* 703: 134699. doi:10.1016/j.scitotenv.2019.134699.

Yurtsever, M. 2019. Tiny, Shiny, and Colorful Microplastics: Are Regular Glitters a Significant Source of Microplastics? *Marine Pollution Bulletin* 146: 678–82. doi:10.1016/j.marpolbul.2019.07.009.

Zhu, J., Yu, X., Zhang, Q., Li, Y., Tan, S., Li, D., Yang, Z., and Wang, J. 2019. Cetaceans and Microplastics: First Report of Microplastic Ingestion by a Coastal Delphinid, Sousa Chinensis. *Science of The Total Environment* 659: 649–54. doi:10.1016/j.scitotenv.2018.12.389.

3 Microplastic in the Air
Sources, Distribution, and Environmental Implications

Somvir Bajar, Neha Yadav, and Kavita Yadav

3.1 INTRODUCTION

Microplastics (MPs) in the environment are an emerging global concern due to their health and environmental impacts. Thompson coined the term "microplastic", which is now widely used to refer to plastics of size smaller than 5 mm (Ahmad et al., 2023). The small size and low density of airborne microplastics (aMPs), one of several contaminants recently discovered in the atmosphere, are a concern for the public, media, government, non-governmental organisations (NGOs), and scientists. The introduction of MPs within the atmosphere has been primarily and rapidly caused by civilisation. Expanding urban areas, the use and disposal of large quantities of plastic products increases the level of microplastic pollution. MPs exist in a variety of forms, sizes, and shapes, including filaments, threads, bits, microbeads, film, foam, and pellets. Different functional zones within the same city have various forms of atmospheric microplastics. The MPs at industrial locations are spherical, fractured, and film-like, whereas those in urban areas are fibrous. In contrast, it was observed that MPs collected from street dust samples contained spherical and film-like particles of varied size (Abbasi et al., 2019; Yao et al., 2022). Since the mid-20th century, there has been significant growth in plastic production and consumption worldwide. With 30% of global plastic output, China leads the way in plastic production, followed by the "United States of America (18%), Europe (17%), the rest of Asia (17%), the Middle East and Africa (7%), Japan (4%), Latin America (4%), and the Commonwealth of Independent States (3%)" (Jahandari, 2023).

Weathering and degradation of plastics result in the formation of MPs. Plastic is photo-oxidised by UV light, which makes it brittle and breaks it up into tiny pieces called MPs. Laundry activities also discharge MPs into the water from plastic-containing garments (Haque et al., 2023). However, 90% of the generated MPs can remain in the surroundings, particularly in non-coastal areas, because of their non-biodegradable nature. Current estimates indicate that the amount of plastic thrown into the terrestrial ecosystem each year is 4–23 times more than the amount dumped into the ocean. Accordingly, these highly developed and mysterious toxins seriously endanger not only people, but also other terrestrial species, aquatic species, microorganisms, and plants (Rose et al., 2023a).

There are laws in some countries to reduce the impacts of plastics, however, the accumulation of plastic trash is continuously affecting the environment. It has been revealed that 10% of plastics manufactured find their way into the oceans, where they persist, build up, and act as a means of transportation for other pollutants. Additionally, 95% of plastic produced ends up in oceans and disturbs the marine ecosystem (Auta et al., 2017). Small plastic components, filaments, and nanoparticles floating in the environment are capable of being carried across great distances by atmospheric currents, which are referred to as airborne microplastics (aMPs) (Jahandari, 2023). Furthermore, these aMPs are ultimately ejected from the atmosphere through dry and wet deposition mechanisms.

Microplastic in the Air: Environmental Implications

According to Chen et al. (2022), aMPs can be as small as 5 mm to several mm. Understanding the impact of their presence, it is becoming imperative to define several effective interventions to reduce the distribution of microplastics into the surroundings and to develop novel strategies to prevent their accumulation in the environment.

This study provides a crude classification of aMPs and their associated adverse impacts on organisms, human health, and ecological systems. It also summarised the challenges associated with the MP mitigation strategies, governmental policies, and future perspectives to minimise their impacts.

3.2 AIRBORNE MICROPLASTICS IN THE ENVIRONMENT

The extent of environmental exposure to airborne microplastics (aMPs) relies on the widespread distribution of these particles. The most significant sources cover the microplastics generated from synthetic fabrics, deterioration of artificial rubber tyres, and urban rubble. According to Boucher et al. (2017), wind transmission is responsible for 7% of oceanic pollution. Based on major sources and methods of the formation of MPs, they can be categorised into two different types: primary and secondary microplastics.

3.2.1 PRIMARY MICROPLASTICS

These MPs are tiny particles of plastic that are used in many commercial applications such as microbeads in skincare products, plastic pellets used in industrial manufacturing, and plastic fibres used in synthetic textiles (Zhang et al., 2017). These primary MPs are further subdivided into two additional categories including microbeads and nurdles. Polyethylene, polypropylene, and polystyrene are the most common materials used to produce microbeads. Microbeads are tiny plastic beads of size <1 mm that are frequently found in personal care and cosmetic items, including body wash, toothpaste, biological and exfoliating scrubs, and health-related research. Although they are intended to add texture, they are also easily able to enter the atmosphere, which can cause environmental pollution. Nurdles are tiny plastic pellets that are used as raw materials to make plastic goods. These MPs are used to produced consumer plastics products on melting down. The pellet loss may happen from leakage during manufacturing or shipping activities and they frequently end up in rivers, oceans, and in the atmosphere (Zhang et al., 2020).

3.2.2 SECONDARY MICROPLASTICS

Secondary microplastics are small particles or fragments resulting in the formation of MPs from larger plastic goods such as bottles, bags, and packaging materials by some physical or chemical change. Due to a variety of environmental conditions, such as sunlight, temperature changes, and mechanical stress, these larger plastic objects deteriorate with time, forming secondary microplastics (Sulistyo et al., 2020).

3.2.3 FRAGMENT AND FIBRE MICROPLASTICS

There are some additional categories of secondary microplastics which include fragment microplastics and fibre microplastics. Fragments MPs are oddly shaped plastic fragments that are generated through the mechanical breakdown of larger plastic products. They can be anything from minuscule particles to a few millimetres in size. These minute plastic fibres may be generated from synthetic fabrics, when they are washed, worn, or subjected to other processes. Over time, weathering can cause fragmentation to microplastics from thin plastic films, and fabrics like nylon and polyester, which are used in packaging and plastic bags. A study conducted to examine washing machine effluent revealed that around 1900 fibres can be produced by a single garment per wash (Dodson et al., 2020).

3.2.4 ADVERSE IMPACT OF MICROPLASTICS ON THE ENVIRONMENT AND PUBLIC HEALTH

A severe issue with the management of plastic waste has emerged in the past few years due to the expanding demand for plastics in both the domestic and industrial sectors. Contamination with microplastics is acknowledged as a major concern due to its potential harmful nature for the entire atmosphere, hydrosphere, and lithosphere. It is dispersed over large areas by the action of winds and waves due to its light weight (Welle et al., 2018; Weis et al., 2020). Plankton, which makes up the lowest trophic level of the marine ecology, have reportedly been found to incorporate microplastics in their biomass. As a part of the food chain, it translocates into other trophic levels and results in bioaccumulation. Moreover, humans consume both plants and animals, resulting in enhanced levels of microplastics being ingested through food (Weis et al., 2020).

Considering the growing issue, it is becoming crucial to determine the origins and dispersion pattern and explore sustainable treatment options to reduce its impact. It is also important to understand the varied outcomes of microplastics on the environment and health of living creatures. Planning, authorisation, sample collection, and additional analysis are the steps that need to be followed for comprehensive assessment of MPs. However, sample collection and subsequent testing are yet to be given a set of standardised operational protocols.

3.3 SOURCES OF MICROPLASTICS

These tiny plastic particles have recently become the focus of growing environmental concern due to their cosmopolitan nature and variable impacts. As far as their origin is concerned, these microplastics sources may be categorised into two main categories including primary and secondary sources (Figure 3.1).

3.3.1 PRIMARY SOURCES OF MICROPLASTICS

MPs which are generated from commercial applications, certain industries, or home applications are referred to as "primary MPs". Various personal care products, including cosmetic products, body products, pesticides, etc. utilise these primary microplastics as ingredients and they are dumped into the environment during their disposal (Fendall et al., 2009; Castaneda et al., 2014). The removal of machine corrosion and painted surfaces, motors, and boat hulls using air-blasting technology also generates MPs. Additionally, the use of acrylic, melamine, or polyester MP scrubbers also increased the emission of MPs into the environment (Hopewell et al., 2009).

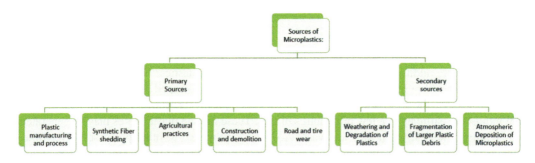

FIGURE 3.1 Classification of microplastics based on origin.

3.3.1.1 Plastic Manufacturing and Process

MPs are produced mostly from the fragmentation of larger plastic items during their production and processing. Plastic materials are shaped and formed into a variety of goods during the plastic process, which can result in the production of waste and byproducts. These waste products, which may take the shape of plastic shavings, particles, or scraps, have the ability to disintegrate into smaller pieces over time to produce MPs. Moreover, a variety of channels, including runoff from production plants, improper waste disposal, and transportation, enable microplastics to infiltrate the environment (Chen et al., 2020).

3.3.1.2 Synthetic Fibre Shedding

Synthetic microfibres are generated from carpets, garments, and other fabrics composed of polyester, nylon, and acrylic. Moreover, a variety of activities, including washing clothes, utilising goods made of synthetic fibres, and disintegration of heavier plastic items, release a number of MPs into the surrounding environment (Browne et al., 2011). However, only 7% of MPs are of the fibre type, suggesting that other types, such as films, spherical forms, and fragments, are more common in suspended air, which are generated from synthetic components (Choi et al., 2024).

3.3.1.3 Road and Tire Wear

Air is contaminated with considerable amounts of plastic particles due to road abrasion and tyre wear. Various studies have revealed that tyre wear is one of the main causes for the generation of MPs. In 2017, IUCN published a report that estimated that up to 28% of MP pollution in the oceans comes from tyre wear. Although there are many different materials used to make rubber tyres and road wear, around 50% of rubber tyres are constructed from synthetic and natural polymers such as carbon black, natural rubber, styrene-butadiene rubber, and others (Sommer et al., 2018; Hagström, 2021).

3.3.1.4 Construction and Demolition

Construction and demolition (C&D) activities contribute to releasing dust and debris into the surroundings, which contributes to microplastics in the environment. The use of plastic-based building materials, such as PVC pipes, plastic insulation, and synthetic roofing materials, can lead to the release of MPs when these materials degrade, break, or are crushed during demolition. Many construction materials, such as paints, coatings, and sealants, also contain MPs as additives for enhancing performance. Deterioration or removal of these material also contributes to aMPs during demolition (Geyer et al., 2017).

3.3.1.5 Agricultural Practices

A number of mechanisms, including the splintering of plastic mulch films, the breakdown of plastic-based agricultural equipment, use of plastic-based fertilisers, and windborne transport from urban and rural areas, may contribute to MPs in the ambient air. Theoretically, these systems release MPs into the atmosphere, where they can spread and travel great distances (Jahandari et al., 2023).

3.3.2 SECONDARY SOURCES

When larger plastic waste particles break down, "secondary microplastics" are formed. Physical, chemical, and biological degradation mechanisms have the potential to weaken and alter the plastic's structural integrity, which results in disintegration (Auta et al., 2017). Various meteorological factors, including solar flux and temperature, can result in plastic fragmentation, size changes, and variations in density (Hopewell et al., 2009; Martinho et al., 2022). Cosmic radiations can encourage bond cleavage in plastic based materials through MP matrix oxidation (Hopewell et al., 2009; Auta et al., 2017). Moreover, in coastal regions, various factors, such as UV radiation, oxygen availability, and mechanical abrasion by waves, sand, and wind, make beaches a more active sites for the disintegration of macroplastics into MPs (Auta et al., 2017).

3.3.2.1 Fragmentation of Larger Plastic Debris

Breakdown of plastic litter in the environment, in urban and rural areas, is a common scenario resulting in the generation of MPs. These plastic wastes can become brittle over time and break apart into smaller fragments, when exposed to environmental stressors. These smaller plastic particles are carried by the winds and transported to different areas, contributing to the generation of aMPs (Browne et al., 2011).

3.3.2.2 Weathering and Degradation of Plastics

Plastics components undergo physical and chemical changes because of variable environmental conditions, which cause them to break down into smaller parts. These smaller fragments are blown away through wind and can be transported long distances. Plastic products undergo photodegradation when they come into contact with UV radiations. Additionally, temperature also affects their chemical bonds and results in rapid fragmentation of larger plastic components. In ambient conditions, plastics can interact with chemicals and airborne pollutants, which lead to chemical changes and result in weakening of the material and an increase in its susceptibility to shattering (Woo et al., 2021).

3.3.2.3 Atmospheric Deposition of Microplastics

The process by which minute plastic particles in the air fall to the Earth's surface because of gravity or other atmospheric factors is referred to as atmospheric deposition of MPs. These MPs of variable size, shape, and composition are easily available in metropolitan regions, industrial settings, agricultural processes, and marine ecosystems, among other places. During their deposition, microplastic pollutants including dust, pollen, and soot particles are present during atmospheric transit, and help with their dispersion and settling. Deposition can happen in both natural and urban settings, and it may impose long-term effects on ecosystems (Allen et al., 2019).

3.4 MECHANISMS OF MICROPLASTICS TRANSPORT IN THE ATMOSPHERE

A variety of MP components are available in surrounding conditions; however, the airborne transfer of MPs is not completely understood. Notably, compared to water systems, air has fewer dispersal boundaries. Moreover, the movement of MPs in the atmosphere is not entirely independent from pollutants existing in water and land, and further, exploration of this aspect is research is required to clarify the mechanisms involved (Dris et al., 2016). Additionally, these microplastics provide a surface to adhere with persistent organic contaminants and aid in their transmission to coastal areas (Figure 3.2) (Padervand et al., 2020).

3.4.1 Atmospheric Transport and Dispersion

When MPs are released into the atmosphere, atmospheric currents can transport them over variable distances. Wind speed, turbulence, and atmospheric conditions are only a few examples of the variables that have an impact on the size, distribution, and direction of dispersion (Hitchcock, 2020; Wang et al., 2020).

Microplastic in the Air: Environmental Implications

FIGURE 3.2 Sources and transport of microplastics.

3.4.2 Function of Wind and Weather

Wind plays an important role in the transportation of MPs due to their light weight and durability. These MPs distribute themselves both locally and far from their initial locations by lifting and moving actions. Moreover, dispersal of MPs in the atmosphere and their eventual deposition are influenced by several meteorological factors including wind speed, wind direction, storms, etc. (Evangeliou et al., 2020).

3.4.3 Aerosolisation and Gas-particle Partitioning

MPs may become aerosolised, which causes them to float in the air as suspended particles. Numerous factors, including physical fragmentation, turbulence, and human activity, contribute to the emergence of this phenomenon. The distribution of MPs between the gas and particle phases in the atmosphere is known as gas–particle partitioning. These particles interact with other particulate pollutants and change their behaviour because of their interactions with gases (Shao et al., 2022).

3.4.4 Deposition and Settling Processes

Deposition refers to the process through which flying MPs settle onto surfaces under the influence of gravity and other factors. These surfaces include a variety of natural and artificial habitats, including soil, water, plants, and buildings. The properties of the receiving surfaces, density, size of the MP particles, and wind speed, all play an influential role on the settling processes. These tiny plastic components can collect after deposition, which could affect ecosystems and human activities (Zhang et al., 2020).

3.4.5 Microbial Degradation of Microplastics

MPs are usually recalcitrant in the environment; however, some identified microbes have the capability to degrade these plastic components. Different enzymes generated by microorganisms are used for the degradation and disintegration of MPs (Othman et al., 2021). These plastic components are less vulnerable to microbial attack than other biodegradable materials. However, microbial degradation has had a significant impact on the transformation of MPs. This microbial colonisation and support for the growth in addition to acting as a carbon source, enables MPs to provide a distinct ecological niche for microorganisms (Yuan et al., 2020).

3.5 SAMPLING AND ANALYSIS METHODS FOR MICROPLASTICS IN THE AIR

MPs are now widespread contaminants in both indoor and outdoor settings, greatly adding to air pollution (Azari et al., 2023). These components in atmospheric fallouts have been discovered for both indoor and outdoor environments using a sampler or vacuum pump, an airborne particles fallout collector, or a rain sampler. These airborne microplastics (aMPs) are distributed and moved because of human, climatic, and meteorological factors. After sampling, the samples are analysed using visual, spectroscopic, or spectrometric techniques (Enyoh et al., 2019). Researchers have been looking into several approaches for detecting and studying aMPs, considering their potential impacts. However, it is difficult to evaluate and interpret the results due to the lack of standardisation and a variety of approaches are employed for the identification of these aMPs. However, each method is associated with difficulties and limitations which are discussed in the present study.

3.5.1 CHALLENGES DURING THE COLLECTION OF MICROPLASTICS SAMPLES

3.5.1.1 Heterogeneity and Sampling Methods

The heterogeneity of airborne particulate matter is one of the main obstacles in understanding aMPs. Organic and inorganic materials of variable sizes, compositions, and forms are associated with airborne particles forming a complex system. Accurate isolation and identification of MPs are challenging due to this complexity. Researchers have been creating and improving sample collection techniques especially made for aMPs to resolve this issue.

3.5.1.2 Challenges in Identification and Quantification

The precise and accurate identification and measurement of MPs after the collection of air samples is the key challenge for MPs. Visual classification is a popular technique for classifying MPs, although it might not be effective for finding smaller MPs (Azari et al., 2023). Microscopy methods can be combined with other analytical tools to improve the precision of MP identification. Another challenge is measuring MPs in air samples. Accurate quantification of smaller plastic particles can be particularly difficult. The assessment of these tiny particles' abundance in the atmosphere is currently complicated in the absence of established quantification methods.

3.5.1.3 Contamination Control Measures

MP sampling requires contamination control procedures to prevent cross-contamination and guarantee reliable results. Even in places where anthropogenic pressure is minimal, MPs are pervasive. Therefore, during handling and processing samples, strict contamination control procedures must be followed. Various contamination control procedures include utilising clean equipment, dressing appropriately, and thoroughly cleaning work areas to reduce the introduction of external MPs. Additionally, samples should be carefully sealed and preserved to avoid contamination during storage and transportation. Any scientific study must include quality assurance and control (Adhikari et al., 2022).

3.5.2 SAMPLING TECHNIQUES FOR AIRBORNE AND GAS-PHASE MICROPLASTICS

Concerns pertaining to potential negative impacts of MPs on public health and the environment have been raised by their presence in air. It is essential to create trustworthy sampling methods and identification techniques to understand the abundance and distribution of these particles. Pre, during, post, and contamination avoidance sampling techniques are some of the major components of sampling and analysis that are discussed in this chapter (Figure 3.3).

Due to their compact size, extensive distribution, and tendency to contaminate samples, airborne microplastics (aMPs) pose several difficulties. In order to efficiently collect and analyse

Microplastic in the Air: Environmental Implications

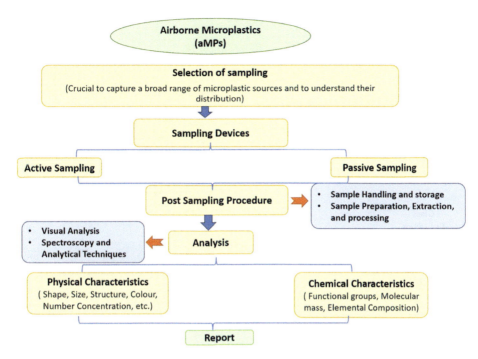

FIGURE 3.3 Sampling and analysis procedure for airborne microplastics.

aMPs, researchers follow a variety of approaches. Some commonly used sampling techniques are explained below.

3.5.2.1 Passive Sampling

Exposing a collecting substrate to the surrounding air during passive sampling allows MPs to accumulate over time (Azari et al., 2023). Sticky surfaces, open-faced filters, and sedimentation plates are examples of typical passive sampling techniques. Analysing the quantity of airborne MPs during wet and dry atmospheric deposition is essential to measuring the final load of MPs released into the surroundings (Rocha-Santos et al., 2022). Complete knowledge of the MPs' accumulation range throughout a certain time frame is crucial for the collection of MPs through passive sampling techniques (Shao et al., 2022). Most of the most popular passive sampling techniques involve collecting wet or dry atmospheric deposition through a funnel into a glass container. The passive fallout from the atmosphere is collected and filtered to remove MPs (Yang et al., 2021). A precise time and place indicator of a number of MPs that disperse outside is provided by passive air deposition samplers (Zhang et al., 2020).

3.5.2.2 Active Sampling

Dynamic sampling techniques actively draw air in and collect MPs via mechanical means. Pumping sampling systems like particle counters and high-volume air samplers are examples of active sampling techniques. These techniques entail pumping a preset volume of air for a predetermined amount of time. Filters capture the granules on their surface or inside of them while allowing the air to circulate into them when it passes through the instrument. The airflow strikes the substrates, causing them to bounce off and leave the particles behind (Li et al., 2020; Abbasi et al., 2022; Shao et al., 2022).

It basically uses a pumping and filtration system in which air is captured by the pump unit and afterwards screened through the filter to isolate the microplastic particles. Different effective

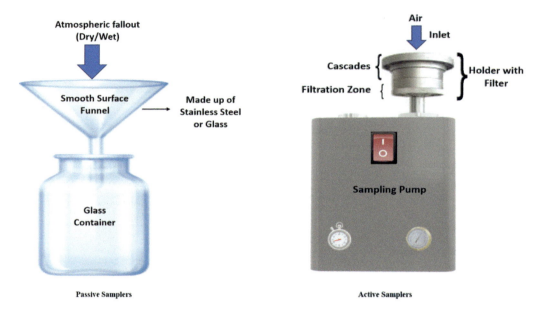

FIGURE 3.4 Sampling devices for airborne microplastics.

quantitative analysis techniques can be performed once the sample is collected to determine the kinds and dimensions of particles (Yang et al., 2021). Active samplers pump air, providing an instance of MP contamination in the air volume as compared to MP debris contamination (Figure 3.4) (Zhang et al., 2020).

3.5.3 Laboratory Analysis and Characterisation of Airborne Microplastics

Laboratory analysis and study are crucial for determining the source distribution and potential negative consequences of aMPs on the environment and health. Both the quantity and quality of plastic particles in the air are detected and quantified using sampling techniques and analytical methods.

3.5.3.1 Analytical Techniques for Microplastics

aMPs can be identified and measured using analytical methods such as pyrolysis gas chromatography-mass spectrometry, XRF, stereomicroscopy, SEM, matrix-assisted laser desorption/ionisation time-of-flight, Raman spectroscopy, FTIR, liquid chromatography-tandem mass spectrometry, and flow cytometry (Chen et al., 2020; Adhikari et al., 2022; Yadav et al., 2023). Because of their small size and extensive environmental spread, gas-phase MP examination poses specific difficulties. Diverse analytical techniques have been developed to locate and measure MPs. Some commonly used techniques are described below.

Fourier Transform Infrared Spectroscopy (FTIR)
The detection of MPs can be done using a particle-based method with FTIR (Rose et al., 2023). The absorption patterns are measured on exposing the MP particles to infrared radiation. This aids in classifying the different plastic polymers that are present in the gas-phase samples of the materials. In recent years, mapping for particle counts and size distribution has been a popular application of FTIR, which has been traditionally used to characterise materials. FTIR provides comprehensive information about a particle's composition and molecular structure by utilising the IR wavelength range (Zhang et al., 2020).

Raman Spectroscopy

The molecular structure of MPs can also be revealed through Raman spectroscopy. This entails shining a laser beam on the sample and measuring the light's inelastic scattering. Based on the distinctive molecular vibrations of each form of plastic, Raman spectroscopy is useful for making this distinction. Raman spectroscopy provides structural data in a non-destructive manner that reveals the identity of the polymer and various types of additives (Zhang et al., 2020; Adhikari et al., 2022).

SEM (Scanning Electron Microscopy)

Researchers can characterise MPs' size, shape, and surface morphology on examining them at extreme magnification using SEM (Ahmad et al., 2023). A combination of EDS (energy-dispersive X-ray spectroscopy) and SEM can reveal the elements present in a sample. A high-powered electron beam is used in the SEM technique to scan the surface of analytical objects. MPs and organic contaminants that may have adhered to their surfaces also be identified using SEM and energy-dispersive X-ray spectroscopy (Adhikari et al., 2022). Projecting a powerful electron beam onto the sample's surface, high-resolution imaging of the particulate can be obtained using SEM. As a result, it is possible to visualise MPs' surfaces clearly and identify their microstructure (Abbasi et al., 2022; Shao et al., 2022).

μ-FTIR (micro-Fourier Transform Infrared Spectroscopy)

μ-FTIR, which combines both features – allowing microscopic inspection of minutely sized plastic-like particles before spectroscopic confirmation on a single platform – makes MPs identification smoother (Woo et al., 2021). This method is an improvement over conventional FTIR that provides greater spatial resolution. FTIR assists in determining the chemical make-up of MPs and possibly their sources on analysing specific MPs from complex matrices (Chen et al., 2020; Bouzid et al., 2022).

Py-GC/MS (pyrolysis-Gas Chromatography/Mass Spectrometry)

Py-GC/MS emerged as a potential technique to identify MPs in a material without using a microscope (Zhang et al., 2020). MPs are heated to high temperatures in an inert environment during pyrolysis, which causes them to thermally degrade into volatile chemicals (Kung et al., 2023). The evolved gases are then analysed with GC/MS to reveal the kind of polymer used and the additives utilised (Bouzid et al., 2022). It determines the kind of plastic present in a sample as well as its concentration (Table 3.1) (Zhang et al., 2020).

3.6 SPATIAL AND TEMPORAL DISTRIBUTION OF AIRBORNE AND GAS-PHASE MICROPLASTICS

Wind and atmospheric currents have the ability to transport airborne microplastics (aMPs) over great distances. Based on factors such as closeness to sources of plastic pollution, wind patterns, urbanisation, and industrial activity, the spatial distribution of aMPs may differ. However, the temporal dispersion of MPs in the air may change under the influence of weather, seasonal cycles, and human impacts.

Transmission, dispersion, and deposition mechanisms propel pollutants through the atmosphere. Additionally, these elements are responsible for the mobility of aMPs. The direction and speed of the ambient wind transfer aMPs from source sites to surrounding and distant places. Deposition, which is the movement of airborne MPs descending to the Earth's surface is influenced by rainfall, scavenging, and deposition, while dispersion is caused by local turbulence and disturbances. The MP's dimensions, form, and length aid the complete movement (transport, dispersion, and deposition) operations (Allen et al., 2019). According to recent studies, MPs can be transmitted through the air in a variety of shapes or configurations (Allen et al., 2019; Enyoh et al., 2019; Liu et al.,

TABLE 3.1

Different Analytical Techniques Used for Identifying Microplastics

S. no.	Analytical Techniques	Description	Advantage	Limitations
1.	**Fourier transform infrared spectroscopy (FTIR)**	Measures the amount of infrared light absorbed by microplastics to analyse their chemical makeup and determine their kind of polymer	Non-destructive, requires little preparation of samples, and can differentiate among various polymer varieties	Restricted specificity for trace analysis; precise identification could need a standard library
2.	**Raman spectroscopy**	Helps identify polymers by providing molecular fingerprinting of microplastics depending upon their vibrational modes		
3.	**Scanning electron microscopy (SEM)**	MPs may be seen at great magnification and clarity using SEM, and their chemical makeup can be analysed using EDS (energy-dispersive X-ray spectroscopy) to help with characterisation and identification	Offers comprehensive morphology of the surface and elemental data, making it appropriate for trace analysis as well as complex samples	Involves the preparation of the sample, which can be costly and time-consuming
4.	**Micro-Fourier transform infrared spectroscopy (µ-FTIR)**	MPs can be directly identified and mapped on sample surfaces using this technique, which is similar to traditional FTIR but modified for microscopic research	Gives spatial data on the distribution of microplastics, which is useful for examining complicated environmental samples	Restricted by the area under analysis, sample preparation can be necessary for the best outcomes
5.	**Pyrolysis-gas chromatography/mass spectrometry (Py-GC/MS)**	Used to isolate and identify the volatile chemicals that are produced when microplastic samples are broken down by pyrolysis	Offers great accuracy, precision, and comprehensive knowledge about the components and structure of the polymer	Potential to cause sample heat deterioration and calls for specific tools and knowledge

2019). However, source and transport assessments of aMPs remain unexplored areas. The kind and volume of emissions alone do not entirely evaluate the air pollution caused by MPs. Additionally, the diffusion and emplacement of MPs are significantly influenced by climate, topography, and meteorological circumstances (Enyoh et al., 2019).

3.6.1 GLOBAL DISTRIBUTION PATTERNS

Global detection of airborne and gas-phase MPs suggests that they are widely dispersed in the atmosphere. MPs are detected in both urban and rural settings, indicating that air currents are responsible for their long-distance transportation. Urban pollution, road dust, industrial emissions, and atmospheric deposition from marine sources are notable sources of aMPs. An essential pathway for the worldwide accumulation of MPs in the aquatic environment and landmasses is MPs being present throughout the atmosphere (Huang et al., 2021). MPs can encourage anaerobic decomposition, which increases CO_2 and methane production, in addition to affecting biological, soil, and microbial community features. Consequently, the pollution of MPs directly contributes to climate change (Shao et al., 2022). Using the online HYSPLIT (Hybrid Single Particle Lagrangian Integrated Trajectory) tool from the NOAA, back trajectories were utilised to assess the origins of masses of air and the possible source range of MPs (Abbasi et al., 2022).

MPs can travel great distances (i.e., 95 km) when dangling in the air (Allen et al., 2019; Shao et al., 2022). Additionally, they can be moved hundreds or even thousands of kilometres before undergoing dry or wet deposition, especially fibrous MPs with comparatively high aspect ratios and slow settling rates. MPs in the atmosphere are a concern for human health due to their small size and persistent nature (Abbasi et al., 2022). MPs also have been discovered in marshes, coastal areas, areas off the coast, and the open ocean through long-range transport.

3.6.2 REGIONAL AND LOCAL VARIABILITY

Depending on the area and local circumstances, the concentration of airborne and gas-phase MPs can change dramatically. Due to increased anthropogenic activities, garbage production, and plastic usage, urban areas typically have higher concentrations. Coastal areas may get MPs from both terrestrial and marine sources, while proximity to coastlines can also affect the levels of these tiny plastic particles. Light weight, durability, and other innate properties aid in the transportation of atmospheric MPs to remote areas where they are deposited through dry or wet deposition. Wind, snowfall, and weathering are significant factors which exist in transporting MPs from sources to ocean or land surfaces (Allen et al., 2019; Zhang et al., 2020). Various regions have distinct atmospheric MP concentrations, which may be impacted by regional geography, local climatic circumstances, the urban heat island effect, and other variables (Li et al., 2020; Yang et al., 2021)

3.6.3 SEASONAL AND DIURNAL TRENDS

MPs concentrations have been found to vary seasonally and during the day, most likely because of weather changes and human activity. For instance, studies have found greater MP concentrations in windy circumstances, which cause plastic particles to resuspend from various surfaces (Allen et al., 2019). Additionally, changes in plastic usage habits, such as an increase in outdoor activity during the warmer months, also have a strong impact on seasonal fluctuations.

Wind, heat, rainfall, and snowfall impact the MP concentration in the atmosphere's lower layers (Allen et al., 2019). The wind's turbulence, velocity, and direction (vertically or horizontally), all affect wide deposition of MPs. The ascent of aMPs is aided by the vertical temperature gradient (Enyoh et al., 2019). The MPs' size and shape, in addition to weather factors like precipitation, snowfall, wind speed, moisture content, air currents, and temperature, all influence the transport and deposition of MPs (Zhang et al., 2020).

3.7 MICROPLASTICS' FATE AND BEHAVIOUR IN THE ATMOSPHERE

Since terrestrial, aquatic, coastal, and atmospheric habitats are interlinked and have diverse source–sink connection networks, the destiny of atmospheric MPs is frequently interconnected with marine and terrestrial MPs. These long-lasting MPs continue to influence both people and the environment. With tremendous propulsion of air, MPs in the external atmosphere are transferred mostly via air movement. Moreover, with their structure, size, low weight, resilience, and buoyancy, MPs are easily carried over long distances through air (Rose et al., 2023). Furthermore, these MPs penetrate soil and oceans through atmospheric deposition (Zhao et al., 2023).

Plastic molecules can linger in the atmosphere for 1–6.4 days (Brahney et al., 2021), which is long enough for trans-Pacific transit to take place. These aMPs are directly introduced into the seas by long-range transport, which functions as a driver (Abbasi et al., 2023). Urban topography, regional weather patterns, and variations in wind direction brought on by thermal cycling can all have an impact on the dispersion and mobility of MPs in the surroundings (Zhao et al., 2023). In 2019, González-Pleiter et al. (2021) employed the Hybridised Single-Particle Lagrangian Incorporated Trajectory (HYSPLIT) tool for the first time to study the movement of environmental MPs and the tracking of pollution sources. Zhao et al. (2023) postulate that ambient air MPs were transported over more than 1000 km across the ocean.

Su et al. (2020) revealed that roughly 141,000 tons of MPs from traffic enter the ocean through the sky each year, which imposes a significant negative influence on coastal ecology. In recent years, MP pollution in terrestrial or aquatic habitats has been significantly linked to air transport (Bergmann et al., 2019). Variables which include the gradient of pollution quantity, wind velocity, prevailing winds, temperature, and humidity affect the dispersion and settling of MPs. This study observed that MPs in the urban environment are washed away and transferred during heavy rains but settle and accumulate on roads and their edges during dry spells.

The destiny of atmospheric MPs that are present indoors and those that are outdoors differs in several ways. Due to the poor air mobility inside, atmospheric MPs cannot easily spread. This is one of reasons for the enrichment of MPs in indoor environments. The higher concentration and smaller size of MPs in indoor environments increase their ability to deeply penetrate the respiratory systems of both humans and animals (Vianello et al., 2019). The size of the MPs in the surrounding environment influences settling and gathering on the ground indoors, which is similar to the behaviour of MPs in outdoor atmospheric conditions (Figure 3.5).

MPs, whether they are deposited on the soil surface or enter rivers to flow with water flow, are responsive to climatic climate factors which break them down into smaller crystallite sizes, even nanoscale, or they are subsequently destroyed and transformed into carbon dioxide (CO_2), water, and methane (Zhao et al., 2023). Under the influence of UV radiation and oxygen in the air, prolonged exposure to sunlight leads to MPs being suspended in the air, floating on the surface of waterbodies, and accumulating on the surface of the land and gradually dissolving (Wang et al., 2018). MPs mistakenly consumed by terrestrial animals and marine species also make their way into the marine and terrestrial food chains. Studies on assessing the biological impacts of MPs on marine species have revealed that these particles can be digested or swallowed by a variety of marine organisms and accumulate in these organisms, disturbing their normal metabolism and disrupting their systemic balance and leading to their death (Franzellitti et al., 2019). Health hazards oriented from MPs in the soil are transported and enriched through the food chain. However, a significant gap has been observed in explaining the influence of MPs on terrestrial ecosystems and transfer factors.

Microplastic in the Air: Environmental Implications

FIGURE 3.5 Fate of airborne microplastics.

3.7.1 Microplastics' Transformation and Ageing

The different changes and ageing processes environmental MPs undergo can have a significant impact on their characteristics and behaviour (Zha et al., 2022). Some of these processes are explained below for a better understanding of the transportation and ageing of MPs.

a. *Weathering:* MPs can undergo physical and chemical changes due to UV light, temperature changes, and moisture. This may result in fragmentation, which over time will reduce the size of the MPs. Natural weathering is responsible for 94% of the MPs that are widespread in the atmosphere (Hu et al., 2023).
b. *Chemical alteration*: Utilising atmospheric constituents and pollutants like ozone, MPs can undergo chemical reactions resulting in the generation of functional groups on their surfaces.
c. *Biological interactions*: Microbes in the air can colonise MPs, producing biofilms in the process and possibly affecting their transit and destiny.
d. *Coating*: MPs can gather physical and chemical coatings from the atmosphere.

3.7.2 Interaction with Atmospheric Gases and Particles

During their transportation and distribution, MPs in the atmosphere have a probable interaction with atmospheric and climatic factors (Zhang et al., 2020). A few among them are listed below.

a. *Aerosol particles*: The atmosphere's aerosol particles resemble MPs which might facilitate their long-distance transit and affect their deposition patterns.
b. *Gases*: Microbeads may soak or adsorb atmospheric air pollutants, altering their stability and function. As a result, gas components may influence their transit and destiny.
c. *Chemical reactions*: The chemical reactions between MPs and atmospheric gases could result in the production of hazardous by-products in the atmosphere, which may further increase the severity.

3.7.3 Microplastics Deposition and Removal Mechanisms

Deposition and removal mechanisms play a crucial role in determining the fate of MPs in air and gas environments (Prata et al., 2022). Various modes of deposition are mentioned below which might have a tendency to contaminate the environment.

a. *Dry deposition*: Due to gravity, MPs can leave the atmosphere. The size and density of the particles affect the pace of dry deposition.
b. *Wet deposition*: Precipitation in the form of rain, snow, or other forms can remove MPs from the atmosphere. For larger MPs, this technique is more efficient for scavenging of MPs from the atmosphere.
c. *Settling and sedimentation*: Surfaces including vegetation, soil, and water bodies are susceptible to the accumulation of MPs. Afterwards, they may merge with terrestrial or aquatic habitats and hold potential impacts on various organisms and the environment through ecological functioning.
d. *Atmospheric transport*: MPs may float around in the air for a long time, allowing them to travel great distances before being deposited. The long-range transportation of MPs make them available in isolated regions.
e. *Human activities*: MPs can be resuspended and released into the atmosphere because of anthropogenic activities including farming, construction, and urban development. In order to evaluate the environmental effects of MPs in the air and create mitigation plans for their adverse effects on ecosystems and human health, it is crucial to comprehend these transformation, interaction, and deposition mechanisms. Ongoing studies in this area are helping to better understand the behaviour of MPs in the atmosphere.

3.8 IMPLICATIONS FOR HUMAN HEALTH AND ECOLOGICAL SYSTEMS

MPs are harmful to organisms including humans and animals due to the degree of susceptibility and level of exposure (Rose et al., 2023a). MPs have gained substantial attention in environmental studies in recent years due to their potential effects on both human health and the ecology. The ability of MPs to travel in the air presents difficulties and critical issues regarding both human exposure and their possible role in climate change. For determining their environmental impact and creating efficient mitigation plans, it is essential to comprehend the distribution and properties of aMPs. These minute plastic fragments are a highly heterogeneous collection of materials since they can be found in a variety of shapes, including microbeads, pellets, pieces, and synthetic threads. With their fine size, these particles build up in the lungs of living things through respiration activities, potentially causing health problems such as inflammation, slowed growth, oxidative stress, reproductive disorders, throat irritation, shortness of breath, obstruction, coughing, and chest pain (Chen et al., 2020; Dong et al., 2020; Shao et al., 2022).

MPs can also be suspended in the atmosphere by wind or air turbulence due to their tiny size and low density, which allows them to stay in surrounding conditions for extended periods of time (Allen et al., 2020; Chen et al., 2020; Batool et al., 2022; Ferrero et al., 2022). As a result, aMPs may also pose a risk to human health due to the possibility of being inhaled and they may cause persistent inflammation, and both benign and malignant lung diseases (Szewc et al., 2021). In addition to possible concerns to human health, there are also associated risks to the ecosystem (Figure 3.6) (Enyoh et al., 2019).

3.8.1 Inhalation and Respiratory Exposure

There has been much focus on the consequences of plastic pollution on marine ecosystems, and concern over the effects of aMPs on human respiratory systems is growing. This study looks at the

Microplastic in the Air: Environmental Implications

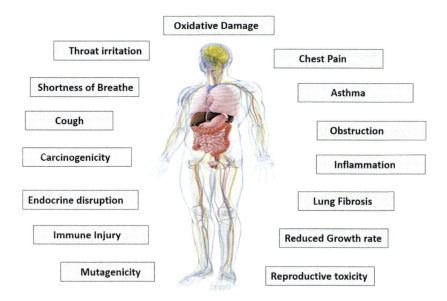

FIGURE 3.6 Impact of airborne microplastics on humans.

potential health risks associated with breathing in aMPs and the need for further exploration to fully understand their impacts. The main mode of exposure to airborne plastic particles is inhalation, which can be hazardous to the respiratory system (Jahandari, 2023)

MPs act as vectors for the co-transportation of pollutants in the environment when they adsorb contaminants. Consequently, transition metals, polycyclic aromatic hydrocarbons (PAHs), and major air pollutants could all be transported by aMPs into the environment. Additionally, co-transported pollutants have been shown to cause mutagenicity, genotoxicity, reproductive toxicity, and carcinogenicity along with additional health issues (Zhang et al., 2020; Ortega and Cort'es-Arriagada, 2023). The respiratory system is an easy route for airborne MPs to enter the body. The same aerodynamic properties as $PM_{2.5}$ may apply to MPs. These airborne particles can enter the deep lung or alveoli by respiration. They differ in concentration, size, and shape, all of which are crucial factors to consider when assessing potential negative consequences to humans (Enyoh et al., 2019).

The aMPs can act as a carrier for the transmission of viruses or other germs to the lungs, where they may infect organs (Prata, 2018). Many variables, including particle size, adsorbed chemicals, concentration, deposition, and clearance rate, influence the associated risk (Pandey et al., 2022). MPs may interact with a variety of species due to their small size, leading to blockage, inflammation, and accumulation in organs after translocation. The digestive system is the primary organ being affected by MPs as a result of possible bioaccumulation and transfer of pollutants through the food chain (Prata, 2018).

3.8.2 Potential Health Effects on Humans

Human exposure during fallout of aMPs may occur by inhalation, skin contact, and open food consumption, with the possibility of bio-persistence and transfer. Immune cell reactions and ingested substances can result in localised inflammation and cancer, especially with impaired metabolism and inadequate clearance systems. The deposition and clearance rates of aMPs affect the potential harm to human organs (Enyoh et al., 2019). Acute bronchial reactions are the first response expressed when particles are inhaled, immune cells produce cytokines, proteases, and reactive oxygen species (ROS) to protect themselves against foreign materials (Prata, 2018). Chronic inflammation may induce cancer because of DNA damage brought on by oxidative stress and particle action, immune system evasion, and pro-inflammatory mediators favouring the development and spread of cancerous cells through angiogenesis and mitogenesis (Enyoh et al., 2019). MPs can be absorbed into human tissues by phagocytosis and cellular adsorption in the respiratory system and gastrointestinal tract, which leads to inflammation, cellular necrosis, and tissue tearing (Enyoh et al., 2019). MPs can also harm the human body chemically by producing ROS, which may impose a wide range of biological impacts and toxicity, including damage to DNA, the immune system, the brain, and histologically (Zhang et al., 2020). The probable impacts introduced through exposure to MPs are detailed below.

a) *Inflammation:* Breathing in aMPs may cause inflammation of the respiratory tract. Numerous respiratory issues, including bronchitis and the escalation of pre-existing illnesses like asthma, can be induced from inflammation in the respiratory system.
b) *Cellular damage:* MPs include a variety of compounds, some of which are poisonous and may cause cellular damage. These particles may release toxic compounds into the lungs when inhaled, possibly resulting in cellular harm. Chronic respiratory conditions and possibly lung cancer may become more likely with prolonged exposure.
c) *Immunological reactions:* According to research, MPs may alter immunological function, potentially making people more prone to respiratory illnesses.
d) *Oxidative stress:* MPs generate oxidative damage in living things, which can harm cells and modify the makeup of cell membranes, reducing the ability to penetrate the membranes and interfering with cells' basic physiological processes. IDH functioning can be inhibited by MPs, which cause oxidative stress and muscle damage in living things (Mao et al., 2022).
e) *Fibrosis:* Prolonged exposure to aMPs might contribute to the development of pulmonary fibrosis, a condition in which lung tissue becomes scarred and less functional. This can lead to impaired lung function and reduced quality of life.
f) *Systemic effects:* There is emerging evidence that MPs, once inhaled, can translocate to other organs and tissues in the body. This raises concerns about potential systemic effects beyond the respiratory system.

3.8.3 Impact on Wildlife and Ecosystem

An essential mechanism for the global deposition of MPs in the hydrosphere and lithosphere is the retention of MPs in the atmosphere (Zhang et al., 2020; Huang et al., 2021). MPs that are in the air could be a source of contamination for terrestrial and aquatic ecosystems (Enyoh et al., 2019).

aMPs are ingested by animals of all sizes, including larger mammals and insects. These particles may eventually clog animals' digestive tracts, which could cause malnutrition or even death. As MPs go up the food chain, they may bioaccumulate and concentrate in higher trophic-level organisms. This could have a cascading effect on entire ecosystems. When entering soils or freshwater sediments, MPs alter the dynamics and emission levels of carbon dioxide (CO_2) and nitrous oxide (N_2O) and decrease the volatilisation of soil ammonia (Zhang et al., 2022). The dry deposition of MPs also contributes to marine plastic pollution. Furthermore, adsorbing air pollutants on aMPs may lead to higher concentrations of greenhouse gases in saltwater (Ding et al., 2021). Some of the major impacts on wildlife caused by MPs are as listed below.

a) *Toxicity:* Harmful compounds and contaminants are frequently found in MPs. These toxins can be consumed and then released into an animal's system, altering physiological processes, and leading to several health problems.
b) *Biodiversity loss:* MPs' penetration of ecosystems can impact the biodiversity and species composition of certain ecosystems. There may be imbalances in the dynamics of ecosystems because certain species are more resistant to exposure to MPs.

Microplastic in the Air: Environmental Implications 45

c) *Changed nutrient cycles:* MPs may disrupt ecosystems' natural nutrient cycles, which may have an impact on plant development and the supply of vital nutrients for all living things.

d) *Destruction of habitat:* The buildup of MPs in ecosystems can result in the degradation of habitat. The quality of soil can be impacted by MPs, which also run the risk of damaging plants and other species that live in the soil.

e) *Unknown long-term effects:* It is as yet unclear how aMPs may affect ecosystems in the long run. There may be slow-onset, cumulative effects that worsen with time.

3.9 MITIGATION AND MANAGEMENT STRATEGIES

The data produced by the detection and quantification of MPs are used by various regulatory bodies to establish and modify regulations for safeguarding people, animals, and essential ecosystem services (Adhikari et al., 2022). Scientists, non-governmental organisations, and the media have recently discovered aMPs and there is currently concern expressed about them (Enyoh et al., 2019).

3.9.1 REGULATORY MEASURES AND POLICIES

Regulatory actions and policies are essential to minimise the impacts of MPs. Governments and international organisations can enact laws that restrict the creation, consumption, and disposal of plastics and promote the use of more environmentally friendly substitutes. Bans or limitations on single-use plastics, extended producer responsibility programmes, and specifications for wastewater treatment facilities to capture MPs are a few examples of these approaches. Microplastics have garnered significant attention from policymakers, prompting the initiation of policy implementations and the exploration of alternative measures to address this issue. These initiatives may introduce elements of uncertainty, necessitating meticulous attention and rigorous enforcement mechanisms (Mitrano and Wohlleben., 2020; Munhoz et al., 2022)

A three-pronged approach is required to successfully mitigate and manage aMPs: reduction at the source, development of cutting-edge filtration equipment, and raising public awareness.

 i. *Source reduction:* Reducing the release of aMPs at the source is one of the key strategies. To ensure effective garbage disposal and recycling, policies for managing plastic waste should be strengthened. This involves enacting programmes for extended producer responsibility (EPR), pushing producers to choose eco-friendly substitutes, and tightening rules on the production of plastic. Reducing the usage of single-use plastics and promoting circular economy models could both make a substantial difference. Better waste management procedures in the plastics-using industries would also reduce the number of MPs and aMPs produced.

 ii. *Innovative air filtration systems:* Cutting-edge air filtration systems are essential for reducing aMPs. High-efficiency particulate air (HEPA) filters can assist capture and restrict the release of MPs into the air and should be used in industrial buildings, automobiles, and enclosed areas. Additionally, the effectiveness of filtering equipment like electrostatic precipitators and cyclone separators can be improved by specially designing them to target MPs. In order to create filtration technologies that are suitable for various environmental conditions and MP emission sources, researchers and engineers need to work together.

 iii. *Public awareness and education:* Increasing public understanding of the problem of aMPs is essential to influencing behaviour and gaining support for legislative initiatives. People can learn more about the causes, effects, and potential solutions of MPs through educational initiatives. To inform people about the effects of plastic pollution on air quality and general well-being, community organisations, schools, and the media should work together. Individuals who have more influence will then be more likely to call for legislative reforms and adopt behaviours that reduce MP and aMP pollution.

iv. *Monitoring and research:* Monitoring and research activities are essential to fully comprehend the breadth and effects of aMPs. Accurate data will be available for risk assessment and policy formulation, if standardised procedures to quantify MP concentrations in air samples are developed. Researchers should explore the potential health impacts of breathing aMPs to fill any knowledge gaps about their long-term repercussions.

v. *International cooperation and policy:* Given the potential for cross-border movement of aMPs, international cooperation is crucial. Countries ought to collaborate to create uniform standards for observing and reducing aMP pollution. The management of aMPs on a global scale could be governed by international conventions and accords.

To minimise and control these microscopic plastic particles, a comprehensive plan that includes source reduction, state-of-the-art filtering technology, public awareness campaigns, stringent monitoring, and international collaboration is required. By implementing these techniques, we may endeavour to produce a cleaner, healthier environment for both the current and future generations. Governments, corporations, researchers, and individuals must properly collaborate to manage this complex challenge and ensure a sustainable future.

3.9.2 Technological Solutions for the Reduction of Microplastics

Technological solutions focus on developing innovative methods to reduce aMP pollution at various stages, including production, use, and disposal. These solutions can involve the development of new materials that are less likely to release aMPs, as well as the implementation of filtration and separation technologies in wastewater treatment plants. A serious global issue is the abundance of MPs in our environment. Addressing the MPs issue has become crucial because of their propensity to survive for hundreds of years and their potential to harm ecosystems and human health. Thankfully, technology is proving to have potential in the battle against MP contamination.

i. *Advanced filtration systems:* Using advanced filtration systems is one of the best strategies to stop the pollution caused by MPs. The goal of these systems is to collect MPs from a variety of sources, including wastewater treatment facilities and stormwater runoff. Microfiltration membranes and high-efficiency particulate air (HEPA) filters are two examples of technologies that can effectively remove MPs and aMPs from the ecosystem.

ii. *Robotic cleanup:* Robotic technologies have been developed to autonomously gather and remove MPs from aquatic settings, where a large fraction of them accumulates. These robots help to successfully clean up our oceans and rivers since they are fitted with sensors and cameras to detect and collect MPs.

iii. *Nanotechnology:* Nanotechnology provides creative ways to reduce the use of MPs. Nanoparticles that can bond to MPs have been created by researchers, making it simpler to remove them from soil and water. To stop the spread of MPs, these nanoparticles might be added to cleaning products or incorporated into filtration systems.

iv. *Biodegradable alternatives:* The creation of biodegradable plastics offers a futuristic answer to the pollution caused by MPs. In order to decrease the persistence of MPs in the environment, these plastics are designed to degrade more quickly than traditional plastics. Plastics made from biodegradable materials can be utilised for a variety of products, including packaging and consumer goods.

v. *Machine learning and AI:* Algorithms based on artificial intelligence and machine learning are essential for locating and following the origins of MPs. Areas where MPs are prevalent can be found and mapped using satellite images, drones, and underwater robots powered by artificial intelligence. This information aids in more precise targeting of cleanup activities.

Microplastic in the Air: Environmental Implications
47

vi. *Bioremediation:* Researchers are investigating the use of specially designed microorganisms that can devour and degrade MPs. The ecosystem could experience a large reduction in MPs with these microorganisms. Their use must be carefully controlled, nevertheless, to prevent unforeseen ecological effects.

vii. *Consumer apps:* Technical solutions aren't just for business and government initiatives. Smartphone apps can help consumers reduce the use of MPs. Users using apps like "Beat the Microbead" can scan product barcodes to check for the presence of MPs in cosmetics and personal care items, giving customers the information they need to make educated decisions.

viii. *Awareness and education initiatives:* Technology is a key tool for spreading knowledge and increasing public awareness of the MP problem. A global audience can be reached using social media, online films, and interactive websites, which can motivate people to use less plastic and take part in cleanup efforts.

ix. *Blockchain tracking:* The manufacture and distribution of plastic items are being monitored via blockchain technology. Consumers can make educated decisions and support businesses who are actively trying to limit their use of plastic by establishing transparent supply chains.

x. *Recycled plastics for 3D printing:* As 3D printing technology develops; recycled plastics will increasingly be used as a feedstock. This decreases the need for new plastic manufacture while simultaneously encouraging the collection and recycling of old plastics, lowering the likelihood that MPs will reach the environment.

These solutions can help us move toward a more sustainable and clean future where MPs do not pose a threat to our ecosystems and public health when combined with increased awareness and responsible consumer behaviour. Therefore, it is essential to keep making investments in and using these technologies to save the world for future generations.

3.9.3 Public Awareness and Educational Initiatives

MPs have grown in importance as a major hazard to wildlife, marine ecosystems, and potentially human health in recent years. These minute pieces of plastic, which are frequently invisible to the unaided eye, can be discovered in rivers, oceans, and even the air we breathe. Public awareness and education campaigns have become crucial weapons in the fight against MPs because of this expanding threat. With the help of these initiatives, people, communities, and businesses will be better equipped to act to reduce the pollution caused by MPs.

i. *Understanding MPs:* It is important to understand about MPs and aMPs before diving into awareness-raising and educational campaigns. MPs can come from a variety of sources. They can be primary, like the microbeads found in personal care products, or secondary, produced when larger plastic objects break down. Therefore, it is essential to understand this if you want to solve the issue successfully.

ii. *Campaigns for public awareness:* These initiatives are essential in educating the public about MPs. These campaigns spread information through a variety of media, including social media, television, and public events. Examples of documentaries that have clarified the effects of plastic pollution, especially MPs, on the environment and human health include "A Plastic Ocean" and "The True Cost". Such visual media can arouse powerful emotions and motivate behaviour.

iii. *Programs at educational institutes:* Educational institutions are important centres for the diffusion of knowledge. Lessons on plastic pollution, aMPs, and MPs are now commonplace in many schools and institutions' curricula. These initiatives not only inform the younger generation about the problem but also inspire them to come up with creative solutions. Students frequently work on efforts like beach clean-ups, plastic recycling programmes, and investigations into the presence of MPs in nearby bodies of water.

iv. *Community empowerment:* Citizens are being empowered to actively participate in data collecting and research on MPs through citizen science projects. People can report and track waste, including MPs, in their areas via apps and websites like "Marine Debris Tracker". This grassroots strategy not only provides important data but also encourages people to feel empowered to resolve the problem locally.

v. *Corporate responsibility:* Companies are becoming more aware of their part in the pollution caused by MPs and are acting to stop it. Many businesses have made commitments to remove MPs from their goods, including apparel and personal care products. These corporate practices are frequently the subject of public awareness campaigns that call on consumers to support eco-friendly brands and make educated decisions.

vi. *Governmental initiatives:* To reduce MP contamination, governments across the globe are passing restrictions. These programmes include restrictions on single-use plastics, bans on microbeads in cosmetics, and financing for studies on the effects of MPs on ecosystems. In order to promote systemic change and enforce ethical manufacturing and disposal processes, government cooperation is essential.

vii. *Community involvement:* Initiatives to reduce MPs are spearheaded by regional communities. A noticeable difference is being made by community recycling programmes and river cleanups. These programmes not only reduce MP pollution but also promote a sense of collective environmental responsibility.

3.10   FUTURE PERSPECTIVES AND RESEARCH DIRECTIONS

There are several directions for future research in the subject of MP analysis in the air and gas environment, which are still developing. To facilitate improved comparisons and data integration, standardising sampling and analytic techniques should be a top priority. The creation of reference materials and procedures for proficiency testing would improve the precision and dependability of results from MP analyses. Despite substantial advancements in airborne MP sampling and analysis, there remain knowledge gaps and areas that need more study. The development of new approaches as well as method standardisation and harmonisation are required.

Understanding the origins, movement, and disposal of aMPs in the gaseous and aerated environments also needs additional study. Informing mitigation methods might benefit from identifying the main sources of MPs and their entry points into the atmosphere. The long-term destiny and degradation of MPs in the atmosphere should also be studied in order to gain knowledge about their persistence and potential effects on ecosystems.

Standardisation of MP sample techniques is crucial for ensuring the comparability of results and advancing research in this area. The development of standardised techniques for air sampling, aMPs separation, identification, and quantification is necessary. This will make it possible for researchers to obtain trustworthy and repeatable data, promoting a better comprehension and evaluation of the environmental consequences of aMPs. Furthermore, new tools and methods must be developed for the analysis of aMPs. Due to their tiny dimensions, MPs cause difficulties in terms of identification and measurement. To fully comprehend the whole extent of aMP pollution, research in this field is essential.

3.10.1   EMERGING AREAS OF RESEARCH

The development of validated techniques for sampling aMPs is a current area of research interest. An investigation into understanding, sampling, and quantification of aMPs and nano-plastics is being carried out by a team at the National Institute of Standards and Technology (NIST). At a municipal recycling centre, they are developing equipment and procedures to sample MPs. For the identification and separation of MP particles from other materials, the NIST team is utilising novel

Microplastic in the Air: Environmental Implications

sampling approaches and combining optical microscopy, scanning electron microscopy (SEM), and Raman microscopy. They want to increase the precision and dependability of MP analysis in air samples by merging these techniques.

3.10.2 Addressing Knowledge Gaps

There is a dearth of comprehensive investigations and a relative lack of clarity in the present research on the dispersion of aMPs. There are still several study gaps in the realm of aMPs. Firstly, it is difficult to precisely estimate these particles' global occurrence due to our poor understanding of their sources, modes of transportation, and atmospheric distribution. Additionally, more research is required to fully understand the possible environmental and health effects of breathing in aMPs.

Structured temporal and spatial studies on the distribution of MPs in the atmosphere are advised for better understanding of the concentrations, types, and occurrence of aMP pollution in various regions. Moreover, such studies can help in identifying the sources, distribution, and fate of atmospheric MPs in various regions. Along with the use of more efficient sampling and analysis procedures, it is advised that the industry standard for monitoring aMP concentration, type, and occurrence be standardised. These knowledge gaps must be filled to appropriately predict the consequences of aMPs on ecosystems and human health.

3.10.3 Collaborative Efforts and International Cooperation

The variations in shape and size of particulate matter, the lack of uniform sampling techniques, the problems of identification and quantification, and the requirement for contamination control measures all provide obstacles for the research of aMPs. Promising solutions, however, are provided by continuous research and improvements in sampling strategies and testing equipment. Researchers can get more precise and comparable data on aMPs by standardising sampling procedures and creating cutting-edge techniques. This information is essential for determining their environmental effects, weighing the dangers of human exposure, and putting appropriate mitigation measures in place to combat aMP contamination.

Establishing standardised monitoring and measurement techniques can be made easier with international cooperation, ensuring that the data collected are similar across locations. Additionally, it makes it possible to exchange technological innovations and best practices for reducing aMP pollution. Additionally, collaborative research projects can reveal the causes and effects of aMPs, assisting policymakers in developing efficient laws. Collaboration across nations can reduce the number of aMPs released into the atmosphere and lessen their harmful effects on ecosystems and public health. Addressing a complex and global environmental issue such as atmospheric microplastics requires a coordinated international approach. The scientific community can overcome the difficulties posed by aMP sampling through ongoing work and cooperation, which will improve knowledge of this developing environmental problem and its effects on ecosystem and human health.

3.11 CONCLUSION

3.11.1 Summary of Key Findings

Microplastics (MPs) are minuscule plastic specks of just under 5 mm in size. They are widely present in various ecosystems and have grown into a severe ecological problem. It is essential to comprehend the abundance and movement of airborne microplastics (aMPs) in the Earth's atmosphere for determining their possible effects on both well-being and the ecosystem. In this study, several monitoring and analytical techniques for MPs are discussed. It is a challenging operation that necessitates careful consideration of numerous different criteria to collect samples and analyse MPs in the air and gas environment. It also highlighted the difficulties and factors to consider while using these approaches and provided some directions for further study. The variety of particulate matter, the absence of standardised sample techniques, the challenges of identification, and the requirement for contamination control measures are only a couple of small special challenges.

3.11.2 IMPORTANCE OF CONTINUED RESEARCH AND MONITORING

It is well known that aMPs have the potential to have a negative influence on both human health and the ecosystem, however, more studies and monitoring are further required to explore the hidden facts of microplastics. Therefore, it has become critical to comprehend the origins, distribution, and potential entry points of these minute plastic particles into our bodies as they are becoming increasingly common in the atmosphere. When inhaled, harmful chemicals and other pollutants carried by aMPs can pose health concerns. The severity of this problem can be more accurately gauged by continual study and monitoring, creating efficient mitigation plans, and formulating wise policy choices that protect both human welfare and the delicate environmental balance. MPs have been extensively researched in freshwater and marine habitats, but academics and the public have paid little attention to MPs in the atmosphere. MPs from the atmosphere enter the body mostly through inhalation and subsequent systemic exposure, leading to toxic reactions, abnormalities in many organs and systems, and even creating a risk of cancer in both humans and animals.

We are currently unable to fully comprehend the worldwide distribution, sources, fate of aMPs, or mechanism of harmful action of atmospheric MPs on ecosystems and humans due to the absence of effective tools for their detection and analysis. Future research on atmospheric MPs should be conducted in depth by the scientific community, particularly in order to investigate appropriate sampling and detection techniques and to provide a uniform industrial standard. Promising solutions, however, are provided by continuous research and improvements in sampling strategies and testing equipment. Researchers can get more precise and comparable data on aMPs by standardising sampling procedures and creating cutting-edge techniques. This information is essential for determining their environmental effects, weighing the dangers of human exposure, and putting appropriate mitigation measures in place to combat MP contamination. The scientific community can overcome the difficulties posed by airborne MP sampling through ongoing work and cooperation, which will improve knowledge of this developing environmental problem and its effects on ecosystem and human health.

REFERENCES

Abbasi, S., Jaafarzadeh, N., Zahedi, A., Ravanbakhsh, M., Abbaszadeh, S., & Turner, A. (2023). Microplastics in the atmosphere of Ahvaz City, Iran. *Journal of Environmental Sciences*, *126*, 95–102.

Abbasi, S., Keshavarzi, B., Moore, F., Turner, A., Kelly, F. J., Dominguez, A. O., & Jaafarzadeh, N. (2019). Distribution and potential health impacts of MPs and microrubbers in air and street dusts from Asaluyeh County, Iran. *Environmental Pollution*, *244*, 153–164.

Abbasi, S., Rezaei, M., Ahmadi, F., & Turner, A. (2022). Atmospheric transport of MPs during a dust storm. *Chemosphere*, *292*, 133456.

Adhikari, S., Kelkar, V., Kumar, R., & Halden, R. U. (2022). Methods and challenges in the detection of MPs and nanoplastics: A mini-review. *Polymer International*, *71*(5), 543–551.

Ahmad, M., Chen, J., Khan, M. T., Yu, Q., Phairuang, W., Furuuchi, M., & Panyametheekul, S. (2023). Sources, analysis, and health implications of atmospheric MPs. *Emerging Contaminants*, *9*(3), 1–16.

Allen, S., Allen, D., Moss, K., Le Roux, G., Phoenix, V. R., & Sonke, J. E. (2020). Examination of the ocean as a source for atmospheric microplastics. *PloS One*, *15*(5), e0232746.

Allen, S., Allen, D., Phoenix, V. R., Le Roux, G., Durántez Jiménez, P., Simonneau, A., & Galop, D. (2019). Atmospheric transport and deposition of MPs in a remote mountain catchment. *Nature Geoscience*, *12*(5), 339–344.

Auta, H. S., Emenike, C. U., & Fauziah, S. H. (2017). Distribution and importance of microplastics in the marine environment: A review of the sources, fate, effects, and potential solutions. *Environment International*, *102*, 165–176.

Azari, A., Vanoirbeek, J. A., Van Belleghem, F., Vleeschouwers, B., Hoet, P. H., & Ghosh, M. (2023). Sampling strategies and analytical techniques for assessment of airborne micro and nano plastics. *Environment International*, *174*, 107885, 1–26.

Batool, I., Qadir, A., Levermore, J. M., & Kelly, F. J. (2022). Dynamics of airborne microplastics, appraisal and distributional behaviour in atmosphere; A review. *Science of The Total Environment*, *806*, 150745.

Bergmann, M., Mützel, S., Primpke, S., Tekman, M. B., Trachsel, J., & Gerdts, G. (2019). White and wonderful? Microplastics prevail in snow from the Alps to the Arctic. *Science Advances*, *5*(8), eaax1157.

Boucher, J., & Friot, D. (2017). *Primary microplastics in the oceans: A global evaluation of sources* (Vol. 10). Gland, Switzerland: Iucn.

Bouzid, N., Anquetil, C., Dris, R., Gasperi, J., Tassin, B., & Derenne, S. (2022). Quantification of microplastics by pyrolysis coupled with gas chromatography and mass spectrometry in sediments: Challenges and implications. *Microplastics*, *1*(2), 229–239.

Brahney, J., Mahowald, N., Prank, M., Cornwell, G., Klimont, Z., Matsui, H., & Prather, K. A. (2021). Constraining the atmospheric limb of the plastic cycle. *Proceedings of the National Academy of Sciences*, *118*(16), e2020719118.

Browne, M. A., Crump, P., Niven, S. J., Teuten, E., Tonkin, A., Galloway, T., & Thompson, R. (2011). Accumulation of microplastic on shorelines woldwide: Sources and sinks. *Environmental Science & Technology*, *45*(21), 9175–9179.

Castañeda, R. A., Avlijas, S., Simard, M. A., & Ricciardi, A. (2014). Microplastic pollution in St. Lawrence river sediments. *Canadian Journal of Fisheries and Aquatic Sciences*, *71*(12), 1767–1771.

Chen, G., Feng, Q., & Wang, J. (2020). Mini-review of microplastics in the atmosphere and their risks to humans. *Science of the Total Environment*, *703*, 135504.

Chen, H., Zou, X., Ding, Y., Wang, Y., Fu, G., & Yuan, F. (2022). Are microplastics the 'technofossils' of the Anthropocene?. *Anthropocene Coasts*, *5*(1), 8.

Choi, D., Jung, S., Lee, J., & Kwon, E. E. (2024). Analysis of microplastics distributed in the environment: Case studies in South Korea. *Energy and Environment*. 0958305X241230616. https://doi.org/10.1177/095830 5X241230616

Ding, Y., Zou, X., Wang, C., Feng, Z., Wang, Y., Fan, Q., & Chen, H. (2021). The abundance and characteristics of atmospheric microplastic deposition in the northwestern South China Sea in the fall. *Atmospheric Environment*, *253*, 118389.

Dodson, G. Z., Shotorban, A. K., Hatcher, P. G., Waggoner, D. C., Ghosal, S., & Noffke, N. (2020). Microplastic fragment and fiber contamination of beach sediments from selected sites in Virginia and North Carolina, USA. *Marine Pollution Bulletin*, *151*, 110869.

Dong, C. D., Chen, C. W., Chen, Y. C., Chen, H. H., Lee, J. S., & Lin, C. H. (2020). Polystyrene microplastic particles: In vitro pulmonary toxicity assessment. *Journal of Hazardous Materials*, *385*, 121575.

Dris, R., Gasperi, J., Saad, M., Mirande, C., & Tassin, B. (2016). Synthetic fibers in atmospheric fallout: A source of microplastics in the environment?. *Marine Pollution Bulletin*, *104*(1–2), 290–293.

Enyoh, C. E., Verla, A. W., Verla, E. N., Ibe, F. C., & Amaobi, C. E. (2019). AMPs: A review study on method for analysis, occurrence, movement and risks. *Environmental Monitoring and Assessment*, *191*, 1–17.

Evangeliou, N., Grythe, H., Klimont, Z., Heyes, C., Eckhardt, S., Lopez-Aparicio, S., & Stohl, A. (2020). Atmospheric transport is a major pathway of microplastics to remote regions. *Nature Communications*, *11*(1), 3381.

Fendall, L. S., & Sewell, M. A. (2009). Contributing to marine pollution by washing your face: Microplastics in facial cleansers. *Marine Pollution Bulletin*, *58*(8), 1225–1228.

Ferrero, L., Scibetta, L., Markuszewski, P., Mazurkiewicz, M., Drozdowska, V., Makuch, P., & Bolzacchini, E. (2022). Airborne and marine microplastics from an oceanographic survey at the Baltic Sea: An emerging role of air-sea interaction?. *Science of The Total Environment*, *824*, 153709.

Franzellitti, S., Canesi, L., Auguste, M., Wathsala, R. H., & Fabbri, E. (2019). MPs exposure and effects in aquatic organisms: A physiological perspective. *Environmental Toxicology and Pharmacology*, *68*, 37–51.

Geyer, R., Jambeck, J. R., & Law, K. L. (2017). Production, use, and fate of all plastics ever made. *Science Advances*, *3*(7), e1700782 (1–5).

González-Pleiter, M., Edo, C., Aguilera, Á., Viúdez-Moreiras, D., Pulido-Reyes, G., González-Toril, E., & Rosal, R. (2021). Occurrence and transport of microplastics sampled within and above the planetary boundary layer. *Science of the Total Environment, 761*, 143213.

Hagström, S. (2021). Fate and transport of microplastic particles in small highway-adjacent streams – A case study in Gothenburg region. Master's thesis in Infrastructure and Environmental Engineering, Chalmers University of Technology, Gothenburg, Sweden. https://odr.chalmers.se/server/api/core/bitstreams/7cd77 2a1-1bd8-42d1-90f5-cc07feedbef3/content

Haque, F., & Fan, C. (2023). Fate and impacts of MPs in the environment: Hydrosphere, pedosphere, and atmosphere. *Environments, 10*(5), 70.

Hitchcock, J. N. (2020). Storm events as key moments of microplastic contamination in aquatic ecosystems. *Science of the Total Environment, 734*, 139436.

Hopewell, J., Dvorak, R., & Kosior, E. (2009). Plastics recycling: challenges and opportunities. *Philosophical Transactions of the Royal Society B: Biological Sciences, 364*(1526), 2115–2126.

Hu, J., Lim, F. Y., & Hu, J. (2023). Characteristics and behaviors of microplastics undergoing photoaging and Advanced Oxidation Processes (AOPs) initiated aging. *Water Research, 232* 119628, 1–12.

Huang, D., Tao, J., Cheng, M., Deng, R., Chen, S., Yin, L., & Li, R. (2021). MPs and nanoplastics in the environment: Macroscopic transport and effects on creatures. *Journal of Hazardous Materials, 407*, 124399.

Jahandari, A. (2023). MPs in the urban atmosphere: Sources, occurrences, distribution, and potential health implications. *Journal of Hazardous Materials Advances, 12*, 100346, 1–12.

Kung, H. C., Wu, C. H., Cheruiyot, N. K., Mutuku, J. K., Huang, B. W., & Chang-Chien, G. P. (2023). The current status of atmospheric micro/nanoplastics research: Characterization, analytical methods, fate, and human health risk. *Aerosol and Air Quality Research, 23*, 220362.

Li, Y., Shao, L., Wang, W., Zhang, M., Feng, X., Li, W., & Zhang, D. (2020). Airborne fiber particles: Types, size and concentration observed in Beijing. *Science of the Total Environment, 705*, 135967.

Liu, K., Wang, X., Wei, N., Song, Z., & Li, D. (2019). Accurate quantification and transport estimation of suspended atmospheric MPs in megacities: Implications for human health. *Environment International, 132*, 105127.

Martinho, S. D., Fernandes, V. C., Figueiredo, S. A., & Delerue-Matos, C. (2022). Microplastic pollution focused on sources, distribution, contaminant interactions, analytical methods, and wastewater removal strategies: A review. *International Journal of Environmental Research and Public Health, 19*(9), 5610.

Mitrano, D. M., & Wohlleben, W. (2020). Microplastic regulation should be more precise to incentivize both innovation and environmental safety. *Nature Communications, 11*(1), 5324.

Munhoz, D. R., Harkes, P., Beriot, N., Larreta, J., & Basurko, O. C. (2022). Microplastics: A review of policies and responses. *Microplastics, 2*(1), 1–26.

Mao, X., Xu, Y., Cheng, Z., Yang, Y., Guan, Z., Jiang, L., & Tang, K. (2022). The impact of MPs pollution on ecological environment: A review. *Frontiers in Bioscience-Landmark, 27*(2), 46.

Ortega, D. E., & Cortés-Arriagada, D. (2023). Atmospheric MPs and nanoplastics as vectors of primary air pollutants – A theoretical study on the polyethylene terephthalate (PET) case. *Environmental Pollution, 318*, 120860.

Othman, A. R., Hasan, H. A., Muhamad, M. H., Ismail, N. I., & Abdullah, S. R. S. (2021). Microbial degradation of MPs by enzymatic processes: A review. *Environmental Chemistry Letters, 19*, 3057–3073.

Pandey, D., Banerjee, T., Badola, N., & Chauhan, J. S. (2022). Evidences of MPs in aerosols and street dust: A case study of Varanasi City, India. *Environmental Science and Pollution Research, 29*(54), 82006–82013.

Padervand, M., Lichtfouse, E., Robert, D., & Wang, C. (2020). Removal of microplastics from the environment. A review. *Environmental Chemistry Letters, 18*(3), 807–828.

Prata, J. C. (2018). Airborne microplastics: Consequences to human health?. *Environmental Pollution, 234*, 115–126.

Prata, J. C., Castro, J. L., da Costa, J. P., Cerqueira, M., Duarte, A. C., & Rocha-Santos, T. (2022). Airborne Microplastics: Concerns Over Public Health and Environmental Impacts. In Rocha-Santos, T., Costa, M., Mouneyrac, C. (eds) *Handbook of Microplastics in the Environment* (pp. 177–201). Cham: Springer International Publishing.

Rocha-Santos, T., Costa, M., & Mouneyrac, C. (2022). *Handbook of MPs in the Environment*. Cham: Springer International Publishing.

Rose, P. K., Jain, M., Kataria, N., Sahoo, P. K., Garg, V. K., & Yadav, A. (2023). Microplastics in multimedia environment: A systematic review on its fate, transport, quantification, health risk, and remedial measures. *Groundwater for Sustainable Development, 20*, 100889.

Rose, P. K., Yadav, S., Kataria, N., & Khoo, K. S. (2023a). Microplastics and nanoplastics in the terrestrial food chain: Uptake, translocation, trophic transfer, ecotoxicology, and human health risk. *TrAC Trends in Analytical Chemistry*, 117249.

Shao, L., Li, Y., Jones, T., Santosh, M., Liu, P., Zhang, M., & BéruBé, K. (2022). AMPs: A review of current perspectives and environmental implications. *Journal of Cleaner Production, 347*, 131048.

Sommer, F., Dietze, V., Baum, A., Sauer, J., Gilge, S., Maschowski, C., & Gieré, R. (2018). Tire abrasion as a major source of microplastics in the environment. *Aerosol and Air Quality Research, 18*(8), 2014–2028.

Sulistyo, E. N., Rahmawati, S., Putri, R. A., Arya, N., & Eryan, Y. A. (2020). Identification of the existence and type of microplastic in code river fish, special region of Yogyakarta. *EKSAKTA: Journal of Sciences and Data Analysis, 1*(1), 85–91.

Su, L., Nan, B., Craig, N. J., & Pettigrove, V. (2020). Temporal and spatial variations of microplastics in roadside dust from rural and urban Victoria, Australia: Implications for diffuse pollution. *Chemosphere, 252*, 126567.

Szewc, K., Graca, B., & Dołęga, A. (2021). Atmospheric deposition of microplastics in the coastal zone: Characteristics and relationship with meteorological factors. *Science of the Total Environment, 761*, 143272.

Vianello, A., Jensen, R. L., Liu, L., & Vollertsen, J. (2019). Simulating human exposure to indoor airborne microplastics using a Breathing Thermal Manikin. *Scientific Reports, 9*(1), 8670.

Wang, J., Zheng, L., & Li, J. (2018). A critical review on the sources and instruments of marine microplastics and prospects on the relevant management in China. *Waste Management & Research, 36*(10), 898–911.

Wang, X., Li, C., Liu, K., Zhu, L., Song, Z., & Li, D. (2020). Atmospheric microplastic over the South China Sea and East Indian Ocean: Abundance, distribution and source. *Journal of Hazardous Materials, 389*, 121846.

Weis, J. S. (2020). Aquatic microplastic research—A critique and suggestions for the future. *Water, 12*(5), 1475.

Welle, F., & Franz, R. (2018). Microplastic in bottled natural mineral water–Literature review and considerations on exposure and risk assessment. *Food Additives & Contaminants: Part A, 35*(12), 2482–2492.

Woo, H., Seo, K., Choi, Y., Kim, J., Tanaka, M., Lee, K., & Choi, J. (2021). Methods of analyzing microsized plastics in the environment. *Applied Sciences, 11*(22), 10640.

Yadav, S., Kataria, N., Khyalia, P., Rose, P. K., Mukherjee, S., Sabherwal, H., & Khoo, K. S. (2023). Recent analytical techniques, and potential eco-toxicological impacts of textile fibrous microplastics (FMPs) and its associated contaminates: A review. *Chemosphere, 326*, 138495.

Yang, H., He, Y., Yan, Y., Junaid, M., & Wang, J. (2021). Characteristics, toxic effects, and analytical methods of MPs in the atmosphere. *Nanomaterials, 11*(10), 2747.

Yao, X., Luo, X. S., Fan, J., Zhang, T., Li, H., & Wei, Y. (2022). Ecological and human health risks of atmospheric MPs (MPs): A review. *Environmental Science: Atmospheres.*

Yuan, J., Ma, J., Sun, Y., Zhou, T., Zhao, Y., & Yu, F. (2020). Microbial degradation and other environmental aspects of MPs/plastics. *Science of the Total Environment, 715*, 136968.

Zha, F., Shang, M., Ouyang, Z., & Guo, X. (2022). The aging behaviors and release of microplastics: A review. *Gondwana Research, 108*, 60–71.

Zhang, Y., Kang, S., Cong, Z., Schmale, J., Sprenger, M., Li, C., & Zhang, X. (2017). Light-absorbing impurities enhance glacier albedo reduction in the southeastern Tibetan Plateau. *Journal of Geophysical Research: Atmospheres, 122*(13), 6915–6933.

Zhang, Y., Kang, S., Allen, S., Allen, D., Gao, T., & Sillanpää, M. (2020). Atmospheric MPs: A review on current status and perspectives. *Earth-Science Reviews, 203*, 103118

Zhang, Y., Wu, H., Xu, L., Liu, H., & An, L. (2022). Promising indicators for monitoring microplastic pollution. *Marine Pollution Bulletin, 182*, 113952 (1–4).

Zhao, X., Zhou, Y., Liang, C., Song, J., Yu, S., Liao, G., & Wu, C. (2023). Airborne microplastics: Occurrence, sources, fate, risks and mitigation. *Science of The Total Environment, 858*, 159943.

4 Fate and Control of Microplastics in Municipal Wastewater Treatment Systems

Linhua Fan, Li Gao, and Arash Mohseni

4.1 INTRODUCTION

Municipal wastewater treatment plants (WWTPs) are known as an important pathway for microplastics (MP) to enter the environment (Murphy et al., 2016). While the WWTPs have been demonstrated as an effective barrier to the MP present in sewage, with a reported removal efficiency in the range of 70%–99.9% (Gatidou, Arvaniti, and Stasinakis, 2019), they may also be a source of MP given the huge volumes of treated wastewater that are discharged into the environment each day (Murphy et al., 2016; Carr, Liu, and Tesoro, 2016). In recent years, many investigations have been carried out to understand the fate and transport of MP in the various sewage treatment processes. Most of the published research work was conducted on conventional activated sludge (CAS)-based systems, with almost no information about the MP removal efficiency of waste stabilisation ponds/lagoons that are utilised widely in some regions of the globe, such as Australia. Moreover, it was observed that many of the previous studies only considered the MP concentrations of the influent and the treated effluent, with relatively limited information about the effect of each treatment step on MP removal. It is known that MP removal depends on the characteristics of microplastic particles and wastewater, and the treatment technologies applied. For effective management of MP at a sewage treatment plant and minimising its MP release to the receiving environment, a detailed understanding of the fate and transport of the microplastics across each treatment system is essential.

This chapter aims to provide an overview of the role of WWTPs in MP control, the general characteristics of MP in sewage such as abundance, polymer type, size distribution, shape and colour, the effect of different sewage treatment systems on MP removal and the knowledge gaps that need to be addressed in MP management. An introduction to a recent study by the authors' group on the MP removal performance of four treatment systems at three WWTPs in Victoria, Australia, is also provided. The key objectives/activities of the investigation were to quantify and characterise the MP in the wastewater samples obtained from the main treatment steps of the three plants over an extended period (two years). Sludge/biosolids samples were also collected from a selected WWTP for MP analysis. The research program included six wastewater samplings at each treatment system. The reported methods of sampling and analysis, and the results can provide more insights into MP management and contribute significantly to the new knowledge in this area.

Fate and Control of Microplastics in Municipal Wastewater

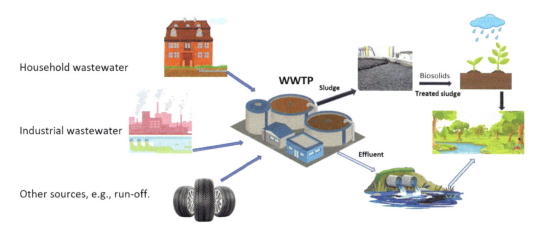

FIGURE 4.1 Schematic diagram of the sources of microplastics in WWTPs.

4.2 WASTEWATER TREATMENT PLANTS AND MICROPLASTICS

WWTPs receive wastewater from industries that use and manufacture microplastics, as well as domestic effluents that contain MP from washing machines, cleaning, and personal care products (Kay et al., 2018) (Figure 4.1). Most of the existing WWTPs were not specifically designed to remove or degrade MP. An elevated concentration of MP downriver from WWTPs has been widely reported (McCormick et al., 2014; Estahbanati and Fahrenfeld, 2016), which was primarily attributed to the residual MP in the treated effluents. The effluents containing the MP can enter oceans directly or indirectly via riverine systems.

Given the high volume of effluents discharged daily, a very large number of MP would enter the receiving environment even at a low MP concentration. Carr, Liu, and Tesoro (2016) reported that at an MP removal efficiency of 99%, one WWTP in Los Angeles (USA) could release 9.3×10^5 MP into the water environment each day. In another study, Ziajahromi et al. (2017) determined that 3.6×10^6 MP were discharged per day from a tertiary WWPT in Sydney, Australia. It is estimated that the WWTPs in Europe alone can release 520 kilotonnes of plastics to the environment each year (Horton and Dixon, 2018). It is worth noting that small-sized MP (e.g., < 20 μm) are not commonly recorded due to the difficulties in quantifying the tiny polymeric particles, and so the total MP release from the WWTPs could be underestimated. According to Xiong et al. (2018), the level of urbanisation and population density of a particular area have a direct influence on the MP level in the receiving environment. The presence of MP in wastewater also has implications for wastewater and sludge treatment operations. For instance, in a review by Zhang and Chen (2020), it was suggested that very small MP could infiltrate the biofilm and produce reactive oxygen species, which exhibited an acute inhibitory effect on the microbes. It was also suggested that the mechanisms involved in MP's impact on wastewater and sludge treatment could be complicated due to various chemicals used in producing the plastics, the adsorption of environmental toxins and the exudation of the chemical additives.

4.3 CHARACTERISTICS OF MICROPLASTICS IN WASTEWATER

It has been reported in some review articles that around 20 types of plastic materials have been detected in the wastewater streams of WWTPs (Kang et al., 2018; Yaseen et al., 2022). The most commonly detected polymers in sewage include polypropylene, polyethene, polystyrene and polyesters due to their high levels of production and use in our lives (Kang et al., 2018). Previous studies have also shown that polyvinyl chloride and polyethene terephthalate are ubiquitous in wastewater streams (Ziajahromi et al., 2017). In municipal wastewater, MP can present in different morphologies and synthetic fibres were frequently reported as the dominant type of MP (Kang et al., 2018; Sutton et al., 2016). The fibres originate mostly from synthetic clothing, which have been detached or pulled apart during washing and appear in domestic laundry effluents (Kang et al., 2018). Fragments are the second most abundant form of MP detected and most likely arise from the degradation of larger pieces of plastic (Sun et al., 2019). Other shapes of MP in municipal wastewater include foams, spheres, pellets, nurdles, flakes and films (Koelmans et al., 2019).

Size is another important characteristic of MP. Sun et al. (2019) stated that more than 70% of MP in WWTP influent streams are larger than 500 μm, whilst more than 90% of the MP in the effluent streams of WWTPs were smaller than 500 μm. However, the size distribution of MP can be markedly different with different catchment conditions, such as the input of MP from industrial sources. Other studies have shown evidence of extremely small MP present in the aquatic environment. According to Sun et al. (2019), 64% of the MP detected in samples from the Atlantic Ocean were smaller than 40 μm, half of which were smaller than 20 μm. Most existing studies investigating microplastics in municipal wastewater treatment systems were from developed countries; there is limited information available for developing countries, particularly in the Mediterranean area, Asia and South America (Gatidou, Arvaniti, and Stasinakis, 2019). As various consumer habits can impact the abundance and type of MP in wastewater, the information obtained from the different regions could be of interest to researchers and water utilities.

4.4 MICROPLASTICS IN WASTEWATER SLUDGE

It is known that most of the MP removed from wastewater would accumulate in the sludge of WWTPs, which is eventually disposed of or reused for beneficial purposes such as land applications. Sludge is, therefore, a significant source of the MP found in soil, which is of concern, especially in the case of agricultural reuse (Collivignarelli et al., 2021). The annual release of MP from WWTPs into the soil would be significantly greater than those entering the oceans. Li et al. (2019) studied the adsorption efficiency of different heavy metals by MP present in wastewater sludge and reported the following affinity scale: Pb > Cd > Zn > Cu > Co > Ni. Considering the potential adsorption of a broad range of harmful chemicals in wastewater and sludge on the MP, it is of great interest to gain knowledge about the abundance and characteristics of MP in the sludge and biosolids.

Studies have been conducted to trace the path of MP into the sludge by collecting and analysing the sludge samples from primary and secondary treatment processes. Based on these studies, commonly, over 95% of MP can be transferred from the influent stream to the sludge stream during wastewater treatment. It is worth noting that various studies may report MP content in different forms of sludge, e.g., primary, secondary, mixed (primary and secondary), dried sludge or biosolids. The abundance of MP in sludge is linked to the characteristics of the wastewater, the adsorption capability of sewage sludge, the treatment process, seasonal fluctuation, and the population density of the catchment area (Liu et al., 2019). Talvitie et al. (2015) observed that fibres were removed best during primary sedimentation. Similar MP characteristics, including type and morphology, are detected in both sludge and wastewater. Studies conducted by Mahon et al. (2017) and Li et al. (2019) showed polyethene, polyesters and PET fibres were the most common types of MP detected in sludge. Compared to wastewater, MP in sludge tend to have a larger average particle size (Sun et al., 2019).

4.5 EFFECT OF WWTPS ON MP REMOVAL

Although almost all existing sewage treatment or wastewater reclamation systems are not designed with MP removal in mind, they are commonly effective in blocking most of the MP from entering the environment through effluent discharge. In general, the MP removal efficiency of WWTPs is

Fate and Control of Microplastics in Municipal Wastewater

dependent upon process variables such as the properties of the wastewater and the MP, treatment technologies and operating conditions. The published studies have shown that most WWTPs could achieve high MP removal (e.g., >90%), particularly those with tertiary treatment processes implemented (Talvitie et al., 2015; Magnusson et al., 2016; Carr, Liu, and Tesoro, 2016; Michielssen et al., 2016; Lares et al., 2018; Gies et al., 2018).

Municipal WWTPs usually include several processing steps, such as preliminary/primary, secondary, post-secondary, and advanced/tertiary treatment. The common unit operations used in the treatment include screening, aeration, coagulation/flocculation, sedimentation, filtration (including membranes), and disinfection. The primary treatment removes large debris with screen meshes of a typical size of 4–6 mm, and some smaller-sized MP may also be removed due to their affinity with the large objects. It has been suggested that an average of 65% of the influent MP could be removed by a typical primary treatment (Burns and Boxall, 2018), in which low-density MP can be skimmed off as they tend to bind and accumulate in the grease layer (Murphy et al., 2016). In the secondary treatment stage, suspended solids and biochemical oxygen demand (BOD) are removed through the microorganisms in aeration tanks. It is expected that some MP can be attached to the solids and then removed via flocculation and sedimentation processes, ending up in the secondary sludge. MP removal rates of 75–99% for secondary treatment have been reported. In advanced (tertiary) treatment, processes such as filtration through media filters or membranes can further improve MP removal before the treated wastewater is disinfected and then discharged/reused. The total MP removal efficiency for tertiary treatment plants was reported as 90–99.9%. It has been noted in some studies that even the membrane-based tertiary treatment did not show complete removal of MP (Talvitie et al., 2017), although the theoretical MP removal is 100% given the pore size of the membranes was significantly smaller than the small-sized MP (e.g., around 1 µm in size).

MP removal efficiency is dependent on the treatment technologies implemented by the WWTPs. Ziajahromi et al. (2017) compared three separate WWTPs in Sydney, which utilised different treatment processes. The first plant consisted of a primary treatment process, the second plant featured both primary and secondary treatment, and the third plant was a tertiary treatment-based process. It was shown that the MP concentration of the plant effluent decreased with additional stages of treatment applied. Almost all the published studies on MP removal in the WWTPs were focused on CAS treatment systems. To date, there is very limited information about the MP's fate and transport in lagoon/pond-based wastewater treatment systems, which are being used as a major treatment option in many countries. In Australia, 60% of the more than 1200 WWTPs use lagoon/pond technology as primary treatment (Hill, Carter, and Kay, 2012; Coggins, Crosbie, and Ghadouani, 2019).

4.6 ANALYTICAL METHODS OF MICROPLASTICS IN WASTEWATER AND SLUDGE

Analysis of the MP present in wastewater or sludge usually involves sample collection, extraction and purification, quantification, and characterisation. As there are currently no standardised procedures for MP analysis, the analytical methods reported in most of the published studies are considerably different. While this can make it difficult to directly compare the results among the various studies, lots of useful information has been generated from the existing studies, and these have laid good foundations for future research in this area.

According to Sun et al. (2019), there are different ways for MP sample collection, including container collection, autosampler collection, separate pumping and filtration, and surface filtration. The collection with containers or autosamplers is simple and convenient, however, this usually means a limited volume of a wastewater sample (e.g., litres or dozens of litres) is collected. As municipal wastewater (e.g., influent of WWTPs) usually has a significantly higher concentration of MP compared with surface water, such methods appear to be suitable and have, therefore, been used most widely in wastewater-related research. Separate pumping and filtration can effectively increase water sample volumes to hundreds of litres or cubic metres. The method is frequently used for collecting samples of the treated effluent of WWTPs (Sun et al. 2019). A sieving procedure after samples were contained has also been used frequently. The mesh size for each sieve differed between studies, with sizes ranging from 20 μm to 5 mm (Carr, Liu, and Tesoro, 2016; Magni et al., 2019). The surface filtration method designed by Carr, Liu, and Tesoro (2016) enables the sampling volume to be further increased to thousands of cubic metres, which was applied for the sampling at the outfall of the treated effluent.

Due to the complexity of wastewater matrices, sample preparation is required to ease MP identification and ensure samples are inherently safe to handle. Sample preparation may involve digestion, staining, and extraction, which could decrease the tendency for misidentification. Common techniques for sample preparation include the addition of hydrogen peroxide (H_2O_2), potassium hydroxide (KOH), or digestive enzymes to break down organic materials that can affect the accurate identification of MP. For sludge sample analysis, most of the reported experimental works used procedures and techniques similar to those for wastewater (Collivignarelli et al., 2021). However, processing techniques involving sludge samples may need specific procedures due to the complex nature of organic materials present in the samples. As such, lab trails may be needed for a certain wastewater/sludge to determine the most suitable sample pretreatment strategy that facilitates MP detection.

Microplastics are commonly characterised by their polymer type, size distribution, colour, and morphology. The most frequently used sieves for size classification include those of 25 μm, 100 μm, and 500 μm (Talvitie et al., 2017; Mintenig et al., 2017; Ziajahromi et al., 2017; Simon, van Alst, and Vollertsen, 2018; Lares et al., 2018). Some characterisation results were based solely on visual identification (Talvitie et al., 2015; Mason et al., 2016; Michielssen et al., 2016); some relied on advanced detection techniques such as micro-Fourier transform infrared spectroscopy (FTIR) or Raman spectroscopy (Murphy et al., 2016; Talvitie et al., 2017; Ziajahromi et al., 2017) in identifying all MP present in samples. Visual identification is simple and rapid; however, it may report false results when the MP under analysis are in doubt. Identifying MP through chemical analysis with FTIR or Raman spectroscopy would be accurate, but the analysis can be tedious and time-consuming, which is not practical when samples contain a large number of MP. For a balanced and practical approach, many studies have incorporated a combined methodology of these two methods (Carr, Liu, and Tesoro, 2016; Mintenig et al., 2017).

Size limitation is a problem associated with the visual identification of MP, as smaller MP particles (e.g., <20 μm particle sizes) can be difficult to sort and identify even under microscopes (Iyare, Ouki, and Bond, 2020). The error rate in visual identification would increase with decreasing size of particles. Moreover, contamination may potentially cause the overestimation of the number of MP in the samples, and strict measures and procedures should, therefore, be implemented to avoid the impact. In general, samples should be contained in glass jars or metal containers instead of plastic containers to avoid contamination by foreign plastics. Other contamination controls include opting for cotton lab coats, thoroughly cleaning laboratory equipment, using a laminar flow cabinet to control airborne plastic particles, and utilising control samples (Koelmans et al., 2019).

4.7 KNOWLEDGE GAPS AND RESEARCH NEEDS

A detailed understanding of the fate and transport of microplastics in wastewater treatment systems can help optimise existing wastewater treatment facilities for achieving the maximum MP removal rate and also help design specific processing systems for MP removal. There is a consensus that the lack of standardisation of MP analysis for water and wastewater, including the procedures of

Fate and Control of Microplastics in Municipal Wastewater

sampling, pre-treatment, quantification, and characterisation, presents a major limitation for all MP research. Standards of MP analysis would enable the comparison of different studies and, hence, more detailed knowledge. To facilitate the standardisation of MP analysis for wastewater and sludge, more work should be conducted to develop more cost-effective and practical methods that can address the shortcomings of the existing methods.

The previous studies investigating the MP abundance in WWTPs were conducted mostly over the short term, and only very few of these studies covered 12 months or greater. Conley et al. (2019) investigated three WWTPs for about 12 months; their results showed that the MP removal rate for two of the WWTPs ranged between 74.8% and 97.1%, while the third plant achieved 95.9–98.1% retention of MP. Their study also looked at the seasonal trends that may impact MP concentration in the WWTPs, however, no clear seasonal effects were shown, and the authors recommended more frequent sampling events throughout the year to validate this observation.

As mentioned earlier, the published studies were predominantly focused on CAS-based systems, with very little knowledge of MP fate and behaviour in lagoon/pond-based WWTPs, and there is a clear demand for more studies on such systems as they are being utilised as the main wastewater treatment in many regions around the globe. Furthermore, most of the previous studies focused on the influent and final effluent of the WWTPs, and detailed surveys of MP content at each treatment step remain limited. This has limited the insightful understanding of MP removal in the treatment processes. It was also observed that the modelling of MP in wastewater treatment systems is lacking. It is expected that more activities will be carried out in this space with the increase in the knowledge of the effect of the treatment processes on MP removal. The research conducted so far has mostly focused on MP particle sizes of greater than 20 µm; understanding the presence of smaller-sized MP (e.g., micron, submicron, and nano size), and their implication for human and environmental health and their treatability during wastewater treatment processes is scarce. This may be related to the fact that their detection and characterisation are much more challenging compared with larger plastic particles.

4.8 A CASE STUDY OF THREE WWTPS ON THEIR EFFICIENCY OF MP REMOVAL

A case study that involved three WWTPs in Victoria, Australia, was conducted to address some of the knowledge gaps outlined in Section 4.7, with a view to improving our understanding of the fate and transport of MP in the various types of sewage treatment process.

4.8.1 Wastewater Treatment Plants Under the Case Study

Wastewater samples were obtained from three water recycling plants (WRPs) denoted as Plants A, B, and C. Brief descriptions of the treatment processes of the three WWTPs are provided below, and further information about the three plants can be found in Table 4.1.

- Plant A is a conventional activated sludge-based treatment plant that produces Class A effluent suitable for environmental discharge; it comprises primary and secondary aeration and tertiary treatment (UF-based).
- Plant B operates a lagoon system producing Class C recycled water. The lagoon system comprises facultative and maturation lagoons followed by storage lagoons.
- Plant C produces Class A and Class C recycled water. The plant treats wastewater primarily with a conventional activated sludge process followed by a membrane (UF)-based system to produce Class A recycled water. A portion of the sewage was treated with a lagoon-based system (with aerated, facultative lagoon and maturation lagoons) to produce Class C water.

TABLE 4.1

Additional Information on the WWTPs Under This Study

WWTP	Catchment (Population)	Process	Inflow (ML/y)	Effluent Quality
A	Residential, commercial, and industrial (~77,000)	Activated sludge-UF	4698 (avg. 2013–17)	Class A
B	Mainly residential, minor commercial, and industrial (~5,700)	Facultative and maturation lagoons	288 (avg. 2011–17)	Class C
C	Significant industrial (~43,000)	Activated sludge-UF (Class A) and AS-Lagoons (Class C)	2291 (avg. 2011–16)	Class A & C

4.8.2 SAMPLING AND ANALYTICAL METHODS APPLIED IN THE STUDY

Six batches of grab wastewater samples were obtained from the three WWTPs over 2019–2021. Wastewater samples of 4–80 L were collected using metal containers (Sandleford, 5 L, 10 L, and 20 L) at the key treatment steps of each system. Two samplings on primary and secondary sludge and biosolids were conducted at Plant A over 2019–2020. For primary and secondary sludge, grab samples were collected, and for biosolids, the samples were the mixtures of the biosolids collected from 10 points (~100 g at each point) of a biosolids pile.

The wastewater samples were first filtered with an MP isolation rig that consists of a series of 20-cm stainless-steel meshes stacking in the order of 500, 200, 100, and 25 μm (from inlet to outlet) for rough fractionation of the MP in the wastewater (Ziajahromi et al., 2017). The retained solids on each mesh were rinsed using deionised water and then transferred into a 50-mL Scotch bottle. The sample of 35 mL was then treated with an equal volume of 30% H_2O_2 at 65°C for 0.5 h for degrading organic matter in the sample. After that, 35 mL of 1M HCl was added to the sample to react for 0.5 h at 65°C to dissolve inorganic salts and digest the non-plastic materials (e.g., cellulose fibres). The sample was then reconstituted in deionised water and filtered through a 25-μm stainless steel mesh. The solids on the mesh were transferred into a Petri dish (d = 80 mm) using 45 mL of deionised water and dried in an oven until no moisture remained. The MP recovery was evaluated with a validation test and was found to be over 90%, which was considered acceptable.

For sludge/biosolid sample processing, each sample (10 g) was placed in a beaker, and 40 mL of 30% hydrogen peroxide and 20 mL of 1 M $FeSO_4$ solutions were added to the beaker. Samples were then digested for 120 min at 65°C at 750 rpm stirring speed on a hot plate stirrer to remove organic materials from the sample matrix (Lusher, McHugh, and Thompson, 2013). After organic degradation, deionised water was poured into the beaker containing the digested sample and agitated by a supersonic shaker for 20 min so that MP floated to the surface. Then, the upper layer of the liquid was decanted into a 25-μm stainless steel mesh, and the content was transferred into a glass Petri dish and dried at 60°C. The dry weight of sludge/biosolids samples was estimated by drying a known amount of sample of each type at 105°C overnight. Measures were implemented to prevent contamination of the samples, and controls were used to quantify the background MP contamination and determine the final count of MP (Fan et al., 2023). In this research, results were reported with the cut-off size of 25 μm for the MP. The shape and colour of the MP were also recorded for the wastewater and sludge samples. A confocal microscope (Leica S9d) fitted with a digital camera was used to count MP on the glass Petri dishes. For identification of MP fibres and fragments, the criteria used previously by Conley et al. (2019) were applied. If a particle under the analysis was under suspicion, a micro-FTIR instrument was utilised to determine the chemical origin of the particle, as described later in this section. In addition to the microscopic analysis, a micro-FTIR

Fate and Control of Microplastics in Municipal Wastewater

imaging system (PerkinElmer Spotlight 400) operated in reflection mode was employed to confirm the material origins of the suspected MP. The details of the instrument settings can be found elsewhere (Fan et al., 2023).

4.9 KEY FINDINGS OF THE CASE STUDY

4.9.1 ABUNDANCE OF MP IN THE PLANT INFLUENT AND FINAL EFFLUENT

The average MP concentrations of plant influent for the three WWTPs over the studied period were 39.6–52.5 MP/L; for the final effluent of the three plants, the average MP concentrations were in the range of 0.17–1.66 MP/L. The lowest MP concentration of the final effluent was achieved by the advanced treatment process utilising membrane filtration. Based on the 15 studies of nine different countries that were summarised by Kang et al. (2018), the MP concentrations of WWTP influent ranged from 15.1 to 640 MP/L, and the MP concentration in the final effluent was in the range of non-detected–65 MP/L. It can be seen that the previously reported MP concentrations in both the plant influent and effluent were highly varied, which would be related to the differences in the wastewater and the treatment processes of those studies. Also, it was noted that the MP cut-off sizes used in those studies varied greatly, from 0.75 to 300 μm. However, the abundance of MP in the influent and final effluent of the three WWTPs under this study was in the reported ranges.

The MP concentration data for the influent of the three WWTPs over the two years were examined to see whether any seasonal trends exist. It was observed that MP concentration was markedly higher in summer (two to three times) than in other seasons. It was shown that the MP concentration was comparable for winter and spring and was marginally lower in autumn. The possible reasons for the observation were presented in a recent publication of the authors' group (Fan et al., 2023), which were related to the rainfall events and the possible seasonal change of population in the catchment areas. There was very limited information in the literature about seasonal trends in the MP concentration of WWTP influent. As introduced early in the chapter, Conley et al. (2019) did not observe a clear seasonal effect on the MP concentration of the influent (an MP cut-off size of 43 μm was used) when studying three WWTPs in South Carolina (USA) over a whole year. It should be noted that WWTPs may have different seasonal trends in terms of MP abundance in plant influent due to the fact that a number of factors related to catchment conditions can affect this. Also, observations in the present and previous studies were based only on a limited number of samples; for a more detailed understanding of the seasonal effect, more frequent samplings should be obtained in future research.

It should also be noted that uncertainties in the different sampling, isolation, and identification methodologies applied in the reported studies made the comparison of the results difficult. For example, the MP size cut-off applied in those studies varied from 0.7 to 300 μm, although several studies utilised 20 or 25 μm, which was similar to that for this study. This highlights the urgent need for the standardisation of MP analytical methodologies for such research.

4.9.2 DOMINANT MP SPECIES IN THE WASTEWATER

It was shown that microfibres were the dominant species in the influents of the three plants, which accounted for 57.3–77.6% of the total MP particles. This was expected due to the usual observation that synthetic fibres were a major source of MP in municipal wastewater (Lares et al., 2018; Gies et al., 2018). The high occurrence of synthetic fibres was related to the fact that a great number of fibres can enter the sewer systems through household washing machine discharge. Fibres were also found to be the main MP species in the final effluent, consisting of 71.9–100% of the total MP. As such, specific strategies to control the fibres could be explored in future studies. For example, this could be achieved through cellulose recovery (e.g., harvesting natural and synthetic fibres) in the preliminary/primary treatment of the WWTPs for resource recovery and mitigating the overall MP in the treated effluent.

The polymer type of each microplastic particle was not specifically determined for each wastewater/sludge sample in this work due to the high number of samples and microplastic particles in the wastewater and sludge. However, random samples were analysed with micro-FTIR, and the results revealed the main polymer types of the MP include polypropylene (PP), polyethene (PE), nylon, polyvinyl chloride (PVC), polystyrene (PS), and polyvinyl alcohol (PVA), which were previously reported as the frequently detected MP material types in sewage (Kang et al., 2018). In terms of colour, the number of blue, black, red, and green MP was found to be significantly higher than that of other colours in the wastewater and sludge samples (data not shown).

4.9.3 MP Removal Efficiency of the WWTPs

The overall removal capacity of microplastics was found to be 99.5%, 97%, and 98.4% for Plants A, B, and C (Class C system), respectively. Although there was only one sampling event on the Class A effluent of Plant C, the MP retention efficacy of 99.6% for the membrane-based treatment system appeared to be consistent with the performance of the Plant A recycled water system. Overall, the two AS-based WWTPs [Plants A and C (Class A system)] performed comparably in terms of MP removal at each stage of the treatment processes (Figure 4.2), with the average efficiency of MP reduction of 62% and 55% for primary treatment, and 77% and 75.3% after secondary treatment, respectively. The results generally followed the trend that MP concentration reduced with an increase in treatment steps, with the highest removal rate achieved with the tertiary treatment processes (e.g., employing membranes).

The two lagoon-based treatment processes under this study showed a promising capacity for MP removal, i.e., 97% and 98.4% for Plants B and C (Class C system), respectively (Figure 4.3). It was shown that both lagoon treatments resulted in a significant MP reduction. As an example of the conventional lagoon system of Plant B, MP concentration decreased by 47% after facultative lagoon treatment, and the accumulative MP reduction efficiency reached 87% after the treatment of the three maturation lagoons in series and 97% after the treated wastewater held up in the storage lagoons for an extended period. Before this work, no published studies were available on the lagoon/

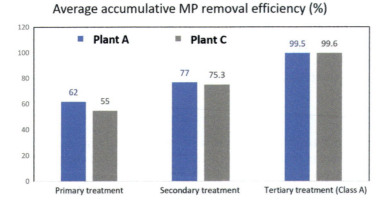

FIGURE 4.2 A comparison of microplastics removal efficiency (mean values) for the two AS-based tertiary treatment systems.

Fate and Control of Microplastics in Municipal Wastewater

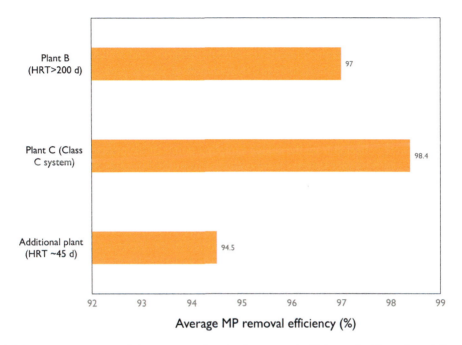

FIGURE 4.3 A comparison of the average microplastics removal efficiency for Plants B and C, and the additional plant investigated.

pond wastewater treatment systems and therefore, the performance of Plants B and C lagoon systems on MP removal could not be compared or benchmarked.

The significant MP reduction for Plant B was attributed to the lengthy retention of the wastewater (high hydraulic retention time (HRT), >200 days, including that of the winter storage lagoons) of the systems that would allow effective separation of MP from the water column via various physical and biological pathways. The suggested effect of wastewater detention time on MP reduction was supported by a later study on another conventional lagoon treatment system in Victoria, which has a similar treatment train to Plant B. The results of the two samplings over 2022 showed the lagoon treatment achieved an average overall MP removal efficiency of 94.5% at the total HRT of 45 days (Figure 4.3). For Plant C (Class C system), it was envisaged that the high efficiency of MP was related to the post-secondary treatment of the wastewater with the lagoon system in which MP was further removed during the month-long detention in the lagoons. The information contributed significantly to the knowledge regarding the effect of wastewater ponds/lagoons on MP retention. To gain a greater understanding of the fate and transport of MP in the conventional lagoon treatment systems, the authors' group conducted a further study using a multimedia model approach to simulate the MP removal in Plant B; the details of the study can be found elsewhere (Um et al., 2023).

Although the membrane-based tertiary treatment processes at Plants A and C (Class A system) showed high MP removal rates, the ultrafiltration processes did not demonstrate a consistent 100% removal of MP, which was unexpected given the theoretical separation efficiency for such membranes (i.e., nominal membrane pore size <0.1 μm). However, the observation was consistent with several previous studies on the MP removal rate of various membrane systems, including microfiltration, ultrafiltration and reverse osmosis, and membrane bioreactor processes (Talvitie et al., 2017; Ziajahromi et al., 2017; Lares et al., 2018; Yang et al., 2019; Bayo, López-Castellanos, and Olmos, 2020; Wang, Lin, and Chen, 2020). To date, there is a lack of research to understand the causes of the incomplete removal of MP observed in membrane processes. Nevertheless, it can be speculated that the presence of a low number of MP in the tertiary treated effluent could be caused by faulty membranes, compromised membrane module integrity, and contamination in the downstream processes (e.g., disinfection, storage), degradation of plastic pipes, and/or contamination during sampling/analysis. This could be an area for future investigations.

4.9.4 ABUNDANCE OF MICROPLASTICS IN SLUDGE AND BIOSOLIDS

Sludge samples were collected from the sludge sampling points of the primary and secondary sedimentation tanks, and biosolids samples were collected from the biosolids pile at the solar pan of Plant A. Table 4.2 shows the results of the total MP concentration (25 μm–5 mm) of the sludge and biosolids. Compared with the previous studies, as summarised in Table 4.3, the relative abundance of MP in the Mt Martha WRP sludge followed a similar trend in that secondary sludge generally had a higher MP concentration than the primary sludge. The MP concentrations of the Mt Martha WRP sludge and biosolids appeared to be in the range reported in the literature, although the MP size ranges under the different studies were slightly different. In terms of morphology, it was observed that the MP in the Plant A sludge and biosolids were primarily fibres (70.3–73.9%), followed by flakes (10.8–18.5%), and fragments (10.5–15.2%), with negligible MP in bead shape. The dominant presence of fibres in the sludge and biosolids was consistent with the high abundance of fibres in the WWTP influent and the results that were reported in other studies, as summarised by Collivignarelli et al. (2021).

TABLE 4.2
Abundance of Microplastics (25 μm–5 mm) in the Sludge/Biosolids Samples of Plant A

Sludge Type	MP Concentration (10^3 MP/kg dry sludge, mean ±SD, n = 4)
Primary sludge	9.4 ± 0.8
Secondary sludge	27.8 ± 2.5
Biosolids	16.6 ± 3.4

TABLE 4.3
Abundance of Microplastics in Different Types of Reported in Some Previous Studies

Sludge Type	MP Content (10^3 MP/kg dry sludge)	Size Range of MP	Mean (10^3 MP/kg dry sludge)	Median (10^3 MP/kg dry sludge)	Number of studies
Primary sludge	5–15	50 μm–5 mm	10	9.5	3
Secondary sludge	4.5–495	10 μm–5 mm	130	23	5
Mixed sludge	1–240	10 μm–5 mm	65	17	11
Biosolids/digested sludge	2.5–171	25 μm–5 mm	70	8.5	5

Source: Li et al. (2019); Collivignarelli et al. (2021).

4.10 CONCLUSIONS AND RECOMMENDATIONS

Microplastic pollution is a growing global concern for water quality and environmental and human health. While sewage treatment plants are an important pathway for microplastics that enter the environment, they are also an effective barrier to block microplastics. This chapter provides an overview of the current knowledge of the fate and control of microplastics in municipal wastewater treatment systems and presents the case study undertaken by the authors' group to address some of the identified knowledge gaps.

The case study showed the average MP concentrations of plant influent for the three WWTPs over the 2 years were in the range of 39.6–52.5 MP/L, with fibres being the dominant species, accounting for 57.3–77.6% of the total MP particles. The results were in agreement that synthetic fibres were a major source of microplastics in municipal sewage, as reported in many previous studies. For the final effluent of the three plants, the average MP concentrations were in the range of 0.17–1.66 MP/L, with the lowest concentration achieved by the advanced treatment process utilising membrane filtration. The abundance of MP in the primary and secondary sludge, and biosolids samples of the selected WWTP under this study was 9.4 ±0.8, 27.8 ±2.5, and 16.6 ±3.4 × 10^3 MP/kg dry solids, respectively. The results followed a similar trend in that secondary sludge generally had a higher MP concentration than primary sludge.

The four treatment systems of the three WWTPs generally exhibited high removal efficiency for MP, ranging from 97 to 99.6%, with the tertiary treatment for Class A water production achieving the highest removal. The two lagoon-based treatment systems were found to give promising removal of MPs (97% and 98.4%, respectively), which were attributed primarily to the lengthy wastewater detention time and the CAS pretreatment of the plant influent. The study also found a seasonal trend of MP concentration in the plant influent that was related to the rainfall events and the possible seasonal change of population in the catchment areas.

The research demonstrated the potential of low-energy and low-cost lagoon/pond wastewater treatment systems for effective control of MP. This new knowledge could help water utilities manage the MP in their sewage treatment systems and mitigate microplastic pollution risk. As synthetic fibres were the main contributor to the MP in sewage, management strategies to control their entrance into the sewer and removal by the treatment systems can be explored. The knowledge of the risks of small MP (e.g., <20 μm) and nano-sized plastics on human health and the environment is scarce. Their removal rate by the existing water treatment technologies is largely unknown; therefore, more work should be done in this area for future guidelines in wastewater treatment and reuse.

REFERENCES

Bayo, Javier, Joaquín López-Castellanos, and Sonia Olmos. 2020. "Membrane bioreactor and rapid sand filtration for the removal of microplastics in an urban wastewater treatment plant." *Marine Pollution Bulletin* 156:111211. doi: 10.1016/j.marpolbul.2020.111211.

Burns, Emily E., and Alistair B. A. Boxall. 2018. "Microplastics in the aquatic environment: Evidence for or against adverse impacts and major knowledge gaps: Microplastics in the environment." *Environmental Toxicology and Chemistry* 37 (11):2776–2796. doi: 10.1002/etc.4268.

Carr, Steve A., Jin Liu, and Arnold G. Tesoro. 2016. "Transport and fate of microplastic particles in wastewater treatment plants." *Water Research (Oxford)* 91:174–182. doi: 10.1016/j.watres.2016.01.002.

Coggins, Liah X., Nicholas D. Crosbie, and Anas Ghadouani. 2019. "The small, the big, and the beautiful: Emerging challenges and opportunities for waste stabilization ponds in Australia." *Wiley Interdisciplinary Reviews. Water* 6 (6):e1383-n/a. doi: 10.1002/wat2.1383.

Collivignarelli, Maria Cristina, Marco Carnevale Miino, Francesca Maria Caccamo, and Chiara Milanese. 2021. "Microplastics in sewage sludge: A known but underrated pathway in wastewater treatment plants." *Sustainability (Basel, Switzerland)* 13 (22):12591. doi: 10.3390/su132212591.

Conley, Kenda, Allan Clum, Jestine Deepe, Haven Lane, and Barbara Beckingham. 2019. "Wastewater treatment plants as a source of microplastics to an urban estuary: Removal efficiencies and loading per capita over one year." *Water Research X* 3:100030. doi: 10.1016/j.wroa.2019.100030.

Estahbanati, Shirin, and N. L. Fahrenfeld. 2016. "Influence of wastewater treatment plant discharges on microplastic concentrations in surface water." *Chemosphere (Oxford)* 162:277–284. doi: 10.1016/j.chemosphere.2016.07.083.

Fan, Linhua, Arash Mohseni, Jonathan Schmidt, Ben Evans, Ben Murdoch, and Li Gao. 2023. "Efficiency of lagoon-based municipal wastewater treatment in removing microplastics." *Science of the Total Environment* 876:162714. doi: 10.1016/j.scitotenv.2023.162714.

Gatidou, Georgia, Olga S. Arvaniti, and Athanasios S. Stasinakis. 2019. "Review on the occurrence and fate of microplastics in Sewage Treatment Plants." *Journal of Hazardous Materials* 367:504–512. doi: 10.1016/j.jhazmat.2018.12.081.

Gies, Esther A., Jessica L. LeNoble, Marie Noël, Anahita Etemadifar, Farida Bishay, Eric R. Hall, and Peter S. Ross. 2018. "Retention of microplastics in a major secondary wastewater treatment plant in Vancouver, Canada." *Marine Pollution Bulletin* 133:553–561. doi: 10.1016/j.marpolbul.2018.06.006.

Hill, R., L. Carter, and R. Kay. 2012. *Wastewater treatment facilities*. Geoscience Australia.

Horton, Alice A., and Simon J. Dixon. 2018. "Microplastics: An introduction to environmental transport processes." *Wiley Interdisciplinary Reviews. Water* 5 (2):e1268-n/a. doi: 10.1002/wat2.1268.

Iyare, Paul U., Sabeha K. Ouki, and Tom Bond. 2020. "Microplastics removal in wastewater treatment plants: A critical review." *Environmental Science Water Research & Technology* 6 (1):2664–2675. doi: 10.1039/d0ew00397b.

Kang, Hyun-Joong, Hee-Jin Park, Oh-Kyung Kwon, Won-Seok Lee, Dong-Hwan Jeong, Byoung-Kyu Ju, and Jung-Hwan Kwon. 2018. "Occurrence of microplastics in municipal sewage treatment plants: A review." *Environmental Health and Toxicology* 33 (3):e2018010– e2018013. doi: 10.5620/eht.e2018013.

Kay, Paul, Robert Hiscoe, Isobel Moberley, Luke Bajic, and Niamh McKenna. 2018. "Wastewater treatment plants as a source of microplastics in river catchments." *Environmental Science and Pollution Research International* 25 (20):20264–20267. doi: 10.1007/s11356-018-2070-7.

Koelmans, Albert A., Nur Hazimah Mohamed Nor, Enya Hermsen, Merel Kooi, Svenja M. Mintenig, and Jennifer De France. 2019. "Microplastics in freshwaters and drinking water: Critical review and assessment of data quality." *Water Research (Oxford)* 155:410–422. doi: 10.1016/j.watres.2019.02.054.

Lares, Mirka, Mohamed Chaker Ncibi, Markus Sillanpää, and Mika Sillanpää. 2018. "Occurrence, identification and removal of microplastic particles and fibers in conventional activated sludge process and advanced MBR technology." *Water Research (Oxford)* 133:236–246. doi: 10.1016/j.watres.2018.01.049.

Li, Xiaowei, Qingqing Mei, Lubei Chen, Hongyuan Zhang, Bin Dong, Xiaohu Dai, Chiquan He, and John Zhou. 2019. "Enhancement in adsorption potential of microplastics in sewage sludge for metal pollutants after the wastewater treatment process." *Water Research (Oxford)* 157:228–237. doi: 10.1016/j.watres.2019.03.069.

Liu, Xiaoning, Wenke Yuan, Mingxiao Di, Zhen Li, and Jun Wang. 2019. "Transfer and fate of microplastics during the conventional activated sludge process in one wastewater treatment plant of China." *Chemical Engineering Journal (Lausanne, Switzerland: 1996)* 362:176–182. doi: 10.1016/j.cej.2019.01.033.

Lusher, A. L., M. McHugh, and R. C. Thompson. 2013. "Occurrence of microplastics in the gastrointestinal tract of pelagic and demersal fish from the English Channel." *Marine Pollution Bulletin* 67 (1–2):94–99. doi: 10.1016/j.marpolbul.2012.11.028.

Magni, Stefano, Andrea Binelli, Lucia Pittura, Carlo Giacomo Avio, Camilla Della Torre, Camilla Carla Parenti, Stefania Gorbi, and Francesco Regoli. 2019. "The fate of microplastics in an Italian Wastewater Treatment Plant." *Science of the Total Environment* 652:602–610. doi: 10.1016/j.scitotenv.2018.10.269.

Magnusson, Kerstin, Karin Eliaeson, Anna Fråne, Kalle Haikonen, Mikael Olshammar, Johanna Stadmark, and Johan Hultén. 2016. *Swedish sources and pathways for microplastics to the marine environment*. IVL Svenska Miljöinstitutet.

Mahon, A. M., B. O'Connell, M. G. Healy, I. O'Connor, R. Officer, R. Nash, and L. Morrison. 2017. "Microplastics in sewage sludge: Effects of treatment." *Environmental Science & Technology* 51 (2):810–818. doi: 10.1021/acs.est.6b04048.

Mason, Sherri A., Danielle Garneau, Rebecca Sutton, Yvonne Chu, Karyn Ehmann, Jason Barnes, Parker Fink, Daniel Papazissimos, and Darrin L. Rogers. 2016. "Microplastic pollution is widely detected in

US municipal wastewater treatment plant effluent." *Environmental Pollution (1987)* 218:1045–1054. doi: 10.1016/j.envpol.2016.08.056.

McCormick, Amanda, Timothy J. Hoellein, Sherri A. Mason, Joseph Schluep, and John J. Kelly. 2014. "Microplastic is an abundant and distinct microbial habitat in an urban river." *Environmental Science & Technology* 48 (20):11863–11871. doi: 10.1021/es503610r.

Michielssen, Marlies R., Elien R. Michielssen, Jonathan Ni, and Melissa B. Duhaime. 2016. "Fate of microplastics and other small anthropogenic litter (SAL) in wastewater treatment plants depends on unit processes employed." *Environmental Science Water Research & Technology* 2 (6):1064–1073. doi: 10.1039/C6EW00207B.

Mintenig, S. M., I. Int-Veen, M. G. J. Löder, S. Primpke, and G. Gerdts. 2017. "Identification of microplastic in effluents of waste water treatment plants using focal plane array-based micro-Fourier-transform infrared imaging." *Water Research (Oxford)* 108:365–372. doi: 10.1016/j.watres.2016.11.015.

Murphy, Fionn, Ciaran Ewins, Frederic Carbonnier, and Brian Quinn. 2016. "Wastewater Treatment Works (WwTW) as a source of microplastics in the aquatic environment." *Environmental Science & Technology* 50 (11):5800–5808. doi: 10.1021/acs.est.5b05416.

Simon, Márta, Nikki van Alst, and Jes Vollertsen. 2018. "Quantification of microplastic mass and removal rates at wastewater treatment plants applying Focal Plane Array (FPA)-based Fourier Transform Infrared (FT-IR) imaging." *Water Research (Oxford)* 142:1–9. doi: 10.1016/j.watres.2018.05.019.

Sun, Jing, Xiaohu Dai, Qilin Wang, Mark C. M. van Loosdrecht, and Bing-Jie Ni. 2019. "Microplastics in wastewater treatment plants: Detection, occurrence and removal." *Water Research (Oxford)* 152:21–37. doi: 10.1016/j.watres.2018.12.050.

Sutton, Rebecca, Sherri A. Mason, Shavonne K. Stanek, Ellen Willis-Norton, Ian F. Wren, and Carolynn Box. 2016. "Microplastic contamination in the San Francisco Bay, California, USA." *Marine Pollution Bulletin* 109 (1):230–235. doi: 10.1016/j.marpolbul.2016.05.077.

Talvitie, Julia, Mari Heinonen, Jari-Pekka Pääkkönen, Emil Vahtera, Anna Mikola, Outi Setälä, and Riku Vahala. 2015. "Do wastewater treatment plants act as a potential point source of microplastics? Preliminary study in the coastal Gulf of Finland, Baltic Sea." *Water Science and Technology* 72 (9):1495–1504. doi: 10.2166/wst.2015.360.

Talvitie, Julia, Anna Mikola, Arto Koistinen, and Outi Setälä. 2017. "Solutions to microplastic pollution – Removal of microplastics from wastewater effluent with advanced wastewater treatment technologies." *Water Research (Oxford)* 123:401–407. doi: 10.1016/j.watres.2017.07.005.

Um, Michelle, Dhakshitha Weerackody, Li Gao, Arash Mohseni, Ben Evans, Ben Murdoch, Jonathan Schmidt, and Linhua Fan. 2023. "Investigating the fate and transport of microplastics in a lagoon wastewater treatment system using a multimedia model approach." *Journal of Hazardous Materials* 446:130694. doi: 10.1016/j.jhazmat.2022.130694.

Wang, Zhifeng, Tao Lin, and Wei Chen. 2020. "Occurrence and removal of microplastics in an advanced drinking water treatment plant (ADWTP)." *Science of the Total Environment* 700:134520. doi: 10.1016/j.scitotenv.2019.134520.

Xiong, Xiong, Kai Zhang, Xianchuan Chen, Huahong Shi, Ze Luo, and Chenxi Wu. 2018. "Sources and distribution of microplastics in China's largest inland lake – Qinghai Lake." *Environmental Pollution (1987)* 235:899–906. doi: 10.1016/j.envpol.2017.12.081.

Yang, Libiao, Kuixiao Li, Song Cui, Yu Kang, Lihui An, and Kun Lei. 2019. "Removal of microplastics in municipal sewage from China's largest water reclamation plant." *Water Research (Oxford)* 155:175–181. doi: 10.1016/j.watres.2019.02.046.

Yaseen, Aarif, Irfana Assad, Mohd Sharjeel Sofi, Muhammad Zaffar Hashmi, and Sami Ullah Bhat. 2022. "A global review of microplastics in wastewater treatment plants: Understanding their occurrence, fate and impact." *Environmental Research* 212 (Pt B):113258–113258. doi: 10.1016/j.envres.2022.113258.

Zhang, Zhiqi, and Yinguang Chen. 2020. "Effects of microplastics on wastewater and sewage sludge treatment and their removal: A review." *Chemical Engineering Journal (Lausanne, Switzerland: 1996)* 382:122955. doi: 10.1016/j.cej.2019.122955.

Ziajahromi, Shima, Peta A. Neale, Llew Rintoul, and Frederic D. L. Leusch. 2017. "Wastewater treatment plants as a pathway for microplastics: Development of a new approach to sample wastewater-based microplastics." *Water Research (Oxford)* 112:93–99. doi: 10.1016/j.watres.2017.01.042.

5 Sources and Occurrence of Microplastics in the Terrestrial Environment

Sarva Mangala Praveena

5.1 INTRODUCTION

Terrestrial ecosystems provide about 99% of human food calories, with only 1% coming from oceans and other aquatic ecosystems. Food and fibre crops are grown on 11% of the Earth's total land area (Pimentel et al., 2012). Microplastics pose a growing hazard to terrestrial ecosystems, as they can infiltrate the soil via multiple routes, including unmanaged plastics debris, windblown refuse, and sewage sludge. Microplastics, which result from the dumping, fragmentation, and degradation of plastic in the land environment, build up in the food chain, posing dangers to ecosystems and human well-being. Plastic waste undergoes degradation on land, resulting in the formation of minuscule particles known as microplastics. These microplastic particles have the ability to infiltrate the food chain by being consumed by organisms that reside in the soil. The infiltration of these particles has harmful consequences for plant health and soil fertility, highlighting the immediate necessity for a thorough comprehension and efficient measures for their management (Wong et al., 2020).

Microplastics are a growing threat to terrestrial ecosystems (de Souza Machado et al., 2018; Rose et al., 2023a). As plastic waste accumulates on land, it undergoes degradation, transforming into microplastics that infiltrate the intricate web of the food chain, thereby posing risks to ecosystems and human health, as emphasized by Mai et al. (2018). Soil organisms, vital components of terrestrial ecosystems, unwittingly ingest these particles, prompting alterations in their behaviour and physiology. Moreover, microplastics introduce chemicals into the soil, leading to contamination of plants and animals (Abdul Khaliq et al., 2017). Recognizing the sources and pathways of microplastics in terrestrial environments is paramount, offering critical insights to inform policies and management practices aimed at curbing plastic pollution. Investigating the intricate interactions between microplastics and soil organisms not only unveils potential ecological effects but also sheds light on changes in soil structure and nutrient cycling. This understanding becomes imperative for safeguarding the well-being of ecosystems, animals, and humans alike, underscoring the essential role of research in mitigating the pervasive impact of microplastics on terrestrial environments (Surendran et al., 2023).

This chapter aims to face the pressing need to address the environmental issue at hand by examining microplastic pollution in terrestrial environments, particularly in soil. The chapter aims to integrate a wide range of facts in order to offer a comprehensive understanding of the sources and occurrence of microplastics in terrestrial ecosystems. The chapter is structured into four sections, which systematically cover important aspects of microplastic pollution in terrestrial environments. These sections include the identification of sources of microplastics, the study of the fate and movement of microplastics, the examination of the impacts of microplastics, and the analysis of case studies that demonstrate the occurrence of microplastics in terrestrial environments. In order to accomplish this extensive synthesis, a meticulous examination of scholarly literature, encompassing journals, books, and conference proceedings, has been conducted. Incorporating ideas from official grey literature, including publications from international organizations and governments, has been carried out to ensure a comprehensive view. This chapter attempts to consolidate these heterogeneous sources of information, thereby providing a valuable resource for numerous stakeholders.

The chapter aims to go beyond scientific research by synthesizing various information sources and become a beneficial resource for different stakeholders. The chapter aims to offer scientists a comprehensive and in-depth comprehension of the intricate dynamics of microplastics in terrestrial ecosystems. This study aims to enhance the current knowledge on microplastic pollution in soil ecosystems by analysing information from multiple academic sources. Its goal is to assist researchers in better understanding the complexities associated with this issue. The chapter is directed toward policymakers with the urgent need for evidence-based guidance on effective management techniques to lessen the impact of microplastics. This chapter moves beyond the realm of intellectual discourse to provide insights that can be used to inform the creation of pragmatic policies that attempt to reduce the sources and pathways of microplastics, thereby contributing to broader environmental sustainability goals.

In addition, non-governmental organizations (NGOs) that are interested in increasing public awareness and conducting educational campaigns consider this chapter to be a significant resource. The consolidated information can be utilized to emphasize the need for tackling microplastic pollution, promoting informed discussions, and advocating for sustainable habits among the general public. Finally, this chapter acknowledges its influence in moulding the educational environment, serving the needs of both undergraduate and postgraduate students. By imparting a basic comprehension of microplastic contamination in land-based ecosystems, particularly in soil, it empowers future environmental guardians with the requisite information to address this widespread problem. The main objective of the chapter is to go beyond the limits of different fields of study, connecting research and practical use, and acting as a versatile source of information for a wide range of people interested in the sustainable management of land-based ecosystems.

5.2 SOURCES OF MICROPLASTICS IN THE TERRESTRIAL ENVIRONMENT

The sources of microplastics in the terrestrial environment are varied and complex (Figure 5.1). Primary sources of microplastics in the terrestrial environment include plastic waste improperly disposed of in the environment due to littering including in construction sites, roadsides, and landfills, synthetic clothing fibres, which shed microfibers when laundered, road tyres, and plastic pellets (Haque & Fan, 2023; Saud et al., 2023). Meanwhile secondary sources of microplastics in the terrestrial environment involve continuous breakdown of larger plastic items into smaller particles due to exposure to environmental factors such as sunlight, wind, and water.

5.2.1 PRIMARY SOURCES OF MICROPLASTICS IN THE TERRESTRIAL ENVIRONMENT

Plastic waste improperly disposed of in the environment can be found in a variety of terrestrial habitats, ranging from forests and grasslands to cities and agricultural farms (Hurley et al., 2020). There are a variety of sources of plastic waste in the terrestrial environment, including both intentional and unintentional actions. One of the primary causes of plastic pollution is the improper disposal of plastic waste. This can occur when plastic debris is improperly disposed of, such as when it is dumped in a soil environment (Okunola et al., 2019). Additionally, inadequate waste management systems are a major contributor to plastic waste in soil environments (Breukelman et al., 2019). Additionally, plastic waste generated from construction sites, along with plastic waste improperly managed in landfill, are also contributors to microplastic pollution in the terrestrial environment (Awoyera & Adesina, 2020; Canopoli et al., 2020).

FIGURE 5.1 Primary and secondary sources of microplastics in the terrestrial environment.

Road transport is another source of plastic littering in the terrestrial environment. Particles generated from vehicle tyres and brakes are also a significant source of microplastics in the terrestrial environment. The process of tyre wear and tear, as well as brake wear, results in the shedding of small particles, which can contribute to the formation of microplastics. Tyre wear and tear occurs due to the friction between the tyres and the road surface, leading to the shedding of small rubber particles called "tyre dust". This tyre dust is composed of a variety of materials, including microplastics, rubber, and metals, and can enter the environment through stormwater runoff, air pollution, and direct deposition (Su et al., 2022). Brake wear, on the other hand, occurs due to the friction between the brake pads and the brake discs. This process generates small particles that contain microplastics and other materials such as copper, iron, and graphite. Brake dust can also be transported through air and water currents, leading to the formation of microplastics in the environment. The process of shredding and release of microplastics from vehicle tyres and brakes is a complex process influenced by several factors, including vehicle speed, traffic density, and road surface type. For instance, high traffic density and low vehicle speed can result in increased tyre wear and tear, leading to higher levels of microplastics in the environment. Similarly, the use of studded tyres can contribute to higher levels of tyre dust and microplastics due to increased friction with the road surface (Worek et al., 2022; Cai et al., 2020; Sommer et al., 2018).

Synthetic clothing fibres have become a primary source of microplastics in the terrestrial environment, with evidence indicating that they shed microfibres during laundering (Yadav et al., 2023; Praveena et al., 2021). When synthetic clothing is laundered in a washing machine, the agitation and water pressure generated in the washing process lead to the shedding of microfibres. The fibres become detached from the clothing and are carried away with the wastewater (De Falco et al., 2018; Pirc et al., 2016). These microfibres then pass through wastewater treatment plants, where they can enter rivers, lakes, and other water bodies. The microfibres can also accumulate in sludge, which is often used as fertilizer in agricultural fields, leading to their introduction into the terrestrial environment. Moreover, these microfibres can be transported over long distances by wind and water, ultimately leading to their deposition in soils, freshwater bodies, and even oceans (Salvador Cesa et al., 2017; Napper & Thompson, 2016).

The primary source of microplastics in the terrestrial environment is plastic pellets, also known as nurdles. These small, pre-production plastic pellets are transported globally in vast volumes for use as raw materials in the production of plastic products. Accidental pellet leaks can occur during transport and handling, leading to plastic litter and the eventual formation of microplastics. In addition, poor waste management practices and inadequate infrastructure in regions where plastic is produced

Occurrence of Microplastics in the Terrestrial Environment

can contribute to the environmental discharge of plastic pellets. These pellets can be carried by wind and water currents and accumulate in ecosystems, posing a threat to the health of fauna and soil (Kallenbach Emilie and Rødland, 2022; Rose et al., 2023b; Balthazar-Silva et al., 2020).

Plastic mulch films are widely used in agricultural practices to enhance crop growth and yield. These thin plastic sheets are applied to the soil surface before planting to control weeds, retain soil moisture, and regulate soil temperature (Kasirajan & Ngouajio, 2012). However, plastic mulch films can also become a primary source of microplastics in the terrestrial environment. The degradation of plastic mulch films into microplastics is a complex process that involves the effects of environmental factors such as sunlight, temperature, and moisture (Praveena et al., 2023). Over time, exposure to these factors causes the plastic mulch films to break down into smaller pieces, which can then be incorporated into the soil and persist for years. Studies have shown that plastic mulch films can release microplastics into the soil at a rate of up to 40 grams per square metre (Zhou et al., 2020).

5.2.2 Secondary Sources of Microplastics in the Terrestrial Environment

Secondary sources of microplastics in the terrestrial environment refer to the fragmentation of larger plastic items due to natural weathering and degradation processes (Rillig, 2012). Figure 5.2 shows

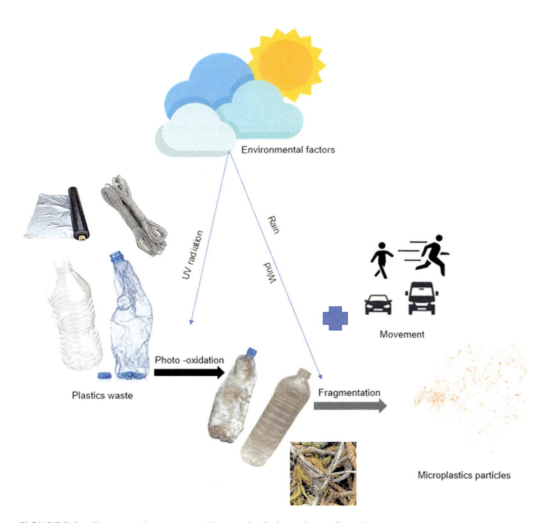

FIGURE 5.2 Fragmentation process of larger plastic items in a soil environment.

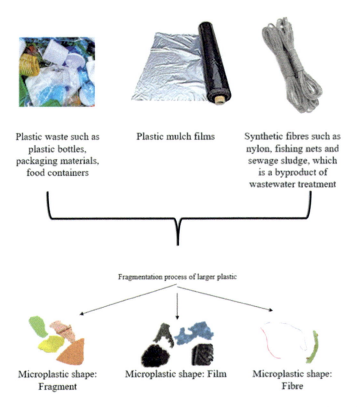

FIGURE 5.3 Fragmentation process of various plastics waste into various microplastic particle shapes.

the fragmentation process of larger plastic items in soil environment. These larger plastic items can come from various primary sources, including plastic waste, plastic packaging, and plastic products that are designed to have a shorter lifespan, such as disposable cutlery and food containers. Once these larger plastic items enter the environment, they are exposed to a variety of environmental factors such as sunlight, heat, wind, and water as well as physical movement (people, vehicle) which can cause them to degrade and fragment into smaller pieces (de Souza Machado et al., 2018).

The gradual breakdown of larger plastic items into smaller particles, including mesoplastics, microplastics, and nanoplastics, is a complex process influenced by various environmental factors (Sa'adu & Farsang, 2023). Understanding this breakdown process is important for comprehending the distribution, transport, and potential impacts of plastic pollution in the environment. As the larger plastic items continue to degrade, they progressively fragment into smaller pieces, forming different size and shape ranges of plastic particles (Figure 5.3):

- Plastic waste such as plastic bottles, packaging materials, food containers, and other plastic items that end up in soil can break down over time, resulting in the release of microplastic fragments into the soil environment (Kadac-Czapska et al., 2023; Sobhani et al., 2020).
- Over time, these plastic mulch films used in agriculture practice can degrade due to exposure to sunlight and physical stresses, resulting in the release of microplastic particles in firm shape into soil (Sa'adu & Farsang, 2023).
- Synthetic fibres such as nylon, fishing nets, and sewage sludge, which is a by-product of wastewater treatment, can contain microplastic particles in fibre shape derived from domestic and industrial sources (L. Zhang et al., 2020).

5.3 FATE AND TRANSPORT OF MICROPLASTICS IN THE TERRESTRIAL ENVIRONMENT

5.3.1 PHYSICAL PROCESS

Understanding the fate and transport of microplastics is crucial for assessing their environmental impacts and developing effective mitigation strategies (Wong et al., 2020; Geyer et al., 2017). Several key factors influence the behaviour of microplastics in terrestrial ecosystems, including their distribution, persistence, and potential interactions with organisms.

Physical processes by climate variables significantly impact the fate and transport of microplastics in the terrestrial environment. The physical characteristics of microplastics, such as their flexibility and buoyancy, can be affected by temperature changes. Increased temperatures can also hasten the biological and chemical processes that interact with microplastics, influencing their rates of disintegration and possible interactions with living organisms (Chamas et al., 2020). The quantity, intensity, and frequency of precipitation events influence microplastics' transport and distribution. Heavy precipitation can mobilize and transport microplastics from various sources, such as urban discharge, to rivers, lakes, and oceans. Precipitation can also influence the erosion of plastic debris from land surfaces, facilitating their movement into water bodies and fostering their further dispersion. Microplastics may become more easily dispersed and transported through wind and water currents as a result of greater degradation and fragmentation brought on by warmer temperatures (Haque & Fan, 2023). Wind plays a significant role in the long-distance transport of microplastics. Strong winds can lift and transport microplastics, particularly those found in open areas such as beaches, fields, and deserts. Wind patterns can determine the spatial distribution of microplastics by conveying them across various regions, including remote and unaffected regions. Microplastics' behaviour is influenced by humidity by changing the characteristics of their surfaces. Microplastics can become more cohesive and prone to aggregation when there is a high concentration of water molecules adhering to them. This grouping may affect how microplastics settle and are deposited on different surfaces, such as soil, vegetation, and water bodies (Chamas et al., 2020; Hurley et al., 2020).

The settling and deposition of microplastics in the terrestrial environment are key processes that influence their fate and transport. Once microplastics are released into the environment, they can undergo various settling and deposition mechanisms, which ultimately determine their distribution and potential impacts (Lu et al., 2023). The settling and deposition mechanisms of microplastics in the terrestrial environment are complex processes that include a combination of physical, chemical, and biological interactions (Saud et al., 2023). Gaining a comprehensive understanding of these mechanisms is of utmost importance in order to accurately evaluate the environmental consequences of microplastics and devise efficient methods to minimize their negative effects. The settling and deposition of microplastics are influenced by several elements, which can be classified into gravitational, hydrodynamic, electrostatic, and biological processes.

Microplastics, especially larger and denser particles, are subject to gravitational forces that cause them to settle and deposit onto different surfaces. Gravity plays a significant role in the vertical movement of microplastics, leading to their accumulation in specific areas such as soils (Koutnik et al., 2021). The surface properties of microplastics, including their size, shape, and roughness, can affect their settling and deposition. Larger microplastics tend to settle more rapidly due to their higher mass, while smaller particles may remain suspended in air or water for longer periods. The shape and roughness of microplastics can influence their interaction with surfaces, affecting their deposition efficiency (Yu & Flury, 2021). In the case of airborne microplastics, aerodynamic forces play a crucial role in their settling and deposition. These forces are influenced by particle size, shape, density, and air turbulence (Yuan & Xu, 2023). Larger, heavier particles with streamlined shapes are more likely to settle closer to their release source, while smaller particles can be transported over longer distances before depositing onto terrestrial surfaces. Hydrodynamic forces, such as water currents and turbulence, influence the settling of microplastics in soils or sediments. Microplastics can be transported by water currents, and when these currents slow down or turbulence decreases, the particles move out of the water column and settle in soils or sediments. Furthermore, vegetation and surface characteristics, such as roughness and hydrophobicity, can affect the deposition of microplastics. Vegetation can act as a physical barrier, intercepting airborne microplastics and causing them to settle onto leaves, branches, or the ground. Surface roughness and hydrophobic properties can enhance the retention and deposition of microplastics in soils or sediments. Similarly, water and soil dynamics influence the movement and transport of microplastic particles, causing them to settle and accumulate in certain areas (Burrows et al., 2020; Yuan & Xu, 2023).

Microplastics can be deposited on surfaces due to the influence of electrostatic interactions. Microplastic deposition can be influenced by the presence of a net charge on natural surfaces in terrestrial environments. Oppositely charged surfaces can exert an attractive force on charged particles. Moreover, the surface charge of microplastics might affect how they interact with nearby particles and surfaces (Lwanga et al., 2022). Microplastics can adsorb onto existing particles, such as sediment grains or organic matter. This process can alter the buoyancy and settling characteristics of microplastics. Adsorption onto particles can also lead to the formation of aggregates, affecting the transport and fate of microplastics in the environment (Al Harraq & Bharti, 2022). Biological settling involves the interaction of microplastics with living organisms and organic matter. Microplastics can be incorporated into soil matrices through the activity of microorganisms and plant roots. The binding of microplastics to organic matter can alter their transport behaviour and affect their availability to biota (Zhang et al., 2021).

5.3.2 Chemical Process

Microplastics can undergo chemical degradation in the soil environment through interactions with soil constituents, such as moisture, oxygen, and reactive minerals. There are a number of chemical reactions that have an impact on the fate and movement of microplastics in the terrestrial environment. Oxidation reactions have the ability to degrade the polymer chains of microplastics, resulting in the formation of smaller and less persistent molecules. Oxidative reactions can be initiated by exposure to environmental stimuli such as sunlight, oxygen, and reactive species. Photodegradation is a process where sun radiation causes oxidation, which in turn breaks down polymer chains and creates smaller pieces. These pieces might possess modified characteristics and heightened vulnerability to additional deterioration (Sutkar et al., 2023). Hydrolysis reactions, triggered by soil pH and moisture, can also contribute to the mineralization of microplastics. The interactions of microplastics with the soil, vegetation, and other elements of the terrestrial ecosystem are largely governed by these processes (Li et al., 2023; Sajjad et al., 2022). Mineralization refers to the conversion of organic compounds into inorganic substances. Microplastics can undergo mineralization processes facilitated by microbial activity in the soil, broken down into smaller fragments through various chemical and biological processes, the eventual mineralization may result in the formation of carbon dioxide, water, and other inorganic substances (Shi et al., 2023).

Microplastics' fate and transport are influenced by their size and shape. Smaller particles are more likely to adsorb chemicals due to their higher surface area to volume ratio, which also makes them easier to transfer via soil pores and water systems. In addition to affecting particle mobility and deposition patterns, irregular forms can alter how microplastics are distributed in the terrestrial environment (Li et al., 2023; Raza et al., 2022). The fate and transport of microplastics can be strongly impacted by the composition, texture, and organic matter concentration of the soil. The ability of various soil types to absorb and hold onto microplastics varies. Soils with higher clay contents

Occurrence of Microplastics in the Terrestrial Environment

have a larger surface area with better adsorption capability. Microplastics and organic materials can potentially interact, changing the dynamics of their stability and transit (Raza et al., 2022).

Microplastics can also release adsorbed chemicals back into the environment (Li et al., 2023). Desorption can occur under certain environmental conditions such as changes in pH, temperature, or the presence of other chemicals. This process can lead to the release of pollutants, potentially affecting the surrounding soil, water, and biota (Burrows et al., 2020). Microplastics can adsorb various chemicals present in the environment onto their surfaces. This includes organic compounds, heavy metals, and persistent organic pollutants. Adsorption can affect the transport of microplastics by changing their buoyancy, stability, and aggregation behaviour. It can also influence the bioavailability and toxicity of the adsorbed chemicals (Haque & Fan, 2023).

Mineralization is a significant process that can contribute to the overall transformation of microplastics in the terrestrial environment. Mineralization refers to the conversion of microplastics into inorganic compounds through chemical processes. It involves the breakdown of the polymer structure of microplastics into smaller, non-polymeric compounds such as carbon dioxide (CO_2), water (H_2O), and mineral residues (Xie et al., 2023). Mineralization can occur through various chemical reactions, including oxidation, hydrolysis, and photodegradation. It is important to note that while mineralization of chemical processes can help mitigate the persistence of microplastics, they may also produce intermediate breakdown products that could still have potential ecological impacts (Xie et al., 2023). However, the extent and rate of these processes can vary depending on factors such as environmental conditions, microplastic characteristics, and the presence of suitable degrading microorganisms (Kaczmarek et al., 2008).

5.4 EFFECTS OF MICROPLASTICS IN THE TERRESTRIAL ENVIRONMENT

5.4.1 ECOLOGICAL EFFECTS OF MICROPLASTICS IN A SOIL ENVIRONMENT

While the consequences of microplastics in aquatic habitats have received a lot of attention, there is also increasing concern about their existence and effects in terrestrial environments (Praveena et al., 2022). Addressing the underlying causes of plastic pollution and implementing sustainable waste management practices are crucial for reducing the negative impacts of microplastics on the terrestrial environment (Schmaltz et al., 2020; Wang et al., 2022).

The physical effects of microplastics on soil structure are among the main issues. The structure and makeup of the soil might change when microplastics build up in it. Microplastics can cause soil to become compacted, pore spaces to be reduced, and water penetration and drainage to be hampered (Guo et al., 2022). This can reduce soil aeration and increase the risk of soil erosion. The habitat and survival of soil-dwelling organisms, such as microorganisms, earthworms, and arthropods, which are essential for nutrient cycling and soil fertility, can be affected by changes in soil structure (Wang et al., 2022).

The direct physical interaction of microplastics with plant root systems is another major concern. In the soil, microplastics can build up and create aggregates that obstruct root growth and development. The nutrition and hydration of plants may be impacted by restricted root development, which also limits nutrient and water absorption. Reduced biomass output, stunted development, and poorer agricultural yields can all result from this. Additionally, the presence of microplastics might prevent the germination of seeds, which has an adverse effect on plant population and ecosystem regeneration (Guo et al., 2022; Wang et al., 2022).

The possible spread of microplastics from plant roots to above-ground tissues is another worry. Microplastics may be absorbed by plants and moved to leaves, stems, and even fruits, according to studies. This raises questions about how people can be exposed to microplastics by eating tainted plant-based food (Wang et al., 2022). In addition, the accumulation of microplastics in plant tissues can impair physiological processes such as photosynthesis, transpiration, and hormone regulation, resulting in stunted plant growth and reduced plant health (Wang et al., 2020).

Microplastics in soil can also affect the availability and cycling of nutrients (Maddela et al., 2023). These microplastic particles have the capacity to absorb and concentrate chemical contaminants, such as heavy metals and persistent organic pollutants, which may prevent plant and soil microbes from effectively utilizing nutrients (Saud et al., 2023; Xie et al., 2023). Additionally, soil microbes, which are essential for nutrient cycling and transformation, might be impacted by microplastics in terms of their activity and composition. Alterations in microbial communities have the potential to reduce soil fertility and jeopardize plant development by causing an imbalance in the mechanisms that cycle nutrients (Wang et al., 2022; Wang et al., 2020; Boots et al., 2019).

5.4.2 Impact of Microplastics in Soil on Human Health

Microplastics present in soil can be taken up by plant roots and subsequently transported to different plant parts, including edible portions such as fruits and vegetables (Wang et al., 2022; Boots et al., 2019). When these contaminated crops are consumed, microplastics can be ingested by humans (Cox et al., 2019). Although research on the health repercussions of ingesting microplastics is still in its infancy, there are worries that tiny particles might harm the gastrointestinal system physically and perhaps release compounds that have been adsorbed by them, having harmful effects (Nam & Ats, 2022; Latunde-Dada et al., 1998). Another route of exposure is by direct skin contact with soil that has been polluted with microplastics. Although it is now thought that dermal absorption of microplastics is rather minimal, there are still concerns regarding potential health implications. Microplastics may irritate sensitive people's skin or induce allergic responses. Furthermore, microplastics that stick to the skin run the risk of moving to the mouth or being accidentally consumed through hand-to-mouth contact (Xie et al., 2023; Qi et al., 2020).

The toxicity and health risks associated with microplastics pollution in the terrestrial environment are of increasing concern (Saud et al., 2023). Through multiple exposure paths, these microplastic particles can build up in soil and pose possible health concerns to people. The potential for microplastics to behave as transporters of chemical contaminants is another major worry. These particles can adsorb and concentrate hazardous compounds, such as heavy metals, persistent organic pollutants (POPs), and other dangerous chemicals found in the environment, due to their large surface area (Burrows et al., 2020). Adsorbed chemicals may be released in the digestive tract after being swallowed, either directly through contaminated food or indirectly through dust inhalation or dermal contact (Durda & Preziosi, 2000). Microplastics are able to cross biological barriers and accumulate in various organs and tissues due to their diminutive size (Makhdoumi et al., 2023; Guerrera et al., 2021). Studies have shown that microplastics can translocate from the gastrointestinal tract to other parts of the body, such as the liver, kidneys, and even the bloodstream (Campanale et al., 2020). As microplastics may interact with vital organs and disrupt normal physiological processes, there is cause for concern regarding potential systemic effects and long-term health consequences. Moreover, the shape, size, and chemical composition of microplastics can influence their toxicity. Certain types of microplastics, such as those containing additives or made from polymers with a higher toxicity level, may produce toxins or stimulate the immune system. Due to their microscopic size, microplastics can also infiltrate cells and potentially cause cellular damage or inflammation (Yang et al., 2022).

5.5 CASE STUDIES OF MICROPLASTICS OCCURRENCE IN TERRESTRIAL ENVIRONMENT

Two case studies that investigated the occurrence of microplastics in agricultural and urban soil samples are used to highlight that the occurrence of microplastics in soil samples depends on the different pollution sources observed in each study area.

5.5.1 CASE STUDY 1: MICROPLASTICS OCCURRENCE IN AGRICULTURAL SOIL

This study involved analysing microplastics occurrence in surface soils from selected agricultural farms in a tropical environment (Klang Valley, Malaysia). Two-step filtration was used to extract microplastics from surface soil samples. Microplastics extraction was conducted using the density separation method using zinc chloride (1.7 g/cm^3) based on the method by Chen et al. (2019) and Rodrigues et al. (2018). For visual identification, all the filter papers were examined using an Olympus CHK Compound Microscope with a magnification of ×40 and photographed. Image J 1.53 (http://imagej.nih.gov), an open-source particle analysis software, was utilized to determine the length of the microplastic particles. Next, these particles were grouped according to shapes; specifically, fibre, fragments, films, pellets, foams, and colour using the classification criteria by Vendel et al. (2017).

The total number of microplastic particles ranged from 1.5 particles/kg to 6.0 particles/kg in surface soils from selected agricultural farms. The mean numbers of microplastic particles in agricultural soils, when averaged across the farms, were similar, ranging between 2.1 and 3.4 particles/kg. This appears proportional to the extent of plastics usage in agricultural activity at these farms. In terms of microplastic particle size, microplastic particles were between 16.73 and 1246.72 μm. These microplastic particles may relate to the extensive breakdown process associated with environmental factors such as sunlight, rainfall, and the movement of people. The largest microplastic particle size observed can be related to a lower breakdown process. For example, this farm used rain shelter to protect vegetables from environmental factors (heavy rainfall, high temperature, flooding, strong winds, sunlight, and relative humidity) and it had less movement of people. Environmental factors may have caused chemical changes in plastics, making them more fragile and breaking them down into smaller particles in the soil environment (Horton et al., 2017).

Figure 5.4 shows microplastic particle shapes found in surface soil from these agricultural farms. The three major microplastic particle shapes identified in these selected agricultural farms are film, fibre, and fragment. According to Lang et al. (2022) and Bläsing and Amelung (2018),

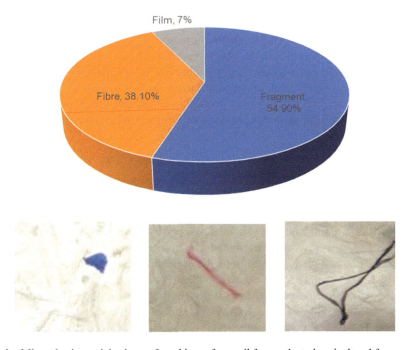

FIGURE 5.4 Microplastic particle shapes found in surface soil from selected agricultural farms.

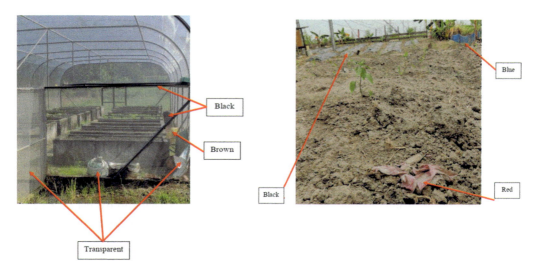

FIGURE 5.5 Plastics usage in selected farms which contributed to microplastic particles in the surface soil sample.

plastic mulching is the main source of film microplastic particles found globally in agricultural soil. Correspondingly, plastic mulching film was extensively found in these farms for use as a rain shelter material. As these farms were in a tropical environment, the plastic films broke down during agricultural activities due to tropical meteorological factors such as high UV radiation, temperature, and rainfall intensity. Fibre microplastic particles can be due to the higher plastic net usage in agricultural applications to protect against sun radiation, rain intensity, birds, and vector insects. Meanwhile fragment particles corresponded with the large unmanaged plastics waste which was breaking down in a brick raised garden planter (Figure 5.5).

Microplastic particle colours found in these agricultural farm soil samples included black coloured microplastic particles which dominated (51.4%), followed in descending order by white (26.2%), blue (3.9%), and red (1.4%) in all the farm soil samples. The black microplastic particle colour was closely connected with the plastic mulching film and nets utilized at all the farms, while white microplastic particles found mostly in fibre and fragment shapes were associated with secondary microplastics, which break down from larger plastic products such as plastic nets and unmanaged plastic waste. The red and blue microplastic particles indicate the decolorized blue plastic sheets and red plastics bags that broke down to form secondary microplastics at this farm.

5.5.2 Case Study 1: Microplastics Occurrence in Urban Soil Samples

This study involved analysing microplastics occurrence in urban soil samples from Cyberjaya (Malaysia). The microplastic particle analysis in urban soil samples was performed using a two-stage filtration procedure with density separation using a saturated sodium chloride (NaCl) solution with a density of 1.2 g/cm^3 following the method developed by Radford et al. (2021) and Lusher et al. (2020). The filter papers were visually examined using an Olympus CHK Compound Microscope with 40× magnification and photographed to facilitate identification. Microplastic particle classification criteria by Lusher et al. (2020) were used to identify plastic and non-plastic particles in these images. The size of the microplastic particles was determined from the photos using the open-source particle analysis software Image J 1.53 (http://imagej.nih.gov). The shape (i.e., fibres, fragments, films, pellets, and foams) and colours of the microplastic particles were grouped according to the classification criteria of Vendel et al. (2017).

Microplastic particles varied widely from 0.3 to 1.15 particles/kg in urban soil samples at all sampling sites in Cyberjaya (Malaysia). The highest percentages of microplastic shapes in the urban soil of Cyberjaya (Malaysia) were fragments (68.7%), followed by films (16.8%), and fibres (14.5%) which were mostly black in colour. The presence of black fragments in urban soils is likely due to distance from the source as sampling locations in this study were near major roads and heavy traffic. These black fragments from tyre wear particles came into contact with road dust and entered the soil. The presence of multicoloured microplastic particles in fragment shape could be related to the high amount of uncontrolled plastic waste disposed of in these land uses, which subsequently breaks down into smaller fragments (Du et al., 2020). This urban soil study in Cyberjaya (Malaysia) has also indicated that microplastic particle size is influenced by soil temperature. This finding suggested that high soil surface temperatures may promote plastic ageing and break-up in tropical environments such as in Malaysia. Acrylonitrile butadiene styrene (ABS), polyethylene glycol terephthalate (PET), polypropylene (PP), polyethylene (PE), and polystyrene (PS) were the predominant polymers found in the urban soil from Cyberjaya (Malaysia). These plastic polymers indicated that the potential sources in these urban soil samples from Cyberjaya (Malaysia) are emitted from packaging materials, beverage bottles such as water bottles, reusable bags, and disposable plastic bags (Figure 5.6).

FIGURE 5.6 Plastic usage practices in selected farms and their contribution to microplastics contamination in surface soil samples.

5.6 CONCLUSION

In conclusion, this chapter has covered terrestrial microplastic pollution, focusing on soil. Microplastics, plastic particles smaller than 5 mm, are becoming important contaminants with widespread environmental and health implications. The terrestrial environment, especially soil, is crucial to ecology because it supports a diversity of species, cycles nutrients, filters water, and stimulates plant development. While most research has focused on maritime habitats, this chapter compiles the latest discoveries and understandings about microplastics in terrestrial ecosystems to fill the knowledge gap. Primary and secondary sources of microplastics in terrestrial ecosystems have been examined in this chapter. The degradation of larger plastic waste releases microplastics into the soil, threatening terrestrial ecosystems. Next, the intricate chemical–physical mechanisms that control microplastic dispersion in these ecosystems were outlined. The effects of soil microplastics on human health and the ecology emphasize the need for a comprehensive solution to this issue. This chapter provides context and examples from Malaysia to demonstrate microplastic pollution in agricultural and urban soil samples. These case studies will increase global awareness of microplastic contamination and help build educated decision-making, environmental management, and public education techniques. This chapter can help scientists, policymakers, NGOs, and students understand microplastic contamination, making it crucial to academic comprehension. It provides the framework for evidence-based decision-making, effective environmental management, and public awareness. The chapter illuminates the sources, fate, and effects of microplastics in terrestrial settings, stimulating research, policy creation, and advocacy to reduce soil ecosystem microplastic pollution. Finally, it highlights the need for a multidisciplinary and collaborative approach to overcoming this Anthropocene environmental crisis.

REFERENCES

Abdul Khaliq, S. J., Al-Busaidi, A., Ahmed, M., Al-Wardy, M., Agrama, H., & Choudri, B. S. (2017). The effect of municipal sewage sludge on the quality of soil and crops. *International Journal of Recycling of Organic Waste in Agriculture*, 6, 289–299. https://doi.org/10.1007/s40093-017-0176-4

Al Harraq, A., & Bharti, B. (2022). Microplastics through the lens of colloid science. *ACS Environmental Au*, 2(1), 3–10. https://doi.org/10.1021/acsenvironau.1c00016

Awoyera, P. O., & Adesina, A. (2020). Plastic wastes to construction products: Status, limitations and future perspective. *Case Studies in Construction Materials*, 12, e00330. https://doi.org/10.1016/j.cscm.2020.e00330

Balthazar-Silva, D., Turra, A., Moreira, F. T., Camargo, R. M., Oliveira, A. L., Barbosa, L., & Gorman, D. (2020). Rainfall and tidal cycle regulate seasonal inputs of microplastic pellets to Sandy beaches. *Frontiers in Environmental Science*, 8. https://doi.org/10.3389/fenvs.2020.00123

Bläsing, M., & Amelung, W. (2018). Plastics in soil: Analytical methods and possible sources. *Science of the Total Environment*, 612, 422–435. https://doi.org/10.1016/j.scitotenv.2017.08.086

Boots, B., Russell, C. W., & Green, D. S. (2019). Effects of microplastics in soil ecosystems: Above and below ground. *Environmental Science and Technology*, 53(19), 11496–11506. https://doi.org/10.1021/acs.est.9b03304

Breukelman, H., Krikke, H., & Löhr, A. (2019). Failing services on urban waste management in developing countries: A review on symptoms, diagnoses, and interventions. *Sustainability*, 11(24), 6977. https://doi.org/10.3390/su11246977

Burrows, S. D., Frustaci, S., Thomas, K. V., & Galloway, T. (2020). Expanding exploration of dynamic microplastic surface characteristics and interactions. *TrAC Trends in Analytical Chemistry*, 130, 115993. https://doi.org/10.1016/j.trac.2020.115993

Cai, R., Zhang, J., Nie, X., Tjong, J., & Matthews, D. T. A. (2020). Wear mechanism evolution on brake discs for reduced wear and particulate emissions. *Wear*, 452–453, 203283. https://doi.org/10.1016/j.wear.2020.203283

Campanale, C., Massarelli, C., Savino, I., Locaputo, V., & Uricchio, V. F. (2020). A detailed review study on potential effects of microplastics and additives of concern on human health. *International Journal of Environmental Research and Public Health, 17*(4). https://doi.org/10.3390/ijerph17041212

Canopoli, L., Coulon, F., & Wagland, S. T. (2020). Degradation of excavated polyethylene and polypropylene waste from landfill. *Science of The Total Environment, 698*, 134125. https://doi.org/10.1016/j.scitotenv.2019.134125

Chamas, A., Moon, H., Zheng, J., Qiu, Y., Tabassum, T., Jang, J. H., Abu-Omar, M., Scott, S. L., & Suh, S. (2020). Degradation rates of plastics in the environment. *ACS Sustainable Chemistry & Engineering, 8*(9), 3494–3511. https://doi.org/10.1021/acssuschemeng.9b06635

Chen, Y., Leng, Y., Liu, X., & Wang, J. (2019). Microplastic pollution in vegetable farmlands of suburb Wuhan, central China. *Environmental Pollution, 257*, 113449.

Cox, K. D., Covernton, G. A., Davies, H. L., Dower, J. F., Juanes, F., & Dudas, S. E. (2019). Human consumption of microplastics. *Environmental Science and Technology, 53*(12), 7068–7074. https://doi.org/10.1021/acs.est.9b01517

De Falco, F., Gullo, M. P., Gentile, G., Di Pace, E., Cocca, M., Gelabert, L., Brouta-Agnésa, M., Rovira, A., Escudero, R., Villalba, R., Mossotti, R., Montarsolo, A., Gavignano, S., Tonin, C., & Avella, M. (2018). Evaluation of microplastic release caused by textile washing processes of synthetic fabrics. *Environmental Pollution, 236*, 916–925. https://doi.org/10.1016/j.envpol.2017.10.057

de Souza Machado, A. A., Kloas, W., Zarfl, C., Hempel, S., & Rillig, M. C. (2018). Microplastics as an emerging threat to terrestrial ecosystems. *Global Change Biology, 24*(4), 1405–1416. https://doi.org/10.1111/gcb.14020

Du, C., Liang, H., & Li, Z. (2020). Pollution characteristics of microplastics in soils in Southeastern Suburbs of Baoding City, China. *International Journal of Environmental Research and Public Health Article, 17*. https://doi.org/10.3390/ijerph17030845

Durda, J. L., & Preziosi, D. V. (2000). Data quality evaluation of toxicological studies used to derive ecotoxicological benchmarks. *Human and Ecological Risk Assessment (HERA), 5*, 747–765. https://doi.org/10.1080/10807030091124176

Geyer, R., Jambeck, J. R., & Law, K. L. (2017). Production, use, and fate of all plastics ever made. *Science Advances*, 25–29. https://doi.org/10.1126/sciadv.1700782

Guerrera, M. C., Aragona, M., Porcino, C., Fazio, F., Laurà, R., Levanti, M., Montalbano, G., Germanà, G., Abbate, F., & Germanà, A. (2021). Micro and nano plastics distribution in fish as model organisms: Histopathology, blood response and bioaccumulation in different organs. *Applied Sciences, 11*(13), 5768. https://doi.org/10.3390/app11135768

Guo, Z., Li, P., Yang, X., Wang, Z., Lu, B., Chen, W., Wu, Y., Li, G., Zhao, Z., Liu, G., Ritsema, C., Geissen, V., & Xue, S. (2022). Soil texture is an important factor determining how microplastics affect soil hydraulic characteristics. *Environment International, 165*, 107293. https://doi.org/10.1016/j.envint.2022.107293

Haque, F., & Fan, C. (2023). Fate and impacts of microplastics in the environment: Hydrosphere, pedosphere, and atmosphere. *Environments, 10*(5). https://doi.org/10.3390/environments10050070

Horton, A. A., Walton, A., Spurgeon, D. J., Lahive, E., & Svendsen, C. (2017). Microplastics in freshwater and terrestrial environments: Evaluating the current understanding to identify the knowledge gaps and future research priorities. *Science of The Total Environment, 586*, 127–141. https://doi.org/10.1016/j.scitotenv.2017.01.190

Hurley, R., Horton, A., Lusher, A., & Nizzetto, L. (2020). Plastic Waste in the Terrestrial Environment. In *Plastic Waste and Recycling* (pp. 163–193). Elsevier. https://doi.org/10.1016/B978-0-12-817880-5.00007-4

Kaczmarek, H., Felczak, A., & Szalla, A. (2008). Studies of photochemical transformations in polystyrene and styrene–maleic anhydride copolymer. *Polymer Degradation and Stability, 93*(7), 1259–1266. https://doi.org/10.1016/j.polymdegradstab.2008.04.011

Kadac-Czapska, K., Knez, E., Gierszewska, M., Olewnik-Kruszkowska, E., & Grembecka, M. (2023). Microplastics Derived from Food Packaging Waste—Their Origin and Health Risks. In *Materials*, 16. https://doi.org/10.3390/ma16020674

Kallenbach Emilie M. F., Rødland, E. S., Buenaventura, N. T., & Hurley, R. (2022). Microplastics in Terrestrial and Freshwater Environments. In M. S. Bank (Ed.), *Microplastic in the Environment: Pattern and Process* (pp. 87–130). Springer International Publishing. https://doi.org/10.1007/978-3-030-78627-4_4

Kasirajan, S., & Ngouajio, M. (2012). Polyethylene and biodegradable mulches for agricultural applications: A review. *Agronomy for Sustainable Development, 32*(2), 501–529. https://doi.org/10.1007/s13593-011-0068-3

Koutnik, V. S., Leonard, J., Alkidim, S., DePrima, F. J., Ravi, S., Hoek, E. M. V., & Mohanty, S. K. (2021). Distribution of microplastics in soil and freshwater environments: Global analysis and framework for transport modeling. *Environmental Pollution, 274*, 116552. https://doi.org/10.1016/j.envpol.2021.116552

Lang, M., Wang, G., Yang, Y., Zhu, W., Zhang, Y., Ouyang, Z., & Guo, X. (2022). The occurrence and effect of altitude on microplastics distribution in agricultural soils of Qinghai Province, northwest China. *Science of the Total Environment, 810*, 152174. https://doi.org/10.1016/j.scitotenv.2021.152174

Latunde-Dada, G. O., Dutra De Oliveira, J. E., Carillo, S. V., Marchini, J. S., & Bianchi, M. L. P. (1998). Gastrointestinal tract and iron absorption: A review. *Brazilian Journal of Food and Nutrition, 9*, 103–125.

Li, Z., Yang, Y., Chen, X., He, Y., Bolan, N., Rinklebe, J., Lam, S. S., Peng, W., & Sonne, C. (2023). A discussion of microplastics in soil and risks for ecosystems and food chains. *Chemosphere, 313*, 137637. https://doi.org/10.1016/j.chemosphere.2022.137637

Lu, X., Wang, X., Liu, X., & Singh, V. P. (2023). Dispersal and transport of microplastic particles under different flow conditions in riverine ecosystem. *Journal of Hazardous Materials, 442*, 130033. https://doi.org/10.1016/j.jhazmat.2022.130033

Lusher, A. L., Munno, K., Hermabessiere, L., & Carr, S. (2020). Isolation and extraction of microplastics from environmental samples: An evaluation of practical approaches and recommendations for further harmonization. *Applied Spectroscopy, 74*, 1049–1065. https://doi.org/10.1177/0003702820938993

Lwanga, E. H., Beriot, N., Corradini, F., Silva, V., Yang, X., Baartman, J., Rezaei, M., van Schaik, L., Riksen, M., & Geissen, V. (2022). Review of Microplastic Sources, Transport Pathways and Correlations with Other Soil Stressors: A Journey from Agricultural Sites into the Environment. In *Chemical and Biological Technologies in Agriculture* (Vol. 9, Issue 1). Springer Science and Business Media Deutschland GmbH. https://doi.org/10.1186/s40538-021-00278-9

Maddela, N. R., Ramakrishnan, B., Kadiyala, T., Venkateswarlu, K., & Megharaj, M. (2023). Do Microplastics and nanoplastics pose risks to biota in agricultural ecosystems? *Soil Systems, 7*(1). https://doi.org/10.3390/soilsystems7010019

Mai, L., Bao, L.-J., Wong, C. S., & Zeng, E. Y. (2018). Chapter 12 – Microplastics in the Terrestrial Environment. In E. Y. Zeng (Ed.), *Microplastic Contamination in Aquatic Environments* (pp. 365–378). Elsevier. https://doi.org/10.1016/B978-0-12-813747-5.00012-6

Makhdoumi, P., Hossini, H., & Pirsaheb, M. (2023). A review of microplastic pollution in commercial fish for human consumption. *Reviews on Environmental Health, 38*(1), 97–109. https://doi.org/10.1515/reveh-2021-0103

Nam, S., & Ats, H. (2022). The occurrence and consequences of microplastics and nanoplastics in fish gastrointestinal tract. *Journal of Survey in Fisheries Sciences, 8*(3), 107–133.

Napper, I. E., & Thompson, R. C. (2016). Release of synthetic microplastic plastic fibres from domestic washing machines: Effects of fabric type and washing conditions. *Marine Pollution Bulletin*. https://doi.org/10.1016/j.marpolbul.2016.09.025

Okunola A, A., Kehinde I, O., Oluwaseun, A., & Olufiropo E, A. (2019). Public and environmental health effects of olastic wastes disposal: A review. *Journal of Toxicology and Risk Assessment, 5*(2). https://doi.org/10.23937/2572-4061.1510021

Pimentel, D., Whitecraft, M., Scott, Z., Zhao, L., Satkiewicz, P., Scott, T. J., Phillips, J., Szimak, D., Singh, G., Gonzalez, D. O., & Moe, T. L. (2012). Ethics of a Sustainable World Population in 100 Years. In R. Chadwick (Ed.), *Encyclopedia of Applied Ethics (Second Edition)* (2nd ed, pp. 173–177). Academic Press. https://doi.org/10.1016/B978-0-12-373932-2.00365-3

Pirc, U., Vidmar, M., Mozer, A., & Kržan, A. (2016). Emissions of microplastic fibers from microfiber fleece during domestic washing. *Environmental Science and Pollution Research, 23*(21), 22206–22211. https://doi.org/10.1007/s11356-016-7703-0

Praveena, S. M., Aris, A. Z., & Singh, V. (2022). Quality assessment for methodological aspects of microplastics analysis in soil. *Trends in Environmental Analytical Chemistry, 34*. https://doi.org/10.1016/j.teac.2022.e00159

Praveena, S. M., Hisham, M. A. F. I., & Nafisyah, A. L. (2023). Microplastics pollution in agricultural farms soils: Preliminary findings from tropical environment (Klang Valley, Malaysia). *Environmental Monitoring and Assessment, 195*(6), 650. https://doi.org/10.1007/s10661-023-11250-5

Praveena, S. M., Syahira Asmawi, M., & Chyi, J. L. Y. (2021). Microplastic emissions from household washing machines: Preliminary findings from Greater Kuala Lumpur (Malaysia). *Environmental Science and Pollution Research*, 28(15), 18518–18522. https://doi.org/10.1007/s11356-020-10795-z

Qi, R., Jones, D. L., Li, Z., Liu, Q., & Yan, C. (2020). Behavior of microplastics and plastic film residues in the soil environment: A critical review. *Science of the Total Environment*, 703, 134722. https://doi.org/10.1016/j.scitotenv.2019.134722

Radford, F., Zapata-Restrepo, L. M., Horton, A. A., Hudson, M. D., Shaw, P. J., & Williams, I. D. (2021). Developing a systematic method for extraction of microplastics in soils. *Analytical Methods*, 13(14), 1695–1705. https://doi.org/10.1039/d0ay02086a

Raza, M., Lee, J.-Y., & Cha, J. (2022). Microplastics in soil and freshwater: Understanding sources, distribution, potential impacts, and regulations for management. *Science Progress*, 105(3), 003685042211266. https://doi.org/10.1177/00368504221126676

Rillig, M. C. (2012). Microplastic in terrestrial ecosystems and the soil? *Environmental Science & Technology*, 46, 6453–6454.

Rodrigues, M. O., Gonçalves, A. M. M., Gonçalves, F. J. M., Nogueira, H., Marques, J. C., & Abrantes, N. (2018). Effectiveness of a methodology of microplastics isolation for environmental monitoring in freshwater systems. *Ecological Indicators*, 89, 488–495. https://doi.org/10.1016/j.ecolind.2018.02.038

Rose, P. K., Jain, M., Kataria, N., Sahoo, P. K., Garg, V. K., & Yadav, A. (2023a). Microplastics in multimedia environment: A systematic review on its fate, transport, quantification, health risk, and remedial measures. *Groundwater for Sustainable Development*, 20, 100889. https://doi.org/10.1016/j.gsd.2022.100889

Rose, P. K., Yadav, S., Kataria, N., & Khoo, K. S. (2023b). Microplastics and nanoplastics in the terrestrial food chain: Uptake, translocation, trophic transfer, ecotoxicology, and human health risk. *TrAC Trends in Analytical Chemistry*, 167, 117249. https://doi.org/10.1016/j.trac.2023.117249

Sa'adu, I., & Farsang, A. (2023). Plastic contamination in agricultural soils: A review. *Environmental Sciences Europe*, 35(1), 13. https://doi.org/10.1186/s12302-023-00720-9

Sajjad, M., Huang, Q., Khan, S., Khan, M. A., Liu, Y., Wang, J., Lian, F., Wang, Q., & Guo, G. (2022). Microplastics in the soil environment: A critical review. *Environmental Technology & Innovation*, 27, 102408. https://doi.org/10.1016/j.eti.2022.102408

Salvador Cesa, F., Turra, A., & Baruque-Ramos, J. (2017). Synthetic fibers as microplastics in the marine environment: A review from textile perspective with a focus on domestic washings. *Science of the Total Environment*, 598, 1116–1129. https://doi.org/10.1016/j.scitotenv.2017.04.172

Saud, S., Yang, A., Jiang, Z., Ning, D., & Fahad, S. (2023). New insights in to the environmental behavior and ecological toxicity of microplastics. Journal of Hazardous Materials Advances, 100298. https://doi.org/10.1016/j.hazadv.2023.100298

Schmaltz, E., Melvin, E. C., Diana, Z., Gunady, E. F., Rittschof, D., Somarelli, J. A., Virdin, J., & Dunphy-Daly, M. M. (2020). Plastic pollution solutions: Emerging technologies to prevent and collectmarineplastic pollution. *Environment International*, 144, 106067. https://doi.org/10.1016/j.envint.2020.106067

Shi, J., Wang, Z., Peng, Y., Fan, Z., Zhang, Z., Wang, X., Zhu, K., Shang, J., & Wang, J. (2023). Effects of microplastics on soil carbon mineralization: The crucial role of oxygen dynamics and electron transfer. *Environmental Science & Technology*, 57(36), 13588–13600. https://doi.org/10.1021/acs.est.3c02133

Sobhani, Z., Lei, Y., Tang, Y., Wu, L., Zhang, X., Naidu, R., Megharaj, M., & Fang, C. (2020). Microplastics generated when opening plastic packaging. *Scientific Reports*, 10(1), 4841. https://doi.org/10.1038/s41598-020-61146-4

Sommer, F., Dietze, V., Baum, A., Sauer, J., Gilge, S., Maschowski, C., & Gieré, R. (2018). Tire abrasion as a major source of microplastics in the environment. *Aerosol and Air Quality Research*, 18(8), 2014–2028. https://doi.org/10.4209/aaqr.2018.03.0099

Su, L., Xiong, X., Zhang, Y., Wu, C., Xu, X., Sun, C., & Shi, H. (2022). Global transportation of plastics and microplastics: A critical review of pathways and influences. *Science of the Total Environment*, 831, 154884. https://doi.org/10.1016/j.scitotenv.2022.154884

Surendran, U., Jayakumar, M., Raja, P., Gopinath, G., & Chellam, P. V. (2023). Microplastics in terrestrial ecosystem: Sources and migration in soil environment. *Chemosphere*, 318, 137946. https://doi.org/10.1016/j.chemosphere.2023.137946

Sutkar, P. R., Gadewar, R. D., & Dhulap, V. P. (2023). Recent trends in degradation of microplastics in the environment: A state-of-the-art review. *Journal of Hazardous Materials Advances*, 11, 100343. https://doi.org/10.1016/j.hazadv.2023.100343

Vendel, A. L., Bessa, F., Alves, V. E. N., Amorim, A. L. A., Patrício, J., & Palma, A. R. T. (2017). Widespread microplastic ingestion by fish assemblages in tropical estuaries subjected to anthropogenic pressures. *Marine Pollution Bulletin, 117*(1–2), 448–455. https://doi.org/10.1016/j.marpolbul.2017.01.081

Wang, F., Wang, Q., Adams, C. A., Sun, Y., & Zhang, S. (2022). Effects of microplastics on soil properties: Current knowledge and future perspectives. *Journal of Hazardous Materials, 424*, 127531. https://doi.org/10.1016/j.jhazmat.2021.127531

Wang, W., Ge, J., Yu, X., & Li, H. (2020). Environmental fate and impacts of microplastics in soil ecosystems: Progress and perspective. *Science of the Total Environment, 708*. https://doi.org/10.1016/j.scitotenv.2019.134841

Wong, J. K. H., Lee, K. K., Tang, K. H. D., & Yap, P.-S. (2020). Microplastics in the freshwater and terrestrial environments: Prevalence, fates, impacts and sustainable solutions. *Science of the Total Environment, 719*, 137512. https://doi.org/10.1016/j.scitotenv.2020.137512

Worek, J., Badura, X., Białas, A., Chwiej, J., Kawoń, K., & Styszko, K. (2022). Pollution from transport: Detection of tyre particles in environmental samples. *Energies, 15*(8). https://doi.org/10.3390/en15082816

Xie, A., Jin, M., Zhou, Q., Fu, L., & Wu, W. (2023). Photocatalytic technologies for transformation and degradation of microplastics in the environment: Current achievements and. *Catalysts, 13*. https://doi.org/10.3390/catal13050846

Yadav, S., Kataria, N., Khyalia, P., Rose, P. K., Mukherjee, S., Sabherwal, H., Chai, W. S., Rajendran, S., Jiang, J.-J., & Khoo, K. S. (2023). Recent analytical techniques, and potential eco-toxicological impacts of textile fibrous microplastics (FMPs) and associated contaminates: A review. *Chemosphere, 326*, 138495. https://doi.org/10.1016/j.chemosphere.2023.138495

Yang, X., Man, Y. B., Wong, M. H., Owen, R. B., & Chow, K. L. (2022). Environmental health impacts of microplastics exposure on structural organization levels in the human body. *Science of the Total Environment, 825*, 154025. https://doi.org/10.1016/j.scitotenv.2022.154025

Yu, Y., & Flury, M. (2021). Current understanding of subsurface transport of micro- and nanoplastics in soil. *Vadose Zone Journal, 20*(2). https://doi.org/10.1002/vzj2.20108

Yuan, Z., & Xu, X.-R. (2023). Chapter Six – Surface Characteristics and Biotoxicity of Airborne Microplastics. In J. Wang (Ed.), *Airborne Microplastics: Analysis, Fate And Human Health Effects* (Vol. 100, pp. 117–164). Elsevier. https://doi.org/10.1016/bs.coac.2022.07.006

Zhang, L., Xie, Y., Liu, J., Zhong, S., Qian, Y., & Gao, P. (2020). An overlooked entry pathway of microplastics into agricultural soils from application of sludge-based fertilizers. *Environmental Science and Technology, 54*(7), 4248–4255. https://doi.org/10.1021/acs.est.9b07905

Zhang, X., Li, Y., Ouyang, D., Lei, J., Tan, Q., Xie, L., Li, Z., Liu, T., Xiao, Y., Farooq, T. H., Wu, X., Chen, L., & Yan, W. (2021). Systematical review of interactions between microplastics and microorganisms in the soil environment. *Journal of Hazardous Materials, 418*(May), 126288. https://doi.org/10.1016/j.jhazmat.2021.126288

Zhou, B., Wang, J., Zhang, H., Shi, H., Fei, Y., Huang, S., Tong, Y., Wen, D., Luo, Y., & Barceló, D. (2020). Microplastics in agricultural soils on the coastal plain of Hangzhou Bay, east China: Multiple sources other than plastic mulching film. *Journal of Hazardous Materials, 388*, 121814. https://doi.org/10.1016/j.jhazmat.2019.121814

6 Analytical Techniques of Microplastic in an Aquatic Environment

Sampling, Extraction, and Identification

Abel Egbemhenghe, Chika J. Okorie, Toluwalase Ojeyemi, Hussein K. Okoro, Ebuka Chizitere Emenike, Bridget Dunoi Ayoku, Kingsley O. Iwuozor, and Adewale George Adeniyi

6.1 INTRODUCTION

Plastic pollution has become a pressing worldwide issue, with over 5000 types of commercial plastics currently produced worldwide (Mendoza and Balcer, 2020). Key plastics such as polyethylene terephthalate and polystyrene, among others, contribute to the vast quantities of plastic waste generated annually (Adeniyi et al., 2023a; Adeniyi et al., 2023c). Unfortunately, a significant portion of plastics, about 79%, ends up in landfills or escapes into the environment, ultimately reaching the oceans through various pathways such as wind and water bodies (Adeniyi et al., 2022a; Geyer et al., 2017). A great deal of marine litter, comprising about 70% of the total, originates from land-based sources (Devriese et al., 2015). This continuous influx of plastic waste into aquatic environments is a matter of concern, with an estimated eight million metric tons of plastic debris entering the oceans each year (Mendoza and Balcer, 2020). Without effective intervention, it is projected that approximately 12,000 million tons of plastic will accumulate in landfills and the natural environment by 2050 (Geyer et al., 2017). Despite efforts at recycling, plastic waste can persist in landfills for many years, further exacerbating the environmental burden (Cole et al., 2011).

The pervasive impact of plastic debris on marine life is evident, with over 600 marine species affected by this pollution (Claessens et al., 2013). The process of degradation breaks down plastics into minute particles due to factors like UV radiation, mechanical abrasion, and biological processes, resulting in the formation of microplastics (MPs) (Dümichen et al., 2017). MPs are defined as plastic particles with a diameter of less than 5 mm, encompassing both resin pellets and small plastic fragments (Mendoza and Balcer, 2020). They are ubiquitously found in marine and freshwater environments and are known to persist for extended periods.

In 2004, Thompson et al. (2004) coined the term "microplastic" to mean small plastic particles. Subsequently, there has been a global rise in the number of studies aimed at understanding the presence and impacts of MPs in aquatic ecosystems (Li et al., 2015). MPs are not confined to marine environments; they also contaminate freshwater ecosystems, appearing in various forms like foams, pellets, and fibres. Due to their small size and buoyancy, detecting and tracking MPs in aquatic environments poses significant challenges, making their distribution and fate complex and not yet

DOI: 10.1201/9781032706573-6

entirely understood (Mendoza and Balcer, 2019). Two major classes of MPs exist: primary and secondary. Primary MPs are intentionally produced and used in industrial settings or incorporated into personal care products, such as microbeads in cosmetics, abrasives for cleaning, and fine powders for various applications (Ilechukwu et al., 2019; Mendoza and Balcer, 2019). Secondary MPs, on the other hand, are generated through the fragmentation and degradation of larger plastic debris under the influence of physical and chemical factors, including sunlight exposure (Gündoğdu et al., 2018). Over time, these larger plastics break down into smaller polymer fragments, eventually becoming microplastics (Dümichen et al., 2017). There are also concerns about the further degradation of microplastics into nanoplastics, which could have more severe environmental consequences (Cole et al., 2011).

The escalating concerns over the environmental, health, and economic impacts of microplastics have prompted increased attention from the public and policymakers. As a result, some countries have introduced policies and regulations to mitigate microplastic pollution, while consumers are increasingly demanding sustainable products. Understanding the sources, environmental effects, and analytical techniques for detecting and identifying microplastics in aquatic environments is crucial to comprehending the magnitude of this issue. Therefore, this chapter focuses on discussing microplastics in aquatic environments, aiming to identify knowledge gaps, propose areas for future research, and highlight the implications of these findings for management and policy decisions. The ultimate goal is to develop effective strategies for reducing the environmental impact of microplastics.

6.2 SOURCES OF MPS IN THE AQUATIC ECOSYSTEM

Microplastics can originate from several sources and can enter aquatic environments through various pathways. The origin of MPs in aquatic environments can be divided based on various classifications. For example, MPs can be classified based on primary and secondary sources, natural and anthropogenic sources, and point and non-point sources. It is crucial to comprehend where microplastics come from to develop effective management strategies to decrease their effect on humans and the environment.

Primary sources of microplastics are the direct release of small plastic particles into the environment. These can include products such as cosmetics, toothpaste, and cleaning products that contain microplastics. Plastic pellets used in the production of plastic products are also a primary source of MPs. Alternatively, secondary sources of MPs are created due to the breakdown of larger plastic materials, such as plastic bags, bottles, and packaging materials. These plastic items are subjected to various environmental elements, such as sunlight and wave action, which lead to their gradual breakdown into smaller fragments.

Microplastics in aquatic environments can also emerge from both natural and anthropogenic sources. Natural sources of microplastics include the breakdown of organic materials such as leaves, wood, and animal waste. These materials can be broken down in the environment and release small fragments, which can contribute to microplastic pollution. Moreover, natural events such as volcanic eruptions, earthquakes, and ocean currents can also generate microplastics. For example, volcanic eruptions can produce tiny plastic particles, and ocean currents can transport plastic waste from one region to another. Alternatively, anthropogenic sources are the result of man's activities, which contribute significantly to the production of microplastics. Some of the major anthropogenic sources of microplastics include plastic products, textiles, personal care products, fishing gear, industrial processes, urban runoff, and agricultural practices.

Microplastics in aquatic environments can also originate from both point and non-point sources. Point sources of microplastics refer to discrete and identifiable sources of pollution, where the release of microplastics occurs at a specific location. Some examples of point sources of microplastics include wastewater treatment plants, plastic manufacturing plants, ports, and marinas. On the other hand, non-point sources of microplastics refer to diffuse sources of pollution, where the release of MPs occurs from a range of sources and it is challenging to identify a single point of origin. Some examples of non-point sources of microplastics include urban runoff, agricultural runoff, and atmospheric deposition.

In general, there are multiple routes through which MPs can infiltrate aquatic media. Everyday items such as our clothes, packaging materials, and vehicle tyres all contribute to this pollutant (Gündoğdu et al., 2018). Synthetic textile fibres released during cloth washing are a great contributor of MPs found in the ocean. Up to 1,900–1,000,000 fibres are released when washing a single garment, over 6,000,000 fibres when washing polyester fabrics, and 700,000 fibres in acrylic fabrics wash (Prata, 2018). MPs can also originate from various sources and enter the aquatic environment through several pathways. For example, industrial plants, littering, and wastewater treatment plant effluents are among the sources of microplastics that increase the contamination of the aquatic environment (Mintenig et al., 2017).

Ideally, effluent treatment plants (ETPs) are designed to act as a mitigation technique for organic materials and nutrients, as well as other contaminants such as MPs in wastewater. Traditional wastewater treatment methods that incorporate primary and secondary treatment processes can effectively remove significant amounts of MPs with high concentrations, up to 90–99%, but the same ETP acts as a route of MPs into the aquatic ecosystem, giving an enormous quantity of wastewater discharged per day (Talvitie et al., 2017). Microplastic particles or microbeads, often made of polyethylene, can be present in household wastewater and cosmetic items like toothpaste, facial cleansers, soaps, and lipsticks. Personal care products serve as one of the many origins of microplastic contamination in domestic water systems (Cristaldi et al., 2020; Gündoğdu et al., 2018). Microbeads, which have an average size of 250 µm, are added to exfoliating washes and toothpaste as they cause less skin irritation and damage compared to natural exfoliants like ground walnut husks and almonds. Cosmetics typically contain between 0.5% and 5% of microbeads, approximately 4000 microbeads are released in a single use (Prata, 2018). Most of the microbeads and fibres are so small that they pass through ETP and are not retained by the plant, so they eventually end up in the watershed (Masura et al., 2015). MPs can be introduced into wastewater through various sources, including plastic resin pellets utilized in transport and manufacturing processes, polystyrene foam waste from fillings and shipping, synthetic fibres discharged by textile industries, and dust created from drilling and cutting plastics (Cristaldi et al., 2020). Moreover, wastewater from effluent treatment plants (ETPs) discharges into the aquatic ecosystems through sewage overflow during heavy rainfall, effluent discharges, and runoff from sewage-based fertilizer application (Hanvey et al., 2017).

Gündoğdu et al. (2018) conducted a study in Turkey to investigate the concentration of MPs in the incoming and outgoing effluent of two effluent treatment plants. The findings revealed that the influent water contained between 1 million and 6.5 million particles per day, while the effluent water contained between 220,000 and 1.5 million particles per day. The study also found that the rate of removal of the MPs ranged from 73% to 79%. This result implies that at least 20% of MPs pass through the treatment plant to the aquatic ecosystems every day. Also, a study conducted in eastern China with the goal of comprehending the fate and management techniques of MPs in effluent treatment plants discovered that the main constituents of the influent MPs were PET, PS, PE, and PP (Lv et al., 2019). The result showed a large amount of MPs, with dominant sizes of >500 µm (40%) and 62.5–125 µm (29%). Proper treatment methods, restricting the use of microbeads, polystyrene foam products, and plastic bags, are some of the recommended source control methods for MPs in ETPs by the study. ETPs therefore play a major role in the spread of MPs in the environment, particularly in aquatic ecosystems. The level of contamination is even more severe in developing countries, especially in Africa, where inadequate wastewater treatment systems exist.

Plastics also get to the ocean through loss or dumping of plastic gear used for fishing activities (Andrady, 2011). These materials degrade with time under the intense sunlight and disintegrate into different fragments of MPs. Mining activities in the sea also contribute to MPs abundance (Adeniyi et al., 2022b; Ighalo et al., 2022). Despite an international agreement banning the disposition of plastic wastes at sea by marine vessels, an estimated 6.5 million tons of plastics were shipped into the oceans in the 1990s by marine vessels (Thiel et al., 2011). Recreational activities in beaches result in huge contamination of this body of water, through which they are transported to the oceans. Approximately 80% of the plastic waste in the oceans is attributed to litter found on beaches (Andrady, 2011). Anthropogenic debris was reported as one of the flotsam categories in the German Bight (North Sea), with plastic particles constituting over 70% (32.4 particles/km^2) of the floating debris in the sea (Thiel et al., 2011). The amount of plastic waste being transported into the oceans is expected to increase in the future due to factors such as wild fishing, recreational and maritime activities, as well as a growing population migrating to coastal areas (Adeniyi et al., 2023b; Adeniyi et al., 2023c).

6.3 ECOTOXICOLOGY OF MPS ON THE AQUATIC ECOSYSTEMS

Ecotoxicological studies provide insights over the adverse effects of a pollutant to the environment (Iwuozor et al., 2021; Iwuozor et al., 2022a; Iwuozor et al., 2022b). Microplastics can have significant environmental impacts on aquatic ecosystems, affecting physical, chemical, biological, ecological, and human health. Microplastics can cause physical harm to aquatic organisms. Small MPs can be eaten by planktonic animals, while larger microplastics can be eaten by larger animals such as fish and sea birds, causing blockages in the digestive system, and even death. Owing to their size, MPs are easily digested by organisms of differing feeding mechanisms such as detritivores, deposit feeders, and filter feeders (Maes et al., 2017; Masura et al., 2015). There have been reports of organisms ingesting microplastics in field and laboratory studies for various marine organisms that include zooplankton, fish, seabirds, crustaceans, bivalves, reptiles, worms, and mammals (Von Moos et al., 2012; Watts et al., 2014). The risks to organisms ingesting MPs can include physiological harm, blockages in the digestive tract, changes in feeding and reproductive behaviours, inflammation, cellular necrosis, and reduced immune response (Gbogbo et al., 2020; Ilechukwu et al., 2019).

The MPs content of the shrimp, *Crangon crangon* L., in aquatic habitats around the North Sea was studied by Devriese et al. (2015), who found that the average microplastic content in the shrimp samples was 0.68 ± 0.55 microplastics per gram of wet weight or 1.23 ± 0.99 microplastics per individual shrimp. The study results indicate that microplastic particles that were larger than 20 μm did not move into the tissues of the shrimp that were studied. However, there is still a potential risk of microplastics being transferred to humans who consume shrimp without removing the intestinal tract.

In one study, adult Pacific oysters were exposed to virgin polystyrene microspheres for 2 months during a reproductive cycle (Sussarellu et al., 2016). The microspheres were 2 and 6 μm in diameter and present at a concentration of 0.023 mg/L. The results showed that the oysters exposed to microspheres had a significant decrease in oocyte number, diameter, and sperm velocity, indicating that the microspheres had a negative impact on feeding modifications and reproductive disruption in oysters (Sussarellu et al., 2016). Watts et al. (2014) conducted research to explore how shore crabs ingest and eliminate MPs via two routes: direct exposure through water and exposure through their diet by feeding on mussels. The microplastic particles were found to remain in the body tissues of shore crabs for up to 14 days after being ingested and up to 21 days after being absorbed through the gills. The study also showed that ventilation is an additional route through which MPs can enter the body of crabs.

Microplastics were found to be present in all nine commercial bivalves tested in China (Li et al., 2015). a variety of MPs including fibres, fragments, and pellets occurred in the tissue of bivalves, with the size of 250 μm being the most common that accounted for an average of 58.8% of the total MPs documented by species. The results showed the spread of MP pollution

Analytical Techniques of Microplastic in an Aquatic Environment

in the commercial bivalves and the potential health risks to human when they consume bivalves contaminated by microplastics. The contaminant can also be transferred from organisms feeding on plastic-contaminated organisms. Farrell and Nelson (2013) found that microplastics can transfer from mussels to crabs. The researchers fed mussels, which were previously exposed to fluorescent polystyrene microspheres, to crabs and observed the highest number of microspheres in the crabs' haemolymph 24 hours after ingestion. Although the amount of MPs remaining in the crabs after 21 hours was small, the study revealed that MPs can be transferred from one trophic level to another, thus increasing concerns for their potential accumulation and implications for the health of animals, the wider food web, and humans.

Microplastics can also entangle and suffocate marine animals, such as sea turtles and marine mammals, leading to injuries and death. Microplastics can also contain toxic chemicals, which can accumulate in the tissues of aquatic organisms, leading to health risks. They can also accumulate on the seafloor and other habitats, altering the physical structure of the environment, as shown in Figure 6.1. This can affect the distribution and abundance of aquatic organisms and modify the habitats available for colonization. MPs can also release toxic chemicals into the environment. The chemicals used in plastic production, such as phthalate esters (Rios Mendoza et al., 2017), brominated flame retardants (BFRs) (Jang et al., 2017), nonylphenol (NP) (Adeyi, 2020), perfluoroalkyl substances (PFAS) (Zeng et al., 2019), and bisphenol A (Liu et al., 2019), can leach out of microplastics and enter the water, potentially harming aquatic organisms and human health. Due to their persistent nature, these compounds can accumulate and be biomagnified up the food chain once absorbed. Microplastics can also have biological impacts on aquatic organisms. Ingestion of microplastics can reduce feeding rates and affect growth and reproduction, leading to population declines. Microplastics can also transport and spread invasive species, as they provide a surface for the attachment and transport of non-native organisms. Microplastics can also affect the behaviour and physiology of aquatic organisms, leading to reduced survival and growth rates.

Microplastics can also have ecological impacts, affecting the structure and function of aquatic ecosystems. Microplastics can alter the physical and chemical properties of water, such as light

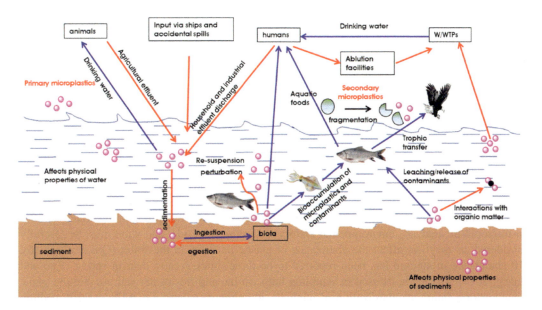

FIGURE 6.1 A visual representation depicting the presence and fate of MPs in aquatic ecosystems. (Adapted with permission from Chaukura et al., 2021, Copyright © 2021 By Authors.)

penetration, oxygen levels, and nutrient cycling, leading to changes in the community structure of aquatic organisms. Microplastics can also affect the primary productivity of aquatic ecosystems, leading to changes in food webs and nutrient cycling. The presence of microplastics can be harmful to human health. Microplastics can accumulate in the tissues of marine life that are ingested by humans, leading to potential health risks. In addition, microplastics can enter the food chain, leading to human exposure to potentially harmful chemicals. The presence of microplastics in sodium chloride (NaCl) was also reported in a laboratory work carried in China (Yang et al., 2015). In this work, 15 brands of sea salts, lake salts, and rock/well salts were collected and measured for MP pollution. MPs measuring less than 200 µm represented most of the fragments and fibres found in the salts, with the abundance in sea salt significantly higher. Microplastics can provide a surface for bacteria and other microorganisms to attach and colonize. This may increase the risk of infection or the spread of antibiotic-resistant bacteria. Microplastics can break down into smaller particles, known as nanoplastics (Akpomie et al., 2022; Emenike et al., 2022; Iwuozor et al., 2022c; Ogunfowora et al., 2021), which can penetrate cells and potentially cause allergic reactions or immune system dysfunction.

6.4 METHODS FOR SAMPLING AND ANALYSING MICROPLASTICS IN AQUATIC ECOSYSTEMS

The difficulty in determining the concentration, density, and dispersal of small plastic particles in the aquatic environment makes microplastics a component of the marine debris that is significantly understudied (Cole et al., 2011). Methods used for sample collection for MP analysis in both the marine and freshwater environment are important for the general estimation of their abundance. The analytical techniques involved in the analysis of microplastics include:

- Sampling technique
- Extraction technique, and
- Identification and quantification technique

6.4.1 SAMPLING TECHNIQUES

Collection of samples is the first step of MP sampling. The choice of method to be used is affected by a range of factors such as the available equipment, aim of the study (qualitative or quantitative), sampling matrix, volumes, and depth (Alimi et al., 2021; de Moura et al., 2019; Prata et al., 2019). The sampling matrix entails the source of the sample, which can be sediment, water, or organisms. Sediment samples are frequently collected from coastal areas such as beaches, estuaries, and the seabed (Qiu et al., 2016). There are different methods to collect microplastics on beaches, such as picking them up directly using forceps, sieving sand, or taking sediment samples. In contrast, collecting samples from the seabed requires a vessel and specialized equipment that is lowered down to the seabed to collect the samples (Prata et al., 2019). According to Qiu et al. (2016), to obtain deep sea sediment samples for microplastic analysis, a box corer can be used while surface samples can be collected using iron spoons or non-plastic sampling spades and then placed in a glass container or wrapped in aluminium foil for later analysis. When collecting water samples for microplastic analysis, environmental factors such as season, tide, and wind must be carefully considered, as they can affect the abundance of plastic particles in the water (Qiu et al., 2016). MP abundance in autumn has been reported to be three times greater than in spring (Kuo and Huang, 2014). The methods used in sampling both marine and freshwater samples are the same, allowing for a future standardization of methods, although distribution of MPs in each medium can be different due to environmental factors such as hydrodynamic source and density (Prata et al., 2019). To collect a large volume of water samples, nets, sieves, or pumps are often used to

Analytical Techniques of Microplastic in an Aquatic Environment

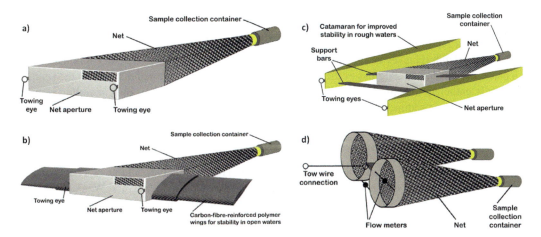

FIGURE 6.2 Tools utilized for collecting microplastics in seawater at various depths: (a) neuston net, (b) manta trawl, (c) catamaran, for sampling MPs on the surface, and (d) bongo nets for capturing MPs in midwater. (Adapted with permission from Silva et al., 2018, Copyright © 2018 Elseiver.)

reduce the in situ size. Near-surface or surface water can be sampled using neuston nets or manta nets, respectively, while bongo nets are coupled nets that collect replicate samples from the water column, as shown in Figure 6.2. However, plankton nets should be towed at low velocities as their small mesh size can break easily (de Moura et al., 2019; Prata et al., 2019). In the collection of organism samples, marine organisms such as mussels, yellowish fish, shrimps, and sea fish are initially captured and then preserved by wrapping them in aluminium foil before being frozen. After freezing, the fish is thawed at room temperature for analysis. The fish is then dissected to extract and analyse the gastrointestinal tract (Qiu et al., 2016).

Currently, there is no universally accepted protocol for sampling techniques and analysis of microplastics. Hence, there are variations among different studies, such as differences in the duration of stirring, temperature of drying, and methods of counting during the processing of environmental samples (Qiu et al., 2016). The absence of a standardized unit for sampling has resulted in presumptions and disparities when contrasting the abundance of microplastics between littoral zones (Hanvey et al., 2017). Characterization and quantification of MPs are associated with several problems which can make inter-study comparisons almost impossible. The disparities between sampling methods can lead to a varieties of sampling units reported, such as mass, volume, area, and length, as well as discrepancies in the minimum and maximum size of microplastics analysed (Hanvey et al., 2017). For instance, most authors have reported MP abundance in particles per square metre (m^2) (Migwi et al., 2020), while others have reported in cubic meters (m^3) (Löder et al., 2017), weight (g) (Missawi et al., 2020), and even litres (L) (Thompson et al., 2004). In general, there are three techniques most commonly used in collecting MPs from the aquatic environment irrespective of the matrix. These include selective sampling, bulk sampling, and volume-reduced sampling.

6.4.2 Selective Sampling

Selective sampling of MPs is employed when the plastic debris is large enough to be seen with the naked eye (particles between 1 and 5 mm) and so can be collected directly from the environment (Silva et al., 2018). This technique has the disadvantage of picking only the obvious particles while the less obvious ones are overlooked, especially when mixed with other debris (Wang and Wang, 2018).

6.4.3 Bulk Sampling

Bulk sampling refers to taking a known volume of environmental sample, mostly from a defined quadrant and depth (Hanvey et al., 2017). In theory, this method should be able to capture all MPs in a particular quadrant regardless of their size, but in practice, only a limited quantity of the sample is typically gathered, thus limiting its effectiveness (Wang and Wang, 2018). Bulk sampling is most suitable in sampling sediment samples, and less frequently, water samples.

6.4.4 Volume-reduced Sampling

Volume-reduced sampling entails the use of fast filtration to reduce the total volume of a bulk sample and retaining only a part of the sample which is selected for analysis (Wang and Wang, 2018). It, however, has the disadvantage of losing a substantial amount of MPs during fast filtration, especially the very small particles (Wang and Wang, 2018). This method is most popular in the collection of water samples.

6.5 EXTRACTION TECHNIQUES

Extraction of the microplastics follows immediately after the sample collection/sampling prior to quantification and identification. The process of isolating, recognizing, and measuring MPs in various samples is crucial in discovering their prevalence, concentration, ingestion by aquatic life, and degradation in marine ecosystems (Gbogbo et al., 2020). Extraction techniques are used to separate MPs from the sample matrix. The extraction technique selected is affected by the type of sample, the size of the microplastics, and the potential for contamination. The methods employed in the extraction of MPs from other debris after sampling include: density separation, filtration, sieving, and digestion. Nevertheless, the best method for extracting MPs from aquatic samples is dependent on the mixture of the matrix and includes density separation using a saturated salt solution and digestion (Gbogbo et al., 2020). Filtration is used together with density separation.

6.5.1 Density Separation

The specific densities of plastics differ based on the type of polymer and the chemicals used in their production (Kershaw et al., 2019). Density separation extracts MPs by taking advantage of these varying densities between the MPs under consideration and the other unwanted particles. This is achieved by thoroughly shaking and settling the mixture materials in a liquid with an average density, and then using the upward force of the liquid to separate the less dense particles from the denser ones (Silva et al., 2018; Wang and Wang, 2018). To extract MPs using density separation, a salt solution of known density is added to the environmental sample, which is then shaken vigorously for a specific duration. This process causes the less dense particles, including MPs, to float to the surface where they can be filtered out and sorted according to their size (Alimi et al., 2021; Thompson et al., 2004). In general, density separation involves four main steps according to Hanvey et al. (2017):

- Adding a liquid solvent with a known density to the sample
- Mixing for a specific period
- A settling equilibrium time and
- Filtration to defined size fractions

Density separation is usually utilized for separating MPs from sediment samples using a sodium chloride (NaCl) solution. However, NaCl solution is only suitable for plastics whose density is less

Analytical Techniques of Microplastic in an Aquatic Environment 93

than that of the saturated salt solution, and is not effective for high-density plastics (Claessens et al., 2013). As a result, NaI solutions have been proposed as an alternative saturated salt solution for the extraction of microplastics from sediment samples (Silva et al., 2018). However, NaI is very costly (70 times costlier than NaCl) and not easily assessable (Claessens et al., 2013). To reduce this cost, Claessens et al. (2013) proposed a method where an elutriation column was developed to allow the separation of the lighter particles from the sediment by creating an upward water flow. This pretreatment greatly reduces the volume of the sediment sample and only 40–80 mL of the high-density NaI solution will be used to extract MPs from 1 kg of sediment sample, instead of 3 L of NaCl solution for a similar amount of sediment. This reduces the needed NaI solution by at least 97% and also saves cost. Other salts that can be used for MP extraction include zinc chloride, calcium chloride, and sodium polytungstate.

6.5.2 SIEVING

The process of sieving makes use of a metal mesh to selectively capture materials that exceed the size of the mesh while allowing smaller materials and water to pass through and be removed from the sample (Wang and Wang, 2018). During sieving, particles that pass through the sieve are discarded, while those retained are collected and sorted (Hidalgo-Ruz et al., 2012). Most of the mesh size typically ranges from 0.035 to 4.75 mm, but this also depends largely on the size range of the desired MPs (Wang and Wang, 2018). This method is, however, less popular, and requires a digestion step, especially where the sample consists of a higher amount of biological materials.

6.5.3 DIGESTION

Digestion separation can remove the inorganic materials contained in the sample, but it cannot remove the organic materials. The presence of these organic materials, such as algae, plankton, and natural debris, in the sample overlays the MPs' spectra, thereby limiting the complete identification of MPs in the subsequent spectroscopic processes (Löder et al., 2017). Therefore, purification of the environmental samples is important to digest these organic materials. Digestion is an extraction technique used to remove the biological materials that are often mixed and confused with plastics (Prata, 2018). As a requirement, the purification and digestion method to be employed should be able to:

- Remove the organic materials effectively and conveniently without influencing the non-natural polymers;
- Enhance the purified samples' concentration using filters with tiny (<1 μm) that is appropriate for spectroscopic measurement; and
- Be less laborious and inexpensive (Löder et al., 2017; Prata, 2018).

Digestion can be carried out using an acid (e.g., HNO_3), alkaline (e.g., KOH), enzyme (e.g., trypsin), or oxidizing agent (e.g., H_2O_2). Combinations of two or more digestion approaches have also been employed (Budimir et al., 2018). Notwithstanding, digestion is dependent on the level of the organic materials present, and when these materials are in a low quantity, researchers often overlook the digestion process.

6.5.3.1 Acid Digestion

In acid digestion, a strong mineral acid such as H_2O_4 or HNO_3 is used at room temperature for a given period of time (overnight or 2 h) (Claessens et al., 2013; Kershaw et al., 2019). Nitric acid (HNO_3) is more effective in the digestion of biological tissue, with the application of heat (Wang and Wang, 2018). Nevertheless, the combination of HNO_3 and $HClO_4$ in the ratio 4:1 v/v, respectively, reached the desired results for MP digestion in aquatic species (Vandermeersch et al., 2015). Unfortunately, this technique is considered unsuitable for digestion because the condition of operating at room temperature can damage some polymers (Kershaw et al., 2019).

6.5.3.2 Alkaline Digestion

The alkalis mostly used for MP digestion include NaOH (Cole et al., 2014) and KOH (Maes et al., 2017). The optimal outcome is achieved by utilizing 40 mL of 10 M potassium hydroxide (KOH) per 0.2 g of the sample that has been dried, and maintaining it at a temperature of 60°C for a duration of 24 hours. Following this, the mixture is neutralized by using hydrochloric acid (HCl), before being subjected to ultrasonication using a bath for a period of 10 minutes, followed by filtration and visual analysis (Kershaw et al., 2019). The use of KOH, however, takes a longer time (up to 3 weeks) to digest if operated at room temperature (Koelmans et al., 2013), while NaOH can destroy the polymer particles in the concentration of 10 M and 60°C (Dehaut et al., 2016).

6.5.3.3 Enzymatic Digestion

This is a technique used to prevent the loss of synthetic polymers caused by chemical attack on the MPs during digestion – a characteristic common to the chemical digestion methods (Löder et al., 2017). In this method, a pre-treatment is first carried out using a detergent (such as 5% or 10% SDS), to enhance the accessibility of the organic materials to the enzymatic treatment (Kershaw et al., 2019). According to Löder et al. (2017), an enzymatic treatment (using plankton sample) can follow the treatment steps below:

- Protease treatment: Protease treatment catalyses the decomposition of protein chains into scattered peptides for easy dissolution.
- Cellulase treatment: This treatment splits the bonds within cellulose and catalyses the decomposition of all cellulose.
- Hydrogen peroxide treatment I: Treatment with H_2O_2 destroys all residual protective coating organic materials.
- Chitinase treatment: Decompose chitin by breaking down the glycosidic bond within the chitin.
- Hydrogen peroxide treatment II: Another application of H_2O_2 to further degrade all organic materials that do not dissolve completely.

6.5.3.4 Oxidative Digestion

Oxidative digestion utilizes oxidizing agents in the separation of microplastics from the material of interest. The most commonly used oxidizing agents include hydrogen peroxide and Fenton reagent (H_2O_2 + Fe) (Alimi et al., 2021). The process of oxidative digestion requires adding hydrogen peroxide and Fe^{2+} catalyst solution to the sample and allowing it to react in a beaker. The addition of H_2O_2 may be continued until no natural organic material is visible. Subsequently, filtration or sieving is carried out before visual analysis (Kershaw et al., 2019; Masura et al., 2015).

6.6 IDENTIFICATION AND QUANTIFICATION

After field collection and laboratory preparation of samples, the MPs of interest need to be carefully and precisely identified. Identification techniques are used to confirm the presence of microplastics and determine their size, shape, and chemical composition. The identification technique selected depends on the type of microplastics being studied and the level of detail required. Examples of identification techniques include microscopy, spectroscopy, and chromatography. Quantification may be done to determine the spatial and temporal distribution of microplastics,

Analytical Techniques of Microplastic in an Aquatic Environment   95

quantify possible organic contaminants, and examine the rates of MP accumulation (Hidalgo-Ruz et al., 2012).

6.6.1 MANUAL COUNTING – OPTICAL MICROSCOPE

Visual identification is the most commonly utilized used and straightforward method for identifying MPs, and it involves the identification of the MPs with the naked eye, based on their colour and size, or with the aid of an optical microscope such as a stereomicroscope (Wang and Wang, 2018). MPs are distinguishable from other particles due to their unique properties, such as colour, shape, and texture, which enables their differentiation from other constituents of a sample (Silva et al., 2018). However, bias has been demonstrated in visual inspection depending on the examiner's subjectivity, with reports of blue fragments having the highest detection probability, while white fragments have the least (Hanvey et al., 2017). Visual counting therefore can result in either overestimation as a result of counting non-plastic particles as plastics, or underestimation due to the bias of the examiner excluding some plastics as non-plastics, especially microplastics (Hanvey et al., 2017). Overestimation can also come from the failure of this technique to fully oxidize the organic matter during the isolation process, which is thus mistakenly included as MPs (Claessens et al., 2011; Gbogbo et al., 2020). For instance, only 75% of fibres and 64% of particles visually identified by Lenz et al. (2015) were confirmed by Raman spectroscopy as MPs. In a different study, SEM analysis confirmed that about 20% of particles initially identified as MPs, with a size smaller than 1 mm through visual identification, were actually composed of aluminium silicate derived from coal ash (Eriksen et al., 2013).

In a bid to minimize the misidentification of MPs by this technique, dye stains have been used to improve identification. Staining dyes are an inexpensive technique used to facilitate the identification of MPs by the naked eye (Prata et al., 2019). Dyes such as Nile Red (Shruti et al., 2021), Rose Bengal (Gbogbo et al., 2020), and Rhodamine B (Tong et al., 2021) have all been used for this purpose. The utilization of Rose Bengal stain to enhance the identification of MPs following a conventional microscopic isolation process was studied by Gbogbo et al. (2020). The following represent the numbers of potential microplastic particles observed in a wetland sample from the coastal region of Ghana using the conventional method, without the use of staining: sediment (3.55 g^{-1}), faecal matter of shorebirds (0.8 g^{-1}), and lagoon water (0.13 mL^{-1}). After staining, these were reduced to: sediment (1.85 g^{-1}), faecal matter (0.35 g^{-1}), and lagoon water (0.09 mL^{-1}). If SEM, FTIR, Raman, and Pyr-GC-MS are not available, utilizing this stain can help to decrease the number of false detections, although further studies have been recommended to ascertain this. Nevertheless, Nile Red is the most commonly used dye and has been reported as the most suitable staining technique for MPs currently. This staining technique requires a brief incubation period and results in a high recovery rate (Prata et al., 2019).

6.6.2 FOURIER TRANSFORM INFRARED SPECTROSCOPY (FTIR)

FTIR is a technique utilized to generate infrared spectra of absorption or emission, and to acquire high-resolution spectral data, which improves the identification of the molecular structure of compounds (Hanvey et al., 2017). This technique is mostly used for the determination of microplastics from an analyte of interest. FTIR can characterize MP particles >20 μm in size (Alimi et al., 2021). It reveals easily the functional group in an MP particle. Attenuated total reflectance-FTIR (ATR-FTIR) improves the details of the MP when determining irregular MPs, a feature in contrast to transmission FTIR (Prata et al., 2019). Micro-FTIR (μFTIR) is a technique utilized for microscopic observation of small-sized particles that appear plastic-like, which can then be spectroscopically confirmed on a single platform by alternating between the object lens and IR beam (Kershaw et al., 2019). μFTIR provides a high-resolution structure of the sample without requiring a pre-selection step, with the ability to detect structures as small as 20 μm (Prata et al., 2019). The utilization of FTIR for the determination of very small microplastics, however, has reduced efficiency, especially when the MPs are less than 20 μm in size.

6.6.3 Raman Spectroscopy

Raman spectroscopy is a type of spectroscopy that utilizes monochromatic light (usually a laser) as a source to interact with a sample through scattering, allowing for analysis of the sample's chemical composition (Käppler et al., 2016). A small fraction of the photons that are scattered undergo a change in energy, providing insight into the vibrations of the molecules in the sample (Käppler et al., 2016). Raman spectroscopy, together with FTIR, are the mostly commonly used techniques for MP identification and characterization. Raman spectroscopy is able to identify MPs as small as <20 μm but >1 μm, works effectively in wet samples, and is capable of generating spatial chemical images of the particles (Alimi et al., 2021; Qiu et al., 2016). With Raman spectroscopy, a non-destructive chemical characterization of MPs can be achieved, which is useful, especially in cases where further analysis of the sample is required (Wang and Wang, 2018). The Raman technique is based on the polarizability of a chemical bond, resulting in excitation of bonds such as aromatic bonds, C–H, and C=C double bonds (Käppler et al., 2016). μRaman is significantly slower in comparison with μFTIR, and requires sample preparation because the remnants of biological materials that fluorescence would essentially influence the result (Kershaw et al., 2019). Raman spectroscopy, however, can be interrupted in its action by the presence of other additives combined with the microplastic (Qiu et al., 2016).

Notwithstanding the minor drawbacks, Raman and FTIR spectroscopes are the best analytical techniques in the recognition and measurement of MPs, and the best result is obtained when the two techniques are incorporated. FTIR spectroscopy will identify MP particles >20 μm, while Raman spectroscopy will identify particles >1 μm.

6.6.4 Scanning Electron Microscopy (SEM)

SEM is a type of microscopy that utilizes a beam of electrons focused on a material, in a bid to produce images with high resolution (Wang and Wang, 2018). SEM can produce images of plastic materials with very high magnification, particularly if the MPs are diverse in size and shape, including but not limited to fibres, spherules, hexagons, and irregular polyhedrons (Silva et al., 2018). The combination of SEM and energy-dispersive X-ray spectroscopy system (EDX) has been reported to produce the best results for microscopic and inorganic chemical constituents inherent in a sample (Eriksen et al., 2013). Despite the success, SEM is not commonly used, particularly when a high volume of samples is to be analysed because it takes a lot of time for adequate analysis of all the samples, requiring a lot of work in the sample preparation (Silva et al., 2018).

6.6.5 Pyrolysis-Gas Chromatography-Mass Spectrometry (Pyr-GC-MS)

Pyr-GC-MS is a destructive method used to identify the polymer type of microplastics by determining the product of their thermal breakdown (Silva et al., 2018). In Pyr-GC-MS, MPs are subjected to thermal decomposition in an inert atmosphere, and the resulting gas is trapped at cryogenic temperatures and separated on a chromatographic column. The separated components are then identified by spectrometry to determine the type of polymer present in the MP sample (Prata et al., 2019). The Pyr-GC-MS method is utilized to investigate the types of polymers and organic plastic additives that are typically challenging to dissolve, extract, or hydrolyse at the beginning (Qiu et al., 2016). Pyr-GC-MS has the advantage of analysing samples without necessarily undergoing pretreatment since it directly examines the polymer sample. Okoffo et al. (2020) recently combined a single-step sequential pressurized liquid extraction (PLE) with a two-stage (double-shot) Pyr-GC-MS for the quantitative extraction and analysis of various plastics. The Pyr-GC-MS method does not require pre-extraction or pre-treatment of samples, making it useful for measuring specific plastics without interference and allowing for uniform reporting of results compared to FTIR and Raman spectroscopy (Okoffo et al., 2020). A significant limitation of the Pyr-GC-MS method is its inability to determine the polymer mass and provide details about the quantity, type, and shape of plastics in the sample (Hanvey et al., 2017).

6.6.6 THERMOGRAVIMETRIC ANALYSIS-DIFFERENTIAL SCANNING CALORIMETRY (TGA-DSC)

TGA-DSC is a thermogravimetric technique that provides an alternative method of analysis to the popular spectroscopes for the determination of MPs. This method involves heating a sample at a specific temperature gradient with a specified heating rate. The difference in heat flow is then determined by comparing the sample to a reference (Majewsky et al., 2016). TGA-DSC is a method that determines the alterations in the attributes of polymers due to their heat resistance (Kershaw et al., 2019). It can also validate the purity of synthetic substances and explore phase transitions (Majewsky et al., 2016). Furthermore, this approach has the benefit of operating directly and needing just a small amount of the sample (1–20 mg) (Dümichen et al., 2015; Majewsky et al., 2016).

6.6.7 THERMAL DESORPTION-GAS CHROMATOGRAPHY-MASS SPECTROMETRY (TDS-GC-MS)

TDS-GC-MS is a method that merges TGA-DSC and Pyr-GC-MS, and is used as an alternative for identifying microplastics. It was proposed by Dümichen et al. (2015), and utilized for the analysis of PE-based MPs in samples. The proposed technique combines TGA-DSC and Pyr-GC-MS to identify MPs and is capable of measuring complex matrices with higher sample masses compared to Pyr-GC-MS. However, further development is required as it has only been tested on PE-spiked environmental samples and its potential needs to be further explored.

6.6.8 HIGH-TEMPERATURE GEL-PERMEATION CHROMATOGRAPHY (HT-GPC)

HT-GPC is another analytical technique for the characterization and quantification of MPs described by Hintersteiner et al. (2015). The study utilized HT-GPC to evaluate the presence of polyolefin microbeads in personal-care products based on aqueous and hydrocarbon substances. The findings demonstrated recovery rates ranging from 92% to 96% for seven distinct personal-care items. Though successful, this technique does not provide any report on the particle size and number, and is only suitable for samples in personal-care products that are formulated with either water or oil-based ingredients as their main solvent or carrier.

SEM, FTIR, Raman spectroscopy, and Pyr-GC-MS are considered the most appropriate methods for identifying MPs due to their capability to analyse the polymer composition of plastics, which minimizes errors and misidentification (Gbogbo et al., 2020). However, the use of visual identification with an optical microscope is still prevalent in most cases, especially in the African perspective. Nevertheless, even when the spectroscopic techniques are assessable, it is still common practice to first visualize the sample with the naked eye and select the most easily detectable particles prior to spectroscopic approaches.

6.7 QUALITY CONTROL/QUALITY ASSURANCE

MP analysis, like any other contaminant study, requires strict adherence to quality control and assurance measures throughout the collection and analysis process (Mendoza and Balcer, 2019). During a field study, there is a risk of contamination that can occur at various stages such as during the process of collecting samples (Olsen et al., 2020). It is important to use validated and acceptable techniques to enhance data quality and enable comparison of the analytical methods employed (Silva et al., 2018). Contaminants such as airborne fibres, which are the most common in the lab, can lead to significant over-reporting of the MPs present in the sample (Mendoza and Balcer, 2019; Wang and Wang, 2018). To mitigate potential contaminants, it is necessary to conduct methodological blank tests during the sampling and handling processes. These contaminants can be further reduced by using filter paper to filter the lab air for a particular duration (Wang and Wang, 2018). Both field and laboratory studies have conducted procedural blanks, replicate samples, spiked blank samples, and matrix spiked samples. However, the lack of a standard reference material with specified concentrations of the target has limited their effectiveness in validating the method, assessing probable estimations, conducting initial quality control, external proficiency essays, and inter-laboratory studies (Silva et al., 2018).

6.8 IMPLICATIONS FOR MANAGEMENT AND POLICY

The growing concern around microplastics in aquatic environments has led to an increased focus on management and policy interventions. Effective management and policy interventions can help to mitigate the impact of MPs on both the aquatic and terrestrial ecosystems, as well as on humans. Some of the implications for management and policy include the following:

- *Source reduction:* One of the most effective ways to manage microplastics is to reduce their sources. This can be achieved through the implementation of regulations and policies that restrict the manufacture and use of non-biodegradable plastics, promote the use of other biodegradable alternatives, and encourage sustainable waste management practices.
- *Improved waste management:* Improving waste management practices, including increasing recycling rates, reducing littering, and implementing effective waste disposal methods, can help to decrease the level of plastic waste that is abandoned in aquatic environments.
- *Education and awareness:* Such campaigns can help to increase public knowledge and understanding of the effects of microplastics on the ecosystem. This can lead to increased support for policy interventions and behaviour change, such as reducing the use of single-use plastics.
- *Standardized monitoring and assessment:* Standardized monitoring and assessment methods can aid in enhancing our knowledge of the distribution and concentration of MPs in aquatic media. This information can be used to inform policy decisions and management interventions.
- *International cooperation:* Treating the problem of MPs in aquatic environments requires international cooperation and collaboration. This includes sharing information and best practices, coordinating research efforts, and developing global agreements and policies.

6.9 GAPS IN KNOWLEDGE AND AREAS FOR FUTURE RESEARCH

While various studies have been geared towards understanding the sources, environmental effects, and analytical techniques for MPs in aquatic environments, several gaps in knowledge and areas for future research remain. Some of these include:

- *Fate and transport:* More studies are required to better comprehend the fate and transport of microplastics in aquatic environments. This includes understanding how microplastics move through different environmental media, such as water, sediment, and biota, and how they are affected by environmental variables such as water flow and sedimentation.
- *Ecological impacts:* While there exist some studies that have shown that MPs can have ecological impacts, additional investigation is necessary to comprehend the exact mechanisms underlying these effects. This includes investigating the effects of chronic exposure to microplastics on aquatic organisms and ecosystems and the potential for long-term impacts.

Analytical Techniques of Microplastic in an Aquatic Environment

- *Human health impacts:* Additional investigation is required to comprehensively comprehend the extent to which MPs negatively affect human health, despite some evidence that suggests their adverse effects. This includes investigating the potential for microplastics to accumulate in human tissues and organs and the potential for health effects from chronic exposure to microplastics.
- *Standardized analytical methods:* At present, there is a lack of universally accepted protocols for analysing MPs in environmental samples. Therefore, further investigation is required to establish and verify standardized methods for sampling, extracting, and identifying MPs across diverse environmental matrices.
- *MPs in freshwater environments*: Various researches on microplastics in aquatic environments have concentrated on marine ecosystems. Based on this, studies geared towards the study of MPs in freshwater environments should be encouraged as they would be helpful in obtaining a better understanding of the sources, environmental impacts, and analytical approaches for MPs in aquatic media.

6.10   CONCLUSION AND SUMMARY

This chapter delves into the diverse sources, environmental consequences, and methods used to collect, isolate, and characterize microplastics (MPs) in aquatic environments. MPs originate from varied origins, including point and non-point sources, natural and anthropogenic sources, as well as primary and secondary sources. These minute plastic particles have substantial adverse effects on aquatic ecosystems, posing physical harm, ingestion risks for numerous organisms, and disruptions to feeding and reproductive behaviours. Moreover, they accumulate toxic chemicals, altering aquatic ecosystem structures and functions, leading to potential ecological and human health impacts through the food chain. The analytical techniques involved in MP analysis encompass sampling techniques, extraction methods, identification techniques, and quantification techniques. Crucial in estimating MP abundance, sampling techniques rely on factors like available equipment, research objectives, the sampling matrix, volumes, and depth. While sampling methods for marine and freshwater environments share similarities, the distribution of MPs can vary due to environmental factors such as hydrodynamics and density. Subsequent to sample collection, extraction techniques like density separation, filtration, sieving, and digestion effectively separate MPs from the sample matrix. Notably, density separation using a saturated salt solution has proven efficient for sediment samples, while cost-effective alternatives like NaI solutions have been proposed. The identification and quantification of MPs are crucial to comprehending their spatial and temporal distribution and potential environmental impacts. While visual identification through optical microscopy is common, it may suffer from biases and misidentification. Spectroscopic techniques, such as Fourier transform infrared spectroscopy (FTIR) and Raman spectroscopy, provide high-resolution structural information for MP identification. Combining multiple techniques enhances the accuracy of MP identification. Despite significant progress, knowledge gaps persist, particularly concerning the long-term effects of MP exposure on both humans and the natural environment. Further research is imperative to fully grasp the potential effects. In summary, addressing the issue of microplastics in aquatic environments requires a holistic approach, considering their sources, environmental effects, and potential impacts on human health. Continuous research, effective management, and policy interventions are vital to mitigating the effects of MPs on both humans and the natural environment.

REFERENCES

Adeniyi, Adewale George, Victor Temitope Amusa, Ebuka Chizitere Emenike, and Kingsley O Iwuozor. 2022a. "Co-carbonization of waste biomass with expanded polystyrene for enhanced biochar production." *Biofuels* no. 14 (6):635–643.

Adeniyi, Adewale George, Victor Temitope Amusa, Kingsley O Iwuozor, and Ebuka Chizitere Emenike. 2023a. "Valorization of waste Biaxially-oriented polypropylene (BOPP) Plastic films by its co-carbonization with Almond leaves." *Environmental Progress & Sustainable Energy* no. 42 (4):e14064.

Adeniyi, Adewale George, Ebuka Chizitere Emenike, Kingsley O Iwuozor, Hussein Kehinde Okoro, and Olusegun Omoniyi Ige. 2022b. "Acid mine drainage: The footprint of the Nigeria mining industry." *Chemistry Africa* no. 5:1907–1920.

Adeniyi, Adewale George, Ebuka Chizitere Emenike, Kingsley O Iwuozor, and Oluwaseyi D Saliu. 2023b. "Solvated polystyrene resin: A perspective on sustainable alternative to epoxy resin in composite development." *Materials Research Innovations* no. 27 (7):490–502.

Adeniyi, Adewale, Kingsley Iwuozor, Ebuka Emenike, Mubarak Amoloye, Emmanuel Aransiola, Fawaz Motolani, and Sodiq Kayode. 2023c. "Prospects and problems in the development of biochar-filled plastic composites: A review." *Functional Composites and Structures* no. 5:012002.

Adeyi, AA. 2020. "Distribution and bioaccumulation of Endocrine Disrupting Chemicals (EDCS) in Lagos Lagoon water, sediment and fish." *Ife Journal of Science* no. 22 (2):057–074.

Akpomie, Kovo G, Kayode A Adegoke, Kabir O Oyedotun, Joshua O Ighalo, James F Amaku, Chijioke Olisah, Adedapo O Adeola, Kingsley O Iwuozor, and Jeanet Conradie. 2022. "Removal of bromophenol blue dye from water onto biomass, activated carbon, biochar, polymer, nanoparticle, and composite adsorbents." *Biomass Conversion and Biorefinery* no. 14:13629–13657.

Alimi, Olubukola S, Oluniyi O Fadare, and Elvis D Okoffo. 2021. "Microplastics in African ecosystems: Current knowledge, abundance, associated contaminants, techniques, and research needs." *Science of The Total Environment* no. 755:142422.

Andrady, Anthony L. 2011. "Microplastics in the marine environment." *Marine Pollution Bulletin* no. 62 (8):1596–1605.

Budimir, Stjepan, Outi Setälä, and Maiju Lehtiniemi. 2018. "Effective and easy to use extraction method shows low numbers of microplastics in offshore planktivorous fish from the northern Baltic Sea." *Marine Pollution Bulletin* no. 127:586–592.

Chaukura, Nhamo, Kebede K Kefeni, Innocent Chikurunhe, Isaac Nyambiya, Willis Gwenzi, Welldone Moyo, Thabo TI Nkambule, Bhekie B Mamba, and Francis O Abulude. 2021. "Microplastics in the aquatic environment—The occurrence, sources, ecological impacts, fate, and remediation challenges." *Pollutants* no. 1 (2):95–118.

Claessens, Michiel, Steven De Meester, Lieve Van Landuyt, Karen De Clerck, and Colin R Janssen. 2011. "Occurrence and distribution of microplastics in marine sediments along the Belgian coast." *Marine Pollution Bulletin* no. 62 (10):2199–2204.

Claessens, Michiel, Lisbeth Van Cauwenberghe, Michiel B Vandegehuchte, and Colin R Janssen. 2013. "New techniques for the detection of microplastics in sediments and field collected organisms." *Marine Pollution Bulletin* no. 70 (1–2):227–233.

Cole, Matthew, Pennie Lindeque, Claudia Halsband, and Tamara S Galloway. 2011. "Microplastics as contaminants in the marine environment: A review." *Marine Pollution Bulletin* no. 62 (12):2588–2597.

Cole, Matthew, Hannah Webb, Pennie K Lindeque, Elaine S Fileman, Claudia Halsband, and Tamara S Galloway. 2014. "Isolation of microplastics in biota-rich seawater samples and marine organisms." *Scientific Reports* no. 4 (1):1–8.

Cristaldi, Antonio, Maria Fiore, Pietro Zuccarello, Gea Oliveri Conti, Alfina Grasso, Ilenia Nicolosi, Chiara Copat, and Margherita Ferrante. 2020. "Efficiency of wastewater treatment plants (WWTPs) for microplastic removal: A systematic review." *International Journal of Environmental Research and Public Health* no. 17 (21):8014.

de Moura, Esperidiana AB, Helio A Furusawa, Marycel EB Cotrim, Emeka E Oguzie, and Ademar B Lugao. 2019. "Microplastics: A Novel Method for Surface Water Sampling and Sample Extraction in Elechi Creek, Rivers State, Nigeria." In *Characterization of Minerals, Metals, and Materials 2019*, 269–281. Springer.

Dehaut, Alexandre, Anne-Laure Cassone, Laura Frère, Ludovic Hermabessiere, Charlotte Himber, Emmanuel Rinnert, Gilles Rivière, Christophe Lambert, Philippe Soudant, and Arnaud Huvet. 2016. "Microplastics in seafood: Benchmark protocol for their extraction and characterization." *Environmental Pollution* no. 215:223–233.

Devriese, Lisa I, Myra D Van der Meulen, Thomas Maes, Karen Bekaert, Ika Paul-Pont, Laura Frère, Johan Robbens, and A Dick Vethaak. 2015. "Microplastic contamination in brown shrimp (Crangon crangon, Linnaeus 1758) from coastal waters of the Southern North Sea and Channel area." *Marine Pollution Bulletin* no. 98 (1–2):179–187.

Dümichen, Erik, Anne-Kathrin Barthel, Ulrike Braun, Claus G Bannick, Kathrin Brand, Martin Jekel, and Rainer Senz. 2015. "Analysis of polyethylene microplastics in environmental samples, using a thermal decomposition method." *Water Research* no. 85:451–457.

Dümichen, Erik, Paul Eisentraut, Claus Gerhard Bannick, Anne-Kathrin Barthel, Rainer Senz, and Ulrike Braun. 2017. "Fast identification of microplastics in complex environmental samples by a thermal degradation method." *Chemosphere* no. 174:572–584.

Emenike, Ebuka Chizitere, Adewale G Adeniyi, Patrick E Omuku, Kingsley Chidiebere Okwu, and Kingsley O Iwuozor. 2022. "Recent advances in nano-adsorbents for the sequestration of Copper from water." *Journal of Water Process Engineering* no. 47 (102715). https://doi.org/10.1016/j.jwpe.2022.102715.

Eriksen, Marcus, Sherri Mason, Stiv Wilson, Carolyn Box, Ann Zellers, William Edwards, Hannah Farley, and Stephen Amato. 2013. "Microplastic pollution in the surface waters of the Laurentian Great Lakes." *Marine Pollution Bulletin* no. 77 (1–2):177–182.

Farrell, Paul, and Kathryn Nelson. 2013. "Trophic level transfer of microplastic: Mytilus edulis (L.) to Carcinus maenas (L.)." *Environmental Pollution* no. 177:1–3.

Gbogbo, Francis, James Benjamin Takyi, Maxwell Kelvin Billah, and Julliet Ewool. 2020. "Analysis of microplastics in wetland samples from coastal Ghana using the Rose Bengal stain." *Environmental Monitoring and Assessment* no. 192 (4):1–10.

Geyer, Roland, Jenna R Jambeck, and Kara Lavender Law. 2017. "Production, use, and fate of all plastics ever made." *Science Advances* no. 3 (7):e1700782.

Gündoğdu, Sedat, Cem Çevik, Evşen Güzel, and Serdar Kilercioğlu. 2018. "Microplastics in municipal wastewater treatment plants in Turkey: A comparison of the influent and secondary effluent concentrations." *Environmental Monitoring and Assessment* no. 190 (11):1–10.

Hanvey, Joanne S, Phoebe J Lewis, Jennifer L Lavers, Nicholas D Crosbie, Karla Pozo, and Bradley O Clarke. 2017. "A review of analytical techniques for quantifying microplastics in sediments." *Analytical Methods* no. 9 (9):1369–1383.

Hidalgo-Ruz, Valeria, Lars Gutow, Richard C Thompson, and Martin Thiel. 2012. "Microplastics in the marine environment: A review of the methods used for identification and quantification." *Environmental Science & Technology* no. 46 (6):3060–3075.

Hintersteiner, Ingrid, Markus Himmelsbach, and Wolfgang W Buchberger. 2015. "Characterization and quantitation of polyolefin microplastics in personal-care products using high-temperature gel-permeation chromatography." *Analytical and Bioanalytical Chemistry* no. 407 (4):1253–1259.

Ighalo, Joshua O, Setyo Budi Kurniawan, Kingsley O Iwuozor, Chukwunonso O Aniagor, Oluwaseun J Ajala, Stephen N Oba, Felicitas U Iwuchukwu, Shabnam Ahmadi, and Chinenye Adaobi Igwegbe. 2022. "A review of treatment technologies for the mitigation of the toxic environmental effects of acid mine drainage (AMD)." *Process Safety and Environmental Protection* no. 157:37–58.

Ilechukwu, Ifenna, Gloria Ihuoma Ndukwe, Nkoli Maryann Mgbemena, and Akudo Ugochi Akandu. 2019. "Occurrence of microplastics in surface sediments of beaches in Lagos, Nigeria." *European Chemical Bulletin* no. 8 (10):371–375.

Iwuozor, Kingsley O, Tunde Aborode Abdullahi, Lawal Adewale Ogunfowora, Ebuka Chizitere Emenike, Ifeoluwa Peter Oyekunle, Fahidat Adedamola Gbadamosi, and Joshua O Ighalo. 2021. "Mitigation of levofloxacin from aqueous media by adsorption: a review." *Sustainable Water Resources Management* no. 7 (6):1–18. doi: https://doi.org/10.1007/s40899-021-00579-9.

Iwuozor, Kingsley O, Kovo G Akpomie, Jeanet Conradie, Kayode A Adegoke, Kabir O Oyedotun, Joshua O Ighalo, James F Amaku, Chijioke Olisah, and Adedapo O Adeola. 2022a. "Aqueous phase adsorption of aromatic organoarsenic compounds: A review." *Journal of Water Process Engineering* no. 49:103059.

Iwuozor, Kingsley O, Ebuka Chizitere Emenike, Fahidat Adedamola Gbadamosi, Joshua O Ighalo, Great C Umenweke, Felicitas U Iwuchukwu, Cyprian O Nwakire, and Chinenye Adaobi Igwegbe. 2022b. "Adsorption of organophosphate pesticides from aqueous solution: A review of recent advances." *International Journal of Environmental Science and Technology* no. 20:5845–5894, In press.

Iwuozor, Kingsley O, Lawal Adewale Ogunfowora, and Ifeoluwa Peter Oyekunle. 2022c. "Review on sugarcane-mediated nanoparticle synthesis: A green approach." *Sugar Tech* no. 24:1186–1197. doi: https://doi.org/10.1007/s12355-021-01038-7.

Jang, Mi, Won Joon Shim, Gi Myung Han, Manviri Rani, Young Kyoung Song, and Sang Hee Hong. 2017. "Widespread detection of a brominated flame retardant, hexabromocyclododecane, in expanded polystyrene marine debris and microplastics from South Korea and the Asia-Pacific coastal region." *Environmental Pollution* no. 231:785–794.

Käppler, Andrea, Dieter Fischer, Sonja Oberbeckmann, Gerald Schernewski, Matthias Labrenz, Klaus-Jochen Eichhorn, and Brigitte Voit. 2016. "Analysis of environmental microplastics by vibrational microspectroscopy: FTIR, Raman or both?" *Analytical and Bioanalytical Chemistry* no. 408 (29):8377–8391.

Kershaw, Peter, Alexander Turra, and Francois Galgani. 2019. "Guidelines for the Monitoring and Assessment of Plastic Litter in the Ocean-GESAMP Reports and Studies No. 99." *GESAMP Reports and Studies.*

Koelmans, Albert A, Ellen Besseling, Anna Wegner, and Edwin M Foekema. 2013. "Plastic as a carrier of POPs to aquatic organisms: A model analysis." *Environmental Science & Technology* no. 47 (14):7812–7820.

Kuo, Fan-Jun, and Hsiang-Wen Huang. 2014. "Strategy for mitigation of marine debris: Analysis of sources and composition of marine debris in northern Taiwan." *Marine Pollution Bulletin* no. 83 (1):70–78.

Lenz, Robin, Kristina Enders, Colin A Stedmon, David MA Mackenzie, and Torkel Gissel Nielsen. 2015. "A critical assessment of visual identification of marine microplastic using Raman spectroscopy for analysis improvement." *Marine Pollution Bulletin* no. 100 (1):82–91.

Li, Jiana, Dongqi Yang, Lan Li, Khalida Jabeen, and Huahong Shi. 2015. "Microplastics in commercial bivalves from China." *Environmental Pollution* no. 207:190–195.

Liu, Xuemin, Huahong Shi, Bing Xie, Dionysios D Dionysiou, and Yaping Zhao. 2019. "Microplastics as both a sink and a source of bisphenol A in the marine environment." *Environmental Science & Technology* no. 53 (17):10188–10196.

Löder, Martin GJ, Hannes K Imhof, Maike Ladehoff, Lena A Löschel, Claudia Lorenz, Svenja Mintenig, Sarah Piehl, Sebastian Primpke, Isabella Schrank, and Christian Laforsch. 2017. "Enzymatic purification of microplastics in environmental samples." *Environmental Science & Technology* no. 51 (24):14283–14292.

Lv, Xuemin, Qian Dong, Zhiqiang Zuo, Yanchen Liu, Xia Huang, and Wei-Min Wu. 2019. "Microplastics in a municipal wastewater treatment plant: Fate, dynamic distribution, removal efficiencies, and control strategies." *Journal of Cleaner Production* no. 225:579–586.

Maes, Thomas, Rebecca Jessop, Nikolaus Wellner, Karsten Haupt, and Andrew G Mayes. 2017. "A rapid-screening approach to detect and quantify microplastics based on fluorescent tagging with Nile Red." *Scientific Reports* no. 7 (1):1–10.

Majewsky, Marius, Hajo Bitter, Elisabeth Eiche, and Harald Horn. 2016. "Determination of microplastic poly-ethylene (PE) and polypropylene (PP) in environmental samples using thermal analysis (TGA-DSC)." *Science of the Total Environment* no. 568:507–511.

Masura, Julie, Joel Baker, Gregory Foster, and Courtney Arthur. 2015. "*Laboratory methods for the analysis of microplastics in the marine environment: Recommendations for quantifying synthetic particles in waters and sediments.*" NOAA Technical Memorandum NOS-OR&R-48.

Mendoza, LM Rios, and Mary Balcer. 2020. *Microplastics in Freshwater Environments.* Elsevier Inc.

Mendoza, Lorena M Rios, and Mary Balcer. 2019. "Microplastics in freshwater environments: A review of quantification assessment." *TrAC Trends in Analytical Chemistry* no. 113:402–408.

Migwi, Francis Kigera, Joanne Atieno Ogunah, and John Mburu Kiratu. 2020. "Occurrence and spatial distribution of microplastics in the surface waters of Lake Naivasha, Kenya." *Environmental Toxicology and Chemistry* no. 39 (4):765–774.

Mintenig, SM, Ivo Int-Veen, Martin GJ Löder, Sebastian Primpke, and Gunnar Gerdts. 2017. "Identification of microplastic in effluents of waste water treatment plants using focal plane array-based micro-Fourier-transform infrared imaging." *Water Research* no. 108:365–372.

Missawi, Omayma, Noureddine Bousserrhine, Sabrina Belbekhouche, Nesrine Zitouni, Vanessa Alphonse, Iteb Boughattas, and Mohamed Banni. 2020. "Abundance and distribution of small microplastics (≤ 3 µm) in sediments and seaworms from the Southern Mediterranean coasts and characterisation of their potential harmful effects." *Environmental Pollution* no. 263:114634.

Ogunfowora, Lawal A, Kingsley O Iwuozor, Joshua O Ighalo, and Chinenye Adaobi Igwegbe. 2021. "Trends in the treatment of aquaculture effluents using nanotechnology." *Cleaner Materials* no. 2:100024. doi: https://doi.org/10.1016/j.clema.2021.100024.

Okoffo, Elvis D, Francisca Ribeiro, Jake W O'Brien, Stacey O'Brien, Benjamin J Tscharke, Michael Gallen, Saer Samanipour, Jochen F Mueller, and Kevin V Thomas. 2020. "Identification and quantification of selected plastics in biosolids by pressurized liquid extraction combined with double-shot pyrolysis gas chromatography–mass spectrometry." *Science of the Total Environment* no. 715:136924.

Olsen, Linn Merethe Brekke, Heidi Knutsen, Sabnam Mahat, Emma Jane Wade, and Hans Peter H Arp. 2020. "Facilitating microplastic quantification through the introduction of a cellulose dissolution step prior to oxidation: Proof-of-concept and demonstration using diverse samples from the Inner Oslofjord, Norway." *Marine Environmental Research* no. 161:105080.

Prata, Joana Correia. 2018. "Microplastics in wastewater: State of the knowledge on sources, fate and solutions." *Marine Pollution Bulletin* no. 129 (1):262–265.

Prata, Joana Correia, João P da Costa, Armando C Duarte, and Teresa Rocha-Santos. 2019. "Methods for sampling and detection of microplastics in water and sediment: A critical review." *TrAC Trends in Analytical Chemistry* no. 110:150–159.

Qiu, Qiongxuan, Zhi Tan, Jundong Wang, Jinping Peng, Meimin Li, and Zhiwei Zhan. 2016. "Extraction, enumeration and identification methods for monitoring microplastics in the environment." *Estuarine, Coastal and Shelf Science* no. 176:102–109.

Rios Mendoza, Lorena M., Satie Taniguchi, and Hrissi K Karapanagioti. 2017. "Chapter 8 – Advanced Analytical Techniques for Assessing the Chemical Compounds Related to Microplastics." In *Comprehensive Analytical Chemistry*, edited by Teresa AP Rocha-Santos and Armando C Duarte, 209–240. Elsevier.

Shruti, VC, Fermín Pérez-Guevara, Priyadarsi D Roy, and Gurusamy Kutralam-Muniasamy. 2021. "Analyzing microplastics with Nile Red: Emerging trends, challenges, and prospects." *Journal of Hazardous Materials* no. 423:127171.

Silva, Ana B, Ana S Bastos, Celine IL Justino, Joao P da Costa, Armando C Duarte, and Teresa AP Rocha-Santos. 2018. "Microplastics in the environment: Challenges in analytical chemistry-A review." *Analytica Chimica Acta* no. 1017:1–19.

Sussarellu, Rossana, Marc Suquet, Yoann Thomas, Christophe Lambert, Caroline Fabioux, Marie Eve Julie Pernet, Nelly Le Goïc, Virgile Quillien, Christian Mingant, and Yanouk Epelboin. 2016. "Oyster reproduction is affected by exposure to polystyrene microplastics." *Proceedings of the National Academy of Sciences* no. 113 (9):2430–2435.

Talvitie, Julia, Anna Mikola, Arto Koistinen, and Outi Setälä. 2017. "Solutions to microplastic pollution–Removal of microplastics from wastewater effluent with advanced wastewater treatment technologies." *Water Research* no. 123:401–407.

Thiel, Martin, Iván A Hinojosa, Tanja Joschko, and Lars Gutow. 2011. "Spatio-temporal distribution of floating objects in the German Bight (North Sea)." *Journal of Sea Research* no. 65 (3):368–379.

Thompson, Richard C, Ylva Olsen, Richard P Mitchell, Anthony Davis, Steven J Rowland, Anthony WG John, Daniel McGonigle, and Andrea E Russell. 2004. "Lost at sea: Where is all the plastic?" *Science* no. 304 (5672):838–838.

Tong, Huiyan, Qianyi Jiang, Xiaocong Zhong, and Xingshuai Hu. 2021. "Rhodamine B dye staining for visualizing microplastics in laboratory-based studies." *Environmental Science and Pollution Research* no. 28 (4):4209–4215.

Vandermeersch, Griet, Lisbeth Van Cauwenberghe, Colin R Janssen, Antonio Marques, Kit Granby, Gabriella Fait, Michiel JJ Kotterman, Jorge Diogène, Karen Bekaert, and Johan Robbens. 2015. "A critical view on microplastic quantification in aquatic organisms." *Environmental Research* no. 143:46–55.

Von Moos, Nadia, Patricia Burkhardt-Holm, and Angela Köhler. 2012. "Uptake and effects of microplastics on cells and tissue of the blue mussel Mytilus edulis L. after an experimental exposure." *Environmental Science & Technology* no. 46 (20):11327–11335.

Wang, Wenfeng, and Jun Wang. 2018. "Investigation of microplastics in aquatic environments: An overview of the methods used, from field sampling to laboratory analysis." *TrAC Trends in Analytical Chemistry* no. 108:195–202.

Watts, Andrew JR, Ceri Lewis, Rhys M Goodhead, Stephen J Beckett, Julian Moger, Charles R Tyler, and Tamara S Galloway. 2014. "Uptake and retention of microplastics by the shore crab Carcinus maenas." *Environmental Science & Technology* no. 48 (15):8823–8830.

Yang, Dongqi, Huahong Shi, Lan Li, Jiana Li, Khalida Jabeen, and Prabhu Kolandhasamy. 2015. "Microplastic pollution in table salts from China." *Environmental Science & Technology* no. 49 (22):13622–13627.

Zeng, Zhuotong, Biao Song, Rong Xiao, Guangming Zeng, Jilai Gong, Ming Chen, Piao Xu, Peng Zhang, Maocai Shen, and Huan Yi. 2019. "Assessing the human health risks of perfluorooctane sulfonate by in vivo and in vitro studies." *Environment International* no. 126:598–610.

7 Advanced Instrumentation for Quantification of Microplastics

Jatinder Singh Randhawa

7.1 INTRODUCTION

Every year, an estimated 8 million tonnes of plastic garbage is released into the ocean, impacting almost 700 marine species (Piazza et al., 2022). Thompson originally used the term "microplastics" in 2004 to refer to the tiny plastic particles in the environment that are less than 5 mm (Thompson et al., 2004). The unintentional discharge of microplastics (MPs) into the environment can exacerbate environmental stresses and have detrimental impacts on plants, ecosystems, and lower trophic levels if improper control is not implemented. By 2060, the amount of secondary microplastics expected to accumulate in the environment is expected to increase by 155–265 million tonnes based on the present trend, with microplastics contributing 13.2% of this total (Lebreton and Andrady, 2019; Sobhani et al., 2020).

Due to its many advantages such as its affordability, durability, and lightweight plastics it plays a big role in offering exceptional comfort in our daily lives (Gu et al., 2020). The environmental contamination caused by plastics is particularly persistent since they are non-biodegradable and cannot disintegrate naturally. Researchers warn that although global plastics output has exceeded 350 million tonnes, it may reach 500 million tonnes by 2025 if prompt action is not taken to curtail it (Geyer et al., 2017). This is based on a study published in Plastics Europe (2019). In air and water currents, they may potentially travel quite far. Substances that come from different sources are categorized as either primary or secondary microplastics. Generally speaking, primary microplastics are plastic particles discharged into rivers and sewage treatment plants. Plastic is broken down and decreased in volume by physical, chemical, and biological processes to create secondary microplastics (Guo and Wang, 2019). Microplastics are most frequently found as pieces, granules, threads, and films (Guo et al., 2020). In air and water currents, they may potentially travel quite far. Substances that come from different sources are categorized as either primary or secondary microplastics. One type of plastic that is easily absorbed by humans through food chains is microplastics that pose a major risk to humans as well as animals (Gaylarde et al., 2020). A few of the many and varied qualities that influence their level of risk include their composition, size, and structure. Pirsaheb et al. (2020)'s research indicates that microscopic, fibre-shaped microplastics pose a greater threat to the environment. Despite the chemicals used in their manufacturing process, the microplastics produced from materials such as polyamide (PA), polypropylene (PP), polyethylene (PE), polyethylene terephthalate (PET), polystyrene (PS), and polyvinyl chloride (PVC) have minimal negative effects on human health (Lithner et al., 2011). For instance, e-waste plastics often include flame retardants (Li et al., 2019) and heavy metals (Pb, Cd, Cr, Hg, and Sb) at concentrations by weight that may exceed hazardous waste standards (Turner et al., 2019; Singh et al., 2020). Research has been conducted on the impact of these microplastics on plants (Zhu et al., 2019). The one that is most commonly

DOI: 10.1201/9781032706573-7

selected to be assessed as a microplastic is PS. One possible reason for this is because the fundamental PS spheres have small and uniform particle sizes, i.e., 1 m. PEs are widely used and persistent in the terrestrial ecosystem as polymers, therefore research on their impacts is frequently conducted as well (Machado et al., 2018). Thus, from the perspective of ecosystem health, uncommon plastic products could have a disproportionately large impact and should be given priority for research. Studies that track microplastics require consistent, dependable standardized sampling and analytical methods (Muller et al., 2020). However, there is currently no defined method in place, and analytical methods for microplastics are still being developed (Uddin et al., 2020). Nowadays, microplastics are widely used in agriculture, and figuring out their types and concentrations has become a hot topic. Numerous research works suggest that microplastics may alter the physical properties of soil, such as the rate of fertility, adsorption–desorption processes, and harmful impacts on terrestrial ecosystems (Brandes et al., 2021; Wang et al., 2022). This chapter summarizes existing techniques for identifying microplastics in real environmental samples and makes predictions about the direction of technology. Microplastics are often mixtures of diverse plastic particles with a wide variety of intricate compositions. The characterization of the chemical composition of microplastics is crucial for both the treatment approaches and the traceability of microplastics (Song et al., 2015). Current and frequently employed methods for the physical and chemical characterization of microplastics are shown in Figure 7.1, including scanning electron microscopy-energy dispersive x-ray (SEM-EDX), Fourier transform infrared spectroscopy, Raman spectroscopy, thermal analysis, and mass spectrometry (Weidner and Trimpin, 2010; Song et al., 2015; Majewsky et al., 2016; Wagner et al., 2017; Araujo et al., 2018). This section has discussed a number of interesting technologies in addition to these conventional techniques.

FIGURE 7.1 Numerous advanced analytical methods for microplastics.

7.2 TECHNIQUES FOR IDENTIFICATION AND QUANTIFICATION OF MICROPLASTICS

7.2.1 MICROSCOPY

According to Talvitie et al. (2017), early research on microplastic counts and physical characterization outlined visual methods for identifying the morphology of the microplastics present in a sample utilizing visual identification instruments such as light microscopes or polarizing microscopy. Based on their external appearance, visual identification techniques usually divide microplastics into four groups: fibres, pieces, pellets, and microbeads. Figure 7.2 shows a variety of physical and chemical characterization techniques for micro- and nanoplastic research.

7.2.2 LIGHT MICROSCOPY

Microplastics a few hundred micrometres in size and can be identified by light microscopy. Since microplastics usually don't have a glossy appearance, their physical response traits like their distinct elasticity or hardness are used to distinguish them (Morgado et al., 2021). Moreover, many kinds of microplastics identified in the environment are beads, fibres, and pieces (Abadi et al., 2021). According to Löder et al. (2015), around 70% of microplastic sample examples are clear. An optical microscope may be used to identify coloured polymers that have had dye added during manufacturing (Dehghani et al., 2017). Colourless or asymmetrical plastic particles in the sub-100 µm size range are challenging to define. Furthermore, it could be more challenging to distinguish microplastics under a microscope if sample particle separation is insufficient. Moreover, microplastics are difficult to detect under a microscope due to the involvement of some biological matter and sediments, which are not eliminated during chemical decomposition of these samples. Transparent particles are misdiagnosed at a rate of over 70%, whereas plastic-like particles are misdiagnosed at a rate above 20%, according to earlier research. Under a microscope, it can be challenging to discern between artificial and natural fibres since they are identical particles with interfering components. Despite the fact that cotton fibres are frequently mistaken for plastic, "destructive tests" have been carried out to find these particles. In order to make up for this, scientists have developed a technique for identifying these particles that involves melting the plastic particles with a heated needle (Shim et al.,

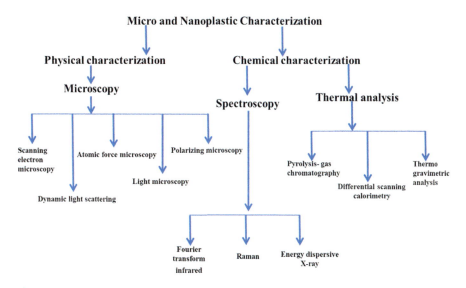

FIGURE 7.2 Physicochemical characterization methods for microplastic detection.

2017; Hendrickson et al., 2018). Whitening agents are commonly used as plastic additives in the textile and plastic synthesis industries. A fluorescence microscope may be used to identify plastics since bleaching chemicals typically exhibit fluorescence. On paper, however, whitening agents are also used. However, since some minerals include compounds that exhibit the self-fluorescence property, identification mistakes may occur (Dehghani et al., 2017).

7.2.3 POLARIZING MICROSCOPY

Polarized light microscopy has been shown to be an efficient method for identifying polyethylene (PE) particles in toxicity testing (Mossotti et al., 2021). The plastic crystal structure may have an effect on the polarized light transmission that can be measured (Abbasi, 2021). Even within the same polymer, there may be variations in the degree of crystallinity due to the kind of polymer and the manufacturing method. But for enough polarized light to pass through, microscopic microplastic samples are required. Samples of opaque microplastic are not amenable to this technique.

7.2.4 LASER DIFFRACTION PARTICLE SIZE

Numerous intricate tools have been developed as a result of material science's explosive progress. The particle size distribution in soil and sediment may be rapidly, accurately, and automatically assessed by employing laser diffraction particle size (Bittelli et al., 2022). This method provides precise, comprehensive, and confirmed results. Critical materials can be retrieved using this largely non-destructive technology (Blott et al., 2004). According to Blott et al. (2004), this approach may be used to examine particles with sizes ranging from 0.04 µm to 2000 µm. However, a variety of contaminants in ambient samples might affect the outcome of the experiment. The secondary goal of the technique is to remove microplastics from the environment. Although this technique has not been used much in the detection of microplastic particle size distribution, it will eventually develop a crucial role in this sector as technology advances.

7.2.5 SCANNING ELECTRON MICROSCOPY

Scanning electron microscopy (SEM) may produce sharp, multi-magnified images even for minuscule particles like nanoplastics (Tunali et al., 2020). Organic particles and microplastics may be easily distinguished from one another according to the texture of the particle surface, which can be verified by high-resolution photos (Hossain et al., 2019). An additional technique for detecting microplastics is transmission electron microscopy (TEM). While user variability may exist in the visual identification of microplastics by TEM, errors can be reduced by employing a tried-and-true method of cross-validating data from many users. Additional analysis is performed by energy-dispersive X-ray spectroscopy (EDS), which shows the elemental conformation of the item. Antioxidants or additive components, such as Al, Si, Na, Mg, and Ca, are gathered as markers and used in EDS microplastic detection (Wagner et al., 2019). By examining a particle's surface elemental makeup, carbon-rich polymers may be identified from inorganic particles (Sabri et al., 2021). However, due to the expensive equipment and laborious sample preparation method, limited quantity of samples can be processed by SEM and TEM.

7.2.6 FLOW CYTOMETRY

In the biological and medical fields, flow cytometry has been extensively applied, according to Adan et al. (2017). Flow cytometry detects substances with a particle size range of 0.5–40 µm by using laser light scattered from particles and recorded in forward- or side-scattering angles (Primpke et al., 2020). However, Kaile et al. (2020) pointed out those measuring microplastics seldom use

Advanced Instrumentation for Quantification of Microplastics

this method. In light of this, flow cytometry is used to analyse the formation of microplastics. They determined this by applying Mie's theory, which uses scattering intensity data to estimate the size of microplastic particles. However, flow cytometry can only identify microplastics with minuscule particle sizes; it cannot differentiate between different kinds of microplastics. Moreover, the accuracy of experimental results is likely to be impacted by some contaminants present in ambient samples. Therefore, to detect microplastics with large particles, flow cytometry has to be enhanced. It is anticipated that flow cytometry will be used in more ways in the future to assist environmental studies (Sorasan et al., 2021).

7.3 SPECTROSCOPY

Spectrophotometry may be used to quantitatively and qualitatively classify microplastics without destroying materials. In spectrophotometry, Fourier transform infrared spectroscopy and Raman spectroscopy are two commonly used methods for microplastic analysis. Microplastics smaller than 20 μm in size may be identified and measured using Raman spectroscopy, whereas those larger than 20 μm can be identified using Fourier transform infrared spectroscopy (Prata et al., 2019). Fourier transform infrared and Raman spectroscopy are often used in conjunction to investigate microplastics (Wright et al., 2019). Nowadays, to count microplastics, spectroscopy is commonly combined with different techniques.

7.3.1 FOURIER TRANSFORM INFRARED SPECTROSCOPY

According to Mecozzi et al. (2016), FTIR spectroscopy is capable of identifying every molecular and functional group found in plastic polymers. FTIR spectroscopy measures the spectrum of infrared (IR) light that the microplastic sample absorbs, which makes it possible to examine the molecular makeup of the several sample forms such as sediment, water, and biota. With absorption peaks corresponding to the frequency of vibration between the bonds of the atoms that make up the material, an infrared spectrum depicts the fingerprint of a microplastic. No two compounds create precisely the same infrared spectrum because every polymer substance is a unique mix of atoms. For this reason, FTIR may be used to identify the chemical structure of a polymer molecule (Chalmers, 2006). The electromagnetic spectrum infrared area is split into three regions: The area of higher energy near infrared (NIR) with wavenumbers between 14,000 and 4000 cm^{-1} (0.78 and 2.5 μm wavelength), which is susceptible to vibration combinations and overtones. Chen et al. (2020) developed FTIR spectroscopy, a technique that combines microscopy with Fourier transform infrared spectroscopy to boost the efficacy of microplastic detection. According to Chen et al. (2020), this novel technique can quantify particles as small as 10 μm and microplastics in environmental specimens. Araujo et al. (2018) assessed studies that employed micro-FTIR and Raman spectroscopy as automated methods for microplastic identification. Researchers also calculated the total analysis time per millimetre by dividing the measured area by the total measurement time they provided. In other cases, integrating the two approaches might result in a more complete MP characterization. FTIR is a non-destructive method with quick findings that is ideal for microplastic investigation since it requires extremely small sample sizes. Researchers worldwide can use this approach since it is widely accessible and reasonably priced. FTIR stands apart from other analytical techniques due to its precise and non-destructive capacity to identify organic molecules (Andoh et al., 2023). Depending on their size, MPs may be detected using different tactics, as shown in Figure 7.3.

7.3.2 RAMAN SPECTROSCOPY

In Raman spectroscopy, every polymer generates a unique spectrum when it is subjected to a laser beam because the molecular and atomic composition of the target affects the frequency of the

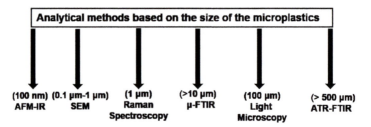

FIGURE 7.3 Commonly used analytical methods for microplastics analysis on the basis of microplastic size.

backscattered light. Similar to FTIR, Raman spectroscopy provides a polymer composition profile for every sample and may be used to identify plastics (Käppler et al., 2015). It may be used to examine the results of algorithms and data libraries. A non-destructive chemical analysis and microscopy component is shared by FTIR and Raman spectroscopy (Kniggendorf et al., 2019). Microplastics as little as a few micrometres in size can be found using Raman spectroscopy since its laser beam diameter is less than that of FTIR. The advantage of the non-contact Raman spectroscopy study is that it preserves the microplastic sample before and after the analysis, which makes it possible to conduct more research in the future. However, it can be more challenging to comprehend the outcomes of each approach in the identification of complicated microplastics due to variations in the reactions and spectra of microplastics between FTIR and Raman spectroscopy. Because colourants and microplastic additives may change the Raman signal, it is challenging to determine the desired polymer by employing this method (Dowarah and Devipriya, 2019). Inaccuracies may also result from worn microplastics' curved surfaces. It also has the same disadvantages as FTIR, including high cost and expert analysis required. Only materials for which a suitable Raman reference spectrum is provided in the database and the possibility of material misunderstanding has been input previously may be distinguished from one another using micro-Raman spectroscopy. The risk of false positives should be considered as it might be difficult to identify the target microplastic (Käppler et al., 2015). PE is a substance that might be confused with SDS or sodium stearate. Researchers combined Raman spectroscopy with microscopy to produce the non-contact, non-destructive analysis technique known as micro-Raman spectroscopy. Microplastics in samples may also be quantitatively analysed using micro-Raman spectroscopy. However, there can only be a maximum of 5,000 micro-Raman spectroscopy particles due to the technological ability to differentiate between particles of different sizes. Consequently, there's a good chance that the quantity of microplastics in samples measured by micro-Raman spectroscopy will be inaccurate (Tsering et al., 2022).

Overall, the fast growth of this sector means that spectroscopy will eventually become an essential technique for the simultaneous chemical characterization and quantification of microplastics.

7.4 MASS SPECTROMETRY

Microplastics can be analysed using liquid and gas chromatography in conjunction with mass spectrometry, which look at the unique products that are created when microplastics are hydrolysed or pyrolysed (Zhou et al., 2018; Li et al., 2021). Because of their excellent sensitiveness, such approaches, especially those involving nanoplastics, have enormous potential for the identification and measurement of microplastics.

7.4.1 Inductively Coupled Plasma Mass Spectroscopy

Furthermore, inductively coupled plasma mass spectroscopy (ICP-MS), a novel mass spectrometry technique for measuring microplastics, was made possible by the advancement of isotope tagging technology. As an alternate method for detecting particles containing carbon, inductively coupled plasma mass spectroscopy (ICP-MS) has recently gained popularity when used in single-particle mode (SP-ICP-MS) (Bolea-Fernandez et al., 2020; Laborda et al., 2021). This innovative method allows for a quantitative analysis of the size and quantity of nanoparticles (Bolea-Fernandez et al., 2020; Sakanupongkul et al., 2024). Nowadays, metal-based nanoparticles in environmental matrices are commonly studied using this technique (Cervantes-Avilés et al., 2021). Using this method, plastic particles as small as 1 μm may be detected at concentrations of 100 particles/mL. This methodology has been utilized in the identification of microplastics present in cosmetics, their extraction from food packaging (Laborda et al., 2021), and their photoaging due to UV radiation (Liu et al., 2023). SP-ICP-MS may be used to track 13C throughout the process in order to measure the concentrations and sizes of model microplastic particles (Laborda et al., 2021). Accordingly, Liu et al. (2021) analysed quantitatively the quantity of particles produced during the photoaging of polystyrene microplastics using SP-ICP-MS. These particles had a wide range of sizes and great environmental importance, down to 7.1×10^6 particles/L. The use of ICP-MS to explore the dynamics of microplastics during the course of ageing shows great potential.

7.4.2 PYROLYSIS GAS CHROMATOGRAPHY–MASS SPECTROMETRY

Pyrolysis gas chromatography is primarily utilized for the chemical characterization of polymers whose large size makes them impossible to characterize using liquid or gas chromatography. These complex polymers are reduced to simpler polymers by the process of pyrolysis, or controlled heat decomposition, which enables mass spectrometry to identify and sort the smaller molecules. This method has long been applied to environmental samples. In recent times, there has been a significant surge in the chemical characterization of microplastics found in environmental samples using this approach (Bouzid et al., 2022). Present-day pyrolysis entails thermally breaking down the analysed material at a temperature ranging from 500 to 1400°C. In these circumstances, the original macromolecule breaks down into characteristic units that are then further separated chromatographically using fused silica capillary columns. The units are then identified by mass spectrometry, primarily with the help of mass spectral libraries or by the selection of characteristic ions of indicator compounds, in the presence of an inert gas. The majority of pre-treatments used to extract (macro)molecules from environmental samples can be applied to this analytical technique, which can perform the analysis directly on the macromolecules and/or polymers whether they are in a liquid or solid state (Picó et al., 2020). This can eliminate the need for the sample to be pretreated. This is what has caused it to be used in this kind of matrix. Since Pyr-GC/MS enables the simultaneous study of several polymers in the entire sample and the initial results are encouraging, it is being suggested more and more for the quantitative investigation of microplastics in environmental matrices. Nevertheless, the existence of left-over mineral and organic compounds remains a significant limitation even if sample purification procedures are becoming increasingly effective in preparing solid materials like sediments. This technique is limited to identifying microplastics that are polystyrene (PS) derivatives. To find molecular markers for different kinds of plastics, more research is required (Watteau et al., 2018).

A different technique for evaluating pyrolysed gases in polymers is pyrolysis–gas chromatography mass spectrometry (Pyr-GC-MS) (Ceccarini et al., 2018). To ascertain whether a sample is plastic, its pyrogram can be compared to the findings of a recognized polymer standard. A small quantity of plastic (particle) samples, ranging from 0.3 to 7 mg is pyrolysed at temperatures over 700°C, a temperature higher than that used in TGA. The samples are then divided and transferred to GC-MS for analysis. The presence of PVC, PS, polyvinyl acetate (PVA), poly(acrylonitrile butadiene styrene), and styrene butadiene rubber (SBR) is shown by bulk analysis of sediment and suspended solid particles.

7.4.3 Liquid Chromatography–Tandem Mass Spectrometry

LC-MSMS is a dependable technique used in pharmacology, medicine, and the study of environmental exposures to minute heat-labile, non-volatile substances that may identify a limited number of polymers (Ng et al., 2020). As an example, microplastics based on PET, polycarbonate, and nylon have been successfully detected and quantitatively analysed in sewage sludge using LC-MS/MS (Zhang et al., 2019; Peng et al., 2020; Zhang et al., 2021). Due to the requirement to depolymerize macromolecules prior to examination, this approach can be damaging. Instead of providing information on the quantity, size, colour, and shape of microplastics and nanoplastics, the detection technique provides information on the mass and number of monomers released during depolymerization. Information on the size profile distribution may also be obtained by applying the approach with size fractionation. Even though combinations of many polymer types are frequently found in microplastics, most polymers lack the characteristic breakdown products needed for mass spectrometry analysis (Peng et al., 2020; Li et al., 2021; Zhang et al., 2021). Thus, it is currently not possible to determine the total amount of microplastics in the environment using mass spectrometry.

7.5 THERMAL ANALYSIS

The characteristics of a material are affected by both time and temperature. One of the most important instruments for material research is thermal analysis, which looks at the functional connection of this shift. Environmental sample materials are heated before being used in thermal analysis to look at microplastics. The high heat absorption of microplastics causes the polymers to gradually transition from a solid to a liquid or gas state as the temperature rises. At a specific temperature, an endothermic peak then develops. Due to the fact that various polymer types have varying thermal stabilities, it is possible to evaluate the composition, type, and additions of microplastics using their unique thermograms (Majewsky et al., 2016).

7.5.1 Differential Scanning Calorimetry

A useful technique for material analysis that looks at the thermal features of plastic polymers is differential scanning calorimetry (DSC) (Tsukame et al., 1997). According to Castañeda et al. (2014), DSC is now often used to identify polyethylene and other significant microplastics in the environment. Despite differential scanning calorimetry's restricted applicability, it is frequently used in combination with other methodologies. In order to support the physical characteristics of polymers, it displays differences in dissolution, crystallization, transition temperatures, and associated enthalpy and entropy (Müsellim et al., 2018). DSC may be used to distinguish between different polymer types since each plastic product has unique properties (Zainuddin and Syuhada, 2020). When DSC detects microplastics with comparable melting points, there are restrictions because of overlapping peaks (Rodriguez Chialanza et al., 2018). It is limited to identifying specific main microplastics, such as PE and PP (Bitter and Lackner, 2021); it is only able to detect the derivatives of these microplastics. To find molecular markers for different kinds of plastics, more research is required (Watteau et al., 2018).

7.5.2 Thermogravimetric Analysis

The determination of microplastics in environmental samples has recently been largely dependent on thermogravimetric analysis (TGA) methods, particularly those that include linking with other analytical tools (also referred as TGA-based techniques) (Becker et al., 2020; La Nasa et al., 2020). TGA and TGA-based procedures are significantly different from the other variations in that they

Advanced Instrumentation for Quantification of Microplastics 113

can assess a sample's weight loss while heating it at a regulated rate. TGA generally has well-known benefits, including simplicity, speed, and adaptability to different sample sizes. The thermal characteristics of polymers have long been studied using TGA. Combining TGA with analytical tools like GC-MS or FTIR spectroscopy may be very effective in characterizing polymers according to the products of their gaseous degradation (Liu et al., 2021).

For instance, it might be challenging to discern between the phase transition signals of polyurethane (PU) and polyethylene terephthalate (PET) (Majewsky et al., 2016; Sun et al., 2019). First of all, certain copolymers are challenging to identify by thermal analysis because of the propensity of polymer branching and other contaminants in microplastics to affect the polymer's transition temperature. Furthermore, because heat analysis ruins environmental samples, it cannot be used to describe the structure and appearance of microplastics. Because of this, thermal analysis is often employed to determine the chemical composition of microplastics or to measure their amount. This can restrict the applications of thermal analysis (Rocha-Santos and Duarte, 2015; Shim et al., 2017; Silva et al., 2018). Appropriate analysis should be the starting point for any MP evaluation. The previously stated TGA-based techniques may provide complicated data, necessitating the need for specialist staff for both operation and analysis. This thus highlights the need for a different approach that is easy to use, reliable, yields easily analysed data, and makes use of readily accessible tools and databases. The detection capabilities of stand-alone TGA have not been utilized in the framework of MP evaluation, as evidenced by its non-application in the identification or quantification of microplastics in environmental samples (Mansa and Zou, 2021).

7.6 EMERGING INNOVATIONS

The current difficulties in identifying microplastics should be overcome by developing new analytical tools and combining cutting-edge detection technologies with already-in-use instruments. The maximum detectable size is one of the problems in microplastic analysis that remains to be resolved. The minimum observable size limit for the current analytical methods is several micrometres. It is more important than ever to comprehend the presence, distribution, destiny, and toxicity of nanoscale plastics since smaller plastic particles can have more harmful consequences. As a result, protocols for nanoplastics sampling, extraction, purification, and concentration should be established with novel identification methods.

7.6.1 Atomic Force Microscopy

Atomic force microscopy (AFM) together with Raman spectroscopy and infrared (IR) might be a useful tool for researching nano- and microplastics. The AFM sensor interacts with material with or without contact and the method produces pictures with nanoscale resolution. The chemical makeup of the desired material can be revealed by Raman spectroscopy or AFM in conjunction with IR (Luo et al., 2021). The AFM cantilever vibrates as a result of the sample thermal expansion brought on by IR absorption. FTIR is used to analyse the ring-down pattern in order to calculate the vibration frequency and amplitude (Luo et al., 2020). AFM and IR spectra of PS beads measuring 100 nm were successfully acquired in prior work. Identifying a single target nanoscale polymer component in an unknown material to focus on using AFM-IR is difficult and time-consuming. One of the labour-intensive analytical steps in conventional microplastic analysis approaches is the search for plastic fragments. The pace at which organic and inorganic particles are removed throughout the separation and purification phases dictates how long the pretreatment procedure will take. In particular, minute, clear plastic particles may go unnoticed while being manually identified. These problems can be solved by automated Raman mapping and plastic fragment tracing with Raman spectroscopy; however, both methods need costly equipment, which is beyond the budget of many microplastics research centres.

7.6.2 Nile Red Staining

An easy staining method might be employed as a workaround for the problem of small, clear particles. Eosin B Hostasol Yellow 3G, Oil Red EGN, and Rose Bengal have little use, despite attempts (Prata et al., 2019). Conversely, Nile Red is an effective fluorescent dye that may be used to precisely colour extremely hydrophobic microplastics, such as 9-diethylamino-5H-benzophenoxazine-5-one, generally used in lipids staining that are physiologically neutral. Nile Red binds selectively to neutral lipids and fluoresces only in a hydrophobic environment. The Nile Red staining method offers quick staining periods (around 10–30 min) and strong recovery rates (up to 96%). If necessary, a quick bleach wash is also carried out. This provides a solid starting point for a more extensive spectroscopic analysis and works well in finding hidden microplastics (Simmerman et al., 2020). By connecting a fluorescent filter to the FTIR microscope after fluorescence microscopy, the identical particle spectrum identification may be instantaneously verified. In situ sample identification may be less likely to overlook microplastics if fluorescence microscopy is combined with FTIR confirmation following Nile Red staining (Sancataldo et al., 2020). Furthermore, it is faster than spectroscopy in recognizing all particles that resemble plastic (Vermeiren et al., 2020). The potential for co-staining natural chemical compounds is one of the main drawbacks of using Nile Red staining on in situ materials. Prior to applying the Nile Red stain, the sample must also be well cleaned. During the pretreatment stage, completely eliminating organic material can be challenging and time-consuming. Efforts have also been made to overcome this by using H_2O_2 to remove organic molecules or, in the instance of Li et al. (2020), a density separation technique including Fenton reagent or NaCl. The wide range of densities found in plastics, however, restricts how efficient these methods may be.

7.6.3 Near-infrared Spectra

Near-IR (NIR) spectroscopic analysis of microplastics has also been investigated. According to Zhang et al. (2018), FTIR studies the spectrum between 600 and 4000 cm^{-1}, whereas NIR analyses the spectra throughout 4000–15,000 cm^{-1}. N–H, O–H, and C–H combinations of X–H chemical vibrations are often used in the investigation of NIR spectra. Additionally, NIR analysis is insensitive and difficult to use for quantitative analysis. Still, it is capable of classifying and analysing large amounts of data from plastic samples rapidly. Therefore, rather than focusing on sample size, it is better to use this analytic method to identify the kind of sample (Paul et al., 2019).

7.6.4 Visible Near Infrared

Vis-NIR spectroscopy, which measures the quantity of light reflected from the sample's surface in the 350–2500 nm wavelength range, is used to calculate the reflectance for each wavelength. This analytical technique may be used to measure microplastics based on the material's chemical composition (Corradini et al., 2019). It is possible to identify microplastics like low-density PET and PVC by using an easily accessible library of vis-NIR spectrum data that includes many polymers that are generally present in the environment. However, biological particles might still be wrongly recognized as plastic since it depends on visual identification, thus, human judgement must be utilized.

7.6.5 Nano Thermal

High spatial resolution local thermal analysis is integrated with nano thermal analysis (nano-TA), a technique for creating AFM images using nano-TA probes. This allows for less than 100 nm spatial precision in the knowledge of the thermal behaviour of materials. When a location of interest is chosen for nano-TA investigation, the probe travels to a fixed spot on the sample surface. The probe exhibits high sensitivity with respect to the microplastic toughness. When the sample surface achieves the glass transition temperature, the tip ceases to heat at a steady pace. Since the specimen softens following state inversion, it is primarily utilized to evaluate the glass transition temperature and investigate the nanoscale surface characteristics of microplastics. Additionally, it is a technique wherein the probe permeates the sample (Luo et al., 2021). Recently, a research study employing nano-TA was carried out to examine how the qualities of microplastics influenced by the advanced oxidation process and the attributes of TiO_2-dyed microplastics impacted the ageing process (Luo et al., 2020; Luo et al., 2021.) Understanding the chemical nature and physical condition of each microplastic region may be aided by knowing the thermal characteristics of particular regions, such as glass temperature, which can be obtained via the nano-TA study of microplastics. The phase transition temperature of semi-crystalline and amorphous polymers, including polycaprolactone (PCL), PET, polymethyl methacrylate (PMMA), LDPE, polyoxymethylene (POM) PS, and related materials, has been studied by a nano-TA analysis technique (Guen et al., 2020).

7.6.6 X-ray Diffraction

Several types of microplastics were examined using X-ray diffraction (XRD) patterns. With a highest peak intensity of 2θ of 25.7°, PET exhibits large peaks and very poor crystalline quality. PP is a powerful, sharply peaked crystalline chemical. Its greatest diffraction peak is 31.2°, and its 2θ drop is 34.3°. PE is naturally crystalline and exhibits highest diffraction peak intensities (i.e., 21.6°, 24.05°, and 27.5°). These three are all prominent peaks. In PS, the broadness of the peaks indicates the poor crystalline nature of the material. At a wavelength, PS achieves its greatest peak intensity of 22.6°. Despite the fact that the PBT microplastic polymer is amorphous and lacks a distinct peak, the pattern is noisy (Thakur et al., 2023). Employing XRD, three different types of microplastics (PE, PVC, and PS) have been analysed. The findings revealed that while PVC had no sharp peak and PS had wide peaks that suggested poor crystalline nature, PE had two prominent peaks at 2θ of 21.1° and 23.4° (Ezeonu et al., 2019; Moura et al., 2023).

7.6.7 Atmospheric Solid Analysis Probe

Although mass spectrometry (MS) has been shown to be a useful tool for characterizing synthetic polymers, its application to the study of single-particle microplastics (MPs) has not yet received much attention. While MPs are currently considered common contaminants, there is a dearth of complete data on their prevalence since techniques that allow for a detailed characterization of MPs are not widely accessible. This work proposes the use of an air solid analysis probe (ASAP) coupled to a tiny quadrupole MS for the chemical analysis of single-particle microplastics, maintaining full compatibility with complementary staining and image analysis techniques. A two-stage ASAP probe temperature schedule was developed in order to eliminate additives and surface contaminants prior to carrying out the actual polymer characterization. The method revealed specific mass spectra with a range of single-particle MPs, such as polyolefins, polyaromatics, polyacrylates, (bio) polyesters, polyamides, polycarbonates, and polyacrylonitriles. It was found that the single-particle size detection limits for polystyrene MPs were 30 and 5 μm, respectively, in full-scan and selective ion recording modes. Furthermore, a multimodal method for microplastic characterization is presented and its results discussed. Using this method, filtered particles are first characterized using staining and fluorescence microscopy, and then each particle is individually selected using a probe to facilitate analysis by ASAP-MS. The method provides a detailed description of microplastic contamination, including details on the amount, size, form, and chemical makeup of the particles. Vitali et al. (2022) conducted a study on microplastics in bioplastic water bottles, which effectively demonstrated the practicality of the created multimodal approach. Advantages and disadvantages of the aforementioned microplastics identification techniques are summarized in Table 7.1.

TABLE 7.1

Advantages and Disadvantages of the Currently Used Microplastics Identification Techniques

S. no.	Identification Techniques	Advantages	Disadvantages
1.	Visual examination	A conventional method for identifying and quantifying microplastics The advantages of visual analysis are easy to use, inexpensiveness, and low chemical threat	Visual inspection is a time-consuming and laborious technique. Furthermore, when microplastic particle sizes are too fine, or pollutants in samples include organic and inorganic particles, the physical examination approach is no longer appropriate The visual analysis approach has a poor level of accuracy and efficiency Visual examination cannot reveal the chemical composition of microplastics Visual analysis is typically employed as an adjunct approach for microplastics investigation
2.	SEM-EDS	A potentially beneficial approach for concurrently analysing the surface shape and elemental composition of microplastics	The pretreatment procedure is complex Work efficiency is low, and it is an expensive method. The amount of microplastics estimated is not particularly accurate. The colour of microplastics cannot be discriminated effectively. This technique is mostly used to detect particular microplastics
3.	FTIR spectroscopy	A type of vibrational spectroscopy that provides information on chemical bonds and functional groups in materials. This technique has been widely employed in qualitative finding and microplastic evaluation	It is only useful for identifying microplastics larger than 20 μm Furthermore, this approach is readily influenced by different circumstances
4.	Raman spectroscopy	Alternative approach to vibration spectroscopy using inelastic light scattering. Furthermore, Raman spectroscopy may be utilized to discover microplastics smaller than 20 μm Furthermore, there is no need to dry or dehydrate materials prior to detection. This method is frequently used in the qualitative detection and evaluation of microplastics	Raman spectroscopy has a comparatively lengthy detection time. Furthermore, Raman spectroscopy has to be enhanced in the study of microplastics

(Continued)

TABLE 7.1 (Continued)
Advantages and Disadvantages of the Currently Used Microplastics Identification Techniques

S. no.	Identification Techniques	Advantages	Disadvantages
5.	Thermal analysis	A technique for analysing components that involves examining their properties as a function of time and temperature This method may be used to determine the chemical composition and bulk concentration of microplastics	The pretreatment of sample procedure is time-consuming. And because this approach is damaging to environmental materials, it cannot be used to analyse the physical characteristics of microplastics
6.	Mass spectrometry	A critical approach for detecting polymers in microplastics Polymer mass spectra can provide useful information on polymer structure, molecular weight, polymerization degree, main functional groups, and end group structure. This approach may be used to analyse chemical characteristics and quantify microplastics in environmental samples	This method's applicability domain is limited Furthermore, this method remains ineffective in quantifying the entire microplastics in the surrounding environment

7.7 CONCLUDING REMARKS AND FUTURE PERSPECTIVES

The identification of microplastics in intricate environmental matrices will involve the application of many combinations of microplastic analytical techniques. The longer it takes to identify microplastics, the smaller they are. Assessing the dangers and effects of microplastics on the environment and human health is increasingly dependent on sub-micron studies. It will be essential to develop new techniques and enhance existing ones in order to decrease the identification time and effort as the need for microplastic contamination monitoring expands. Developing precise and practical identification methods will also be essential for identifying and quantifying nanoplastics in environmental sample data. Future studies should concentrate on developing a completely or partially automated analytical method that can combine chemical analysis to identify the individual plastic components with image analysis-based approaches to identify the physical characteristics (such as shape and size) of microplastics. Furthermore, synthetic cellulose fibres, known as polyester in Europe and rayon in the United States, pollute the environment in a fashion that is remarkably comparable to that of microplastics. Consequently, while researching microplastics, this should also be taken into consideration. In order to avoid this mistake, the characteristics that separate microplastics from comparable materials should be investigated. It is also important for research to keep developing new dyeing methods, nanotechnology, and analytical tools to find and remove microplastics from environmental sample data.

REFERENCES

Zahra Taghizadeh Rahmat Abadi, Behrooz Abtahi, Hans-Peter Grossart Hans-Peter Grossart, Saber Khodabandeh. "Microplastic content of Kutum fish, Rutilus frisii kutum in the southern Caspian Sea." *Science of the Total Environment* 752 (2021): 141542.

Abbasi, Sajjad. "Prevalence and physicochemical characteristics of microplastics in the sediment and water of Hashilan Wetland, a national heritage in NW Iran." *Environmental Technology & Innovation* 23 (2021): 101782.

Aysun Adan, Günel Alizada, Yağmur Kiraz, Yusuf Baran, Ayten Nalbant. Flow cytometry: Basic principles and applications. *Critical Reviews in Biotechnology* 37, no. 2 (2017): 163–176.

Andoh, Collins Nana, Francis Attiogbe, Nana Osei Bonsu Ackerson, Mary Antwi, and Kofi Adu-Boahen. "Fourier Transform Infrared Spectroscopy: An analytical technique for microplastic identification and quantification." *Infrared Physics & Technology* 136 (2023): 105070.

Araujo, Catarina F., Mariela M. Nolasco, Antonio M. P. Ribeiro, and Paulo J. A. Ribeiro-Claro. "Identification of microplastics using Raman spectroscopy: Latest developments and future prospects." *Water Research* 142 (2018): 426–440.

Becker, Roland, Korinna Altmann, Thomas Sommerfeld, and Ulrike Braun. "Quantification of microplastics in a freshwater suspended organic matter using different thermoanalytical methods–outcome of an interlaboratory comparison." *Journal of Analytical and Applied Pyrolysis* 148 (2020): 104829.

Bittelli, Marco, Sergio Pellegrini, Roberto Olmi, Maria Costanza Andrenelli, Gianluca Simonetti, Emilio Borrelli, and Francesco Morari. "Experimental evidence of laser diffraction accuracy for particle size analysis." *Geoderma* 409 (2022): 115627.

Bitter, Hajo, and Susanne Lackner. "Fast and easy quantification of semi-crystalline microplastics in exemplary environmental matrices by differential scanning calorimetry (DSC)." *Chemical Engineering Journal* 423 (2021): 129941.

Blott, Simon J., Debra J. Croft, Kenneth Pye, Samantha E. Saye, and Helen E. Wilson. "Particle size analysis by laser diffraction." *Geological Society, London, Special Publications* 232, no. 1 (2004): 63–73.

Bolea-Fernandez, Eduardo, Ana Rua-Ibarz, Milica Velimirovic, Kristof Tirez, and Frank Vanhaecke. "Detection of microplastics using inductively coupled plasma-mass spectrometry (ICP-MS) operated in single-event mode." *Journal of Analytical Atomic Spectrometry* 35, no. 3 (2020): 455–460.

Bouzid, Nadia, Christelle Anquetil, Rachid Dris, Johnny Gasperi, Bruno Tassin, and Sylvie Derenne. "Quantification of microplastics by pyrolysis coupled with gas chromatography and mass spectrometry in sediments: Challenges and implications." *Microplastics* 1, no. 2 (2022): 229–239.

Brandes, Elke, Martin Henseler, and Peter Kreins. "Identifying hot-spots for microplastic contamination in agricultural soils a spatial modelling approach for Germany." *Environmental Research Letters* 16, no. 10 (2021): 104041.

Castañeda, Rowshyra A., Suncica Avlijas, M. Anouk Simard, and Anthony Ricciardi. "Microplastic pollution in St. Lawrence river sediments." *Canadian Journal of Fisheries and Aquatic Sciences* 71, no. 12 (2014): 1767–1771.

Ceccarini, Alessio, Andrea Corti, Francesca Erba, Francesca Modugno, Jacopo La Nasa, Sabrina Bianchi, and Valter Castelvetro. "The hidden microplastics: new insights and figures from the thorough separation and characterization of microplastics and of their degradation byproducts in coastal sediments." *Environmental Science & Technology* 52, no. 10 (2018): 5634–5643.

Cervantes-Avilés, Pabel, and Arturo A. Keller. "Incidence of metal-based nanoparticles in the conventional wastewater treatment process." *Water Research* 189 (2021): 116603.

Chalmers, John M. "Infrared spectroscopy in analysis of polymers and rubbers." *Encyclopedia of Analytical Chemistry: Applications, Theory and Instrumentation* (2006). https://doi.org/10.1002/9780470027 318.a2015

Chen, Yiyang, Dishi Wen, Jianchuan Pei, Yufan Fei, Da Ouyang, Haibo Zhang, and Yongming Luo. "Identification and quantification of microplastics using Fourier-transform infrared spectroscopy: Current status and future prospects." *Current Opinion in Environmental Science & Health* 18 (2020): 14–19.

Corradini, Fabio, Harm Bartholomeus, Esperanza Huerta Lwanga, Hennie Gertsen, and Violette Geissen. "Predicting soil microplastic concentration using vis-NIR spectroscopy." *Science of the Total Environment* 650 (2019): 922–932.

de Souza Machado, Anderson Abel, Chung Wai Lau, Jennifer Till, Werner Kloas, Anika Lehmann, Roland Becker, and Matthias C. Rillig. "Impacts of microplastics on the soil biophysical environment." *Environmental Science & Technology* 52, no. 17 (2018): 9656–9665.

Dehghani, Sharareh, Farid Moore, and Razegheh Akhbarizadeh. "Microplastic pollution in deposited urban dust, Tehran metropolis, Iran." *Environmental Science and Pollution Research* 24 (2017): 20360–20371.

Dowarah, Kaushik, and Suja P. Devipriya. "Microplastic prevalence in the beaches of Puducherry, India and its correlation with fishing and tourism/recreational activities." *Marine Pollution Bulletin* 148 (2019): 123–133.

Ezeonu, S., E. Egonnaya, and C. Nweze. "Electrical and structural properties of polystyrene/graphite composite by direct mixing method." *Advances in Physics Theories and Applications* 81 (2019): 11–15.

Gaylarde, Christine C., José Antonio Baptista Neto, and Estefan Monteiro da Fonseca. "Nanoplastics in aquatic systems-are they more hazardous than microplastics?." *Environmental Pollution* 272 (2021): 115950.

Geyer, Roland, Jenna R. Jambeck, and Kara Lavender Law. "Production, use, and fate of all plastics ever made." *Science Advances* 3, no. 7 (2017): e1700782.

Gu, Yuwei, Julia Zhao, and Jeremiah A. Johnson. "Polymer networks: From plastics and gels to porous frameworks." *Angewandte Chemie International Edition* 59, no. 13 (2020): 5022–5049.

Guen, Eloise, Petr Klapetek, Robert Puttock, Bruno Hay, Alexandre Allard, Tony Maxwell, Pierre-Olivier Chapuis et al. "SThM-based local thermomechanical analysis: Measurement intercomparison and uncertainty analysis." *International Journal of Thermal Sciences* 156 (2020): 106502.

Guo, Jing-Jie, Xian-Pei Huang, Lei Xiang, Yi-Ze Wang, Yan-Wen Li, Hui Li, Quan-Ying Cai, Ce-Hui Mo, and Ming-Hung Wong. "Source, migration and toxicology of microplastics in soil." *Environment International* 137 (2020): 105263.

Guo, Xuan, and Jianlong Wang. "The chemical behaviors of microplastics in marine environment: A review." *Marine Pollution Bulletin* 142 (2019): 1–14.

Hendrickson, Erik, Elizabeth C. Minor, and Kathryn Schreiner. "Microplastic abundance and composition in western Lake Superior as determined via microscopy, Pyr-GC/MS, and FTIR." *Environmental Science & Technology* 52, no. 4 (2018): 1787–1796.

Hossain, Mohammed R., Miao Jiang, QiHuo Wei, and Laura G. Leff. "Microplastic surface properties affect bacterial colonization in freshwater." Journal of Basic Microbiology 59, no. 1 (2019): 54–61.

Kaile, Namrata, Mathilde Lindivat, Javier Elio, Gunnar Thuestad, Quentin G. Crowley, and Ingunn Alne Hoell. "Preliminary results from detection of microplastics in liquid samples using flow cytometry." *Frontiers in Marine Science* 7 (2020): 552688.

Käppler, Andrea, Frank Windrich, Martin G. J. Löder, Mikhail Malanin, Dieter Fischer, Matthias Labrenz, Klaus-Jochen Eichhorn, and Brigitte Voit. "Identification of microplastics by FTIR and Raman microscopy: a novel silicon filter substrate opens the important spectral range below 1300 cm− 1 for FTIR transmission measurements." *Analytical and Bioanalytical Chemistry* 407 (2015): 6791–6801.

Kniggendorf, Ann-Kathrin, Christoph Wetzel, and Bernhard Roth. "Microplastics detection in streaming tap water with Raman spectroscopy." *Sensors* 19, no. 8 (2019): 1839.

La Nasa, Jacopo, Greta Biale, Daniele Fabbri, and Francesca Modugno. "A review on challenges and developments of analytical pyrolysis and other thermoanalytical techniques for the quali-quantitative determination of microplastics." *Journal of Analytical and Applied Pyrolysis* 149 (2020): 104841.

Laborda, Francisco, Celia Trujillo, and Ryszard Lobinski. "Analysis of microplastics in consumer products by single particle-inductively coupled plasma mass spectrometry using the carbon-13 isotope." *Talanta* 221 (2021): 121486.

Lebreton, Laurent, and Anthony Andrady. "Future scenarios of global plastic waste generation and disposal." *Palgrave Communications* 5, no. 1 (2019): 1–11.

Li, Chengjun, Yan Gao, Shuai He, Hai-Yuan Chi, Ze-Chen Li, Xiao-Xia Zhou, and Bing Yan. "Quantification of nanoplastic uptake in cucumber plants by pyrolysis gas chromatography/mass spectrometry." *Environmental Science & Technology Letters* 8, no. 8 (2021): 633–638.

Li, Huiru, Mark J. La Guardia, Hehuan Liu, Robert C. Hale, T. Matteson Mainor, Ellen Harvey, Guoying Sheng, and Jiamo Fu. "Brominated and organophosphate flame retardants along a sediment transect encompassing the Guiyu, China e-waste recycling zone." *Science of the Total Environment* 646 (2019): 58–67.

Li, Xiaowei, Lubei Chen, Yanyan Ji, Man Li, Bin Dong, Guangren Qian, John Zhou, and Xiaohu Dai. "Effects of chemical pretreatments on microplastic extraction in sewage sludge and their physicochemical characteristics." *Water Research* 171 (2020): 115379.

Lithner, Delilah, Åke Larsson, and Göran Dave. "Environmental and health hazard ranking and assessment of plastic polymers based on chemical composition." *Science of the Total Environment* 409, no. 18 (2011): 3309–3324.

Liu, Yi, Ruojia Li, Jianping Yu, Fengli Ni, Yingfei Sheng, Austin Scircle, James V. Cizdziel, and Ying Zhou. "Separation and identification of microplastics in marine organisms by TGA-FTIR-GC/MS: A case study of mussels from coastal China." *Environmental Pollution* 272 (2021): 115946.

Liu, Ziyi, Yanjie Zhu, Shangsi Lv, Yuxian Shi, Shuofei Dong, Dong Yan, Xiaoshan Zhu, Rong Peng, Arturo A. Keller, and Yuxiong Huang. "Quantifying the dynamics of polystyrene microplastics UV-aging process." *Environmental Science & Technology Letters* 9, no. 1 (2021): 50–56.

Liu, Hang, Xian Zhang, Bin Ji, Zhimin Qiang, Tanju Karanfil, and Chao Liu. "UV aging of microplastic polymers promotes their chemical transformation and byproduct formation upon chlorination." *Science of The Total Environment* 858 (2023): 159842.

Löder, Martin G. J., and Gunnar Gerdts. "Methodology used for the detection and identification of microplastics— A critical appraisal." In: Bergmann, M., Gutow, L., Klages, M. (eds), *Marine Anthropogenic Litter* (2015): 201–227. Springer, Cham. https://doi.org/10.1007/978-3-319-16510-3_8

Luo, Hongwei, Yahui Xiang, Yaoyao Zhao, Yu Li, and Xiangliang Pan. "Nanoscale infrared, thermal and mechanical properties of aged microplastics revealed by an atomic force microscopy coupled with infrared spectroscopy (AFM-IR) technique." *Science of the Total Environment* 744 (2020): 140944.

Luo, Hongwei, Yahui Xiang, Yu Li, Yaoyao Zhao, and Xiangliang Pan. "Photocatalytic aging process of Nano-TiO2 coated polypropylene microplastics: Combining atomic force microscopy and infrared spectroscopy (AFM-IR) for nanoscale chemical characterization." *Journal of Hazardous Materials* 404 (2021): 124159.

Luo, Hongwei, Yifeng Zeng, Yaoyao Zhao, Yahui Xiang, Yu Li, and Xiangliang Pan. "Effects of advanced oxidation processes on leachates and properties of microplastics." *Journal of Hazardous Materials* 413 (2021): 125342.

Majewsky, Marius, Hajo Bitter, Elisabeth Eiche, and Harald Horn. "Determination of microplastic polyethylene (PE) and polypropylene (PP) in environmental samples using thermal analysis (TGA-DSC)." *Science of the Total Environment* 568 (2016): 507–511.

Mansa, Rola, and Shan Zou. "Thermogravimetric analysis of microplastics: A mini review." *Environmental Advances* 5 (2021): 100117.

Mecozzi, Mauro, Marco Pietroletti, and Yulia B. Monakhova. "FTIR spectroscopy supported by statistical techniques for the structural characterization of plastic debris in the marine environment: Application to monitoring studies." *Marine Pollution Bulletin* 106, no. 1–2 (2016): 155–161.

Morgado, Vanessa, Luís Gomes, Ricardo JN Bettencourt da Silva, and Carla Palma. "Validated spreadsheet for the identification of PE, PET, PP and PS microplastics by micro-ATR-FTIR spectra with known uncertainty." *Talanta* 234 (2021): 122624.

Mossotti, Raffaella, Giulia Dalla Fontana, Anastasia Anceschi, Enrico Gasparin, and Tiziano Battistini. "Preparation and analysis of standards containing microfilaments/microplastic with fibre shape." *Chemosphere* 270 (2021): 129410.

Moura, Diana S., Carlos J. Pestana, Colin F. Moffat, Jianing Hui, John T. S. Irvine, and Linda A. Lawton. "Characterisation of microplastics is key for reliable data interpretation." *Chemosphere* 331 (2023): 138691.

Müller, Yanina K., Theo Wernicke, Marco Pittroff, Cordula S. Witzig, Florian R. Storck, Josef Klinger, and Nicole Zumbülte. "Microplastic analysis are we measuring the same? Results on the first global comparative study for microplastic analysis in a water sample." *Analytical and Bioanalytical Chemistry* 412, no. 3 (2020): 555–560.

Müsellim, Ece, Mudassir Hussain Tahir, Muhammad Sajjad Ahmad, and Selim Ceylan. "Thermokinetic and TG/DSC-FTIR study of pea waste biomass pyrolysis." *Applied Thermal Engineering* 137 (2018): 54–61.

Ng, Keng Tiong, Helena Rapp-Wright, Melanie Egli, Alicia Hartmann, Joshua C. Steele, Juan Eduardo Sosa-Hernández, Elda M. Melchor-Martínez, et al. "High-throughput multi-residue quantification of contaminants of emerging concern in wastewaters enabled using direct injection liquid chromatography-tandem mass spectrometry." *Journal of Hazardous Materials* 398 (2020): 122933.

Paul, Andrea, Lukas Wander, Roland Becker, Caroline Goedecke, and Ulrike Braun. "High-throughput NIR spectroscopic (NIRS) detection of microplastics in soil." *Environmental Science and Pollution Research* 26 (2019): 7364–7374.

Peng, Chu, Xuejiao Tang, Xinying Gong, Yuanyuan Dai, Hongwen Sun, and Lei Wang. "Development and application of a mass spectrometry method for quantifying nylon microplastics in environment." *Analytical Chemistry* 92, no. 20 (2020): 13930–13935.

Piazza, Veronica, Abdusalam Uheida, Chiara Gambardella, Francesca Garaventa, Marco Faimali, and Joydeep Dutta. "Ecosafety screening of photo-fenton process for the degradation of microplastics in water." *Frontiers in Marine Science* 8 (2022): 791431.

Picó, Yolanda, and Damià Barceló. "Pyrolysis gas chromatography-mass spectrometry in environmental analysis: Focus on organic matter and microplastics." *TrAC Trends in Analytical Chemistry* 130 (2020): 115964.

Pirsaheb, Meghdad, Hooshyar Hossini, and Pouran Makhdoumi. "Review of microplastic occurrence and toxicological effects in marine environment: Experimental evidence of inflammation." *Process Safety and Environmental Protection* 142 (2020): 1–14.

Prata, Joana Correia, João P. Da Costa, Armando C. Duarte, and Teresa Rocha-Santos. "Methods for sampling and detection of microplastics in water and sediment: A critical review." *TrAC Trends in Analytical Chemistry* 110 (2019): 150–159.

Sebastian Primpke, Richard K Cross, Svenja M Mintenig, Marta Simon, Alvise Vianello, Gunnar Gerdts, Jes Vollertsen. "Toward the systematic identification of microplastics in the environment: evaluation of a new independent software tool (siMPle) for spectroscopic analysis." *Applied Spectroscopy*, 74 no. 9 (2020): 1127–1138.

Primpke, Sebastian, Marisa Wirth, Claudia Lorenz, and Gunnar Gerdts. "Reference database design for the automated analysis of microplastic samples based on Fourier transform infrared (FTIR) spectroscopy." *Analytical and Bioanalytical Chemistry* 410 (2018): 5131–5141.

Rocha-Santos, Teresa, and Armando C. Duarte. "A critical overview of the analytical approaches to the occurrence, the fate and the behavior of microplastics in the environment." *TrAC Trends in Analytical Chemistry* 65 (2015): 47–53.

Rodríguez Chialanza, Mauricio, Ignacio Sierra, Andrés Pérez Parada, and Laura Fornaro. "Identification and quantitation of semi-crystalline microplastics using image analysis and differential scanning calorimetry." *Environmental Science and Pollution Research* 25 (2018): 16767–16775.

Sabri, N. H., A. Muhammad, NH Abdul Rahim, A. Roslan, and AR Abu Talip. "Feasibility study on co-pyrolyzation of microplastic extraction in conventional sewage sludge for the cementitious application." *Materials Today: Proceedings* 46 (2021): 2112–2117.

Sakanupongkul, Apinya, Kalyanee Sirisinha, Rattaporn Saenmuangchin, and Atitaya Siripinyanond. "Analysis of microplastic particles by using single particle inductively coupled plasma mass spectrometry." *Microchemical Journal* 199 (2024): 110016.

Sancataldo, Giuseppe, Giuseppe Avellone, and Valeria Vetri. "Nile Red lifetime reveals microplastic identity." *Environmental Science: Processes & Impacts* 22, no. 11 (2020): 2266–2275.

Shim, Won Joon, Sang Hee Hong, and Soeun Eo Eo. "Identification methods in microplastic analysis: A review." *Analytical Methods* 9, no. 9 (2017): 1384–1391.

Silva, Ana B., Ana S. Bastos, Celine I. L. Justino, João P. da Costa, Armando C. Duarte, and Teresa AP Rocha-Santos. "Microplastics in the environment: Challenges in analytical chemistry-A review." *Analytica Chimica Acta* 1017 (2018): 1–19.

Simmerman, Claire B., and Jill K. Coleman Wasik. "The effect of urban point source contamination on microplastic levels in water and organisms in a cold-water stream." *Limnology and Oceanography Letters* 5, no. 1 (2020): 137–146.

Singh, Narendra, Huabo Duan, and Yuanyuan Tang. "Toxicity evaluation of E-waste plastics and potential repercussions for human health." *Environment International* 137 (2020): 105559.

Sobhani, Zahra, Yongjia Lei, Youhong Tang, Liwei Wu, Xian Zhang, Ravi Naidu, Mallavarapu Megharaj, and Cheng Fang. "Microplastics generated when opening plastic packaging." *Scientific Reports* 10, no. 1 (2020): 4841.

Song, Young Kyoung, Sang Hee Hong, Mi Jang, Gi Myung Han, and Won Joon Shim. "Occurrence and distribution of microplastics in the sea surface microlayer in Jinhae Bay, South Korea." *Archives of Environmental Contamination and Toxicology* 69 (2015): 279–287.

Sorasan, Carmen, Carlos Edo, Miguel González-Pleiter, Francisca Fernández-Piñas, Francisco Leganés, Antonio Rodríguez, and Roberto Rosal. "Generation of nanoplastics during the photoageing of low-density polyethylene." *Environmental Pollution* 289 (2021): 117919.

Sun, Jing, Xiaohu Dai, Qilin Wang, Mark C. M. Van Loosdrecht, and Bing-Jie Ni. "Microplastics in wastewater treatment plants: Detection, occurrence and removal." *Water Research* 152 (2019): 21–37.

Talvitie, Julia, Anna Mikola, Arto Koistinen, and Outi Setälä. "Solutions to microplastic pollution–Removal of microplastics from wastewater effluent with advanced wastewater treatment technologies." *Water Research* 123 (2017): 401–407.

Thakur, Babita, Jaswinder Singh, Joginder Singh, Deachen Angmo, and Adarsh Pal Vig. "Identification and characterization of extracted microplastics from agricultural soil near industrial area: FTIR and X-ray diffraction method." *Environmental Quality Management* 33, no. 1 (2023): 173–181.

Thompson, Richard C., Ylva Olsen, Richard P. Mitchell, Anthony Davis, Steven J. Rowland, Anthony W. G. John, Daniel McGonigle, and Andrea E. Russell. "Lost at sea: where is all the plastic?." *Science* 304, no. 5672 (2004): 838–838.

Tsering, Tenzin, Mirka Viitala, Maria Hyvönen, Satu-Pia Reinikainen, and Mika Mänttäri. "The assessment of particle selection and blank correction to enhance the analysis of microplastics with Raman microspectroscopy." *Science of the Total Environment* 842 (2022): 156804.

Tsukame, Takahiro, Yasushi Ehara, Yasuko Shimizu, Michio Kutsuzawa, Hideki Saitoh, and Yoshio Shibasaki. "Characterization of microstructure of polyethylenes by differential scanning calorimetry." *Thermochimica Acta* 299, no. 1–2 (1997): 27–32.

Tunali, Merve, Edwin Nnaemeka Uzoefuna, Mehmet Meric Tunali, and Orhan Yenigun. "Effect of microplastics and microplastic-metal combinations on growth and chlorophyll a concentration of Chlorella vulgaris." *Science of the Total Environment* 743 (2020): 140479.

Turner, Andrew, Claire Wallerstein, and Rob Arnold. "Identification, origin and characteristics of bio-bead microplastics from beaches in western Europe." *Science of the Total Environment* 664 (2019): 938–947.

Uddin, Saif, Scott W. Fowler, and Talat Saeed. "Microplastic particles in the Persian/Arabian Gulf-A review on sampling and identification." *Marine Pollution Bulletin* 154 (2020): 111100.

Vermeiren, P., C. Muñoz, and K. Ikejima. "Microplastic identification and quantification from organic rich sediments: A validated laboratory protocol." *Environmental Pollution* 262 (2020): 114298.

Vitali, Clementina, Hans-Gerd Janssen, Francesco Simone Ruggeri, and Michel W. F, Nielen. "Rapid single particle atmospheric solids analysis probe-mass spectrometry for multimodal analysis of microplastics." *Analytical Chemistry* 95, no. 2 (2022): 1395–1401.

Wagner, Jeff, Zhong-Min Wang, Sutapa Ghosal, Chelsea Rochman, Margy Gassel, and Stephen Wall. "Novel method for the extraction and identification of microplastics in ocean trawl and fish gut matrices." *Analytical Methods* 9, no. 9 (2017): 1479–1490.

Wagner, Jeff, Zhong-Min Wang, Sutapa Ghosal, Margaret Murphy, Stephen Wall, Anna-Marie Cook, William Robberson, and Harry Allen. "Nondestructive extraction and identification of microplastics from freshwater sport fish stomachs." *Environmental Science & Technology* 53, no. 24 (2019): 14496–14506.

Wang, Fayuan, Quanlong Wang, Catharine A. Adams, Yuhuan Sun, and Shuwu Zhang. "Effects of microplastics on soil properties: Current knowledge and future perspectives." *Journal of Hazardous Materials* 424 (2022): 127531.

Watteau, Francoise, Marie-France Dignac, Adeline Bouchard, Agathe Revallier, and Sabine Houot. "Microplastic detection in soil amended with municipal solid waste composts as revealed by transmission electronic microscopy and pyrolysis/GC/MS." *Frontiers in Sustainable Food Systems* 2 (2018): 81.

Weidner, Steffen M., and Sarah Trimpin. "Mass spectrometry of synthetic polymers." *Analytical Chemistry* 82, no. 12 (2010): 4811–4829.

Wright, Stephanie L., Joseph M. Levermore, and Frank J. Kelly. "Raman spectral imaging for the detection of inhalable microplastics in ambient particulate matter samples." *Environmental Science & Technology* 53, no. 15 (2019): 8947–8956.

Wright, Stephanie L., Joseph M. Levermore, and Frank J. Kelly. "Raman spectral imaging for the detection of inhalable microplastics in ambient particulate matter samples." *Environmental Science & Technology* 53, no. 15 (2019): 8947–8956.

Zainuddin, Zarlina, and Syuhada. "Study of analysis method on microplastic identification in bottled drinking water." *Macromolecular Symposia* 391, no. 1 (2020): 1900195.

Zhang, Junjie, Lei Wang, and Kurunthachalam Kannan. "Quantitative analysis of polyethylene terephthalate and polycarbonate microplastics in sediment collected from South Korea, Japan and the USA." *Chemosphere* 279 (2021): 130551.

Zhang, Junjie, Lei Wang, Rolf U. Halden, and Kurunthachalam Kannan. "Polyethylene terephthalate and polycarbonate microplastics in sewage sludge collected from the United States." *Environmental Science & Technology Letters* 6, no. 11 (2019): 650–655.

Zhang, Jixiong, Kuangda Tian, Chunli Lei, and Shungeng Min. "Identification and quantification of microplastics in table sea salts using micro-NIR imaging methods." *Analytical Methods* 10, no. 24 (2018): 2881–2887.

Zhou, Liling, Tiecheng Wang, Guangzhou Qu, Hanzhong Jia, and Lingyan Zhu. "Probing the aging processes and mechanisms of microplastic under simulated multiple actions generated by discharge plasma." *Journal of Hazardous Materials* 398 (2020): 122956.

Zhou, Qian, Haibo Zhang, Chuancheng Fu, Yang Zhou, Zhenfei Dai, Yuan Li, Chen Tu, and Yongming Luo. "The distribution and morphology of microplastics in coastal soils adjacent to the Bohai Sea and the Yellow Sea." *Geoderma* 322 (2018): 201–208.

Zhu, Fengxiao, Changyin Zhu, Chao Wang, and Cheng Gu. "Occurrence and ecological impacts of microplastics in soil systems: A review." *Bulletin of Environmental Contamination and Toxicology* 102 (2019): 741–749.

8 Microplastics in the Aquatic Environment
Analytical and Mitigation Methods

Md Abdullah Al Noman, S. Shanthakumar, Shishir Kumar Behera, S. Srinivasan, Eldon R. Rene, and Tejas R. Atri

8.1 INTRODUCTION

In aquatic ecosystems, plastic pollution takes various forms, ranging from conspicuous debris to almost invisible plastic fragments known as microplastics (MPs) (Chaukura et al., 2021). They infiltrate the aquatic biome in a multitude of ways, posing a significant threat to flora and fauna. Analyzing MPs in aquatic ecosystems necessitates a range of analytical methods that are specifically designed to detect, quantify, and characterize them (Sun et al., 2019). However, these methods have their own challenges and limitations, encompassing aspects like sample collection, identification, and protocols. A variety of techniques, such as visual identification, microscopy, spectroscopy, sieve filtration, and density separation, are deployed to scrutinize MPs (Prata et al., 2019). These methods enable researchers to understand their presence and attributes. However, the analytical landscape is far from uniform, and each technique possesses its strengths and weaknesses (Lv et al., 2021). Sample collection proves to be a formidable challenge due to the heterogeneous distribution of MPs in aquatic environments. Moreover, MPs' diminutive size and minimal concentration levels make them elusive targets. Fragmentation and degradation further complicate the matter, as do the persistent issues of distinguishing MPs from natural particles (Picó & Barceló, 2021).

There are various existing and potential mitigation strategies that have already been proposed to curb the spread and adverse effects of MPs. Preventing MPs from entering the environment is a primary mitigation strategy (Napper et al., 2021). This approach involves minimizing the use of plastics that readily degrade into MPs. Instead, replacing these materials with eco-friendly alternatives and promoting sustainable consumption habits should be followed (Chaukura et al., 2021). A significant portion of MPs enter the aquatic biome through wastewater discharge. Enhancement of wastewater treatment facilities to capture and remove MPs can be done (Kiran et al., 2020). A step forward would be to properly manage the sewage sludge in order to prevent its accumulation in the environment. Additionally, extended producer responsibilities schemes can hold manufacturers accountable for the entire lifecycle of their products, thus incentivizing them to reduce plastic pollution (Chaukura et al., 2021).

Emerging technologies, such as MP capture devices, biodegradable alternatives, and novel degradation methods such as nanozymes and enzymatic degradation, offer exciting prospects for managing and mitigating MP pollution (Li et al., 2019). Ecological studies provide valuable insights into the effects of MPs on aquatic ecosystems and guide targeted mitigation efforts (Napper et al., 2021). Policy makers also have a crucial role to play by implementing and enforcing strict regulations and policies (Onyena et al., 2021).

MPs in the Aquatic Environment: Analytical & Mitigation Methods   125

While the study of MPs in aquatic environments has progressed significantly, the analysis and development of effective mitigation strategies need to be further strengthened. Analytical challenges arise from the absence of universally accepted and standardized protocols for MP analysis, hindering the comparability of different results obtained (Prata et al., 2019). Current analysis methods often focus on specific size ranges or types of MPs, potentially leading to underestimations of contamination levels (Napper et al., 2021). On the mitigation front, assessing the real-world effectiveness of existing strategies and understanding their long-term environmental effects remain rather limited (Chaukura et al., 2021).

In response to these knowledge gaps, this chapter sets forth two primary objectives. First, a rigorous evaluation of existing MP detection methods, focusing on their performance and limitations was conducted. By comparing the accuracy, precision, and practicality of these methods, a comprehensive assessment is provided to assist researchers and practitioners in selecting appropriate techniques for precise MP quantification and characterization. Second, effective mitigation approaches for combating MP pollution are also elaborated.

8.2 SAMPLING METHODS FOR MICROPLASTICS IN THE AQUATIC ENVIRONMENT

8.2.1 SAMPLING METHODS OF MICROPLASTICS

Research on the abundance and characteristics of MPs has been carried out in different aquatic environments including seawater, freshwater, groundwater, and wastewater in different parts of the world over the last couple of decades; however, no standard approaches have been observed in the sampling methods, separation, purification, detection, and quantification (Gupta et al., 2023; Kye et al., 2023). Different sampling methods, separation techniques, and analysis tools were applied in these studies to detect and characterize MPs in different aquatic media such as water, sediment, and biota of different trophic levels (Ivleva et al., 2017; Stock et al., 2019; Wang & Wang, 2018; Yusuf et al., 2022; Zhao et al., 2022). The characteristics of MPs such as density, size, shape, and chemical properties including adsorption of chemicals and the ambient settings such as water density, waves, wind, and currents influence the distribution of MPs in the aquatic environment and the sampling methods (Gupta et al., 2023; Sajad et al., 2023). Since MPs are buoyant and persistent, they float and sustain in the aquatic environment. That is why sampling of MP is widely performed from the surface water at different depths as there is no standardized depth mentioned in the existing literature (Mai et al., 2018; Yu et al., 2016). However, taking samples only from surface water may mislead the research findings because MPs having additives and higher density ones sink to the bottom of the water body (Reisser et al., 2015). Therefore, researchers also consider sediment and aquatic biota as potential sources of MPs.

8.2.2 SAMPLING METHODS OF MICROPLASTICS FROM WATER

The most widely used tools for MP sampling in surface water are manta trawls and neuston nets. Bongo nets and benthic trawls are for the midwater level and seabed, respectively (Cole et al., 2011; Lee et al., 2014). Moreover, plankton nets, multiple opening–closing nets, and continuous plankton recorders were also reported as major sampling apparatuses of MPs in water columns (Crawford & Quinn, 2017; Silva et al., 2018). Water intake pumps and water collection bottles are also applied occasionally for MP sampling from surface water and water columns (Cole et al., 2011). The size of the mesh and the opening area of the sampling net play a crucial role in MP sampling from water. The usual size of the trawl mesh is between 100 and 500 µm, but the most widely used size is 300 µm for MP sampling (Enders et al., 2015). In a recent research, an aquatic drone (Jellyfishbot®) was utilized to sample MPs from surface water to develop a standardized method of sampling. This research compared the findings with a manta net and reported similar findings for shapes, colors, and abundance of MPs from the surface water of rivers and oceans in the specified sampling site (Pasquier et al., 2022).

8.2.3 SAMPLING METHODS OF MICROPLASTICS FROM SEDIMENTS

The porous media/sediments have been considered as the long-standing sink of MPs (Woodall et al., 2014). Sediment is generally sampled from the shoreline or water bottom/sea floor. To observe MP contamination, sediments have been widely sampled on the beach or along the shoreline across the globe due to the convenient nature of sampling (da Costa et al., 2023; Fok & Cheung, 2015; Yu et al., 2016; Zhang et al., 2016). Due to the lack of a standardized sampling procedure, beach sampling varies based on the depth/volume of the sample, tide lines, area, etc. Shovels, trowels, tweezers, spatulas, and spoons are commonly used to sample from the beach (Tampang & Viswanathan, 2022; Klein et al., 2018; Löder & Gerdts, 2015; Nor & Obbard, 2014). In some research, the sampling depth mostly varied between 1 and 5 cm (Tampang & Viswanathan, 2022; Hidalgo-Ruz et al., 2012; Venkatramanan et al., 2022). For sampling sediments from the bottom of the water body/subtidal region, different sampling tools such as corers, Van Veen sediment grab, metal collection container, and stainless-steel grab were used (Bakir et al., 2023; Ling et al., 2017; Markic et al., 2023). MP distribution in the subtidal region is very heterogeneous, which is why it is advised to perform sampling in several replicates, especially when using a corer or grab (Wang & Wang, 2018). Plastic tools are not recommended as they can lead to the overestimation of MPs.

8.2.4 SAMPLING METHODS OF MICROPLASTICS FROM AQUATIC BIOTA

The aquatic biota can be used for monitoring MP contamination since they can uptake MPs. MPs were detected in numerous biotas in the laboratory and natural settings (Rezania et al., 2018; Weber et al., 2018; Yasaka et al., 2022). Fish, marine worms, turtles, bivalves, seabirds, plankton, etc. are the most commonly monitored aquatic species for MP contamination. After dissecting, MP is obtained in several tissues and organs of these biotas (Mai et al., 2018). Different sampling approaches such as trawls, bongo nets, manta nets, grasps, creels, and electrofishing can be used to catch biota from aquatic environments. The sample shall be preserved as quickly as possible by freezing or by using preservatives such as formalin, formaldehyde, or ethanol (Stock et al., 2019). To ensure representative sampling and effective monitoring, it is advised to consider temporal and spatial distribution, sizes, sex, age, length, and weight of biota (Lusher et al., 2020; Stock et al., 2019; Wesch et al., 2016).

8.2.5 PRECAUTIONS TO BE TAKEN WHILE SAMPLING MPS

The study of MPs from environmental samples involves many steps and long processes, starting from sampling, pretreatment, extraction, analysis (accuracy and consistency), and interpretation of the results. There is great potential for contamination while dealing with MP in any of these steps. Therefore, it is very important to perform MP sampling and analysis cautiously to avoid any contamination. Some precautions such as avoiding plastic tools, wearing cotton lab coats and nitrile/latex gloves, and properly covering samples are recommended to avoid external contamination during sampling, handling, and lab analysis.

Before conducting sampling, all sampling bottles and sampling tools should be rinsed properly with MP-free water/Milli-Q water. In the sampling spot, cotton clothes shall be worn, and sample collection equipment needs to be non-plastic (glass, stainless steel, natural fibers). Samples should be covered properly to avoid external contamination. It is recommended to use sufficient replicates and an appropriate quantity of samples to have representative findings. Some previous studies have also considered using a field blank to quantify the level of background contamination during the sampling phase (Ghanadi et al., 2024; Wang & Wang, 2018).

Before performing laboratory analysis, all laboratory apparatus shall be rinsed with Milli-Q water and oven dried if necessary. It is recommended to limit the number of users in the lab and a regular cleaning schedule must be followed to avoid background contamination. During sample analysis with different analytical apparatus, there should be a reagent blank solution as a quality control approach (Dris et al., 2024; Ghanadi et al., 2024; Liu et al., 2024). Some digestive agents (H_2O_2, HNO_3, $HClO_4$) can damage or dissolve certain plastic polymers which can result in an underestimation of MPs. Therefore, it is advised to determine the effective concentration and digestion conditions (temperature, time) that will lead to correct MP estimation (Lusher et al., 2017).

For living biota samples, quick preservation should be followed using recommended preservatives or by freezing the sample because defecation can result in inaccurate MP detection (Stock et al., 2019).

8.2.6 SAMPLE PRETREATMENT AND EXTRACTION

When MPs are extracted from environmental samples, they must be separated with the highest purity and yield. The majority of MP separation procedures are dependent on the size, density, and presence of naturally occurring organic matter on the MP surface. Typical MP pretreatment methods include sieving, density separation, flotation, filtering, drying, chemical oxidation, digestion, and others that are included in Table 8.1.

Pure water samples are mostly pretreated using glass fiber or aluminum oxide filters to separate the MP particles (Stock et al., 2019). Density separation, which uses floating and fluidization, is one of the prevalent methods for sediments containing MPs. The denser salt solution is prepared using $NaCl/NaI/ZnCl_2$ and mixed with sediments which allows the MP particles to float. After that, the MP particles are separated by filtering the supernatant (Banik et al., 2022; Hidalgo-Ruz et al., 2012; Polt et al., 2023). Natural organic materials adhering to the surface of the plastic particles are invariably present in high concentrations in samples taken from the environment, which can make it difficult to accurately identify and characterize the MPs. To eliminate organic materials, MPs extracted from water and sediment samples are subjected to chemical or enzymatic digestion. Several chemicals have been utilized for the biomaterial dissolving process, which commonly makes use of oxidizers (H_2O_2), acids (HNO_3, HCl), or alkaline (NaOH, KOH) chemicals (Wang & Wang, 2018). Enzymatic digestion involves combining MP materials with a variety of enzymes, including cellulose, lipase, proteinase, amylase, and chitinase (Li et al., 2018).

In the case of biota samples, the digestive tract of larger organisms (fish, amphibians, birds) was examined and some studies analyzed entire organisms which are smaller in size (shrimps, macroinvertebrates, crustaceans). Several techniques have been designed to separate MPs from biota, including dissection, homogenization, and digestion with enzymes or chemicals (Lusher et al., 2020; O'Connor et al., 2019; Pastorino et al., 2023; Rendell-Bhatti et al., 2023). Acid digestion, which uses powerful oxidizing acids like nitric acid, perchloric acid, hydrochloric acid, or a combination of the aforementioned, is one of the most often used techniques for biota samples to break down biological tissues (Wang & Wang, 2018).

8.3 ANALYTICAL METHODS FOR MICROPLASTICS IN THE AQUATIC ENVIRONMENT

A range of visual, thermal, and spectral techniques, described in Figure 8.1, have been applied to perform qualitative and quantitative analysis of MPs after extraction, separation, and purification from the raw samples. Visual characterization (optical microscopy and scanning electron microscopy) is a primary and convenient way of characterizing the physical properties of MPs. Thermal (Py-GC-MS, TGA-GC-MS) and spectral (FTIR and Raman spectroscopy) methods are also commonly applied to

TABLE 8.1

Typical Microplastic Pretreatment Methods

Sample Type	Location	Pretreatment and Extraction	Analytical Methods	Results	References
Water, sediment, zooplankton, tadpoles, and fish	Lake Balma, Italy	Mechanical agitation, filtration, digestion	Stereomicroscopy and FTIR	Sediment: 1.33 ± 0.67 items/m³; Water and Zooplankton: 0 items/m³; Fish: 0.45 items/g (lower balma)	(Pastorino et al., 2023)
Fish, invertebrate, sediment	Wadden Sea Coastline, Germany	Density separation, filtration, digestion	Fluorescence microscopy and µRaman spectroscopy	Biota: 0 to 248.1 items/g; Sediment: 0 to 8128 part/kg	(Polt et al., 2023)
Lamprey Species, Sediment	Teith and Tweed river, UK	Sieving, filtration, chemical oxidation, density separation	µFTIR	Lamprey: 1.00 to 27.47 particles/g sediment	(Rendell-Bhatti et al., 2023)
Sediment	Kuakata Beach, Bangladesh	Density separation, sieving, drying and digestion	FTIR and Stereomicroscope	232 ± 52 items/kg	(Banik et al., 2022)
Water, Snail	Qing River, China	Filtration, density separation, digestion,	LD-IR	Water: 22–40 particles/L; Snail: 28 particles/individual	(An et al., 2022)
Sediment	Marina di Vecchiano, Italy	Homogenization, Sieving, microwave digestion	Py-GC-MS	0.003–0.2 g/350 g	(La Nasa et al., 2021)
Sediment	Aveiro beach, Portugal	Density separation, filtration	ATR-FTIR and stereoscopy	22 items/kg sand	(Prata et al., 2020)
Water, Sediment and Fish	Poyang Lake, China	Filtration, drying, density separation, dissection, digestion	Stereoscopic microscope and Raman spectrometry	Water: 5–34 items/L; Sediment: 4–506 items/kg; Crucians: 0–18 items/individual	(Yuan et al., 2019)
Water, sediments and biota	Levantine Basin, Lebanon	Filtration, density separation, digestion	Stereomicroscopy and micro-Raman spectroscopy	Water: 6.7 MPs/m³; Sediments: 4.68 MPs/g; Biota: 2.9–8.3 MPs/ individual	(Kazour et al., 2019)
Sediment and Water	Southern beaches, Sri Lanka	Sieving, density separation	Raman ATR-FTIR	Sediment: 6–738 MPs/m²; Water: 0–29 MPs/m³	(Koongolla et al., 2018)

MPs in the Aquatic Environment: Analytical & Mitigation Methods

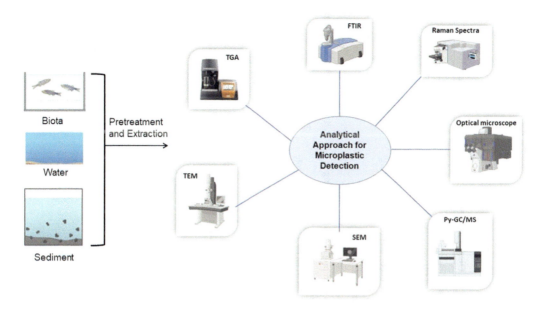

FIGURE 8.1 Methods of microplastics analysis in water, sediment, and biota samples.

detect and quantify MPs more accurately. The advantages and disadvantages of these methods are summarized in Table 8.2.

8.3.1 Visual Characterization

The conventional method of identification of MPs based on size, shape, and color usually uses light microscopy (Wang & Wang, 2018). Different polymer species fall into one of the three size categories: 100 μm (62–100 μm), 100–500 μm, or > 1000 μm. The identification of the polymer is not revealed and particles <100 μm are difficult to detect using this method (Desforges et al., 2014). In a recent research, the fragments and fiber-type MPs were detected in edible marine fish species by optical microscopy (Srisiri et al., 2024). Polymer identification was not possible in that study using optical microscopy, except low-resolution particle-type determination.

Scanning electron microscopy (SEM) is a method that scans the surface of the MP particle with a high-intensity electron beam and reveals its morphological properties. SEM provides clear pictures with a resolution of up to 0.1 μm and exceptionally high magnification that can distinguish between MPs and organic particles. SEM can precisely determine MP particles of varied sizes and forms (Ribeiro-Claro et al., 2017; Xiang et al., 2022). Its application together with EDS increases the possibility of chemical element analysis, plastic polymer type identification in the samples, and relative abundance estimation (Soursou et al., 2023). A recent study detected the size, shape, and surface morphology of MPs from surface water and sediment using SEM at 400× magnification. Due to the high resolution of SEM, weathered MP particles with cracked/deformed surfaces were clearly detected (Zhao et al., 2024). The same study also reported the proportion of chemical composition (Al, Fe, Cu, Ca, Si and Cl) present in MPs by EDS measurement.

Transmission electron microscopy (TEM) is another promising electron microscopy technique that may provide images with atomic-scale resolution. However, the polymers' amorphous nature and simple composition (organic nature) render the electron beam ineffective. That is why TEM has been rarely applied for MP analysis despite it being very efficient in determining nanomaterials. However, staining MPs with heavy elements may provide efficient detection (Mariano et al., 2021).

TABLE 8.2

Advantages and Disadvantages of Different Analytical Methods for Detecting MPs

Analytical Methods	Analyzed Parameters	Advantages	Disadvantages	References
Optical microscopy	Size, shape, morphology	- Inexpensive method - Simple operation - Non-destructive method - Widely used for MPs detection	- Low resolution - Potential for human error - No chemical composition detection - Not very clear surface morphology	(Dong et al., 2023; Schwaferts et al., 2019; Srisiri et al., 2024; Wang & Wang, 2018)
SEM-EDS	Size, shape, surface morphology, and chemical composition	- High resolution image - Clear surface morphology and shape - Proportion of chemical composition can be detected - Widely used for MPs detection	- Sample needs preprocessing (coating with Au/Pt) - Quantification is difficult - Expensive and time consuming	(Schwaferts et al., 2019; Ye et al., 2022; Zhao et al., 2024)
TEM	Size, shape, morphology, and aggregation	- High resolution image	- MPs need preprocessing (staining required) - Difficult to analyze weathered sample - Quantification is difficult - Limited application for MPs detection	(Li et al., 2020; Schwaferts et al., 2019a)
FTIR	Chemical composition, Spectral analysis	- High sensitivity and accuracy - Can accurately detect MPs polymer type - Consist of standard spectral library - Provide information about physiochemical weathering of MPs - µFTIR is more sensitive and automatic	- Time consuming - Expensive - Sample needs preprocessing (cleaning) - Water can induce interference	(Rocha-Santos & Duarte, 2015; Wang & Wang, 2018; Ye et al., 2022)
Raman spectroscopy	Detect plastic type, Spectral analysis	- Non-destructive chemical characterization - High sensitivity and precision - Standard spectral library - Better spatial resolution than FTIR	- Complex MPs mixture can produce overlapping spectra - Long processing time - Sample preparation - Interferences by fluorescence	(Jung et al., 2021; Rocha-Santos & Duarte, 2015; Ye et al., 2022)

(Continued)

TABLE 8.2 (Continued)
Advantages and Disadvantages of Different Analytical Methods for Detecting MPs

Analytical Methods	Analyzed Parameters	Advantages	Disadvantages	References
Py-GC/MS	Detect chemical composition of polymers and organic additives	- High sensitivity and accuracy - Both polymer and organic additives can be analyzed in one run - Can avoid background contamination - Minimal sample pretreatment - Very small quantity of sample is required for analysis (0.5 g)	- Destructive method - Very time consuming (30–100 min per sample) - High temperature - Manual sample placement in the crucible	(Wang & Wang, 2018; Ye et al., 2022)
TED-GC-MS	Qualitative and quantitative analysis of polymers and organic additives.	- Special pretreatment not required - Excellent for detecting mass of polymers and organic additives - High reproducibility - Effective than Py-GC-MS - Able to analyze large quantity of sample (100 mg)	- Destructive - Expensive - High temperature	(Ainali et al., 2021; Schwaferts et al., 2019; Wu et al., 2020)

However, the structure and chemical composition of MP polymers can be impacted by these stains (Campagnolo et al., 2021). TEM analysis was carried out in another study to observe the surface morphology of silver-stained nanoparticles and their interaction with polystyrene microplastics in simulated surface water (Li et al., 2020). Though TEM analysis was carried out at 80 kV, the images were not very sharp (Li et al., 2020). A change in agglomeration between two MPs was observed with time, indicating potential challenges of using TEM for environmental MP samples.

8.3.2 SPECTROMETRIC CHARACTERIZATION

The principle of Fourier transform infrared spectroscopy (FTIR) is based on the information on the specific chemical bond of the identifiers. The interaction between molecular vibration and infrared radiation produces spectral maps with different peak patterns for specific bonds. As a result, it is possible to identify the polymers that make up particles, and by comparing those materials to those in the standard library, the goal of detection can be accomplished (Mai et al., 2018; Xu et al., 2019). Therefore, the unique fingerprint produced by FTIR can detect the chemical composition of the MPs. For MP analysis, there are two different FTIR types: microscopic equipped FTIR and mercury cadmium telluride (MCT) single mode. Materials with heterogeneous MP components can be identified using microscope-equipped FTIR, however, the MCT single mode can only examine single and bulk samples by detecting the sample with an ATR tip (Lee & Chae, 2021). The detection limit of µ-FTIR was reported as 20 µm in several studies (Lee & Chae, 2021; Müller et al., 2020). More cutting-edge methods, such as nano-FTIR, which combines an optical microscope with a broadband infrared source, are advised to be used for particles smaller than the diffraction extent (Mattsson et al., 2021). Drying the sample is crucial since FTIR is readily disrupted by water and results in poor horizontal resolution, and complicated spectra when detecting wet substances. Different plastic polymers such as low-density polyethylene (LDPE), high-density polyethylene (HDPE), polystyrene (PS), etc. were detected in water, sediment, and fish samples by their characteristic blending deformation at the specific region of the FTIR spectra (Chatterjee et al., 2024).

Raman spectroscopy is a photon scattering method where light from a powerful laser source is scattered by molecules. Raman scattering happens when a sample gets exposed to a monochromatic laser light source; certain molecules absorb the light while others scatter it (Lee & Chae, 2021). Several research studies have reported Raman spectroscopy-based MP detection (Chakraborty et al., 2022; Dey, 2023; Mikac et al., 2023). A study was conducted based on Raman spectroscopy to detect two degradable and three non-degradable MPs. Polybutylene adipate terephthalate (PBAT), polybutylene succinate (PBS), polyethylene (PE), polypropylene (PP), and polystyrene (PS) polymers were detected based on their characteristic Raman spectra and corresponding functional groups (Li et al., 2024). The key advantages of Raman spectroscopy over other analytical techniques are its ability to investigate microscopic particles as small as 1 µm and its superior responsiveness to non-polar plastic functional groups (Lenz et al., 2015). Rigorous sample purification is strongly advised to reduce spurious signals while using Raman micro-spectrometry. In comparison to FTIR, Raman offers a greater spatial resolution for MPs. This is due to Raman's ability to direct the incident laser beam to concentrate into a 10 µm spot diameter and gather the scattered light from a broad angle.

8.3.3 THERMAL CHARACTERIZATION

Pyrolysis-GC-MS (Py-GC-MS) has been reported for the detection of MPs in water, sand, and sediment samples, typically in conjunction with other analytical techniques to acquire more comprehensive information on the MPs separated from these matrices (Castelvetro et al., 2021; Ibrahim et al., 2021; Sorolla-Rosario et al., 2023). Py-GC-MS makes use of the thermal disintegration of MPs in an inert environment to reveal the structural makeup of macromolecules. Depending on the environmental matrix, this approach also offers direct injection, which necessitates little sample preparation. Research on MP occurrence and exposure has been hampered by the protracted discussion about the categorization of sizes and global standardization. This method circumvents the issue of potentially underestimating plastics by providing information on the mass of polymers in the samples regardless of the particle size (Okoffo et al., 2019). Py-GC/MS has the ability to concurrently analyze polymer type and organic additives, which is a significant advantage (Mai et al., 2018). Fries et al. (2013) conducted studies using Py-GC/MS to characterize the polymer types and organic additives in marine sediments. The authors reported six different polymer types [polyethylene (PE), polypropylene, polystyrene, polyamide, chlorinated PE and chlorosulfonated PE] and several organic additives including 2,4-di-tert-butylphenol and different phthalates (Fries et al., 2013). The calibration of different organic additives was found to be time-consuming in that study.

TED-GC-MS (thermal extraction and desorption coupled with GC-MS) is another technique that uses TGA to extract the degradation product of MPs by thermal decomposition. The degraded plastic products are then analyzed using thermal desorption in a GC-MS for about 3 hours. This technique allows the use of relatively higher masses (up to 100 mg) than Py-GC-MS and is suitable for identifying and quantifying complex heterogeneous matrices (Dümichen et al., 2017; Elert et al., 2017). Polymer composites (polyethylene, polypropylene, polystyrene) and their decomposition products were quantified and analyzed in a river sample without peak overlay (Dümichen et al., 2019). These authors, in a previous study, also successfully identified the decomposition products of polyethylene spiked in environment samples using TED-GC-MS (Dümichen et al., 2015).

Some studies also mentioned the use of chromatographic techniques in addition to visual, thermal, and spectrometric methods. However, the use of liquid chromatographic techniques for MP analysis is less common because of the difficulties it brings, including sample extraction. High-pressure liquid chromatography (HPLC) has, however, been used in investigations in conjunction with a variety of detectors, including diode array, UV, and fluorescence (Soursou et al., 2023).

8.4 MITIGATION MEASURES FOR MICROPLASTICS IN AQUATIC ENVIRONMENTS

Inevitable usage of plastics results in MPs spreading all over the environment. The removal, degradation, and mitigation methods for MPs and their efficiency are demonstrated in Figure 8.2 and Table 8.3, respectively.

8.4.1 PHYSICAL METHOD

8.4.1.1 Granular Filtration

Usually, MPs are filtered through vacuum by granular filtration and membrane filtration. Activated carbon, glass beads, and quartz sands were used as granular media for filtration. The flow rate, granular media, and characteristics of the fluid have significant impacts on separation efficiency. Strong interactions between MP particles may make them more suited for granular filtration because of their comparable wettability, reverse surface charge, and heterogeneous shape (Wu et al., 2013; Zhang et al., 2021).

8.4.1.2 Membrane Filtration

In membrane filtration, the membrane functions as a selective barrier, permitting the passage of specific compounds while preventing the passage of others. Microfiltration, ultrafiltration, nanofiltration, and reverse osmosis membranes are the different kinds of membranes based on pore size. Particles larger than 0.08–2 μm, 0.005–0.02 μm, and 0.002 μm may be removed using microfiltration, ultrafiltration, and nanofiltration, respectively. In order to remove MPs, a variety of factors must be considered, including membrane characteristics, transmembrane pressure, filtration time and mode,

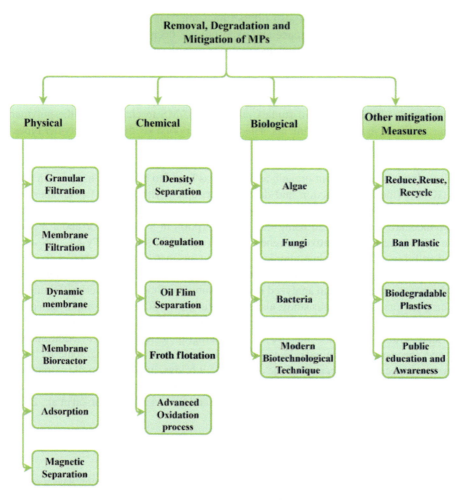

FIGURE 8.2 Removal, degradation, and mitigation methods for MPs.

flow rate, and MP size. Membrane filtration's benefits include low maintenance, high throughput, and great efficiency. Membrane fouling and the need for a thorough chemical pretreatment are drawbacks of this method (Poerio et al., 2019).

8.4.1.3 Dynamic Membrane

Dynamic membrane (DM) has emerged as a promising wastewater treatment technique due to its affordability, ease of cleaning, and lower energy consumption. DM technology removes microparticles between 1.65 μm and 516 μm by gravity. The creation of stable and light DM layers is aided by DM, making it a viable solution for the elimination of low-density and non-degradable MPs. Yet, DM filtration's flaws also stem from membrane performance fluctuations. Thick fouling layers decrease the membrane performance since the filtration resistance is driven primarily by the cake layer. The MPs may pass through the filter before creating a cake (Li et al., 2018).

8.4.1.4 Membrane Bioreactor

Tertiary wastewater treatment methods such as membrane bioreactors (MBRs), disc filters (DFs), rapid sand filters (RSFs), and dissolved air flotation (DAF) are used in WWTPs to remove MPs from wastewater effluent (Talvitie et al., 2017). Biological catalysts are combined with membrane

TABLE 8.3

Efficiency of the Different Methods Used for MPs Removal and Degradation

Method	Technique	Media	Source	MPs	Efficiency %	Reference
Physical	Granular filtration	Sand	WWTPs	PE, PS, PP, and PET	79	(Funck et al., 2021)
		Biochar + sand		PS	>95	(Wang et al., 2020)
	Microfiltration	SiC		Nylon fiber, PVC	98.5	(Luogo et al., 2022)
	Ultrafiltration	ZrO_2		Nylon fiber, PVC	99.2	
	Membrane bioreactor		WWTPs	PES, PE, PA, PP	99.4	(Lares et al., 2018)
	Adsorption	Zn/Al LDHs		PS	100	(Tiwari et al., 2020)
		Chitin-GO		PS	89.8	(Sun et al., 2020)
		$CaCO_3$		PMMA, PVAc	99	(Batool & Valiyaveettil, 2020)
	Magnetic separation	CNTs		PE, PET, PA	>80	(Tang et al., 2021)
Chemical	Electrocoagulation	Al	Microbeads	PE	99	(Perren et al., 2018)
	Oil film separation	Castor oil		PP, PS, PMMA, and PET-G	95	(Mani et al., 2019)
		Canola oil		PVC	98	(Crichton et al., 2017)
		Olive oil		PE, PU	90	(Scopetani et al., 2020)
				PS, PC	97	
				PVC, PET	95	
	Dissolved air floatation			PES, PAC, PVC, PS, PP	95	(Talvitie et al., 2017)
	Advanced oxidation processes	Photo-Fenton		PP	96	(Dos Santos et al., 2023)
				PVC	94	
Biological	Fungi	*Aspergillus* sp. and *Lysinibacillus* sp.	Granules	LDPE	29.5	(Esmaeili, et al., 2013)
	Bacteria	*Microbacterium paraoxydans*		LDPE	61	(Rajandas et al., 2012)
		Pseudomonas aeruginosa		LDPE	50.5	

Abbreviations: PES, polyester; PE, polyethylene; PA, polyamide; PP, polypropylene; PMMA, poly(methyl methacrylate); PVAc, poly(vinyl acetate); PET, polyethylene terephthalate; PVC, polyvinyl chloride; PS, polystyrene; PET-G, polyethylene terephthalate glycol; PU, polyurethane; PAC, polyacrylates; LDPE, low-density polyethylene separation in the MBR. Bioreactor, membrane filtration, and pretreatment are the steps that make up the MBR process. Biodegradation of organic materials is a key aspect of MBR's efficacy in removing MPs. The MBR had the greatest removal effectiveness (99.4%) for MPs, indicating that it was the most effective method for MP removal (Lares et al., 2018).

8.4.1.5 Adsorption

Zn-Al layered double hydroxides (LDHs) adsorbent has been used as a potential adsorbent for complete removal of nano-scale plastic debris from freshwater systems at a pH of 4.0 (Tiwari et al., 2020). Similarly, a sponge made from chitin and graphene oxide adsorbent used for the adsorption of different types of MPs showed up to 89.8% removal. While the hydrogen bond interactions, electrostatic interactions, and π–π interactions were the key driving forces for the adsorption of MPs, the intra-particle diffusion played a significant role in the entire adsorption process (Sun et al., 2020). $CaCO_3$ coprecipitation guarantees 99% removal of polymer nanoparticles from water within 30 minutes (Batool & Valiyaveettil, 2020). The adsorption of MPs on $CaCO_3$ is facilitated by the negative surface charge of synthetic polymers. Although this method is effective, it is challenging to achieve selective adsorption to isolate MPs.

8.4.1.6 Magnetic Separation

Magnetic fields are utilized to separate MPs from the sewer system. Long-range magnetic force enables increased separation efficiency, reduced waste sludge production, and high throughput. The three steps involved in this process include (i) preparing the magnetic seeds, (ii) increasing their ability to aggregate, and (iii)optimizing the separation efficiency (Wan et al., 2011). A separation efficiency of more than 80% was observed when MPs were separated using magnetic carbon nanotubes (Tang et al., 2021; Zhang et al., 2021).

8.4.2 CHEMICAL METHODS

8.4.2.1 Density Separation

Density separation is a very basic method used for the removal of MPs from sediments. After being stirred, the MPs that are lighter than the medium will rise to the top of the suspension and get separated from the sediment. Thus, it is essential for density separation that flotation media densities be variable. NaCl, NaBr, NaI, $ZnBr_2$, $CaCl_2$, $3Na_2WO_4.9WO_3$, and NaH_2PO_4 are all examples of common media used for density separation. Nevertheless, since this approach was not used to remove high-density MPs, such as PET from sediments, comparison between this and other technologies may not be accurate (Zhang et al., 2021). Density flotation is exclusively used in sample analysis, despite its good application. Due to their high cost and toxicity, denser salts need special handling during the preparation of the density flotation solution and its subsequent storage (Fuller & Gautam, 2016). There seems to be a realistic ceiling on density separation provided by the size fractions measurement. MPs should not be extracted from their natural habitats via density separation since it may be difficult to recycle the extraction solution from water. The effectiveness of removal might be constrained by induced gravity separation.

8.4.2.2 Coagulation

Suspended MPs require an increase in the size of the particles for physical separation because of their tiny size. When sol aggregates act as an unstable colloid, flocculation may be used to filter it out of water used for drinking and wastewater treatment (Nakazawa et al., 2018). Several different types of coagulants, both synthetic and natural, are available for use in the clearance of MPs. There was a minimal correlation between ionic strength, turbidity, natural organic matter, and removal efficiency, perhaps because of the low impact on floc characteristics. Microfibers and other larger MP particles might readily bind to floccules, improving the sedimentation property. Small (10–20 μm)

MPs in the Aquatic Environment: Analytical & Mitigation Methods 137

spherically shaped, smooth MP particles escape in large numbers during coagulation (Zhang et al., 2020). Coagulation is often used in conjunction with other physical separation processes, including filtering, membrane technology, and froth flotation for MP elimination. Granular activated carbon filtration was more effective in removing MPs than sedimentation and coagulation. Throughout the coagulation process, it is important to consider how different reagents could alter the water's quality. The electrocoagulation (EC) process is cheaper to run, friendly to the environment, and produces less sludge. Almost 90% of microbeads were removed over a wider pH range, demonstrating EC's superior effectiveness in removing MPs from wastewater (Perren et al., 2018; Zhang et al., 2021).

8.4.2.3 Oil Film Separation

Separation by oil films relies on hydrophobicity rather than density. A removal of 95% and 98%, respectively, were attained using castor oil and canola oil (Crichton et al., 2017; Mani et al., 2019). In this process, the oil film was separated by freezing samples at − 4°C in polytetrafluoroethylene cylinders and MPs were recovered (Scopetani et al., 2020; Truc & Lee, 2019). This low-cost, low-risk method depends on oleophilic properties rather than density. For samples with high biogenic concentrations, further digestion is needed before shaking the oil and environmental matrix to separate the oil film. If any traces of oil remains after seprating the oil film, it is recommended to use ethyl alcohol and hexane to remove the trace oil.

8.4.2.4 Froth Floatation

In froth flotation process, the hydrophobic materials float as bubble aggregations, whereas hydrophilic ones are settled as underflow. Consequently, hydrophobic MP segments attract froth flotation. Oil, grease, algae, and surfactants are effectively treated by this technique from sewage effluents using functionalized polymers (Bolto & Xie, 2019). The size and density of MPs determine the bubble size and air flow capacity needed for good adhesion and debris transportation. Advanced final-stage wastewater treatment technology including reduced 95% of MPs from WWTPs (Talvitie et al., 2017). Uncertain conditions such as surface deterioration, biomass fouling, surface contamination, and morphological features might impede the separation of MPs.

8.4.2.5 Advanced Oxidation Processes

Advanced oxidation processes (AOPs) are allowed for carbonizing organic pollutants using reactive oxygen species and to convert the organic compound into a simple one. Industrial wastewater treatment uses homogeneous AOPs, however, the restricted pH range, sludge production, and catalyst inactivation during the process are some of the drawbacks (Xu et al., 2019). Heterogeneous AOPs have been employed to improve the degradation efficiency of cosmetic MPs using combined carbocatalytic oxidation and hydrothermal hydrolysis (Kang et al., 2019).

8.4.3 Biological Methods

Biological removal or degradation involves algae, fungi, and bacteria and also includes emerging modern biotechnological techniques. Degradation results, and changes in physical and chemical properties have been confirmed by various spectroscopy and chromatography instruments.

8.4.3.1 Algae

Algal degradation includes corrosion, hydrolysis, penetration, fouling, etc. Microalgae colonize on plastic surfaces in aquatic media to degrade MPs with enzymes. The surface of low-density polyethylene was found to be colonized by *Oscillatoria subbrevis* and *Phormidium lucidum*, and was subsequently degraded without pretreatment. Algae form biofilms on the surface of polymers as they degrade. The ability to build biofilms on MPs was demonstrated by a number of cyanobacterial strains, including those from the genera *Microcystis, Rivularia, Pleurocapsa, Synechococcus,*

Prochlorothrix, *Leptolyngbya calothrix*, and *Scytonema* (Bryant et al., 2016; Didier et al., 2017; Dussud et al., 2018).

8.4.3.2 Fungi

Fungi survive with wide habitat adaptation. They also tolerate various toxic metals and chemicals. They can convert complex polymers into simple monomers with their intracellular and extracellular enzymes, including oxidases and hydrolases, and natural biosurfactants such as hydrophobins. *Mucor circinelloides* and *Aspergillus flavus* degraded low-density polyethylene which was isolated from a municipal landfill (Kunlere et al., 2019). LDPE films exposed to UV radiation was degraded by *Aspergillus* species and *Lysinibacillus* species at rates of 29.5% and 15.8%, respectively (Esmaeili et al., 2013).

8.4.3.3 Bacteria

Numerous habitats, including polluted sediments, wastewater, sludge, compost, and municipal landfills, produce bacteria that can degrade MPs. Microbes that thrive in polluted environments frequently learn how to activate the enzyme mechanism that breaks down MPs. Lipases, esterases, laccases, amidases, cutinases, hydrolases, and carboxylesterases were the enzymes involved in degradation of MPs. The microbial degradation, usually a slow process, is influenced by temperature, pH, substrate characteristics, and surfactants. To promote biodegradation, pretreatment is advised, including the use of chemical oxidizing agents, thermooxidation, nitric acid, and UV irradiation of MPs (Anand et al., 2023). The bacterial consortium had greater degradation efficiency and community stability than pure cultures. Within two months of incubation, MPs were degraded by *Microbacterium paraoxydans* and *Pseudomonas aeruginosa* by almost 61.0% and 50.5%, respectively (Rajandas et al., 2012). A study conducted using *Pseudomonas chlororaphis*, *Pseudomonas stutzeri*, and *Vibrio* sp. consortia showed the addition of starch to enhance the degradation of polypropylene (Cacciari et al., 1993).

8.4.3.4 Modern Biotechnological Techniques

The pervasiveness of MPs requires their effective bioremediation. Genetic engineering approaches allow us to modify the genetic components of bacteria and improve their biodegradation effectiveness. Recombinant DNA technologies, gene cloning, and genetic manipulation have improved microbial bioremediation of hydrocarbons (Kumar et al., 2020). *Ideonella sakaiensis* 201-F6 produces enzymes that break down polyethylene terephthalate to terephthalic acid and non-hazardous ethylene glycol. Cloning the genes of *Ideonella sakaiensis* 201-F6 allows other bacterial strains to break down polyethylene terephthalate into non-hazardous monomers (Moog et al., 2019). Genetically engineered strains of *E. coli* BL21 and *P. chrysosporium* produced manganese-dependent peroxidase and laccase enzymes. Meanwhile, it has been reported that the degradation of polyethylene terephthalate is improved by enzymes that have been genetically modified. (Paço et al., 2019; Sharma et al., 2018).

8.5 CONCLUSIONS

The concentration, type, and age of the MPs, the species and developmental stage of the bioassay species employed, as well as any potential interactions between the MPs and organisms, all influence how MPs affect the environment. The evaluation of the environmental implications of MPs can be facilitated by standardizing and establishing consensus on the identification and quantification procedures. The establishment of regulatory controls that can improve environmental quality won't be possible until we are able to the adequately characterize the threats that MPs pose to the environment. Besides, as a part of the prevention and mitigation measures, it is critical to educate all stakeholders on the effects of MPs and improper plastic waste management. At the local, national,

MPs in the Aquatic Environment: Analytical & Mitigation Methods

regional, and international levels, strict regulations are needed to limit the use and consumption of plastics and to create incentives for their reduction in use and waste.

REFERENCES

Ainali, N. M., Kalaronis, D., Kontogiannis, A., Evgenidou, E., Kyzas, G. Z., Yang, X., Bikiaris, D. N., & Lambropoulou, D. A. (2021). Microplastics in the environment: Sampling, pretreatment, analysis and occurrence based on current and newly-exploited chromatographic approaches. *Science of The Total Environment*, *794*, 148725. https://doi.org/10.1016/j.scitotenv.2021.148725

An, L., Cui, T., Zhang, Y., & Liu, H. (2022). A case study on small-size microplastics in water and snails in an urban river. *Science of The Total Environment*, *847*, 157461. https://doi.org/10.1016/j.scitotenv.2022.157461

Anand, U., Dey, S., Bontempi, E., Ducoli, S., Vethaak, A. D., Dey, A., & Federici, S. (2023). Biotechnological methods to remove microplastics: a review. *Environmental Chemistry Letters*, *21*(3), 1787–1810. https://doi.org/10.1007/s10311-022-01552-4

Bakir, A., Doran, D., Silburn, B., Russell, J., Archer-Rand, S., Barry, J., Maes, T., Limpenny, C., Mason, C., Barber, J., & Nicolaus, E. E. M. (2023). A spatial and temporal assessment of microplastics in seafloor sediments: A case study for the UK. *Frontiers in Marine Science*, *9*(January). https://doi.org/10.3389/fmars.2022.1093815

Banik, P., Hossain, M. B., Nur, A.-A. U., Choudhury, T. R., Liba, S. I., Yu, J., Noman, M. A., & Sun, J. (2022). Microplastics in sediment of Kuakata Beach, Bangladesh: Occurrence, spatial distribution, and risk assessment. *Frontiers in Marine Science*, *9*, 860989.

Batool, A., & Valiyaveettil, S. (2020). Coprecipitation – An efficient method for removal of polymer nanoparticles from water. *ACS Sustainable Chemistry and Engineering*, *8*(35), 13481–13487. https://doi.org/10.1021

Bolto, B., & Xie, Z. (2019). The use of polymers in the flotation treatment of wastewater. *Processes*, *7*(6). https://doi.org/10.3390/PR7060374

Bryant, J. A., Clemente, T. M., Viviani, D. A., Fong, A. A., Thomas, K. A., Kemp, P., Karl, D. M., White, A. E., & DeLong, E. F. (2016). Diversity and activity of communities inhabiting plastic debris in the North Pacific Gyre. *mSystems*, *1*(3), 1-19. https://doi.org/10.1128/msystems.00024-16

Cacciari, I., Quatrini, P., Zirletta, G., Mincione, E., Vinciguerra, V., Lupattelli, P., & Sermanni, G. G. (1993). Isotactic polypropylene biodegradation by a microbial community: Physicochemical characterization of metabolites produced. *Applied and Environmental Microbiology*, *59*(11), 3695–3700. https://doi.org/10.1128/aem.59.11.3695-3700.1993

Campagnolo, L., Petersen, E., Riediker, M., Dini, L., Mariano, S., Tacconi, S., Fidaleo, M., & Rossi, M. (2021). Micro and nanoplastics identification: Classic methods and innovative detection techniques. *Frontiers in Toxicology*, *3*, 636640. https://doi.org/10.3389/ftox.2021.636640

Castelvetro, V., Corti, A., La Nasa, J., Modugno, F., Ceccarini, A., Giannarelli, S., Vinciguerra, V., & Bertoldo, M. (2021). Polymer identification and specific analysis (PISA) of microplastic total mass in sediments of the protected marine area of the Meloria Shoals. *Polymers*, *13*(5), 796.

Chakraborty, I., Banik, S., Biswas, R., Yamamoto, T., Noothalapati, H., & Mazumder, N. (2022). Raman spectroscopy for microplastic detection in water sources: A systematic review. *International Journal of Environmental Science and Technology*, *2*, 10435–10448. https://doi.org/10.1007/s13762-022-04505-0

Chaukura, N., Kefeni, K. K., Chikurunhe, I., Nyambiya, I., Gwenzi, W., Moyo, W., Nkambule, T.T.I., Mamba, B.B., & Abulude, F.O. (2021). Microplastics in the aquatic environment – the occurrence, sources, ecological impacts, fate, and remediation challenges. *Pollutants*, *1*(2), 95–118. https://doi.org/10.3390/pollutants1020009

Chatterjee, N. H., Manna, S., Ray, A., Das, S., Rana, N., Banerjee, A., Ray, M., & Ray, S. (2024). Microplastics contamination in two species of gobies and their estuarine habitat of Indian Sundarbans. *Marine Pollution Bulletin*, *198*, 115857. https://doi.org/10.1016/j.marpolbul.2023.115857

Cole, M., Lindeque, P., Halsband, C., & Galloway, T. S. (2011). Microplastics as contaminants in the marine environment: A review. *Marine Pollution Bulletin*, *62*(12), 2588–2597.

Crawford, C. B., & Quinn, B. (2017). Microplastic collection techniques. In: Crawford, C. B., & Quinn, B. (eds.), *Microplastic Pollutants* (pp. 179–202). Elsevier Science, Amsterdam, The Netherlands.

Crichton, E. M., Noël, M., Gies, E. A., & Ross, P. S. (2017). A novel, density-independent and FTIR-compatible approach for the rapid extraction of microplastics from aquatic sediments. *Analytical Methods, 9*(9), 1419–1428. https://doi.org/10.1039/C6AY02733D

da Costa, M. B., Otegui, M. B. P., Zamprogno, G. C., Caniçali, F. B., dos Reis Cozer, C., Pelletier, E., & Graceli, J. B. (2023). Abundance, composition, and distribution of microplastics in intertidal sediment and soft tissues of four species of Bivalvia from Southeast Brazilian urban beaches. *Science of The Total Environment, 857*, 159352. https://doi.org/10.1016/j.scitotenv.2022.159352

Desforges, J.-P. W., Galbraith, M., Dangerfield, N., & Ross, P. S. (2014). Widespread distribution of microplastics in subsurface seawater in the NE Pacific Ocean. *Marine Pollution Bulletin, 79*(1–2), 94–99.

Dey, T. (2023). Microplastic pollutant detection by Surface Enhanced Raman Spectroscopy (SERS): A mini-review. *Nanotechnology for Environmental Engineering, 8*(1), 41–48. https://doi.org/10.1007/s41204-022-00223-7

Didier, D., Anne, M., & Alexandra, T. H. (2017). Plastics in the North Atlantic garbage patch: A boat-microbe for hitchhikers and plastic degraders. *Science of the Total Environment, 599–600*, 1222–1232. https://doi.org/10.1016/j.scitotenv.2017.05.059

Dong, H., Wang, X., Niu, X., Zeng, J., Zhou, Y., Suona, Z., Yuan, Y., & Chen, X. (2023). Overview of analytical methods for the determination of microplastics: Current status and trends. *TrAC Trends in Analytical Chemistry, 167*, 117261. https://doi.org/10.1016/j.trac.2023.117261

Dos Santos, N. de O., Busquets, R., & Campos, L. C. (2023). Insights into the removal of microplastics and microfibres by advanced oxidation processes. *Science of The Total Environment, 861*, 160665. https://doi.org/10.1016/j.scitotenv.2022.160665

Dris, R., Beaurepaire, M., Bouzid, N., Stratmann, C., Nguyen, M. T., Bordignon, F., Gasperi, J., & Tassin, B. (2024). Sampling and analyzing microplastics in rivers: What methods are being used after a decade of research? In *Microplastic Contamination in Aquatic Environments* (pp. 65–91). Elsevier. https://doi.org/10.1016/B978-0-443-15332-7.00013-2

Dümichen, E., Barthel, A. K., Braun, U., Bannick, C. G., Brand, K., Jekel, M., & Senz, R. (2015). Analysis of polyethylene microplastics in environmental samples, using a thermal decomposition method. *Water Research, 85*, 451–457. https://doi.org/10.1016/j.watres.2015.09.002

Dümichen, E., Eisentraut, P., Bannick, C. G., Barthel, A.-K., Senz, R., & Braun, U. (2017). Fast identification of microplastics in complex environmental samples by a thermal degradation method. *Chemosphere, 174*, 572–584.

Dümichen, E., Eisentraut, P., Celina, M., & Braun, U. (2019). Automated thermal extraction-desorption gas chromatography mass spectrometry: A multifunctional tool for comprehensive characterization of polymers and their degradation products. *Journal of Chromatography A, 1592*, 133–142.

Dussud, C., Meistertzheim, A. L., Conan, P., Pujo-Pay, M., George, M., Fabre, P., Coudane, J., Higgs, P., Elineau, A., Pedrotti, M. L., Gorsky, G., & Ghiglione, J. F. (2018). Evidence of niche partitioning among bacteria living on plastics, organic particles and surrounding seawaters. *Environmental Pollution, 236*, 807–816. https://doi.org/10.1016/j.envpol.2017.12.027

Elert, A. M., Becker, R., Duemichen, E., Eisentraut, P., Falkenhagen, J., Sturm, H., & Braun, U. (2017). Comparison of different methods for MP detection: What can we learn from them, and why asking the right question before measurements matters? *Environmental Pollution, 231*, 1256–1264.

Enders, K., Lenz, R., Stedmon, C. A., & Nielsen, T. G. (2015). Abundance, size and polymer composition of marine microplastics ≥ 10 μm in the Atlantic Ocean and their modelled vertical distribution. *Marine Pollution Bulletin, 100*(1), 70–81.

Esmaeili, A., Pourbabaee, A. A., Alikhani, H. A., Shabani, F., & Esmaeili, E. (2013). Biodegradation of low-density polyethylene (LDPE) by mixed culture of *Lysinibacillus xylanilyticus* and *Aspergillus niger* in Soil. *PLoS ONE, 8*(9), e71720. https://doi.org/10.1371/journal.pone.0071720

Fok, L., & Cheung, P. K. (2015). Hong Kong at the Pearl River Estuary: A hotspot of microplastic pollution. *Marine Pollution Bulletin, 99*(1–2), 112–118.

Fries, E., Dekiff, J. H., Willmeyer, J., Nuelle, M. T., Ebert, M., & Remy, D. (2013). Identification of polymer types and additives in marine microplastic particles using pyrolysis-GC/MS and scanning electron microscopy. *Environmental Sciences: Processes and Impacts, 15*(10), 1949–1956. https://doi.org/10.1039/c3em00214d

Fuller, S., & Gautam, A. (2016). A procedure for measuring microplastics using pressurized fluid extraction. *Environmental Science and Technology, 50*(11), 5774–5780. https://doi.org/10.1021/acs.est.6b00816

Funck, M., Al-Azzawi, M. M. S., Yildirim, A., Knoop, O., Schmidt, T. C., Drewes, J. E., & Tuerk, J. (2021). Release of microplastic particles to the aquatic environment via wastewater treatment plants: The impact of sand filters as tertiary treatment. *Chemical Engineering Journal*, *426*, 130933. https://doi.org/10.1016/J.CEJ.2021.130933

Ghanadi, M., Joshi, I., Dharmasiri, N., Jaeger, J. E., Burke, M., Bebelman, C., Symons, B., & Padhye, L. P. (2024). Quantification and characterization of microplastics in coastal environments: Insights from laser direct infrared imaging. *Science of the Total Environment*, *912*. https://doi.org/10.1016/j.scitotenv.2023.168835

Gupta, D. K., Choudhary, D., Vishwakarma, A., Mudgal, M., Srivastava, A. K., & Singh, A. (2023). Microplastics in freshwater environment: Occurrence, analysis, impact, control measures and challenges. *International Journal of Environmental Science and Technology*, *20*(6), 6865–6896. https://doi.org/10.1007/s13762-022-04139-2

Hidalgo-Ruz, V., Gutow, L., Thompson, R. C., & Thiel, M. (2012). Microplastics in the marine environment: A review of the methods used for identification and quantification. *Environmental Science & Technology*, *46*(6), 3060–3075.

Ibrahim, Y. S., Hamzah, S. R., Khalik, W. M. A. W. M., Yusof, K. M. K. K., & Anuar, S. T. (2021). Spatiotemporal microplastic occurrence study of Setiu wetland, South China sea. *Science of the Total Environment*, *788*, 147809.

Ivleva, N. P., Wiesheu, A. C., & Niessner, R. (2017). Microplastic in aquatic ecosystems. *Angewandte Chemie International Edition*, *56*(7), 1720–1739. https://doi.org/10.1002/anie.201606957

Jung, S., Cho, S. H., Kim, K. H., & Kwon, E. E. (2021). Progress in quantitative analysis of microplastics in the environment: A review. *Chemical Engineering Journal*, *422*, 130154. https://doi.org/10.1016/J.CEJ.2021.130154

Kang, J., Zhou, L., Duan, X., Sun, H., Ao, Z., & Wang, S. (2019). Degradation of cosmetic microplastics via functionalized carbon nanosprings. *Matter*, *1*(3), 745–758. https://doi.org/10.1016/j.matt.2019.06.004

Kazour, M., Jemaa, S., Issa, C., Khalaf, G., & Amara, R. (2019). Microplastics pollution along the Lebanese coast (Eastern Mediterranean Basin): Occurrence in surface water, sediments and biota samples. *Science of The Total Environment*, *696*, 133933. https://doi.org/10.1016/j.scitotenv.2019.133933

Kiran, B. R., Kopperi, H., & Venkata Mohan, S. (2022). Micro/nano-plastics occurrence, identification, risk analysis and mitigation: challenges and perspectives. *Reviews in Environmental Science and Bio/technology. 21(1)*, 169–203.

Klein, S., Dimzon, I. K., Eubeler, J., & Knepper, T. P. (2018). Analysis, Occurrence, and Degradation of Microplastics in the Aqueous Environment. In: Wagner, M., Lambert, S. (eds) *Freshwater Microplastics*. The Handbook of Environmental Chemistry, vol 58, Springer, Cham.

Koongolla, J. B., Andrady, A. L., Terney Pradeep Kumara, P. B., & Gangabadage, C. S. (2018). Evidence of microplastics pollution in coastal beaches and waters in southern Sri Lanka. *Marine Pollution Bulletin*, *137*, 277–284. https://doi.org/10.1016/j.marpolbul.2018.10.031

Kumar, M., Xiong, X., He, M., Tsang, D. C. W., Gupta, J., Khan, E., Harrad, S., Hou, D., Ok, Y. S., & Bolan, N. S. (2020). Microplastics as pollutants in agricultural soils. *Environmental Pollution*, 265, 114980. https://doi.org/10.1016/J.ENVPOL.2020.114980

Kunlere, I. O., Fagade, O. E., & Nwadike, B. I. (2019). Biodegradation of low density polyethylene (LDPE) by certain indigenous bacteria and fungi. *International Journal of Environmental Studies*, *76*(3), 428–440. https://doi.org/10.1080/00207233.2019.1579586

Kye, H., Kim, J., Ju, S., Lee, J., Lim, C., & Yoon, Y. (2023). Microplastics in water systems: A review of their impacts on the environment and their potential hazards. *Heliyon*, *9*(3), e14359. https://doi.org/10.1016/j.heliyon.2023.e14359

La Nasa, J., Biale, G., Mattonai, M., & Modugno, F. (2021). Microwave-assisted solvent extraction and double-shot analytical pyrolysis for the quali-quantitation of plasticizers and microplastics in beach sand samples. *Journal of Hazardous Materials*, *401*, 123287. https://doi.org/10.1016/j.jhazmat.2020.123287

Lares, M., Ncibi, M. C., Sillanpää, M., & Sillanpää, M. (2018). Occurrence, identification and removal of microplastic particles and fibers in conventional activated sludge process and advanced MBR technology. *Water Research*, *133*, 236–246. https://doi.org/10.1016/j.watres.2018.01.049

Lee, H., Shim, W. J., & Kwon, J.-H. (2014). Sorption capacity of plastic debris for hydrophobic organic chemicals. *Science of the Total Environment*, *470*, 1545–1552.

Lee, J., & Chae, K.-J. (2021). A systematic protocol of microplastics analysis from their identification to quantification in water environment: A comprehensive review. *Journal of Hazardous Materials*, *403*, 124049.

Lenz, R., Enders, K., Stedmon, C. A., Mackenzie, D. M. A., & Nielsen, T. G. (2015). A critical assessment of visual identification of marine microplastic using Raman spectroscopy for analysis improvement. *Marine Pollution Bulletin*, *100*(1), 82–91.

Li, F., Liu, D., Guo, X., Zhang, Z., Martin, F. L., Lu, A., & Xu, L. (2024). Identification and visualization of environmental microplastics by Raman imaging based on hyperspectral unmixing coupled machine learning. *Journal of Hazardous Materials*, *465*, 133336. https://doi.org/10.1016/j.jhazmat.2023.133336

Li, J., Liu, H., & Chen, J. P. (2018). Microplastics in freshwater systems: A review on occurrence, environmental effects, and methods for microplastics detection. *Water Research*, *137*, 362–374.

Li, L., Geng, S., Wu, C., Song, K., Sun, F., Visvanathan, C., Xie, F., & Wang, Q. (2019). Microplastics contamination in different trophic state lakes along the middle and lower reaches of Yangtze River Basin. *Environmental Pollution*, *254*, 112951.

Li, L., Xu, G., Yu, H., & Xing, J. (2018). Dynamic membrane for micro-particle removal in wastewater treatment: Performance and influencing factors. *Science of the Total Environment*, *627*, 332–340. https://doi.org/10.1016/j.scitotenv.2018.01.239

Li, P., Zou, X., Wang, X., Su, M., Chen, C., Sun, X., & Zhang, H. (2020). A preliminary study of the interactions between microplastics and citrate-coated silver nanoparticles in aquatic environments. *Journal of Hazardous Materials*, *385*, 121601. https://doi.org/10.1016/j.jhazmat.2019.121601

Ling, S. D., Sinclair, M., Levi, C. J., Reeves, S. E., & Edgar, G. J. (2017). Ubiquity of microplastics in coastal seafloor sediments. *Marine Pollution Bulletin*, *121*(1), 104–110. https://doi.org/10.1016/j.marpolbul.2017.05.038

Liu, S., Li, Y., Wang, F., Gu, X., Li, Y., Liu, Q., Li, L., & Bai, F. (2024). Temporal and spatial variation of microplastics in the urban rivers of Harbin. *Science of the Total Environment*, *910*. https://doi.org/10.1016/j.scitotenv.2023.168373

Löder, M. G. J., & Gerdts, G. (2015). Methodology used for the detection and identification of microplastics - A critical appraisal. *Marine Anthropogenic Litter*, 201–227

Luogo, B. D. P., Salim, T., Zhang, W., Hartmann, N. B., Malpei, F., & Candelario, V. M. (2022). Reuse of water in laundry applications with micro- and ultrafiltration ceramic membrane. *Membranes*, *12*(2), 223. https://doi.org/10.3390/membranes12020223

Lusher, A., Welden, N., Sobral, P., & Cole, M. (2017). Sampling, isolating and identifying microplastics ingested by fish and invertebrates. *Analytical Methods*, *9*. https://doi.org/10.1039/C6AY02415G

Lusher, A. L., Welden, N. A., Sobral, P., & Cole, M. (2020). Sampling, isolating and identifying microplastics ingested by fish and invertebrates. In *Analysis of Nanoplastics and Microplastics in Food* (pp. 119–148). CRC Press.

Lv, L., Yan, X., Feng, L., Jiang, S., Lu, Z., Xie, H., Xie, H., Sun, S., Chen, J., & Li, C. (2021). Challenge for the detection of microplastics in the environment. *Water Environment Research*, *93*(1), 5–15.

Mai, L., Bao, L.-J., Shi, L., Wong, C. S., & Zeng, E. Y. (2018). A review of methods for measuring microplastics in aquatic environments. *Environmental Science and Pollution Research*, *25*(12), 11319–11332. https://doi.org/10.1007/s11356-018-1692-0

Mani, T., Frehland, S., Kalberer, A., & Burkhardt-Holm, P. (2019). Using castor oil to separate microplastics from four different environmental matrices. *Analytical Methods*, *11*(13), 1788–1794. https://doi.org/10.1039/c8ay02559b

Mariano, S., Tacconi, S., Fidaleo, M., Rossi, M., & Dini, L. (2021). Micro and nanoplastics identification: Classic methods and innovative detection techniques. *Frontiers in Toxicology*, *3*, 636640.

Markic, A., Bridson, J. H., Morton, P., Hersey, L., Budiša, A., Maes, T., & Bowen, M. (2023). Microplastic pollution in the intertidal and subtidal sediments of Vava'u, Tonga. *Marine Pollution Bulletin*, *186*, 114451. https://doi.org/10.1016/j.marpolbul.2022.114451

Mattsson, K., da Silva, V. H., Deonarine, A., Louie, S. M., & Gondikas, A. (2021). Monitoring anthropogenic particles in the environment: Recent developments and remaining challenges at the forefront of analytical methods. *Current Opinion in Colloid & Interface Science*, *56*, 101513.

Mikac, L., Rigó, I., Himics, L., Tolić, A., Ivanda, M., & Veres, M. (2023). Surface-enhanced Raman spectroscopy for the detection of microplastics. *Applied Surface Science*, *608*, 155239. https://doi.org/10.1016/j.apsusc.2022.155239

Moog, D., Schmitt, J., Senger, J., Zarzycki, J., Rexer, K. H., Linne, U., Erb, T., & Maier, U. G. (2019). Using a marine microalga as a chassis for polyethylene terephthalate (PET) degradation. *Microbial Cell Factories, 18*(1), 1–15. https://doi.org/10.1186/s12934-019-1220-z

Müller, Y. K., Wernicke, T., Pittroff, M., Witzig, C. S., Storck, F. R., Klinger, J., & Zumbülte, N. (2020). Microplastic analysis – Are we measuring the same? Results on the first global comparative study for microplastic analysis in a water sample. *Analytical and Bioanalytical Chemistry, 412*(3), 555–560.

Nakazawa, Y., Matsui, Y., Hanamura, Y., Shinno, K., Shirasaki, N., & Matsushita, T. (2018). Identifying, counting, and characterizing superfine activated-carbon particles remaining after coagulation, sedimentation, and sand filtration. *Water Research, 138*, 160–168. https://doi.org/10.1016/j.watres.2018.03.046

Napper, I.E., Baroth, A., Barrett, A.C., Bhola, S., Chowdhury, G.W., Davies, B.F.R., Duncan, E.M., Kumar, S., Nelms, S.E., Niloy, M.N.H., Nishat, B., Maddalene, T., Thompson, R.C., & Koldewey, H. (2021) The abundance and characteristics of microplastics in surface water in the transboundary Ganges River, *Environmental Pollution, 274*, 116348, https://doi.org/10.1016/j.envpol.2020.116348.

Nor, N. H. M., & Obbard, J. P. (2014). Microplastics in Singapore's coastal mangrove ecosystems. *Marine Pollution Bulletin, 79*(1–2), 278–283.

O'Connor, J. D., Mahon, A. M., Ramsperger, A., Trotter, B., Redondo-Hasselerharm, P. E., Koelmans, A. A., Lally, H. T., & Murphy, S. (2019). Microplastics in freshwater biota: A critical review of isolation, characterization, and assessment methods. *Global Challenges, 4*(6) *1800118.*

Okoffo, E. D., O'Brien, S., O'Brien, J. W., Tscharke, B. J., & Thomas, K. V. (2019). Wastewater treatment plants as a source of plastics in the environment: A review of occurrence, methods for identification, quantification and fate. *Environmental Science: Water Research & Technology, 5*(11), 1908–1931.

Onyena, A. P., Aniche, D. C., Ogbolu, B. O., Rakib, M. R., Uddin, J., & Walker, T. R. (2022). Governance strategies for mitigating microplastic pollution in the marine environment: a review. *Microplastics, 1*, 15–46.

Paço, A., Jacinto, J., da Costa, J. P., Santos, P. S. M., Vitorino, R., Duarte, A. C., & Rocha-Santos, T. (2019). Biotechnological tools for the effective management of plastics in the environment. *Critical Reviews in Environmental Science and Technology, 49*(5), 410–441. https://doi.org/10.1080/10643 389.2018.1548862

Pasquier, G., Doyen, P., Carlesi, N., & Amara, R. (2022). An innovative approach for microplastic sampling in all surface water bodies using an aquatic drone. *Heliyon, 8*(11). https://doi.org/10.1016/j.heliyon.2022. e11662

Pastorino, P., Anselmi, S., Esposito, G., Bertoli, M., Pizzul, E., Barceló, D., Elia, A. C., Dondo, A., Prearo, M., & Renzi, M. (2023). Microplastics in biotic and abiotic compartments of high-mountain lakes from Alps. *Ecological Indicators, 150*, 110215. https://doi.org/10.1016/j.ecolind.2023.110215

Perren, W., Wojtasik, A., & Cai, Q. (2018). Removal of microbeads from wastewater using electrocoagulation. *ACS Omega, 3*(3), 3357–3364. https://doi.org/10.1021/acsomega.7b02037/asset/images/ao-2017-0203 7a_m013.gif

Picó, Y., & Barceló, D. (2021). Analysis of microplastics and nanoplastics: How green are the methodologies used?, *Current Opinion in Green and Sustainable Chemistry, 31*, 100503, https://doi.org/10.1016/ j.cogsc.2021.100503

Poerio, T., Piacentini, E., & Mazzei, R. (2019). Membrane processes for microplastic removal. *Molecules, 24*(22). https://doi.org/10.3390/molecules24224148

Polt, L., Motyl, L., & Fischer, E. K. (2023). Abundance and distribution of microplastics in invertebrate and fish species and sediment samples along the German Wadden Sea coastline. *Animals, 13*(10). https://doi. org/10.3390/ani13101698

Prata, J. C., da Costa, J. P., Lopes, I., Duart,e A. C., & Rocha-Santos, T. (2019). Effects of microplastics on microalgae populations: A critical review. *Science of the Total Environment, 665*, 400–405. https:// doi:10.1016/j.scitotenv.2019.02.132.

Prata, J. C., Reis, V., Paço, A., Martins, P., Cruz, A., da Costa, J. P., Duarte, A. C., & Rocha-Santos, T. (2020). Effects of spatial and seasonal factors on the characteristics and carbonyl index of (micro) plastics in a sandy beach in Aveiro, Portugal. *Science of The Total Environment, 709*, 135892. https://doi.org/ 10.1016/j.scitotenv.2019.135892

Rajandas, H., Parimannan, S., Sathasivam, K., Ravichandran, M., & Su Yin, L. (2012). A novel FTIR-ATR spectroscopy based technique for the estimation of low-density polyethylene biodegradation. *Polymer Testing, 31*(8), 1094–1099. https://doi.org/10.1016/j.polymertesting.2012.07.015

Reisser, J., Slat, B., Noble, K., Du Plessis, K., Epp, M., Proietti, M., De Sonneville, J., Becker, T., & Pattiaratchi, C. (2015). The vertical distribution of buoyant plastics at sea: An observational study in the North Atlantic Gyre. *Biogeosciences, 12*(4), 1249–1256.

Rendell-Bhatti, F., Bull, C., Cross, R., Cox, R., Adediran, G. A., & Lahive, E. (2023). From the environment into the biomass: Microplastic uptake in a protected lamprey species. *Environmental Pollution, 323*, 121267. https://doi.org/10.1016/j.envpol.2023.121267

Rezania, S., Park, J., Din, M. F. M., Taib, S. M., Talaiekhozani, A., Yadav, K. K., & Kamyab, H. (2018). Microplastics pollution in different aquatic environments and biota: A review of recent studies. *Marine Pollution Bulletin, 133*, 191–208.

Ribeiro-Claro, P., Nolasco, M. M., & Araújo, C. (2017). Characterization of microplastics by Raman Spectroscopy. *Comprehensive Analytical Chemistry, 75*, 119–151. https://doi.org/10.1016/BS.COAC.2016.10.001

Rocha-Santos, T., & Duarte, A. C. (2015). A critical overview of the analytical approaches to the occurrence, the fate and the behavior of microplastics in the environment. *TrAC Trends in Analytical Chemistry, 65*, 47–53. https://doi.org/10.1016/J.TRAC.2014.10.011

Sajad, S., Allam, B. K., Mushtaq, Z., & Banerjee, S. (2023). Microplastic detection and analysis from water and sediment: A review. *Macromolecular Symposia, 407*(1), 2100367. https://doi.org/10.1002/masy.202100367

Schwaferts, C., Niessner, R., Elsner, M., & Ivleva, N. P. (2019). Methods for the analysis of submicrometer- and nanoplastic particles in the environment. *TrAC Trends in Analytical Chemistry, 112*, 52–65. https://doi.org/10.1016/j.trac.2018.12.014

Scopetani, C., Chelazzi, D., Mikola, J., Leiniö, V., Heikkinen, R., Cincinelli, A., & Pellinen, J. (2020). Olive oil-based method for the extraction, quantification and identification of microplastics in soil and compost samples. *Science of the Total Environment, 733*, 139338. https://doi.org/10.1016/j.scitotenv.2020.139338

Sharma, B., Dangi, A. K., & Shukla, P. (2018). Contemporary enzyme based technologies for bioremediation: A review. *Journal of Environmental Management, 210*, 10–22. https://doi.org/10.1016/j.jenvman.2017.12.075

Silva, A. B., Bastos, A. S., Justino, C. I. L., da Costa, J. P., Duarte, A. C., & Rocha-Santos, T. A. P. (2018). Microplastics in the environment: Challenges in analytical chemistry – A review. *Analytica Chimica Acta, 1017*, 1–19. https://doi.org/10.1016/j.aca.2018.02.043

Sorolla-Rosario, D., Llorca-Porcel, J., Pérez-Martínez, M., Lozano-Castelló, D., & Bueno-López, A. (2023). Microplastics' analysis in water: Easy handling of samples by a new thermal extraction desorption-gas chromatography-mass spectrometry (TED-GC/MS) methodology. *Talanta, 253*, 123829. https://doi.org/10.1016/j.talanta.2022.123829

Soursou, V., Campo, J., & Picó, Y. (2023). A critical review of the novel analytical methods for the determination of microplastics in sand and sediment samples. *TrAC Trends in Analytical Chemistry, 166*, 117190. https://doi.org/10.1016/j.trac.2023.117190

Srisiri, S., Haetrakul, T., Dunbar, S. G., & Chansue, N. (2024). Microplastic contamination in edible marine fishes from the upper Gulf of Thailand. *Marine Pollution Bulletin, 198*, 115785. https://doi.org/10.1016/j.marpolbul.2023.115785

Stock, F., Kochleus, C., Bänsch-Baltruschat, B., Brennholt, N., & Reifferscheid, G. (2019). Sampling techniques and preparation methods for microplastic analyses in the aquatic environment – A review. *TrAC Trends in Analytical Chemistry, 113*, 84–92. https://doi.org/10.1016/j.trac.2019.01.014

Sun, J., Dai, X., Wang, Q., van Loosdrecht, M. C. M., Ni, & B.-J. (2019). Microplastics in wastewater treatment plants: Detection, occurrence and removal. *Water Research, 152*, 21–37. https://doi.org/10.1016/j.watres.2018.12.050

Sun, C., Wang, Z., Chen, L., & Li, F. (2020). Fabrication of robust and compressive chitin and graphene oxide sponges for removal of microplastics with different functional groups. *Chemical Engineering Journal, 393*. https://doi.org/10.1016/j.cej.2020.124796

Talvitie, J., Mikola, A., Koistinen, A., & Setälä, O. (2017). Solutions to microplastic pollution – Removal of microplastics from wastewater effluent with advanced wastewater treatment technologies. *Water Research, 123*, 401–407. https://doi.org/10.1016/j.watres.2017.07.005

Tampang, A. M. A. A, & Viswanathan, P. M. (2022). Occurrence, distribution and sources of microplastics in beach sediments of Miri coast, NW Borneo. *Chemosphere, 305*, 135368. https://doi.org/10.1016/j.chemosphere.2022.135368

Tang, Y., Zhang, S., Su, Y., Wu, D., Zhao, Y., & Xie, B. (2021). Removal of microplastics from aqueous solutions by magnetic carbon nanotubes. *Chemical Engineering Journal*, *406*, 126804. https://doi.org/10.1016/j.cej.2020.126804

Tiwari, E., Singh, N., Khandelwal, N., Monikh, F. A., & Darbha, G. K. (2020). Application of Zn/Al layered double hydroxides for the removal of nano-scale plastic debris from aqueous systems. *Journal of Hazardous Materials*, *397*, 122769. https://doi.org/10.1016/j.jhazmat.2020.122769

Truc, N. T. T, & Lee, B. K. (2019). Sustainable hydrophilization to separate hazardous chlorine PVC from plastic wastes using H_2O_2/ultrasonic irrigation. *Waste Management*, *88*, 28–38. https://doi.org/10.1016/j.wasman.2019.03.033

Venkatramanan, S., Chung, S. Y., Selvam, S., Sivakumar, K., Soundhariya, G. R., Elzain, H. E., & Bhuyan, Md. S. (2022). Characteristics of microplastics in the beach sediments of Marina tourist beach, Chennai, India. *Marine Pollution Bulletin*, *176*, 113409. https://doi.org/10.1016/j.marpolbul.2022.113409

Wan, T. J., Shen, S. M., Siao, S. H., Huang, C. F., & Cheng, C. Y. (2011). Using magnetic seeds to improve the aggregation and precipitation of nanoparticles from backside grinding wastewater. *Water Research*, *45*(19), 6301–6307. https://doi.org/10.1016/j.watres.2011.08.067

Wang, W., & Wang, J. (2018). Investigation of microplastics in aquatic environments: An overview of the methods used, from field sampling to laboratory analysis. *TrAC Trends in Analytical Chemistry*, *108*, 195–202. https://doi.org/10.1016/j.trac.2018.08.026

Wang, Z., Sedighi, M., & Lea-Langton, A. (2020). Filtration of microplastic spheres by biochar: Removal efficiency and immobilisation mechanisms. *Water Research*, *184*, 116165. https://doi.org/10.1016/j.watres.2020.116165

Weber, A., Scherer, C., Brennholt, N., Reifferscheid, G., & Wagner, M. (2018). PET microplastics do not negatively affect the survival, development, metabolism and feeding activity of the freshwater invertebrate *Gammarus pulex*. *Environmental Pollution*, *234*, 181–189.

Wesch, C., Bredimus, K., Paulus, M., & Klein, R. (2016). Towards the suitable monitoring of ingestion of microplastics by marine biota: A review. *Environmental Pollution*, *218*, 1200–1208.

Woodall, L. C., Sanchez-Vidal, A., Canals, M., Paterson, G. L. J., Coppock, R., Sleight, V., Calafat, A., Rogers, A. D., Narayanaswamy, B. E., & Thompson, R. C. (2014). The deep sea is a major sink for microplastic debris. *Royal Society Open Science*, *1*(4), 140317.

Wu, M., Yang, C., Du, C., & Liu, H. (2020). Microplastics in waters and soils: Occurrence, analytical methods and ecotoxicological effects. *Ecotoxicology and Environmental Safety*, *202*, 110910. https://doi.org/10.1016/j.ecoenv.2020.110910

Wu, N., Wyart, Y., Liu, Y., Rose, J., & Moulin, P. (2013). An overview of solid/liquid separation methods and size fractionation techniques for engineered nanomaterials in aquatic environment. *Environmental Technology,* *2*(1), 55–70. https://doi.org/10.1080/09593330.2013.788073

Xiang, S., Xie, Y., Sun, X., Du, H., & Wang, J. (2022). Identification and quantification of microplastics in aquaculture environment. *Frontiers in Marine Science*, *8*, 1–10. https://doi.org/10.3389/fmars.2021.804208

Xu, J.-L., Thomas, K. V, Luo, Z., & Gowen, A. A. (2019). FTIR and Raman imaging for microplastics analysis: State of the art, challenges and prospects. *TrAC Trends in Analytical Chemistry*, *119*, 115629. https://doi.org/10.1016/j.trac.2019.115629

Xu, X., Zong, S., Chen, W., & Liu, D. (2019). Comparative study of Bisphenol A degradation via heterogeneously catalyzed H_2O_2 and persulfate: Reactivity, products, stability and mechanism. *Chemical Engineering Journal*, *369*, 470–479. https://doi.org/10.1016/j.cej.2019.03.099

Yasaka, S., Pitaksanurat, S., Laohasiriwong, W., Neeratanaphan, L., Jungoth, R., Donprajum, T., & Taweetanawanit, P. (2022). Bioaccumulation of microplastics in fish and snails in the Nam Pong River, Khon Kaen, Thailand. *Environment Asia*, *15*(1), 81–93. https://doi.org/10.14456/ea.2022.8

Ye, Y., Yu, K., & Zhao, Y. (2022). The development and application of advanced analytical methods in microplastics contamination detection: A critical review. *Science of The Total Environment*, *818*, 151851. https://doi.org/10.1016/j.scitotenv.2021.151851

Yu, X., Peng, J., Wang, J., Wang, K., & Bao, S. (2016). Occurrence of microplastics in the beach sand of the Chinese inner sea: The Bohai Sea. *Environmental Pollution*, *214*, 722–730.

Yuan, W., Liu, X., Wang, W., Di, M., & Wang, J. (2019). Microplastic abundance, distribution and composition in water, sediments, and wild fish from Poyang Lake, China. *Ecotoxicology and Environmental Safety*, *170*, 180–187. https://doi.org/10.1016/j.ecoenv.2018.11.126

Yusuf, A., Sodiq, A., Giwa, A., Eke, J., Pikuda, O., Eniola, J. O., Ajiwokewu, B., Sambudi, N. S., & Bilad, M. R. (2022). Updated review on microplastics in water, their occurrence, detection, measurement, environmental pollution, and the need for regulatory standards. *Environmental Pollution*, *292*, 118421. https://doi.org/10.1016/j.envpol.2021.118421

Zhang, K., Su, J., Xiong, X., Wu, X., Wu, C., & Liu, J. (2016). Microplastic pollution of lakeshore sediments from remote lakes in Tibet plateau, China. *Environmental Pollution*, *219*, 450–455.

Zhang, Y., Diehl, A., Lewandowski, A., Gopalakrishnan, K., & Baker, T. (2020). Removal efficiency of micro- and nanoplastics (180 nm–125 µm) during drinking water treatment. *Science of the Total Environment*, *720*. https://doi.org/10.1016/j.scitotenv.2020.137383

Zhang, Y., Jiang, H., Bian, K., Wang, H., & Wang, C. (2021). A critical review of control and removal strategies for microplastics from aquatic environments. *Journal of Environmental Chemical Engineering*, *9*(4). https://doi.org/10.1016/j.jece.2021.105463

Zhao, K., Wei, Y., Dong, J., Zhao, P., Wang, Y., Pan, X., & Wang, J. (2022). Separation and characterization of microplastic and nanoplastic particles in marine environment. *Environmental Pollution*, *297*, 118773. https://doi.org/10.1016/j.envpol.2021.118773

Zhao, W., Li, J., Liu, M., Wang, R., Zhang, B., Meng, X. Z., & Zhang, S. (2024). Seasonal variations of microplastics in surface water and sediment in an inland river drinking water source in southern China. *Science of The Total Environment*, *908*, 168241. https://doi.org/10.1016/j.scitotenv.2023.168241

9 Analytical and Detection Techniques for Microplastics

Anu Kumari, Meenu Yadav, and Rachna Bhateria

9.1 INTRODUCTION

The word "plastics" is derived from the Greek word "plastikos," which means a flexible and moldable material. Plastics were first introduced in 1907 and have gained immense popularity in the industry and modern life since their subsequent mass production in the mid-1950s (Miranda et al., 2019). The synthetic organic polymer known as plastic is a hydrophobic organic polymer made up of long monomer chains joined by covalent bonds. It is known for being flexible, robust, steady, and affordable. A wide range of products, including water bottles, food packaging, electronics, clothing, medical supplies, and building materials, use plastics. According to the World Economic Forum, 14–18% of plastic garbage is recycled, 25% is burnt, and 56% is deposited in landfills and oceans (WEF, 2020). Plastic garbage is currently generated at a rate of roughly 2.1 billion metric tonnes per year, and is expected to increase to 3.4 billion metric tonnes by 2050 (Khan et al., 2022).

Plastics can break down into smaller and smaller fragments as they age in the environment, mostly due to physical-chemical degradation and microbial deterioration. On the basis of fragment size, plastic can be classified into macroplastics (>5 mm), microplastics (<5 mm), and nanoplastics (size range from 1–100 nm) (Yang et al., 2020). The term "microplastics" was first introduced in 2004 (Law and Thompson, 2014). Plastic debris (microplastics) can remain for millions of years in oceans, rivers, lakes, and land due to its non-biodegradable nature. The most common plastic polymers that are the main source of microplastics are polypropylene (PP), polystyrene (PS), polyethylene (PE), thermoplastic polyester (PET), and polyvinyl chloride (PVC). Numerous items in our daily lives, such as makeup, single-use clothes, kitchenware, tapes, rugs, wraps, shampoo bottles, and wire insulation, contain microplastics (Ziani et al., 2023).

Microplastics are categorized as primary or secondary microplastics based on their origin. Primary microplastics are polymers with dimensions of 5 mm or less and secondary microplastics are produced by the weathering and fragmentation of plastic debris in the environment, such as plastic bags (Ma et al., 2019). The monitoring of microplastics in various ecosystems (oceans, rivers, lakes, terrestrial, and atmosphere) has received a lot of attention recently. According to estimates, approximately 15–31% of the microplastics discharged into the environment are primary microplastics, whereas 70–80% are secondary microplastics. Additionally, a number of physical characteristics, such as density (light/heavy), flexibility (hard/soft), or shape (fragments, pellets, filaments, and granules), are used to categorize microplastics (Sheriff et al., 2023).

Microplastics are made up of a variety of materials that differ not only in particle attributes such as size and shape but also in chemical composition. Toxic compounds associated with microplastics via adsorption processes have an impact on the biotic and abiotic environments (Campanale et al., 2020). Desorption mechanisms allow dangerous compounds to be discharged into the environment

DOI: 10.1201/9781032706573-9

after ingestion, potentially causing toxicity and/or accumulating in the food chain. Once in the ecosystem, microplastics interact with the numerous species there, potentially altering their behavior. Microplastics have been shown in the literature to have minimal effects on a variety of species, including bacteria, yeast, protozoa, and nematodes (Heinlaan et al., 2020), while other studies show different toxic effects of microplastics and related contaminants in invertebrates (Foley et al., 2018) and vertebrates (Miranda et al., 2019).

Microplastics enter into the human body mostly by food, inhalation, and skin contact. The identification of microplastics and nanoplastics has a significant impact on biological barrier diffusion, epithelial adherence, and bioaccumulation in deep tissues (Prata et al., 2020). The study of microplastics is very important for tracing microplastics and for their better removal from the environment (Zhang et al., 2018). As a result, microplastics analysis is an essential precursor and basis for treating microplastic pollution. Microplastics, consisting of polymers with high molecular masses, pose challenges for direct investigation using common analytical methods such as liquid chromatography (LC), gas chromatography (GC), and mass spectrometry (MS). The size of microplastics hinders their entry into chromatographic columns, leading to clogging and incomplete separation. The complex mixtures of polymers, along with their non-volatile nature, make them unsuitable for effective separation and analysis by GC. Additionally, the high molecular masses of these polymers complicate ionization and fragmentation in mass spectrometry, rendering it challenging to obtain clear information about their structure. As a result, alternative techniques such as spectroscopy and microscopy are often employed for the direct observation and identification of microplastics. Over the past 10 years, significant scientific research work has been conducted globally to address this problem (Huppertsberg et al., 2018). In light of this, microplastic analysis in this chapter is divided into physical characterization (visual analysis, dynamic light scattering, and laser diffraction) and chemical composition identification which includes analytical techniques and microscopy that are further classified as shown in Figure 9.1. In this study, microplastics analysis techniques and their

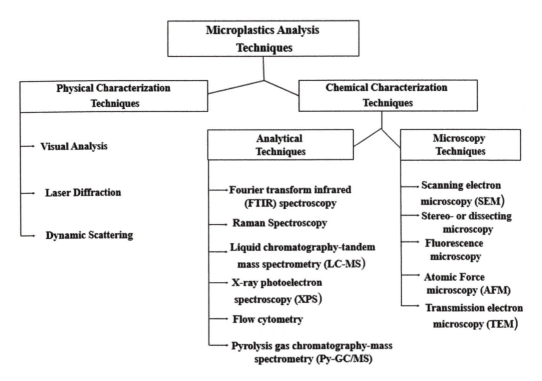

FIGURE 9.1 Different microplastic analysis techniques.

Analytical and Detection Techniques of Microplastics

applications are based on these two factors. The main goal of this chapter is to review unique and efficient microplastics detection techniques for better identification, detection, and quantification.

9.2 SAMPLING

To effectively study microplastics, appropriate sampling techniques need to be employed to capture these plastic particles from different environmental media such as water, sediment, soil, and biota. The choice of sampling method depends on the specific research objectives and the environmental components being investigated, as shown in Table 9.1.

Microplastics in the air are a growing problem. Airborne microplastics can travel considerable distances in the open air. Human exposure by microplastic inhalation and ingestion may be especially hazardous indoors due to the abundance of plastic objects, and their limited recycling. The data on microplastics remain sparse, although there is a considerable body of literature on ambient particles in indoor and outdoor air. These are commonly collected using various porosity filters. It should be noted that glass fiber filters are frequently utilized in this application, and this matrix may enmesh and obfuscate microplastics, complicating subsequent spectroscopic examination (Zhang et al., 2020). Sediment sampling involves collecting sediment cores using coring devices, and the sediments can be further processed for microplastic extraction. Soil sampling can be done using coring or excavation techniques, sieving, and density separation methods which are employed for microplastics isolation. For water sampling, commonly used techniques include surface water grab sampling, in which water samples are collected using bottles or buckets, and towed nets such as manta trawls and neuston nets. These methods allow for the collection of large volumes of water, which can be subsequently filtered to concentrate microplastic particles. Biota sampling involves the collection of organisms, such as fish, shellfish, and zooplankton, which can ingest or accumulate

TABLE 9.1
Comparison of Different Microplastics Samples

Sample Type	Collection Method	Techniques	Disadvantages	References
Water sample	Collect water in a container or use a plankton net for smaller organisms	Filtration, visual inspection, Fourier transform infrared spectroscopy (FTIR), Raman spectroscopy	Low concentrations, interference from natural particles, limited detection of smaller particles	(Lwanga et al. 2017)
Air sample	Use air samplers such as high-volume air samplers or passive air samplers	Filtration, FTIR, scanning electron microscopy (SEM)	Low concentrations, potential loss during collection, interference from other airborne particles	(Zhang et al. 2020)
Sediment sample	Collect sediment using sediment grabs or coring techniques	Density separation, sieving, FTIR, SEM	High background contamination, complex matrix, time-consuming sample preparation	(Horton et al. 2017)
Biota sample	Collect organisms (e.g., fish, mussels) using nets, traps, or direct sampling	Digestion, density separation, FTIR, SEM	Low extraction efficiency, potential damage to organisms, identification of microplastics within tissues	(Prata et al. 2019)

microplastics. These organisms are typically dissected, and their gastrointestinal tracts or tissues are analyzed for microplastic content (Prata et al., 2019).

9.3 SEPARATION OF MICROPLASTICS FROM SAMPLES

Separation techniques for obtaining microplastics from sample analysis include digestion, density separation, sieving, and filtration. Separation methods may vary depending on the sample type, the presence of interfering substances, and the analytical techniques employed for microplastic analysis. Digestion involves removing organic matter from the sample and leaving behind microplastics. Digestion methods must be compatible with the analytical targets to obtain reliable results. Enzymatic digestion involves enzymes breaking down organic matter, while acid digestion dissolves organic materials while microplastics remain unaffected. The resulting solution is then filtered or centrifuged to separate microplastics. Chemical digestion is a common pretreatment technique used to reduce potential interferences in environmental samples containing microplastics. Complex matrices can be broken down by a variety of chemicals and reagents, including acids, bases, oxidizing agents, and enzymes (Rist et al., 2017).

Density separation techniques separate microplastics from denser materials in the sample, such as sediment or soil particles. Flotation involves using a saline solution with a specific density higher than microplastics, allowing microplastics to float to the surface. Heavy liquid separation uses dense liquids with densities higher than that of microplastics, allowing microplastics to sink to the bottom while other particles remain suspended or float. Sodium chloride is inexpensive, simple to obtain, and safe for the environment. For the separation of microplastics, saturated NaCl solution, which has a density of 1.2 g cm^3, is frequently used as a density flotation solution to separate microplastics (Green et al., 2018).

Another technique for removing microplastics from digested solutions is filtering and purification. Filtration methods separate microplastics from water samples or further concentrate them after digestion or density separation steps. For the filtering and eventual identification of microplastics, selecting the right filter membrane is essential. Cellulose nitrate membrane filters, nylon membrane filters, glass fiber filters, cellulose acetate membrane filters, and polycarbonate membrane filters are a few of the materials utilized to make filter membranes and have been used in several investigations. Sieving involves using various specific pore sizes to separate microplastics from larger particles. Membrane filtration uses filters with specific pore sizes to retain microplastics while allowing other particles and solutes to pass through (Radford et al., 2021).

Microplastic separation utilizing oils is also gaining popularity these days. Recent studies have reported the separation of microplastics using castor and olive oil. This method is simple and relies on solvent extraction. For six different types of plastic (PE, PS, PET, PVC, polycarbonate, and polyurethane) a high recovery rate of MPs with a value of 90%–97% was observed. The high MPs recovery rate, environmental friendliness, and resource-efficient separation are caster oil's key benefits (Scopetani et al., 2020). Table 9.2 discusses various treatment methods for the analysis of microplastics with their advantages and disadvantages.

9.4 PHYSICAL MICROPLASTIC ANALYSIS TECHNIQUES

Microplastic particle size, shape, color, morphology, corrosion degree, and aging degree can all be determined by physical characterization analysis. Visual analysis, dynamic light scattering analysis, and laser diffraction particle size analysis are the main physical characterization study approaches commonly used for microplastics analysis and detection. One of the most popular techniques for examining the physical characteristics of microplastics is visual inspection. The pretreated samples must first be examined with the naked eye or a microscope as the first stage in the visual analysis approach. The microplastics can then be broadly classed and numbered according to their color,

Analytical and Detection Techniques of Microplastics

TABLE 9.2

Comparison of Various Treatment Methods Used for Microplastics Analysis

Treatment Method	Solution Used	Recovery Rate	Advantages	Disadvantages	References
Digestion method	Strong base and acids	98–100%	Cost effective and efficient method	Causes the decomposition of some polymers	(Olsen et al. 2020)
Density separation	NaCl, ZnCl, Distilled water	91–99%	Rapid, reproducible and low cost	Requires solution recovery due to high cost, toxicity, and corrosiveness	(Li et al. 2019)
Flotation	Distilled water	90%	Can separate MPs from oil	Time-consuming	(Zhang et al. 2018)
Filtration	Filter papers of different types	–	Traditional method and easy to use	MPs stick with filter paper and requires an additional step such as washing	(Radford et al. 2021)
Oil-based extraction	Olive and coaster oil	99%	High recovery rate for MPs extraction	Poor separation efficiency for fluoropolymers	(Scopetani et al. 2019)

shape, and size. The microplastics are then removed using tweezers. Visual analysis can be done with the naked eye or under a microscope. Only microplastics with particle sizes of 1–5 mm can be distinguished by the naked eye, however, microscopic inspection can reveal microplastics with particle sizes of at least 100 microns (Dris et al., 2015). The accuracy of visual analysis to identify microplastics can improve by combining it with fluorescence staining. The visual analysis approach is advantageous in that it is easy to use, inexpensive, and doesn't provide a significant chemical risk when in use. However, visual analysis is unable to reveal the chemical makeup of microplastics. It also requires a lot of time and labor. The microplastics experimental results are influenced by various factors such as environmental media and impurities in samples, which further reduces the accuracy and efficiency of the visual analysis method. For instance, when microplastics are very small in amount or environmental samples contain additional particle pollutants, the visual analysis approach is difficult to apply. As a result, the visual analysis methodology is often used in conjunction with other microplastics analysis techniques rather than as a stand-alone technique (Liu et al., 2022).

Laser diffraction is an effective physical technique for identifying and analyzing microplastics. It offers a reliable and efficient method for characterizing microplastics based on their size distribution. The principle of laser diffraction involves the interaction of a laser beam with microplastic particles suspended in a liquid medium. When a laser beam passes through the sample, it interacts with the microplastics, causing scattering of the light. The scattered light is then detected at different angles using a collection optics system. Based on the intensity and angular distribution of the scattered light, the size distribution of the microplastics can be determined (Rakesh et al., 2021). The microplastic analysis by laser diffraction involves several steps: sample preparation, laser illumination, scattered light collection, intensity measurement, and data analysis. Sample preparation involves suspending microplastics in a liquid medium to ensure proper particle dispersion, allowing for accurate measurements. Laser illumination directs a laser beam through the sample, resulting in scattering of light. The scattered light is collected at different angles using a collection optics system, with multiple detectors employed to capture it simultaneously. The intensity of the scattered light is measured for each angle, directly related to the size of microplastic particles, allowing for size estimation. Data analysis is then performed using mathematical algorithms to determine the size distribution of microplastics in the sample, providing information about their abundance and size range (Enders et al., 2015). Although this technique has not been widely used to detect the spread of microplastic particles, it will eventually be crucial to the detection of the distribution of microplastic particles. It provides rapid and non-destructive measurements, allowing for real-time monitoring of microplastic pollution (Lusher et al., 2017). Additionally, laser diffraction can handle a wide size range of particles, from a few micrometers to hundreds of micrometers, making it suitable for analyzing microplastics in various environmental samples.

Dynamic scattering is another physical technique used for the identification and analysis of microplastics based on the phenomenon of dynamic light scattering. It provides valuable insights into the size and size distribution of microplastic particles in environmental samples. It is also called quasi-elastic light scattering and is commonly use to study the Brownian motion of macromolecules in solution. It is used for the analysis of particles size in range from 0.3 nm to 10,000 nm. Dynamic scattering works by analyzing the fluctuations in the intensity of light scattered by microplastic particles suspended in a liquid medium. When a laser beam is directed through the sample, it interacts with the microplastics, causing the scattered light to fluctuate. These fluctuations are then analyzed to extract information about the size distribution of the microplastics. Microplastic analysis by dynamic scattering involves several steps, including sample preparation, laser illumination, scattered light detection, dynamic light scattering analysis, and data interpretation. Sample preparation involves suspending microplastics in a liquid medium, while laser illumination involves interacting with the particles and scattering light. The scattered light is collected and detected at a specific angle using a photodetector, and the intensity is measured over time. Dynamic light scattering analysis analyzes fluctuations in the intensity, providing information on the size distribution of microplastics. Data interpretation helps determine the size range, abundance, and other characteristics of microplastics present in the sample (Lwanga et al., 2017). It offers several advantages for microplastic identification and analysis. It is a non-destructive and rapid technique that allows for real-time monitoring of microplastic populations. Additionally, it provides information regarding the size distribution of the particles, which is crucial for understanding their environmental impact and potential risks. It requires only a small volume of sample, typically in the milliliter range, making it suitable for situations where the sample volume is limited or precious (Gerolin et al., 2020).

This technique has limited resolution for distinguishing particles that are very close in size, especially in cases where the size distribution is narrow or bimodal. It does not provide direct information about the shape or morphology of microplastics, which can be relevant for certain applications. Dynamic scattering measurements can be affected by optical interferences, such as sample turbidity or the presence of colored impurities. These interferences can lead to erroneous results or limit the applicability of the technique for highly turbid or complex sample matrices. The commonly used physical methods for microplastic analysis and comparison of these techniques are shown in Table 9.3.

9.5 ANALYTICAL TECHNIQUES

Chemical characterization of microplastics involves the analysis of their composition, additives, and associated chemicals. Understanding the chemical properties of microplastics is crucial for assessing their potential ecological and human health impacts, as well as for identifying potential sources and degradation processes (Garaba et al., 2018). Several analytical techniques are employed to chemically characterize microplastic response; here some commonly used methods are discussed. Analyzing microplastics requires specialized analytical methods that can accurately detect, quantify,

TABLE 9.3
Comparison of Various Physical Characterization Techniques for Microplastic Analysis

Technique	Laser Diffraction	Visual Analysis	Dynamic Scattering	References
Principle	Measures particle size based on light diffraction patterns	Relies on visual observation	Measures particle size based on light scattering	(Liu et al. 2022)
Parameters	Particle size distribution, concentration, morphology	Presence/absence, morphology, color, shape, texture	Particle size distribution, concentration	(Gerolin et al. 2020)
Detection limit	Typically, suitable for larger particles	Limited to visible range of microplastics	Suitable for a wide range of microplastic sizes	(Huang et al. 2023)
Speed	Rapid analysis	Time-consuming analysis	Rapid analysis	(Liu et al. 2021)
Applications	Bulk analysis of larger microplastics in environmental samples	Rapid screening for presence/absence of microplastics in visually clear samples	Quantitative analysis of microplastics in environmental samples	(Sorasan et al. 2021)

and characterize these small particles in various environmental matrices. A detailed comparison of different analytical techniques used for microplastics analysis is provided in Table 9.4.

9.5.1 Fourier Transform Infrared Spectroscopy

Fourier transform infrared (FTIR) spectroscopy is extensively used to investigate polymer chemical bonding. This approach is useful for identifying carbon-based polymers. Plastics have a different bonding configuration that creates unique spectra which are used to distinguish them from other organic and inorganic polymers. Plastics can also be detected using well-established polymer spectrum libraries. FTIR is used to identify colored microplastics whose pigments fluoresce and can interact with Raman spectra. Furthermore, FTIR identification exposes the polymer composition, which can provide information about the sample's origin (Veerasingam et al., 2021). It can also be used to study the oxidation state of microplastics by analyzing the composition of oxygen bonds (e.g., carbonyl groups). The interaction between infrared light and the molecular bonds present in the microplastic samples is crucial to the FTIR method for the characterization of microplastics. Specific infrared wavelengths are absorbed as a result of this interaction, and they can be utilized to determine the kinds of polymers contained in the samples.

FTIR involves sample preparation, spectral measurement, infrared radiation interaction, absorption spectrum acquisition, Fourier transform and spectrum analysis, and polymer identification. Microplastic samples are collected from environmental sources and prepared for analysis, with a mixture of different polymer types. The samples are mounted on a suitable substrate and placed in the FTIR instrument. The resulting spectrum represents the absorption of infrared radiation by the microplastic sample at different wavelengths, with distinctive absorption bands corresponding to the vibrational modes of the polymer molecules. The specific polymer types present in the microplastic sample are then identified by matching the absorption bands in the sample spectrum to those in the reference spectra (Araujo et al., 2018).

FTIR gives vital information on the polymers found in microplastics, allowing researchers to identify and quantify various polymer compositions. It can also provide insights into additional

154 Microplastic Pollution

TABLE 9.4

Comparison of Different Analytical Techniques Used for Analysis of Microplastics

Analytical Technique	Principle	Sample Preparation	Advantages	Limitations	References
FTIR	Absorption of infrared light by chemical bonds	Extraction, filtration, drying, spectroscopic analysis	High sensitivity for polymer identification	Limited to larger particle sizes, inability to differentiate similar polymers	(Akhbarizadeh et al. 2021)
Raman spectroscopy	Scattering of laser light by molecular vibrations	Minimal sample preparation, direct analysis	High sensitivity, rapid analysis	Limited penetration depth, fluorescence interference	(Sullivan et al. 2020)
LC-MS	Separation by liquid chromatography followed by mass spectrometric analysis	Extraction, chromatographic separation, mass spectrometric analysis	High sensitivity, capability for target compound analysis	Limited to soluble microplastics, complex sample matrix	(O'Brien et al. 2021)
XPS	Measurement of emitted electrons after X-ray excitation	Sample cleaning, surface analysis	Surface-specific analysis, elemental composition	Limited depth of analysis, requires vacuum conditions	(Torres et al. 2021)
Flow cytometry	Analysis of particle size and fluorescence signals using laser-based detection	Sample filtration, staining, flow cytometric analysis	High throughput, rapid analysis	Limited to larger microplastics, limited polymer identification	(Bianco et al. 2023)
Py-GC/MS	Pyrolysis of the sample followed by gas chromatography and mass spectrometry	Sample heating, pyrolysis, gas chromatographic separation, mass spectrometric analysis	Comprehensive analysis of polymer fragments	Limited to thermal-stable microplastics, extensive sample preparation	(Lusher et al. 2020)

properties such as chemical modifications, degradation, or the presence of additives. It is a non-destructive technique and offers several advantages for microplastic analysis, including its ability to identify a wide range of polymer types simultaneously and provide information about their chemical structure. However, it also has some limitations, such as difficulties in detecting very small microplastics and differentiating between polymers with similar spectra. The spatial resolution of the FITR spectroscopic spectrum is 10–20 μm, it can only detect microplastics with a particle size over 20 μm (Akhbarizadeh et al., 2021).

9.5.2 RAMAN SPECTROSCOPY

Raman spectroscopy is widely used for the recognition and detection of microplastics. It provides valuable information about the chemical composition and molecular structure of microplastic particles. The principle of Raman spectroscopy is based on the Raman effect, which involves the scattering of photons when they interact with molecular vibrations or rotations within a sample. When a laser beam is directed onto a microplastic particle, a small fraction of the scattered light undergoes a shift in energy (known as the Raman shift) due to the interaction with the molecular vibrations of the plastic material. This shift in energy provides information regarding the molecular composition and structure of the microplastic, enabling its detection and analysis (Catarino et al., 2017).

Raman spectroscopy provides several advantages for the analysis of microplastics, including its non-destructive nature, high specificity, and the ability to analyze samples directly without extensive sample preparation. It can be used to identify different types of plastics, detect additives or contaminants, and analyze the physical and chemical properties of microplastic particles. However, Raman spectroscopy has significant limitations when it comes to detecting microplastics. The first limitation of Raman spectroscopy's usage in the examination of microplastics in samples is that it cannot be utilized to detect samples with fluorescence (Sullivan et al., 2020). Additionally, Raman spectra produced by impurities adsorbed on the surface of microplastics and additives in microplastics and polymer Raman spectra may overlap, making it difficult to identify microplastics.

Briefly, the detection of microplastics employing Raman spectroscopy needs to be enhanced. Raman spectroscopy has recently seen a number of advancements. Raman spectroscopy technology is combined with optimized software to achieve efficient detection of microplastics with sizes of particles ranging from 1 μm to 500 μm (Araujo et al., 2018). In the future, Raman spectroscopy will be essential in determining the presence of microplastics in the real world.

9.5.3 LIQUID CHROMATOGRAPHY–TANDEM MASS SPECTROMETRY (LC-MS)

Liquid chromatography–tandem mass spectrometry (LC-MS) is a powerful analytical technique used for the detection and analysis of microplastics. It combines liquid chromatography (LC) separation abilities with the sensitive and focused mass spectrometry (MS) detection, enabling the identification and quantification of microplastic particles based on their chemical composition and structure (Avio et al., 2015). LC-MS is a method for detecting microplastics by utilizing sample preparation, liquid chromatography separation, and mass spectrometry. Samples are collected from environmental sources and pretreated to isolate microplastics. Different LC techniques, such as size exclusion chromatography (SEC) or reversed-phase chromatography, are used to separate microplastic particles based on size, hydrophobicity, or other properties. The eluted particles are introduced into a mass spectrometer for detection and identification. In tandem mass spectrometry (MS/MS), multiple stages of mass analysis are performed, including ionization, fragmentation, and mass-to-charge ratio analysis. The chemical composition and structure of the microplastics can be recognized by comparing mass spectra with databases or reference standards (O'Brien et al., 2021).

LC-MS offers several advantages for detecting microplastics, including high sensitivity and selectivity, chemical identification, quantitative analysis, and multiclass analysis. It allows for low concentration detection in complex environmental samples, provides insights into plastic types and sources, and can detect and analyze various plastic polymers, including polyethylene, polypropylene, polystyrene, and polyethylene terephthalate. However, it is important to note that the detection and identification of microplastics using LC-MS can be challenging due to the complexity of environmental samples, the potential presence of matrix interferences, and the lack of comprehensive libraries or reference standards for all types of microplastics (Napper and Thompson, 2020).

9.5.4 X-RAY PHOTOELECTRON SPECTROSCOPY (XPS)

X-ray photoelectron spectroscopy (XPS), also known as electron spectroscopy for chemical analysis (ESCA), is a technique widely used for the identification and detection of microplastics. It provides crucial details about the elemental composition, chemical bonding, and surface properties of microplastic samples. By utilizing XPS, researchers can gain insights into the types of plastics presents, their sources, and their potential interactions with the environment. The principle behind XPS involves irradiating a sample with monochromatic X-rays, typically generated by an X-ray source such as a magnesium or aluminum anode. When the X-rays interact with the sample's surface, they can ionize the atoms or molecules present, resulting in the emission of photoelectrons. The emitted photoelectrons are then detected and their kinetic energies are measured. The energy of the emitted electrons corresponds to the binding energy of the electrons in the sample's atoms, providing information about the elemental composition and chemical bonding (Käppler et al., 2016).

In the context of microplastics, XPS allows researchers to identify and distinguish plastic particles from other organic and inorganic materials. The characteristic binding energies of carbon (C 1s), oxygen (O 1s), and nitrogen (N 1s) atoms can be measured, providing information about the specific polymer types present in the microplastic sample. These data can be used to differentiate between different types of plastics, such as polyethylene (PE), polypropylene (PP), polyethylene terephthalate (PET), and others (Van et al., 2014). Furthermore, XPS can provide insights into the surface properties of microplastics, such as their oxidation state and the presence of surface coatings or contaminants. This information is valuable for understanding the potential interactions of microplastics with the environment and other substances.

XPS is used for the identification and detection of microplastics in various environmental samples. It is employed to identify different types of microplastics in sediment samples from an urban estuary and to determine the presence of various polymer types, such as polypropylene, polyethylene, and polystyrene, indicating different sources of contamination. These examples highlight the utility of XPS in the identification and characterization of microplastics. However, it is worth noting that XPS is primarily a surface-sensitive technique, and therefore, it may not provide information about the bulk composition or internal structure of microplastics. Complementary techniques, such as Fourier-transform infrared spectroscopy (FTIR) and Raman spectroscopy, are often employed alongside XPS to gain a more comprehensive understanding of microplastic samples (He et al., 2023).

9.5.5 FLOW CYTOMETRY

Flow cytometry is an advanced analytical technique that is increasingly being utilized for the detection and characterization of microplastics. Originally developed for cell analysis, flow cytometry offers several advantages in terms of high-throughput analysis, multi-parametric detection, and size-based discrimination, making it a promising tool for the detection and quantification of microplastics in various environmental samples (Karthik et al., 2018).

The flow cytometry method for detecting microplastics involves several steps: sample preparation, particle suspension, instrument setup, particle analysis, and data analysis. Samples are collected from various sources, with pretreatment steps required to isolate microplastics. The sample is suspended in a fluid, and the flow cytometer is calibrated and configured to detect and analyze microplastic particles. The particles are hydrodynamically focused and passed through laser beams, allowing for real-time analysis of scattered and fluorescent signals. Data analysis is performed using specialized software, allowing for the identification and classification of microplastics based on size, shape, and fluorescence characteristics (Bianco et al., 2023).

Flow cytometry is a powerful tool for detecting microplastics, offering high-throughput analysis, size-based discrimination, multiparametric analysis, and high sensitivity. It can detect microplastics at low concentrations, providing comprehensive information about particles, and can be enhanced with fluorescent probes or antibodies targeting specific microplastic characteristics. This method also offers high specificity, making it an ideal method for detecting microplastics in complex mixtures. Despite its advantages, there are certain limitations associated with flow cytometry for microplastic detection, including the need for proper calibration, standardization, and the potential interference from natural particles or organic matter in environmental samples.

Analytical and Detection Techniques of Microplastics

9.5.6 Pyrolysis Gas Chromatography–Mass Spectrometry (Py-GC/MS)

This is an effective analytical method for identifying and characterizing microplastics. Microplastics are small plastic particles, typically less than 5 mm in size, that have become a serious environmental issue as a result of their pervasive presence in ecosystems. Py-GC/MS combines the principles of pyrolysis, gas chromatography (GC), and mass spectrometry (MS) to investigate the chemical composition of microplastics. The technique requires subjecting the microplastic sample to high temperatures in a pyrolysis chamber, causing the polymers to thermally degrade into smaller fragments. These pyrolysis products are then separated and detected using a GC-MS system (Lusher et al., 2020).

The gas chromatograph separates the pyrolysis products based on their volatility and polarity, using a capillary column with a stationary phase. The separated compounds are subsequently into the mass spectrometer, which ionizes the molecules and measures their mass-to-charge ratio (m/z). By comparing the mass spectra of the pyrolysis products with reference databases or using other identification techniques, the chemical composition of the microplastics can be determined. Py-GC/MS offers several advantages for microplastic analysis. First, it allows for the identification of a wide range of polymers, including common plastics such as polyethylene (PE), polypropylene (PP), polyethylene terephthalate (PET), and polystyrene (PS). This capability is crucial because microplastics can originate from various sources, including consumer products, industrial processes, and environmental degradation. Second, the technique provides quantitative information about the abundance of different polymer types within a sample. This information is valuable for assessing the environmental impact of microplastics and understanding their sources and pathways of release into the environment. Overall, Py-GC/MS is a valuable tool for the analysis of microplastics, offering the ability to identify and quantify different polymer types. It enables researchers to gain insights into the sources, distribution, and environmental impact of microplastics, contributing to the development of effective mitigation strategies (Lusher et al., 2020).

9.6 MICROSCOPY TECHNIQUES

Microplastics have become a significant environmental concern due to their widespread presence in various ecosystems. Analyzing microplastics requires specialized microscopic techniques to accurately identify, quantify, and characterize these particles. Several advanced microscopic methods are employed for microplastic analysis (scanning electron microscopy, stereo- or dissecting microscopy, fluorescence microscopy, atomic force microscopy, and transmission electron microscopy).

9.6.1 Scanning Electron Microscopy (SEM)

Scanning electron microscopy (SEM) is a potent imaging technique widely used for the detection, characterization, and analysis of microplastics. SEM allows for high-resolution imaging and provides valuable morphological information about microplastic particles (size, shape, surface features, and composition). SEM operates by scanning a sample surface with a focused electron beam. Different signals are produced when the beam interacts with the material, including secondary electrons, backscattered electrons, and distinctive X-rays. These signals are collected and analyzed to create detailed images and obtain information about the sample's elemental composition (Minor et al., 2020).

SEM involves sample preparation, mounting and coating, instrument setup, imaging, and analysis. Microplastic samples are collected from various sources and undergo pretreatment steps to isolate particles. The sample is then mounted on a conductive substrate and coated with a thin layer of conductive material to enhance conductivity and prevent charging artifacts. The SEM instrument is calibrated and configured based on the sample's requirements, optimizing parameters like electron beam energy, aperture size, and working distance. The electron beam interacts with the sample, generating signals that are collected by detectors for high-resolution images and information about the microplastics' morphology, size, and surface characteristics. It offers high-resolution imaging of microplastic particles, allowing for surface features, shape, and size distribution observation. It also enables elemental analysis, characterization, and differentiation of polymer types. SEM can accommodate large sample sizes and can be combined with other techniques like FTIR or Raman spectroscopy for complementary information on microplastic chemical composition. SEM has disadvantages for microplastic detection, including time-consuming sample preparation, limited sample depth, and high instrument cost and expertise. It also lacks direct observation of microplastics' internal structure. Additionally, SEM's operation may require expertise (Sekudewicz et al., 2021).

9.6.2 STEREO- OR DISSECTING MICROSCOPY

Stereo- or dissecting microscopy is a versatile technique commonly used for the detection and characterization of microplastics. Also known as low-magnification or macroscopic microscopy, this method allows for the visual examination of samples in three dimensions, providing valuable information about the size, shape, color, and surface features of microplastic particles. The principle of stereomicroscopy for microplastic detection utilizes two separate optical paths to create a three-dimensional image of the sample. This is achieved by capturing two slightly different views of the sample from two distinct angles. The human brain then combines these views to perceive a three-dimensional image. Stereomicroscopes typically have lower magnification compared to other microscopy techniques, but they provide a larger field of view and greater depth of focus, making them suitable for the examination of macroscopic samples, including microplastic particles (Kalaronis et al., 2022).

It is used for microplastic detection by collecting samples from various sources, undergoing pretreatment, mounting them on transparent substrates, setting up the instrument for optimal imaging conditions, and observing the sample under the microscope. Microplastic particles are identified based on size, shape, color, and surface characteristics, with three-dimensional visualization capabilities. Images or videos of the particles are captured for further analysis and documentation. It offers three-dimensional visualization of microplastic particles, enabling better understanding of the size, shape, and surface properties. Its large field of view enables efficient screening of larger samples, while its greater depth of focus enables clear visualization of particles with uneven surfaces or complex structures. Stereomicroscopy is non-destructive, requiring no sample preparation or chemical treatments, preserving the integrity of microplastic particles for further analysis. Stereomicroscopes typically have lower magnification capabilities compared to compound microscopes. They are primarily designed for low to moderate magnification, usually up to around 50× or 100×. Stereomicroscopes provide a three-dimensional (3D) view of the specimen, which is beneficial for tasks such as dissection or manipulation. However, the depth perception provided by stereomicroscopes is limited and may not accurately represent the true depth of the specimen. This can be a disadvantage when precise depth information is crucial for analysis or measurement. Stereomicroscopes are primarily designed for the observation of opaque or thick specimens. Stereomicroscopes are primarily used for macroscopic or large-scale observations (Hurley et al., 2018).

9.6.3 FLUORESCENCE MICROSCOPY

Fluorescence microscopy is a valuable technique for the detection and characterization of microplastics. By exploiting the unique fluorescent properties of certain dyes or naturally occurring autofluorescence, fluorescence microscopy enables the visualization and identification of microplastic particles with high specificity and sensitivity. The principle of fluorescence microscopy for microplastic detection utilizes the absorption of light at a specific wavelength by fluorescent dyes or autofluorescent materials. When excited by light of the appropriate wavelength, these substances emit light at a longer wavelength, referred to as fluorescence. By labeling microplastic particles with fluorescent dyes or exploiting their inherent autofluorescence, fluorescence microscopy allows for the selective visualization and discrimination of microplastics from background materials (Tadsuwan et al., 2022). It includes the detection of microplastics by collecting samples from various sources, such as water bodies, sediments, or biological tissues. These samples can undergo pretreatment steps, such as filtration or digestion, to isolate and concentrate the particles. Microplastics can be labeled with fluorescent dyes or intrinsic autofluorescence, which can be detected without external labeling. The microscope is configured with excitation and emission filters specific to the microplastics' properties, and the emitted fluorescence is captured by the detector. Images are acquired using digital cameras or imaging software, revealing microplastic particles as bright fluorescent objects against a dark background. Image analysis software can quantify parameters like particle size, shape, and fluorescence intensity (Crew et al., 2020).

It offers several advantages for microplastic detection, including high specificity, visualization of small particles, multiplexing capability, and non-destructive analysis. It uses fluorescent labeling or intrinsic autofluorescence to detect and differentiate microplastics from other materials, allowing for selective detection and differentiation. Additionally, it allows for the simultaneous use of multiple fluorescent dyes with different emission spectra, enabling the detection and differentiation of different types or sizes of microplastics in a single sample. Furthermore, fluorescence microscopy is non-destructive, preserving the integrity of microplastic particles for further analysis. Fluorescence microscopy has numerous benefits, however, there are several significant drawbacks that have been found. Polymers that cannot reflect the fluorescence light cannot be identified through this technique (Brandon et al., 2020).

9.6.4 Atomic Force Microscopy (AFM)

Atomic force microscopy (AFM) is a powerful technique used for the analysis and detection of microplastics. AFM provides high-resolution imaging capabilities and allows for the characterization of microplastic particles in terms of their size, shape, surface topography, and mechanical properties. This technique has been widely utilized in microplastic research due to its ability to deliver detailed information at the nanoscale level. The detection of microplastics using AFM involves several steps, including probe selection, imaging, size and shape analysis, surface characterization, surface topography, mechanical properties, and mechanical properties. The choice of probe depends on the specific properties of the microplastics being analyzed, such as size and surface characteristics. The cantilever bends as a result of being scanned in a raster pattern, revealing the presence and morphology of microplastics. The size and shape of microplastic particles are determined through AFM images, providing quantitative data on dimensions. Surface characterization is also essential for understanding the potential interactions and behavior of microplastics in different environments. Mechanical properties, such as stiffness, elasticity, and adhesion properties, can be probed by measuring the forces required to deform or manipulate them. Overall, AFM is a valuable tool for detecting microplastics and understanding their physical behavior in various environments (Bonfanti et al., 2021).

AFM provides high-resolution imaging capabilities, allowing for the direct visualization of microplastics. It can reveal the size, form, and surface features of microplastic particles with sub-nanometer resolution, enabling detailed characterization and identification. It can assess the surface characteristics of microplastics, such as roughness, texture, and the presence of coatings or contaminants. This information is valuable for understanding the potential interactions of microplastics with other substances and organisms, as well as their behavior in different environments. However, it also has some limitations and disadvantages like time-consuming, costly, and if the microplastics are buried within a complex matrix or embedded in a larger particle, it may be challenging to access and detect them using AFM (Sharma et al. 2020).

9.6.5 TRANSMISSION ELECTRON MICROSCOPY (TEM)

Transmission electron microscopy (TEM) is an effective technique used for microplastics detection and analysis. TEM provides high-resolution imaging and detailed structural information at the nanoscale level. It allows for the direct visualization of microplastic particles and offers valuable insights into their size, shape, morphology, and composition. It is a different potential kind of electron microscopy with particular advantages when looking into small-sized microplastics. It offers high-resolution images at spatial resolutions which are used for the study of the structure, chemical composition, atomic dimensions, and morphology of microplastic samples (Liu et al., 2022). Its ability to deliver all of the previously described information on microplastics is dependent on the electron beam that is released from the very thin foil of the TEM apparatus. After interacting with the sample, the electron beam is converted into elastically or inelastically scattered electrons. Detectors can gather this signal and produce a detailed image of the object being investigated. In comparison to SEM instruments, this instrument operates at greater voltages (80 kV) (Hu et al., 2022). Additionally, this electron microscope can be connected to image-processing software and EDS to determine the elemental composition of microplastics (Bonfanti et al., 2021).

TEM offers extremely high resolution, allowing the visualization of nanoscale structures and details. It can achieve sub-Angstrom resolution, enabling the examination of atomic arrangements and molecular structures. TEM provides high magnification capabilities, often ranging from 50,000× to 10 million times (Egerton, 2011). TEM is particularly suitable for the analysis of thin specimens, such as biological samples, nanoparticles, and thin films. The electron beam passes through the specimen, making it possible to observe internal structures and obtain valuable information about their composition and morphology. TEM provides various imaging techniques, such as bright-field imaging, dark-field imaging, and high-resolution imaging. These techniques allow for the visualization of different features and properties of the sample, including defects, crystalline structures, and compositional variations. Preparing samples for TEM can be complex and time-consuming. It often involves cutting or thinning the specimen to a thickness of a few nanometers and requires specialized techniques such as ion milling. TEM operates under high vacuum conditions, which can limit the types of samples that can be examined. Biological samples, for example, need to be dehydrated and stabilized before analysis, which may introduce artifacts or alter their natural state. The applications, detection limit, advantages, and limitations of the different microscopic techniques used for microplastics analysis are discussed in Table 9.5.

9.7 NEW APPROACHES AND ADVANCED IDENTIFICATION METHODS FOR MICROPLASTIC ANALYSIS

Microplastic identification is a rapidly evolving field, and researchers are continuously developing new approaches and advanced methods to detect and characterize microplastic particles. Enhanced microplastic identification methods for microplastic include microfluidics, isotope ratio mass spectrometry (IRMS), hyperspectral imaging, and DNA barcoding. Microfluidics-based approaches have emerged as promising tools for microplastic identification. These systems utilize miniaturized devices and microchannels to manipulate and analyze microplastic samples. Microfluidics offers advantages such as low sample volume requirements, high throughput, and integration with various detection techniques. IRMS utilizes stable isotope analysis to identify the sources of microplastics and differentiate them from natural organic materials. Hyperspectral imaging utilizes the unique spectral signatures of different types of microplastics to distinguish and classify them based on their

TABLE 9.5
Comparison of Different Microscopy Techniques Used for Detection of Microplastics

Methods	Application	Detection Limit	Advantages	Disadvantages	References
Scanning electron microscopy (SEM)	To study morphology of microplastic	Resolution: 2 nm at 2 kV	High-resolution imaging, elemental analysis	Time-consuming, limited sample depth, time-consuming	(Fang et al. 2021)
Stereo- or dissecting microscopy	Morphology analysis and characterization of microplastics	<100 μm	Three-dimensional visualization, fast, inexpensive	Need for coupling with other analytical techniques	(Kalaroni et al. 2019)
Fluorescence microscopy	Size, shape, and color determination and quantification of microplastics	<5 μm	Fast, inexpensive, simple, no need for special treatment	Nonfluorescence microplastics cannot be detected	(Tadsuwan et al. 2022)
Atomic force microscopy (AFM)	Topography analysis	Resolution <0.3 nm	Very high resolution, no need for pretreatment, 3-D image	Need for coupling with other analytical techniques	(Bonfanti et al. 2021).
Transmission electron microscope (TEM)	Morphology analysis	Resolution limit <0.1 nm	Very high resolution, elemental analysis	Time-consuming as time needed for sample pretreatment	(Li et al. 2021)

chemical composition and morphology. DNA barcoding applies DNA sequencing techniques to identify microplastics by their associated microbial communities or DNA fragments, allowing for source tracking and potential identification of specific polymer types.

9.8 CONCLUSION

In conclusion, precise microplastics are essential, and identification and detection techniques are the foundation for the treatment of microplastic contamination. Numerous techniques for microplastic analysis have been developed in recent years. The microplastic analysis technology is divided into two categories based on the information gathered: physical characterization technology and chemical composition identification technology. Physical characterization technology primarily consists of visual analysis, dynamic light scattering analysis, and laser diffraction particle size analysis. Additionally, analytical and microscopy methods are frequently employed for the chemical characterization investigation of microplastics.

Even though there have been significant advancements in microplastic analysis technology recently, techniques still have limitations when it comes to their actual use. Additionally, relying on a single approach may yield inaccurate data on microplastics. Additionally, relying on a single approach makes the analysis less accurate due to intrusion from false-positive or false-negative signals. To assure the accuracy of the data gathered on microplastics in the environment, a variety of analytical procedures are often used. Furthermore, the development of effective microplastics sample techniques is critical for the investigation of microplastics. Microplastics are characterized by their very small particle size, different shapes, and complex compositions. Additionally, microplastics are easily harmed by outside forces. As a result, it is challenging to fully distinguish between microplastics and their environmental components. Currently, there is no efficient or unified method for removing microplastics from a complex environmental matrix. This not only affects the accuracy of microplastic analysis results but also makes using a variety of microplastic analysis techniques in real-world contexts challenging.

Therefore, it is crucial to create analytical techniques for microplastic samples and analyses that are quick, precise, affordable, and useful. The flotation method, one of several microplastic sample techniques, has the potential to eventually become a special way to remove and recycle microplastics from the environment. More importantly, using in vitro selection to eliminate some low-efficiency methods for identifying particular microplastics is highly likely to become a prominent sector for microplastic analysis in the future. Novel techniques, such as dynamic light scattering analysis, flow cytometry, laser fraction particle size analysis, and confocal laser scanning microscopy, play an important role in the future analysis of microplastics and even nanoplastics due to the integration of biomedicine, material science, environmental medicine, and environmental analysis.

REFERENCES

Akhbarizadeh, Razegheh, Sina Dobaradaran, Mehdi Amouei Torkmahalleh, Reza Saeedi, Roza Aibaghi, and Fatemeh Faraji Ghasemi. "Suspended fine particulate matter (PM2. 5), microplastics (MPs), and polycyclic aromatic hydrocarbons (PAHs) in air: Their possible relationships and health implications." *Environmental Research* 192 (2021): 110339.. https://doi.org/10.1016/j.envres.2020.110339

Araujo, Catarina F., Mariela M. Nolasco, Antonio M. P. Ribeiro, and Paulo J. A. Ribeiro-Claro. "Identification of microplastics using Raman spectroscopy: Latest developments and future prospects." *Water Research* 142 (2018): 426–440. https://doi.org/10.1016/j.watres.2018.05.060

Avio, Carlo Giacomo, Stefania Gorbi, and Francesco Regoli. "Experimental development of a new protocol for extraction and characterization of microplastics in fish tissues: First observations in commercial species from Adriatic Sea." *Marine Environmental Research* 111 (2015): 18–26. https://doi.org/10.1016/j.marenvres.2015.06.014

Bianco, Angelica, Luca Carena, Nina Peitsaro, Fabrizio Sordello, Davide Vione, and Monica Passananti. "Rapid detection of nanoplastics and small microplastics by Nile-Red staining and flow cytometry." *Environmental Chemistry Letters* 21, no. 2 (2023): 647–653. https://doi.org/10.1007/s10311-022-01545-3

Bonfanti, Patrizia, Anita Colombo, Melissa Saibene, Giulia Motta, Francesco Saliu, Tiziano Catelani, Dora Mehn et al. "Microplastics from miscellaneous plastic wastes: Physico-chemical characterization and impact on fish and amphibian development." *Ecotoxicology and Environmental Safety* 225 (2021): 112775. https://doi.org/10.1016/j.ecoenv.2021.112775

Bradney, Lauren, Hasintha Wijesekara, Kumuduni Niroshika Palansooriya, Nadeeka Obadamudalige, Nanthi S. Bolan, Yong Sik Ok, Jörg Rinklebe, Ki-Hyun Kim, and M. B. Kirkham. "Particulate plastics as a vector for toxic trace-element uptake by aquatic and terrestrial organisms and human health risk." *Environment International* 131 (2019): 104937. https://doi.org/10.1016/j.envint.2019.104937

Brandon, Jennifer A., Alexandra Freibott, and Linsey M. Sala. "Patterns of suspended and salp-ingested microplastic debris in the North Pacific investigated with epifluorescence microscopy." *Limnology and Oceanography Letters* 5, no. 1 (2020): 46–53.

Campanale, Claudia, Carmine Massarelli, Ilaria Savino, Vito Locaputo, and Vito Felice Uricchio. "A detailed review study on potential effects of microplastics and additives of concern on human health." *International Journal of Environmental Research and Public Health* 17, no. 4 (2020): 1212. https://doi.org/10.3390/ijerph17041212

Catarino, Ana I., Richard Thompson, William Sanderson, and Theodore B. Henry. "Development and optimization of a standard method for extraction of microplastics in mussels by enzyme digestion of soft tissues." *Environmental Toxicology and Chemistry* 36, no. 4 (2017): 947–951. https://doi.org/10.1002/etc.3608

Crew, Alex, Irene Gregory-Eaves, and Anthony Ricciardi. "Distribution, abundance, and diversity of microplastics in the upper St. Lawrence River." *Environmental Pollution* 260 (2020): 113994. https://doi.org/10.1016/j.envpol.2020.113994

Dris, Rachid, Johnny Gasperi, Vincent Rocher, Mohamed Saad, Nicolas Renault, and Bruno Tassin. "Microplastic contamination in an urban area: A case study in Greater Paris." *Environmental Chemistry* 12, no. 5 (2015): 592–599. https://doi.org/10.1071/EN14167

Egerton, Ray F. *Electron energy-loss spectroscopy in the electron microscope.* Springer Science & Business Media, 2011.

Enders, Kristina, Robin Lenz, Colin A. Stedmon, and Torkel G. Nielsen. "Abundance, size and polymer composition of marine microplastics≥ 10 μm in the Atlantic Ocean and their modelled vertical distribution." *Marine Pollution Bulletin* 100, no. 1 (2015): 70–81. https://doi.org/10.1016/j.marpolbul.2015.09.027. lusher

Fang, Cheng, Zahra Sobhani, Xian Zhang, Luke McCourt, Ben Routley, Christopher T. Gibson, and Ravi Naidu. "Identification and visualisation of microplastics/nanoplastics by Raman imaging (iii): algorithm to cross-check multi-images." *Water Research* 194 (2021): 116913.

Fang, Cheng, Yunlong Luo, Xian Zhang, Hongping Zhang, Annette Nolan, and Ravi Naidu. "Identification and visualisation of microplastics via PCA to decode Raman spectrum matrix towards imaging." *Chemosphere* 286 (2022): 131736. https://doi.org/10.1016/j.chemosphere.2021.131736

Foley, Carolyn J., Zachary S. Feiner, Timothy D. Malinich, and Tomas O. Höök. "A meta-analysis of the effects of exposure to microplastics on fish and aquatic invertebrates." *Science of the Total Environment* 631 (2018): 550–559. https://doi.org/10.1016/j.scitotenv.2018.03.046

Garaba, Shungudzemwoyo P., and Heidi M. Dierssen. "An airborne remote sensing case study of synthetic hydrocarbon detection using short wave infrared absorption features identified from marine-harvested macro-and microplastics." *Remote Sensing of Environment* 205 (2018): 224–235. https://doi.org/10.1016/j.rse.2017.11.023

Gerolin, Cristiano Rezende, Fabiano Nascimento Pupim, André Oliveira Sawakuchi, Carlos Henrique Grohmann, Geórgia Labuto, and Décio Semensatto. "Microplastics in sediments from Amazon rivers, Brazil." *Science of the Total Environment* 749 (2020): 141604.

Green, Dannielle S., Louise Kregting, Bas Boots, David J. Blockley, Paul Brickle, Marushka Da Costa, and Quentin Crowley. "A comparison of sampling methods for seawater microplastics and a first report of the microplastic litter in coastal waters of Ascension and Falkland Islands." *Marine Pollution Bulletin* 137 (2018): 695–701. https://doi.org/10.1016/j.marpolbul.2018.11.004

He, Jinsong, Zhuojun Jiang, Xiao Fu, Fan Ni, Fei Shen, Shirong Zhang, Zhang Cheng, Yongjia Lei, Yanzong Zhang, and Yan He. "Unveiling interactions of norfloxacin with microplastic in surface water by 2D FTIR correlation spectroscopy and X-ray photoelectron spectroscopy analyses." *Ecotoxicology and Environmental Safety* 251 (2023): 114521. https://doi.org/10.1016/j.ecoenv.2023.114521

Heinlaan, Margit, Kaja Kasemets, Villem Aruoja, Irina Blinova, Olesja Bondarenko, Aljona Lukjanova, Alla Khosrovyan et al. "Hazard evaluation of polystyrene nanoplastic with nine bioassays did not show particle-specific acute toxicity." *Science of the Total Environment* 707 (2020): 136073. https://doi.org/10.1016/j.scitotenv.2019.136073

Horton, Alice A., Alexander Walton, David J. Spurgeon, Elma Lahive, and Claus Svendsen. "Microplastics in freshwater and terrestrial environments: Evaluating the current understanding to identify the knowledge gaps and future research priorities." *Science of the Total Environment* 586 (2017): 127–141. https://doi.org/10.1016/j.scitotenv.2017.01.190

Hu, Rui, Kaining Zhang, Wei Wang, Long Wei, and Yongchao Lai. "Quantitative and sensitive analysis of polystyrene nanoplastics down to 50 nm by surface-enhanced Raman spectroscopy in water." *Journal of Hazardous Materials* 429 (2022): 128388. https://doi.org/10.1016/j.jhazmat.2022.128388

Huang, Zike, Bo Hu, and Hui Wang. "Analytical methods for microplastics in the environment: A review." *Environmental Chemistry Letters* 21, no. 1 (2023): 383–401.

Huppertsberg, Sven, and Thomas P. Knepper. "Instrumental analysis of microplastics—benefits and challenges." *Analytical and Bioanalytical Chemistry* 410 (2018): 6343–6352. https://doi.org/10.1007/s00216-018-1210-8

Hurley, Rachel R., Amy L. Lusher, Marianne Olsen, and Luca Nizzetto. "Validation of a method for extracting microplastics from complex, organic-rich, environmental matrices." *Environmental Science & Technology* 52, no. 13 (2018): 7409–7417. https://doi.org/10.1021/acs.est.8b01517

Kalaronis, Dimitrios, Nina Maria Ainali, Eleni Evgenidou, George Z. Kyzas, Xin Yang, Dimitrios N. Bikiaris, and Dimitra A. Lambropoulou. "Microscopic techniques as tools for the determination of microplastics and nanoplastics in the aquatic environment: A short review." *Green Analytical Chemistry* (2022): 100036. https://doi.org/10.1016/j.greeac.2022.100036

Käppler, Andrea, Dieter Fischer, Sonja Oberbeckmann, Gerald Schernewski, Matthias Labrenz, Klaus-Jochen Eichhorn, and Brigitte Voit. "Analysis of environmental microplastics by vibrational microspectroscopy: FTIR, Raman or both?." *Analytical and Bioanalytical Chemistry* 408 (2016): 8377–8391.

Karthik Ramamurthy, Radhakrishnan Subhadra Robin, Ramachandran Purvaja, Dipnarayan Ganguly, Iyyanar Anandavelu, Rajendran Raghuraman, Hariharan Gopalakrishna, Addepalli Ramakrishna, and Ramesh Ramachandran. "Microplastics along the beaches of southeast coast of India." *Science of the Total Environment* 645 (2018): 1388–1399.

Khan, Shujaul Mulk, Zahoor Ul Haq, Noreen Khalid, Zeeshan Ahmad, and Ujala Ejaz. "Utilization of three indigenous plant species as alternative to plastic can reduce pollution and bring sustainability in the environment." In *Natural Resources Conservation and Advances for Sustainability*, pp. 533–544. Elsevier, 2022. https://doi.org/10.1016/B978-0-12

Law, Kara Lavender, and Richard C. Thompson. "Microplastics in the seas." *Science* 345, no. 6193 (2014): 144–145. https://doi.org/10.1126/science.1254065

Li, Jianlong, Zhuozhi Ouyang, Peng Liu, Xiaonan Zhao, Renren Wu, Chutian Zhang, Chong Lin, Yiyong Li, and Xuetao Guo. "Distribution and characteristics of microplastics in the basin of Chishui River in Renhuai, China." *Science of The Total Environment* 773 (2021): 145591.

Li, Qinglan, Jiangtong Wu, Xiaopeng Zhao, Xueyuan Gu, and Rong Ji. "Separation and identification of microplastics from soil and sewage sludge." *Environmental Pollution* 254 (2019): 113076. https://doi.org/10.1016/j.envpol.2019.113076

Liu, Weiyi, Jinlan Zhang, Hang Liu, Xiaonan Guo, Xiyue Zhang, Xiaolong Yao, Zhiguo Cao, and Tingting Zhang. "A review of the removal of microplastics in global wastewater treatment plants: Characteristics and mechanisms." *Environment International* 146 (2021): 106277.

Liu, Yuxuan, Yonghua Wang, Na Li, and Shengnan Jiang. "Avobenzone and nanoplastics affect the development of zebrafish nervous system and retinal system and inhibit their locomotor behavior." *Science of The Total Environment* 806 (2022): 150681. https://doi.org/10.1016/j.scitotenv.2021.150681

Lusher, A. L., N. A. Welden, P. Sobral, and MJAoN Cole. "Sampling, isolating and identifying microplastics ingested by fish and invertebrates." *Analytical Methods* 9, no. 9 (2017): 1346–1360. https://doi.org/10.1039/C6AY02415G

Lusher, Amy L., Inger Lise N. Bråte, Keenan Munno, Rachel R. Hurley, and Natalie A. Welden. "Is it or isn't it: The importance of visual classification in microplastic characterization." *Applied Spectroscopy* 74, no. 9 (2020): 1139–1153.

Lwanga, Esperanza Huerta, Hennie Gertsen, Harm Gooren, Piet Peters, Tamás Salánki, Martine van der Ploeg, Ellen Besseling, Albert A. Koelmans, and Violette Geissen. "Incorporation of microplastics from litter into burrows of Lumbricus terrestris." *Environmental Pollution* 220 (2017): 523-531.

Ma, Piao, mu Wei Wang, Hui Liu, yu Feng Chen, and Jihong Xia. "Research on ecotoxicology of microplastics on freshwater aquatic organisms." *Environmental Pollutants and Bioavailability* 31, no. 1 (2019): 131–137. https://doi.org/10.1080/26395940.2019.1580151

Minor, Elizabeth C., Roselynd Lin, Alvin Burrows, Ellen M. Cooney, Sarah Grosshuesch, and Brenda Lafrancois. "An analysis of microlitter and microplastics from Lake Superior beach sand and surface-water." *Science of The Total Environment* 744 (2020): 140824. https://doi.org/10.1016/j.scitotenv.2020.140824

Miranda, Tiago, Luis R. Vieira, and Lúcia Guilhermino. "Neurotoxicity, behavior, and lethal effects of cadmium, microplastics, and their mixtures on Pomatoschistus microps juveniles from two wild populations exposed under laboratory conditions—implications to environmental and human risk assessment." *International Journal of Environmental Research and Public Health* 16, no. 16 (2019): 2857. https://doi.org/10.3390/ijerph16162857

Napper, Imogen Ellen, and Richard C. Thompson. "Plastic debris in the marine environment: History and future challenges." *Global Challenges* 4, no. 6 (2020): 1900081. https://doi.org/10.1002/gch2.201900081

O'Brien, Stacey, Elvis Dartey Okoffo, Cassandra Rauert, Jake W. O'Brien, Francisca Ribeiro, Stephen D. Burrows, Tania Toapanta, Xianyu Wang, and Kevin V. Thomas. "Quantification of selected microplastics in Australian urban road dust." *Journal of Hazardous Materials* 416 (2021): 125811.

Olsen, Linn Merethe Brekke, Heidi Knutsen, Sabnam Mahat, Emma Jane Wade, and Hans Peter H. Arp. "Facilitating microplastic quantification through the introduction of a cellulose dissolution step prior to oxidation: Proof-of-concept and demonstration using diverse samples from the Inner Oslofjord, Norway." *Marine Environmental Research* 161 (2020): 105080. https://doi.org/10.1016/j.marenvres.2020.105080

Prata, Joana Correia, João P. da Costa, Armando C. Duarte, and Teresa Rocha-Santos. "Methods for sampling and detection of microplastics in water and sediment: A critical review." *TrAC Trends in Analytical Chemistry* 110 (2019): 150–159. https://doi.org/10.1016/j.trac.2018.10.029

Prata, Joana Correia, João P. da Costa, Isabel Lopes, Armando C. Duarte, and Teresa Rocha-Santos. "Environmental exposure to microplastics: An overview on possible human health effects." *Science of the Total Environment* 702 (2020): 134455. https://doi.org/10.1016/j.scitotenv.2019.134455

Radford, Freya, Lina M. Zapata-Restrepo, Alice A. Horton, Malcolm D. Hudson, Peter J. Shaw, and Ian D. Williams. "Developing a systematic method for extraction of microplastics in soils." *Analytical Methods* 13, no. 14 (2021): 1695–1705. https://doi.org/10.1039/D0AY02086A

Rakesh Subramanian, Veeraswamy Davamani, Murugaragavan Ramasamy, Thanakkan Ramesh, and Sheeba Shrirangasami. *Pharma Innov. Journal* 10, 8 (2021): 1412–1417.

Rist, Sinja, Anders Baun, and Nanna B. Hartmann. "Ingestion of micro-and nanoplastics in Daphnia magna–Quantification of body burdens and assessment of feeding rates and reproduction." *Environmental pollution* 228 (2017): 398–407. https://doi.org/10.1016/j.envpol.2017.05.048

Scopetani, Costanza, David Chelazzi, Juha Mikola, Ville Leiniö, Reijo Heikkinen, Alessandra Cincinelli, and Jukka Pellinen. "Olive oil-based method for the extraction, quantification and identification of microplastics in soil and compost samples." *Science of the Total Environment* 733 (2020): 139338.

Scopetani, C., S. Pflugmacher, and J. Pellinen. "A method for the extraction of microplastics from solid samples using olive oil." In *Poster Session Presented at 16 Th International Conference on Environmental Science and Technology Rhodes, Greece, 4 to 7 September 2019*. 2019.

Sekudewicz, Ilona, Agnieszka Monika Dąbrowska, and Marcin Daniel Syczewski. "Microplastic pollution in surface water and sediments in the urban section of the Vistula River (Poland)." *Science of The Total Environment* 762 (2021): 143111. https://doi.org/10.1016/j.scitotenv.2020.143111

Sharma, Madhu D., Anjana I. Elanjickal, Juili S. Mankar, and Reddithota J. Krupadam. "Assessment of cancer risk of microplastics enriched with polycyclic aromatic hydrocarbons." *Journal of Hazardous Materials* 398 (2020): 122994.

Sheriff, Ishmail, Mohd Suffian Yusoff, and Herni Binti Halim. "Microplastics in wastewater treatment plants: A review of the occurrence, removal, impact on ecosystem, and abatement measures." *Journal of Water Process Engineering* 54 (2023): 104039. https://doi.org/10.1016/j.jwpe.2023.104039

Sorasan, Carmen, Carlos Edo, Miguel González-Pleiter, Francisca Fernández-Piñas, Francisco Leganés, Antonio Rodríguez, and Roberto Rosal. "Generation of nanoplastics during the photoageing of low-density polyethylene." *Environmental Pollution* 289 (2021): 117919. https://doi.org/10.1016/j.envpol.2021.117919

Sullivan, G. L., J. Delgado Gallardo, E. W. Jones, P. J. Hollliman, T. M. Watson, and Sarper Sarp. "Detection of trace sub-micron (nano) plastics in water samples using pyrolysis-gas chromatography time of flight mass spectrometry (PY-GCToF)." *Chemosphere* 249 (2020): 126179. https://doi.org/10.1016/j.chemosphere.2020.126179

Tadsuwan, Katekanya, and Sandhya Babel. "Unraveling microplastics removal in wastewater treatment plant: A comparative study of two wastewater treatment plants in Thailand." *Chemosphere* 307 (2022): 135733. https://doi.org/10.1016/j.chemosphere.2022.135733

Torres, Fernando G., Diana Carolina Dioses-Salinas, Carlos Ivan Pizarro-Ortega, and Gabriel E. De-la-Torre. "Sorption of chemical contaminants on degradable and non-degradable microplastics: Recent progress and research trends." *Science of the Total Environment* 757 (2021): 143875.

Van Cauwenberghe, Lisbeth, and Colin R. Janssen. "Microplastics in bivalves cultured for human consumption." *Environmental Pollution* 193 (2014): 65–70. https://doi.org/10.1016/j.envpol.2014.06.010

Veerasingam, S., M. Ranjani, Ramadoss Venkatachalapathy, Andrei Bagaev, Vladimir Mukhanov, Daria Litvinyuk, M. Mugilarasan et al. "Contributions of Fourier transform infrared spectroscopy in microplastic pollution research: A review." *Critical Reviews in Environmental Science and Technology* 51, no. 22 (2021): 2681–2743. https://doi.org/10.1080/10643389.2020.1807450.

WEF. (2020). Plastics, the Circular Economy and Global Trade, World Economic Forum Report. www3.weforum.org/docs/WEF_Plastics_the_Circular_Economy_and_Global_Trade_2020.pdf

Yang, Yun, Xiangru Zhang, Jingyi Jiang, Jiarui Han, Wanxin Li, Xiaoyan Li, Kenneth Mei Yee Leung, Shane A. Snyder, and Pedro J. J. Alvarez. "Which micropollutants in water environments deserve more attention globally?." *Environmental Science & Technology* 56, no. 1 (2021): 13–29. https://doi.org/10.1021/acs.est.1c04250.

Zhang, G. S., and Y. F. Liu. "The distribution of microplastics in soil aggregate fractions in southwestern China." *Science of the Total Environment* 642 (2018): 12–20. https://doi.org/10.1016/j.scitotenv.2018.06.004

Zhang, Shaoliang, Xiaomei Yang, Hennie Gertsen, Piet Peters, Tamás Salánki, and Violette Geissen. "A simple method for the extraction and identification of light density microplastics from soil." *Science of the Total Environment* 616 (2018): 1056–1065. https://doi.org/10.1016/j.scitotenv.2017.10.213

Zhang, Yulan, Shichang Kang, Steve Allen, Deonie Allen, Tanguang Gao, and Mika Sillanpää. "Atmospheric microplastics: A review on current status and perspectives." *Earth-Science Reviews* 203 (2020): 103118. https://doi.org/10.1016/j.earscirev.2020.103118

Ziani, Khaled, Corina-Bianca Ioniță-Mîndrican, Magdalena Mititelu, Sorinel Marius Neacșu, Carolina Negrei, Elena Moroșan, Doina Drăgănescu, and Olivia-Teodora Preda. "Microplastics: A real global threat for environment and food safety: A state of the art review." *Nutrients* 15, no. 3 (2023): 617. https://doi.org/10.3390/nu15030617

10 Microplastic Contamination in Water, Sediment, and Biota in Mangrove Forests

Rungpilin Jittalerk and Sandhya Babel

10.1 INTRODUCTION

The mangrove forests act as a buffer between terrestrial and marine ecosystems (Deng et al., 2021; Qian et al., 2021). They have become a potential sink for land- and marine-based waste (Deng et al., 2021). Microplastics are brought to the mangrove forests along with an enormous volume of waste. High vegetation density supports the accumulation of marine litter and microplastics (Martin et al., 2019; Zhou et al., 2020). The degradation of plastic debris results in the widespread presence of microplastics in the mangrove forest (Deng et al., 2021). The specialized characteristics of mangrove forests, such as aerial root systems, contribute as natural traps of microplastics (Luo et al., 2021). Microplastic pollution in the mangrove forest is not only associated with the surrounding environment (sediment and water) but also with mangrove biota. Their habitats have been encroached upon by small indigestible threats (microplastics). So far, studies have mainly reported on the microplastic abundance in sediment, water, and biota in mangrove forests. Most studies have found the distinguished prevalence of microfibers among other shapes in mangrove biota (Deng et al., 2021; Qian et al., 2021). Fibrous microplastics, mostly originating from textile industries, wastewater treatment plants, and domestic laundry, potentially have detrimental effects on animals (Yadav et al., 2023). Apart from these sources, the presence of microplastics in the mangrove forests can be attributed to various sources, such as terrestrial, atmospheric, and marine sources. Tracing sources of microplastics enhances the efficiency of plastic waste management and prevention. Plastic products are in demand and are basic elements in daily life. In order to solve plastic and microplastic pollution, integral management from origin to destination is necessary. Presently, adequate policies and legislation to enable citizens to acknowledge and realize this global concern are vague (Rose et al., 2023a). Moreover, mitigation strategies are occasionally feasible in realistic scenarios. Although technology in wastewater treatment plants can prevent a large proportion of domestic microplastics from being released into the environment, it has not been deliberately developed for this goal (Rose et al., 2023b). The effect of microplastic pollution is beyond animal and human health. Future economy and commercialization could be ruined by contaminated products. Microplastic pollution is destroying the health of and affects living organisms at all trophic levels including producers, consumers, and decomposers (Rose et al., 2023a). However, recent research on trophic transfer and toxicity, mainly conducted in laboratories, remains insufficient to uncover the actual situation in nature (Rose et al., 2023b). The microplastic examination in the mangrove ecosystem is still in its infancy and has mainly been reported in living organisms at the consumer level and the environment (e.g., water and sediment). Furthermore, although mangrove forests play several significant roles, there are few studies investigating microplastic contamination in mangrove forests compared to other ecosystems. This chapter aims to provide the current view of microplastic contamination in mangrove forests. It highlights the current status

DOI: 10.1201/9781032706573-10

of microplastic contamination in mangrove forests, mainly focusing on (1) microplastic pollution in mangrove forests, (2) potential pathways of microplastics to mangrove forests, (3) impacts of microplastics on mangrove animals and human subsistence, and (4) feasible solutions to prevent microplastic pollution.

10.2 MICROPLASTIC POLLUTION IN MANGROVE FORESTS

Mangrove forests are typically found on subtropical and tropical shores (John et al., 2022), covering about 132,000 km² (Martin et al., 2019). Mangrove forests play several significant roles, such as providing nurseries for juveniles (Abu El-Regal and Ibrahim, 2014), habitats for marine and terrestrial biota (Deng et al., 2021), and resources for food and housing (Luo et al., 2021). Acting as a protective shield for natural disasters, such as tsunamis and hurricanes, and the prevention of shoreline erosion, are part of mangrove actions (Luo et al., 2021; John et al., 2022). With the special growing zone, several mangrove forests have become potential receptacles for enormous volumes of waste from terrestrial and marine sources. As shown in Figure 10.1, secondary microplastics (fragments and sheets) are present in mangrove forests, which may be generated by accumulated plastic waste. Mangrove forests, whether unused or urbanized, are vulnerable to the invasion of microplastics (Nabizadeh et al., 2019; Trindade et al., 2023). Microplastics from both land- and marine-based waste are brought to the mangrove forests through wind, tides, and currents (Deng et al., 2021). Although studies on the investigation of microplastics in mangrove forests have recently increased, there are few studies compared to those on marine ecosystem. Most studies mainly focus on sediment, followed by water and biota. However, since a standard method for microplastic analysis has not been clarified, several methods have been applied. Different procedures can affect the results.

10.2.1 MICROPLASTICS IN MANGROVE SEDIMENT

A large burial rate of microplastics in mangrove forests was detected in sediment (Martin et al., 2020). Several studies have reported microplastic investigations in mangrove sediment (Deng et al., 2021). As shown in Table 10.1, the abundance of microplastics in mangrove sediment has been reported in both low and high concentrations and the concentration varied among regions and countries. Different debris loads (Rose and Webber, 2019) and natural factors, such as the carrying capacity of rivers and oceans (Li et al., 2020a), mangrove density, and sediment properties (Deng et al., 2021) can affect the microplastic contamination in each mangrove area. The abundance of microplastics in mangrove sediment reported in studies from Indonesia is much lower than in other countries, such as Iran and China. Cordova et al. (2021) found the contamination of microplastics at 28.09 ± 10.28 items/kg dry weight (d.w.) in sediment from Muara Angke Wildlife Reserve in Indonesia. Improved

FIGURE 10.1 Microplastics detected in the Bang Pu mangrove forest, Thailand.

Microplastic Contamination in the Biota of Mangrove Forests

TABLE 10.1

Microplastic Contamination in Mangrove Sediment

Location	Microplastic Abundance	Predominant Shape	Predominant Polymer Type	References
Southern Iran	2,169 items/kg d.w.	Fiber	Polystyrene, polypropylene, polyethylene terephthalate	(Maghsodian et al. 2021)
Beibu Gulf, China	640 to 6,360 items/kg	–	Polypropylene	(Li et al. 2020a)
Six mangroves, Southern China	227 ± 173 to 2249 ± 747 items/kg	Fiber	Polypropylene, polyethylene	(Li et al. 2020b)
Jinjiang Estuary, China	490 ± 127.3–1170 ± 99.0 items/500 g d.w.	Fiber	Polyethylene terephthalate, polypropylene, polyethylene	(Deng et al. 2020)
Maowei Sea, China	520 ± 8 to 940 ± 17 items/kg	–	Polypropylene	(Li et al. 2019)
Tallo River, Makassar, Indonesia	16.67 ± 20.82 to 150 ± 36.06 items/kg	Fragment, line	Polyethylene	(Wicaksono et al. 2021)
Muara Angke Wildlife Reserve, Indonesia	28.09 ± 10.28 items/kg d.w.	Foam	Polystyrene	(Cordova et al. 2021)
East Surabaya, Indonesia	73.9 ± 157 items/kg d.w. (ND) to 598 items/kg	Fragment, foam	Polypropylene	(Ni'am et al. 2022)
Six mangrove habitats, Ciénaga Grande de Santa Marta, Colombian Caribbean	31 to 2863 items/kg d.w.	Film	Polystyrene, Nylon	(Garcés-Ordóñez et al. 2019)
Saija and Timbiqui River estuaries, the Colombian Pacific	48 to 832 items/m^2	Film, filament	–	(Garcés-Ordóñez et al. 2023)
Sundarbans mangrove forest, Bangladesh	1.57 × 10^5 items/kg	Fragment	Polyamide	(Nawar et al. 2023)
Moheshkhali channel, Bay of Bengal, Bangladesh	6.66 to 138.33 items/m^2	Film	Polypropylene, polyethylene, polystyrene, polyamide	(Nahian et al. 2023)
Ulhas River Estuary, India	96.67 ± 12.02 to 130 ± 5.77 items/kg d.w.	Fragment	Polypropylene	(Kumkar et al. 2021)
Mangalavanam, Kerala, India	1275 ± 532 items/kg d.w.	Fragment	Polypropylene	(Kannankai et al. 2022)
Butuan Bay, Philippines	8 items/150 g d.w.	Fiber	Polypropylene	(Navarro et al. 2022)
St. Lucia, uMgeni, Durban Harbour, and Isipingo, South Africa	18.5 ± 34.4 to 143.5 ± 93.0 items/500 g	Fiber	Polyethylene, polypropylene (with water)	(Govender et al. 2020)
Mauritius	107.4 ± 76.42 to 140.2 ± 85.38 items/kg	Fragment, film	–	(Seeruttun et al. 2023)

Abbreviations: d.w., dry weight; ND, not detected.

river and coastal cleanup programs in Jakarta possibly reduced the contamination of microplastics in the environment, which may cause a low concentration of microplastics in the mangrove areas (Cordova et al., 2021). China is one of the largest global plastic producers, accounting for 32% in 2021 (Plastics Europe, 2022), and also the top contributor to plastic waste (Liu and Liu, 2023). In 2020, most studies on the investigation of microplastics in wetland ecosystems were from China (61.7%) (Luo et al., 2021). Microplastic pollution in the mangrove forests in China ranges from 8.3 to 7,900 items/kg d.w. (Deng et al., 2021). A study by Li et al. (2020a) detected microplastics in sediment from mangrove habitats in the Beibu Gulf in China that varied from 640 to 6,360 items/kg. Furthermore, mangrove forests in southern Iran were also contaminated with a high abundance of microplastics at 2,169 items/kg d.w. (Maghsodian et al., 2021). The world's largest mangrove forest, Sundarbans, was reported to be contaminated with an extremely high abundance of microplastics in sediment (1.57×10^5 items/kg) (Nawar et al., 2023). Apart from regional differences in debris load, natural factors, and different methods employed, sampling strategies, sample treatment, and microplastic characterization can also affect the results.

Not only urban mangrove forests have been invaded by microplastic pollution, but also remote mangrove forests. Microplastics were detected in the mangrove forests of the Saija and Timbiqui river estuaries in the remote Colombian Pacific with a range of 48–832 items/m² (Garcés-Ordóñez et al., 2023). They found that the concentration of microplastics in the high-rainfall season (764–832 items/m²) was higher than in the low-rainfall season (48–92 items/m²). Microplastics from terrestrial sources, such as runoffs from urban and agricultural areas, after rainfall possibly increase the concentration of microplastics in the high-rainfall season (Deng et al., 2021; Garcés-Ordóñez et al., 2023). Microplastics ranging from 6.66 to 138.33 items/m² were detected in shore sediments of the Moheshkhali channel surrounded by mangrove forests in Bangladesh (Nahian et al., 2023). This channel is the receptacle of an enormous volume of terrestrial waste carried by the Bakkhali and Matamohori rivers. In addition, a study by Seeruttun et al. (2023) reported that mangrove forests in a small island, Mauritius, have been plagued by plastic waste due to the location within the trade wind belts and the outer rim of ocean gyres.

In addition, as presented in Table 10.1, microplastic fragments were widely detected in mangrove sediment in many studies. The high-density microplastics sink into surface sediment, while microplastics with a lower density than seawater float (Deng et al., 2021). However, biofilm formation by microbes may affect the buoyancy and settlement of microplastics (Qian et al., 2021). A study by Pete et al. (2022) examined biofilm formation on polyethylene (PE) microplastics by *Anabaena* sp., *Synechococcus elongatus*, and *Alcanivorax borkumensis*. They found that only *A. borkumensis* led to the settlement of microplastics. Furthermore, various types of plastic waste accumulating in the mangrove area generate microplastics in different shapes. Fragments, irregular forms of microplastics, are mainly generated through the degradation of large plastic debris and retained in sediment with their setting mechanism (Cordova et al., 2021; Seeruttun et al., 2023). Microplastic fragments may originate from plastic goods, such as containers, furniture, and toys (Ta and Babel, 2020).

PE and polypropylene (PP) are frequently reported as the dominant polymer types of microplastics contaminated in mangrove sediment in many studies (Table 10.1). The demand for PE and PP plastic for European plastic converters has increased, reaching 49.2% in 2021 (Plastics Europe, 2022). Both polymer types are part of the most commonly used polymer types together with polyvinyl chloride (PVC), polystyrene (PS), and polyethylene terephthalate (PET), accounting for 90% of global plastic production (Phuong et al., 2016). As the potential interception of terrestrial- and marine-based waste, accumulated plastic waste in the mangrove forest reflects the plastic consumption of humans. Both PE and PP plastic waste are ubiquitous in many mangrove areas around the world.

As depicted in Figure 10.2, unique attributes of mangrove forests, such as respiratory roots, play significant roles in the accumulation of plastic debris. Li et al. (2019) found that the concentration of microplastics in the rhizospheres was significantly higher in the non-rhizospheres. Pneumatophores of mangrove trees advocate the accumulation and preservation of microplastics in sediment (Duan et al., 2021). Moreover, the distribution of microplastics in sediment is possibly influenced by the height and density of mangrove forests (Zhou et al., 2020).

Microplastic Contamination in the Biota of Mangrove Forests

FIGURE 10.2 Plastics and microplastics trapped by pneumatophores in the Bang Pu mangrove forest, Thailand.

10.2.2 Microplastics in Mangrove Water

Water currents and weathering lead to the accessibility of microplastics in the mangrove forests (John et al., 2022). As shown in Table 10.2, the investigation of microplastics in mangrove surface water was examined in some countries. The variability of microplastic contamination in mangrove water is also influenced by regional differences and microplastic analysis methods. Moreover, debris loads and natural factors, such as hydrodynamics, vegetation density, and the level of aquatic primary productivity in each mangrove area (Rose and Webber, 2019; Deng et al., 2021) impact microplastic contamination. Additionally, as mentioned earlier, sampling strategies can affect the results. For example, water sample collection was conducted using a manta trawl (Rose and Webber, 2019), a neuston net (Aliabad et al., 2019), and a steel water sampler (Li et al., 2020a). Recently, a study by Nawar et al. (2023) reported an extremely high concentration of microplastics (2.66×10^3 items/L) in Sundarbans, the world's largest mangrove forest, in Bangladesh. To the extent of our knowledge, it has the highest abundance of microplastics in mangrove water compared to other studies. Surprisingly, microplastic contamination in the river mouth in Bangladesh appeared at a much lower concentration. Nahian et al. (2023) reported that microplastics in surface water at the Moheshkhali channel in Bangladesh, which carries tons of land-based waste, ranged from 0 to ~0.1 items/m^3 which was only detected at the connected narrow tributaries. Besides, surface water in mangrove habitats of the Ulhas estuary in India was contaminated with microplastics ranging from 0.23 ± 0.02 to 0.41 ± 0.05 items/L (Kumkar et al., 2021). A study by Li et al. (2020a) reported microplastic contamination in water from mangrove habitats in the Beibu Gulf in China. They reported the concentration of microplastics ranging

172   Microplastic Pollution

from 399 to 5,531 items/m³ in tidal water from mangrove habitats. In South America, Trindade et al. (2023) reported microplastics-contaminated water from mangrove estuarine waters around Todos os Santos Bay, the largest and most important navigable bay in Brazil. They reported a maximum number of microplastics at 32,800 items/m³ with an average concentration of 5,180 items/m³. The mangrove forests of this bay are surrounded by several anthropogenic activities with high populations and industrial complexes (Paes et al., 2022; Trindade et al., 2023). Moreover, the mangrove soils of Todos os Santos Bay have recently been stated as the most polluted mangrove soil in the world, with a maximum of 31,087 items/kg (Paes et al., 2022).

Fragments predominated in mangrove water in several studies, followed by fibers (Table 10.2). Fragments and fibers possibly originate from the degradation of large pieces of plastic debris and sewage of cloth laundry, respectively (Deng et al., 2021). Similar to sediment, a predominance of PE and PP microplastics was frequently observed in mangrove water in most studies.

Weak turbulence of tidal water leads to the sinking of microplastics (Zhang et al., 2020). Biofilm formation by fouling organisms and heteroaggregates of sticky microplastics may also cause the

TABLE 10.2
Microplastic Contamination in Mangrove Water

Location	Microplastic Abundance	Predominant Shape	Predominant Polymer Type	References
Chabahar Bay, Iran	0.14 ± 0.06 items/m³	Fiber	Polyethylene, polypropylene	(Aliabad et al. 2019)
Beibu Gulf, China	399 to 5,531 items/m³	–	Polyethylene, polypropylene	(Li et al. 2020a)
Yunxiao Mangrove Reserve, China	275 items/m³	Fragment, fiber	Polypropylene, polyethersulfone, polyethylene terephthalate	(Pan et al. 2020)
Tallo River, Makassar, Indonesia	0.74 ± 0.46 to 3.41 ± 0.13 items/m³	Fragment, line	Polyethylene	(Wicaksono et al. 2021)
Saija and Timbiqui River estuaries, the Colombian Pacific	0.06 to 0.53 items/m³	Foam, fragment	–	(Garcés-Ordóñez et al. 2023)
Kingston Harbour, Jamaica	0.76 items/m³	Fragment	Polyethylene, polypropylene	(Rose and Webber 2019)
Ulhas River Estuary, India	0.23 ± 0.02 to 0.41 ± 0.05 items/L	Fragment	Surlyn ionomer	(Kumkar et al. 2021)
Sundarbans mangrove forest, Bangladesh	2.66×10^3 items/L	Fragment	Polyamide	(Nawar et al. 2023)
Moheshkhali channel, Bay of Bengal, Bangladesh	0 to ~0.1 items/m³	Film	Polypropylene, polyethylene	(Nahian et al. 2023)
Penang, Malaysia	201 ± 21.214 to $1,407 \pm 124.265$ items/L	Foam, fragment	Polyethylene	(Tan and Zanuri 2023)
Sitio Pulo-Manila Bay, Philippines	449 items in collected samples	Fragment	Polyethylene, polypropylene	(Galicia et al. 2023)
Todos os Santos Bay, Brazil	637 to 32.8×10^3 items/m³	Fiber, fragment	–	(Trindade et al. 2023)
St. Lucia, uMgeni, Durban Harbour and Isipingo, South Africa	11.9 ± 11.2 to 50.6 ± 56.0 items/10,000 L	Fiber	Polyethylene, polypropylene (with sediment)	(Govender et al. 2020)

Microplastic Contamination in the Biota of Mangrove Forests 173

settlement of floating microplastics (Rummel et al., 2017). On the other hand, strong hydrodynamics can resuspend microplastics settling in sediment during rainy or flood periods (Li et al., 2022).

10.2.3 Microplastics in Mangrove Biota

Mangrove forests provide habitats and breeding grounds for terrestrial and aquatic vertebrates and invertebrates (Arceo-Carranza et al., 2021). Mangrove biota is also important for food consumption in many countries. As mentioned earlier, microplastics have ubiquitously distributed in water and sediment in mangrove forests around the world. This can lead to microplastic contamination in biota since their habitats are contaminated. The investigation of microplastics contaminated in mangrove biota is very sparse compared to marine organisms. As shown in Table 10.3, microplastic abundance in mangrove biota has been reported in studies from Indonesia, Mexico, Peru, China, Iran, and India.

TABLE 10.3
Microplastic Contamination in Mangrove Biota

Location	Studied Species	Microplastic Abundance	Predominant Shape	Predominant Polymer Type	References
Pramuka Island, Jakarta Bay, Indonesia	Crabs	327.56 items/individual	Fiber	–	(Patria et al. 2020)
Isla del Carmen, Mexico	Crabs	1.3 ± 1.2 items/g (all soft tissues)	Fiber	–	(Capparelli et al. 2022)
Tumbes, Peru	Crabs	475 items (gills) and 446 items (digestive tracts) from 30 crabs	Fiber	Polyethylene terephthalate, polyethylene vinyl acetate	(Aguirre-Sanchez et al. 2022)
Hong Kong	Crabs	2,966 items from 49 crabs	–	Polyethylene, Polyethylene terephthalate, Rayon (9 identified items)	(Not et al. 2020)
Hangzhou Bay and Yangtze Estuary, China	Mudskippers	5.3 ± 2.4 items/individual (gut) 1.2 ± 1.2 items/individual (gill)	Fiber, fragment	–	(Su et al. 2019)
Five mangrove forests, Iran	Mudskippers	15 items from 14 mudskippers	Fiber	Polystyrene, Polypropylene, polyethylene terephthalate	(Maghsodian et al. 2021)
Ulhas River estuary, India	Mudskippers	3.75 ± 1.26 to 6.11 ± 1.17 items/individual (gut)	Filament	Low-density polyethylene	(Kumkar et al. 2021)
Beibu Gulf, China	Sea snails	95.6 ± 5.0 items/kg	–	Polypropylene, polyethylene	(Li et al. 2020a)
Pramuka Island, Jakarta Bay, Indonesia	Periwinkle snail	75.5 items/individual	Fiber	–	(Patria et al. 2020)
Zhanjiang, China	Fish	2.83 ± 1.84 items/individual	Fiber	Polyethylene	(Huang et al. 2020)
KwaZulu-Natal, South Africa	Fish	0.79 ± 1.00 items/fish	Fiber	Rayon	(Naidoo et al. 2020)

Burrowing decapods, the crustaceans associated with mangrove forests, are mainly sesarmid crabs and fiddler crabs (Kristensen, 2008). They play essential roles in leaf litter dynamics, energy flow, and the productivity of the mangrove ecosystem (Arceo-Carranza et al., 2021). Their bioturbation can also maintain nutrient cycling (Kristensen, 2008). Mangrove crabs are significant bioindicators for contaminants in mangrove sediment. Moreover, they are important commercial crabs in some countries such as Thailand (Yeesin and Bautip, 2021) and Peru (Aguirre-Sanchez et al., 2022). The high abundance of microplastics contaminated in mangrove crabs (*Metopograpsus quadridentata*) from Indonesia was detected at 327.56 items/individual (Patria et al., 2020). Moreover, Capparelli et al. (2022) detected the contamination of microplastics in fiddler crabs (*Minuca rapax*) and their burrows. They also found that the distribution of microplastics in sediment has been modified by the construction and maintenance of the burrows of the crabs. Fibers were observed as the predominant shape of microplastics in mangrove crabs from Indonesia (Patria et al., 2020), Hong Kong (Not et al., 2020), Mexico (Capparelli et al., 2022), and Peru (Aguirre-Sanchez et al., 2022). The dominant polymer type of microplastics in mangrove crabs has been scarcely reported. A study by Not et al. (2020) reported the polymer composition, consisting of PE, PET, and rayon, for nine identified items. Though the shapes of microplastics can help in understanding the characteristics of plastic wastes that are sources of microplastics, polymer identification provides efficient results to trace the origins of microplastics. Polymer types effectively separate microplastics from incompletely digested organic matter or false microplastics and confirm the chemical composition of microplastics (Tirkey and Upadhyay, 2021). Comprehensive characteristics of detected microplastics are needed to examine the sources of microplastics in mangrove crabs and other species.

Mudskippers are amphibious fish found in mangrove swamp and intertidal mud (Swennen et al., 1995). Moreover, they are valuable and consumed as food in China, Taiwan, Korea, the Philippines, Bangladesh (Ansari et al., 2014), and India (Kumkar et al., 2021). Microplastic contamination has been reported in mudskippers from China (Su et al., 2019), India (Kumkar et al., 2021), and Iran (Maghsodian et al., 2021). A high abundance of microplastics was detected in Dussumier's mudskippers (*Boleophthalmus dussumieri*) from India, with a range from 3.75 ± 1.26 to 6.11 ± 1.17 items/individual (Kumkar et al., 2021). The life cycle of mudskippers associates mangrove water and sediment (Swennen et al., 1995; Kumkar et al., 2021). Kumkar et al. (2021) reported the association between microplastic contamination in mudskippers and their adjacent environment. Moreover, mudskippers construct their burrow by mouth excavation (Tran et al., 2020), which makes them vulnerable to microplastics contaminated in the surrounding environment. A predominance of microplastic fibers was observed in mudskippers. Interestingly, some species of mudskippers, such as *B. dussumieri*, possibly trap other shapes of microplastics by the specific configuration of the teeth on pharyngeal plates (Kumkar et al., 2021). However, this specific characteristic does not appear in *Periophthalmodon schlosseri* and *Periophthalmus chrysospilos* (Tran et al., 2021). Polymer types of microplastics in mudskippers varied with regional differences, such as PS, PP, PET (Maghsodian et al., 2021), and low-density polyethylene (LDPE) (Kumkar et al., 2021).

The contamination of microplastics in many marine vertebrates, especially fish, has widely been investigated from mangrove forests. Huang et al. (2020) detected microplastics ranging from 0.6 to 8.0 items/individual in 30 fish species from Zhanjiang mangrove wetland in China. Juvenile fish from South Africa were also contaminated with microplastics (Naidoo et al., 2020). They found that microplastics were detected in 91 of 174 fish samples. However, it is notable that the abundance of microplastics in juvenile fish from the protected mangrove forests was comparable to juvenile fish from the urbanized mangrove forest. Naidoo et al. (2020) explained that mangrove forests may receive microplastics from runoff, and domestic and industrial effluent, which may cause accumulation in the protected mangrove forest that has limited water exchange with the

Microplastic Contamination in the Biota of Mangrove Forests   **175**

ocean. Besides, microplastic fibers were predominantly observed in mangrove fish from China and South Africa.

Some studies have reported the abundance of microplastics in mangrove invertebrates. The concentration of microplastics in sea snails (*Ellobium chinense*) from China (Li et al., 2020a) was higher than in periwinkle snails (*Littoraria scabra*) from Indonesia (Patria et al., 2020). The potential source of microplastic uptake in *E. chinense* is likely related to microplastics contaminated in pore water, which provides food for sea snails during ebb tide (Li et al., 2020a). Moreover, *E. chinense* is an endangered species on the International Union for Conservation of Nature (Shin et al., 2021). Although *L. scabra* is not eaten by humans, it is prey for the higher trophic levels, such as cephalopods and fish (Patria et al., 2020).

Microplastics transferred by the prey–predator relationship may be a potential pathway for the translocation of microplastics through the food chain. A study by Sarker et al. (2022) reported on the bioaccumulation of microplastics across different trophic levels from secondary consumers to quaternary consumers in the Sundarbans mangrove forest in Bangladesh. However, some studies revealed that the excretion ability of predators can reduce the accumulation of microplastics. Farrell and Nelson (2013) found that microplastics were retained in shore crabs (*Carcinus maenas*) for only 21 days. Similarly, a study by Santana et al. (2017) reported that swimming crabs (*Callinectes ornatus*) and puffer fish *(Spheoeroides greeleyi)* excreted microplastics after 10 days of depuration. However, the complex food web in the mangrove forests with high biodiversity is still vulnerable to the distribution, accumulation, and translocation of microplastics.

10.2.4 MICROPLASTIC ANALYSIS IN MANGROVE BIOTA

Due to the absence of a standard method for microplastic analysis, different methodologies were applied in previous studies. A flow diagram for microplastic analysis in mangrove biota is shown in Figure 10.3. Microplastic analysis for mangrove biota consists of seven main procedures comprising sample collection, sample preparation, extraction, density separation, filtration, characterization, and polymer identification.

10.2.4.1 Sample Collection

Sample collection for mangrove biota is similar to that for other animals in other ecosystems. Sampling strategies depend on the objectives. For example, mangrove crabs can be collected by hand picking (Not et al., 2020) or purchasing from local markets (Aguirre-Sanchez et al., 2022). Not et al. (2020) examined microplastic intake in four different crab species in mangrove habitats, while Aguirre-Sanchez et al. (2022) investigated microplastic contamination in mangrove crabs sourced from the main local markets in Tumbes City, Peru. Sample collection for other mangrove animals, such as fish or amphibious fish, can be conducted in several ways. For example, fish can be collected by dip netting (Naidoo et al., 2020), boat trawling, drift nets, cage nets (Huang et al., 2020), netting (Maghsodian et al., 2021), or collaborating with local fishers (Kumkar et al., 2021). There are very few studies on microplastic contamination in mangrove snails. Li et al. (2020a) applied a snail survey for the sampling method of sea snails.

Some studies specified anesthetized methods for mangrove animals, such as 4% formaldehyde (Maghsodian et al., 2021) or tricaine methane sulfonate (Kumkar et al., 2021) for mudskippers and the euthanasia method using 99% anhydrous alcohol for juvenile fish (Naidoo et al., 2020). However, 10% formaldehyde solution and 100% alcohol can damage some polymers (Lusher et al., 2017). In addition, these methods should be approved by related organizations.

Suitable containers for transportation are required. For instance, sea snails were immediately stored in a cryogenic storage tank at −4°C and transported to the laboratory (Li et al., 2020a). Dussumier's mudskippers were kept in steel containers with tightly fitting lids (Kumkar et al.,

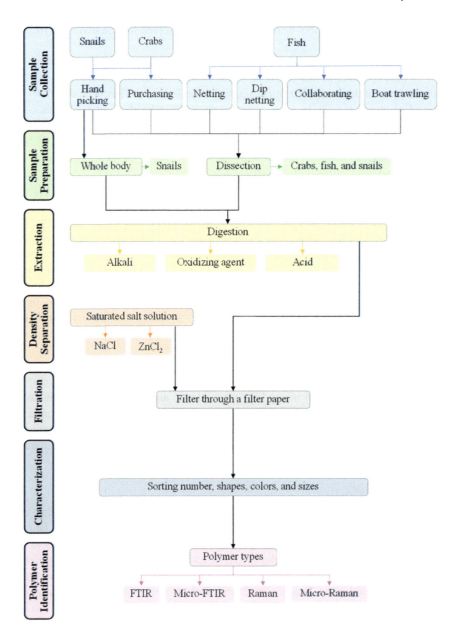

FIGURE 10.3 Microplastic analysis in mangrove biota.

2021). Proper containers should be considered to avoid microplastic contamination from the container itself. Moreover, if preservation is needed, samples should be desiccated or frozen (Lusher et al., 2017). For example, Not et al. (2020) stored mangrove crab samples at −18°C until further processing. Thus, sample collection for mangrove biota depends on the species and available instruments.

10.2.4.2 Sample Preparation
After removing external dirt, mangrove animals are dissected to separate target organs. Most field studies mainly focus on digestive tracts to assess the plastic consumption of animals (Lusher et al.,

Microplastic Contamination in the Biota of Mangrove Forests

2017). The digestive tracts of mangrove fish were frequently examined, followed by gills and other organs (Huang et al., 2020; Kumkar et al., 2021; Maghsodian et al., 2021). The presence of microplastics in mangrove crabs was investigated in the digestive tracts and gill ventilation (Not et al., 2020; Aguirre-Sanchez et al., 2022). A study by Capparelli et al. (2022) examined all soft tissues of mangrove crabs. For mangrove snails, only soft tissues (Patria et al., 2020) or the whole body with shells (Li et al., 2020a) were used in previous studies. Therefore, sample preparation is based on studied organs. Mostly, dissection is needed for fish and crabs.

10.2.4.3 Extraction

For enumerating microplastics in mangrove biota, target organs have to be digested to remove organic matter (Lusher et al., 2017). Alkali digestion with 10% potassium hydroxide (KOH) was applied to mudskipper samples (Kumkar et al., 2021; Maghsodian et al., 2021). This method was also used to digest the target organs of fish (Huang et al., 2020) and mangrove crabs (Aguirre-Sanchez et al., 2022). According to a study by Dehaut et al. (2016), 10% KOH provided efficient digestion for mussels, crabs, and fish without polymer degradation except for cellulose acetate. Besides, an oxidizing agent was applied in some studies. Hydrogen peroxide (H_2O_2) was used in some studies on microplastic contamination in mangrove crabs (Not et al., 2020; Capparelli et al., 2022). Moreover, Fenton's reagent, the mixture of 30% H_2O_2 and 0.05 M iron (II) solution, can be used for gills and guts digestion of crabs (Vermeiren et al., 2023). Though excessive foaming of H_2O_2 may cause the loss of microplastics (Lusher et al., 2017), two-step digestion with 10% KOH and Fenton's reagent provided the best performance for plankton digestion (Alfonso et al., 2021). Due to their habitats and the feeding behavior of some species, such as mangrove crabs, two-step digestion is necessary to remove sediment contents or plant tissues (e.g., branch and leaves detritus). A study by Patria et al. (2020) used nitric acid (HNO_3) in low quantity to digest mangrove crabs and snails. Although HNO_3 provides highly efficient digestion, it can destroy some polymer types, such as PE and PS (Lusher et al., 2017). Thus, the digestion for mangrove biota samples should suit species and stomach content according to their feeding behavior. Moreover, the damage to microplastics of the digestion method should be considered.

10.2.4.4 Density Separation

Most studies performed density separation to segregate microplastics from the remaining organic matter. This procedure is not necessary if organic matter is totally degraded (Capparelli et al., 2022). In order to separate microplastics from incompletely digested organic matter, some studies applied sodium chloride (NaCl) for segregating microplastics (Huang et al., 2020; Patria et al., 2020). Although NaCl solution (1.2 g/cm³) is cost-effective and reliable, it cannot separate PVC and PET (Tirkey and Upadhyay, 2021). Karami et al. (2017) found that sodium iodide (NaI), having a density of 1.5 g/cm³, can recover both polymer types and provide an efficient recovery rate for other polymer types (e.g., PS, LDPE, and PP). However, NaI is much more expensive than common salt like NaCl. Kedzierski et al. (2017) proposed several reuses of NaI solution without density reduction and chemical contamination (except for mass loss). Moreover, a two-step density separation with NaCl followed by NaI can reduce the demand for NaI and overcome the limitation of NaCl (Nuelle et al., 2014). It should be noted that the combined method takes more time. Similarly, the combination of NaCl and zinc chloride ($ZnCl_2$) was used to perform density separation of microplastics in mangrove crabs (Aguirre-Sanchez et al., 2022). Despite $ZnCl_2$ being efficient, it is hazardous and corrosive (Tirkey and Upadhyay, 2021). Moreover, the cost of $ZnCl_2$ is also high. Nevertheless, a study by Rodrigues et al. (2020) reported that the reuse of $ZnCl_2$ can be performed at least five times. Mangrove animals have different feeding behaviors and habitats that expose them to different types of microplastics. Low-density microplastics floating in surface water mostly are exposed to mangrove fish. Density separation with NaCl is possibly sufficient. On the other hand, various types of plastic debris accumulated in sediment generate microplastics with different densities and polymer types. Animals living on the ground of mangrove forests, such as mangrove crabs, could receive a wide range of microplastics. Two-step density separation or highly efficient chemicals (e.g., NaI and $ZnCl_2$) are viable choices. Therefore, the proper density separation method depends on the feeding behavior and species of mangrove biota.

10.2.4.5 Filtration

Filtration using a funnel with a vacuum pump system is simply applied to separate solid and liquid using filter paper as a medium (Tirkey and Upadhyay, 2021). Different types and pore sizes of filter papers were used in previous studies. Maghsodian et al. (2021) used Whatman Schleicher and Schuell filter paper with a pore size of < 2 μm for the filtration of digested samples containing chemical solution. For samples that employ density separation, the supernatant is collected and filtered through a filter paper (Huang et al., 2020; Kumkar et al., 2021). Chemical removal before filtration is necessary for some chemicals. NaI can blacken the cellulose filter (Tirkey and Upadhyay, 2021). The collected supernatant from NaI solution should be rinsed to remove the remaining chemical. After filtration, filter papers are dried by leaving them at room temperature (Maghsodian et al., 2021) or placed in an oven (Naidoo et al., 2020) or desiccator (Kumkar et al., 2021). A Petri dish is used to store filter paper (Li et al., 2020a; Kumkar et al., 2021; Maghsodian et al., 2021). Aluminum foil can be used to cover a Petri dish for airborne contamination (Maghsodian et al., 2021).

10.2.4.6 Characterization

Microplastic appearance can be characterized into four categories comprising source, shape, size, and color (Rodríguez-Seijo and Pereira, 2017). Most previous studies on microplastic contamination in mangrove biota mainly report shapes, sizes, and colors. The shape categories of microplastics are divided into four main shapes comprising fragment, film, fiber, and pellet (Huang et al., 2020; Kumkar et al., 2021; Maghsodian et al., 2021). The sizes of microplastics are reported as varied in different studies. Capparelli et al. (2022) detected that all microplastics contaminated in fiddler crabs were less than 0.45 mm. Similarly, Kumkar et al. (2021) found only fiber microplastics in Dussumier's mudskippers, ranging from 0.7 to 11.3 mm. Some studies classified microplastics in different size ranges depending on methods and differentiating ability (Rodríguez-Seijo and Pereira, 2017). Aguirre-Sanchez et al. (2022) divided microplastics into four size ranges (1–5, 0.5–1, 0.25–0.5, 0.002–0.25 mm). Huang et al. (2020) classified microplastics into five size ranges (0.02–1, 1–2, 2–3, 3–4, and 4–5 mm). The color categories can be done by visual inspection. However, visual sorting can lead to misidentification with incompletely digested organic matter, such as shell fragments (Tirkey and Upadhyay, 2021). It is difficult to differentiate small-size microplastics from organic matter. In order to identify and separate small-size microplastics, fluorescence staining is applied to microplastic identification (Tirkey and Upadhyay, 2021). Not et al. (2020) used Rose Bengal to track microplastics. Though fluorescence staining can track the presence of microplastics, the color categories cannot be examined due to the stain. Besides, scanning electron microscopy coupled with energy-dispersive spectroscopy can be used for the surface morphology of microplastics (cracks, fissures, pits, and adherent particles) (Kumkar et al., 2021).

10.2.4.7 Polymer Identification

The identification of the polymer composition of microplastics can provide a powerful method to separate microplastics from false microplastics and define the chemical characterization of microplastics (Tirkey and Upadhyay, 2021). Different detection techniques were used in studies on microplastic contamination in the mangrove biota, such as micro-Raman spectroscopy (Li et al., 2020a), Raman spectroscopy (Maghsodian et al., 2021), Fourier transform infrared (FTIR) spectroscopy (Naidoo et al., 2020; Not et al., 2020; Kumkar et al., 2021; Maghsodian et al., 2021), and micro-FTIR (Huang et al., 2020). FTIR is the most common analytical technique employed for microplastic analysis (Rose et al., 2023b). However, some studies did not report on polymer identification. Due to the limitation of a lack of instruments, Aguirre-Sanchez et al. (2022) identified polymer types in a very low numbers of microplastics compared to the abundance of microplastics in samples. They found only two main polymer types (PET and polyethylene vinyl acetate). Similarly, Not et al. (2020) were unable to identify most microplastics, which may have been caused by the small size of microplastics and the interruption of Rose Bengal or remaining organic matter. Near-infrared spectroscopy (NIR) provides better performance than FTIR (Tirkey and Upadhyay, 2021). However, NIR spectroscopy has been used in a few studies. Among the analytical techniques, pyrolysis gas chromatography–mass spectrometry, FTIR, and Raman spectroscopy provide reliability on accurate chemical identification (Rose et al., 2023a). Thus, the selected techniques for polymer identification should suit microplastics in samples and access small-size microplastics to provide comprehensive results of microplastic uptake in mangrove biota.

10.3 POTENTIAL PATHWAYS FOR MICROPLASTICS

Microplastics enter mangrove forests through different potential pathways. As shown in Figure 10.4, microplastics probably enter mangrove forests through three main sources, comprising terrestrial, atmospheric, and marine sources.

Terrestrial sources of microplastics mainly include domestic wastewater, wastewater treatment plants, and agricultural runoff (Deng et al., 2021). Dense populations and nearby industrial zones probably release wastewater contaminated with microplastics into mangrove areas (Kumkar et al., 2021). Some microplastics, such as fibers shed from textiles and cloths, can pass through the filters during wastewater treatment (Govender et al., 2020). Most microfibers originate from domestic and commercial textile washing, textile industries, and wastewater treatment plants (Yadav et al., 2023). Although the removal capacity of wastewater treatment plants is over 70% and can be enhanced to 99.9% when combined with a tertiary stage, the huge daily discharge cannot be completely overlooked (Hassan et al., 2023; Rose et al., 2023b). Nevertheless, mismanaged and free-dumped plastic waste can be effective sources of microplastic origin. A study by Pan et al. (2020) found that the nearness to a local hub of economic activities, industries, and population increased the abundance of microplastics in surface water from the Zhangjiang River. Microplastic contamination in urbanized mangrove forests is likely related to population density, and inefficient wastewater and solid waste management, while preserved mangrove forests that allow local fisheries are possibly affected by the activities of inhabitants (Trindade et al., 2023). Moreover, currents from the ocean may transport microplastics from adjacent lands to the mangrove areas

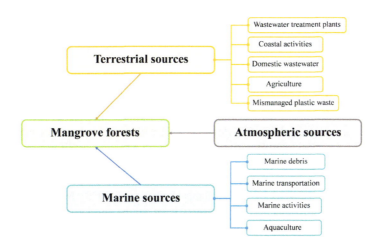

FIGURE 10.4 Pathways for microplastic contamination in mangrove forests.

(Rose and Webber, 2019). Agricultural runoff, such as agricultural films or fragments, can be another source of microplastics (Deng et al., 2021; Qian et al., 2021). Film mulching leaves some plastic fractions that break down into soil via ultraviolet (UV) radiation and mechanical tillage (Rose et al., 2023a). Sewage sludge, employed as a soil conditioner, contains escaping microplastics from wastewater treatment plants that can affect crop production (Hassan et al., 2023). Another pathway of microplastics is coastal activities. A study by Aliabad et al. (2019) revealed that ports with the accumulation of fishing waste, such as broken vessels, contained a higher abundance of microplastics. Additionally, paint flakes from boats and vessel cleaning can be sources of microplastics in the surrounding areas of ports, including mangrove forests. Moreover, mismanaged plastic litter from land-based sources can be driven to mangrove forests by tidal currents and waves (Martin et al., 2019) and generate microplastics through degradation (Deng et al., 2021).

Atmospheric microplastics possibly enter mangrove forests through rainfall (Qian et al., 2021). A study by Abbasi (2021) detected microplastic contamination in precipitation collected during monsoon in Iran that ranged from ~20.83 to 156.25 items/m^2 per 10 min. Moreover, microplastics can be brought by typhoons, which dramatically increase after cyclones (Lo et al., 2020).

The marine sources of microplastics consist of aquacultures, marine activities, and transportation. Ta and Babel (2020) reported the contamination of microplastics in water and sediment from aquaculture zones in the Chao Phraya River Estuary, Thailand. It is possible that microplastics released from estuaries may be brought to the mangrove forests located at the shoreline near the estuaries. Another source is marine activities. Currents can drive microplastics released from marine tourism activities to the mangrove forests (John et al., 2022). Besides, microplastics can be leaked from marine transportation (Deng et al., 2021). Primary microplastics, such as cylindrical virgin resin pellets, are widely used for transportation (Joint Group of Experts on the Scientific Aspects of Marine Environmental Protection, 2015). Furthermore, floating marine debris from the ocean may be translocated to mangrove forests by currents (Martin et al., 2019). The accumulation of marine debris can cause the occurrence of microplastics in mangrove forests (Luo et al., 2021).

10.4 IMPACTS OF MICROPLASTICS

10.4.1 IMPACTS ON MANGROVE BIOTA

A study by Brennecke et al. (2015) found the translocation of microplastics ranging between 180–250 μm from the digestive tracts to the hepatopancreas in the fiddler crab (*Uca rapax*). Villegas et al. (2022) proposed that the synergistic effects of malathion with microplastics led to a high mortality rate in the fiddler crabs (*Minuca ecuadoriensis*). They reported that microplastics alone did not kill the fiddler crabs after 5 days of exposure. Conversely, Wang et al. (2021a) found that microplastics mostly accumulated in the hepatopancreas of predatory crabs (*Charybdis japonica*). The crabs can excrete microplastics by their defense mechanism, but it was not endured after surpassing the limit concentrations in the hepatopancreas, which caused severe hepatic injury (Wang et al., 2021a). This defense mechanism was also detected in mudskippers. Kumkar et al. (2021) proposed that the special teeth on the pharyngeal plate of *B. dussumieri* can filter other shapes of microplastics except fibers to access its digestive tract. However, this may be due to the papilliform teeth of mudskippers in species of the genus *Boleophthalmus*. Although the defense mechanism efficiently blocks other shapes, fibrous microplastics are sufficient to harm animal health since they have a large surface area and high adsorption capacity of contaminants (Yadav et al., 2023). Moreover, other mudskipper species, such as giant mudskippers (*P. schlosseri*), have canine and large teeth on the small and flat pharyngeal plate (Tran et al., 2021), which microplastics possibly can pass through and access their gut. Apart from that, large-size microplastics can cause physical damage, such as obstructions inside digestive organs (Yuan et al., 2022).

Besides, heavy metals were detected in mangrove crabs and giant mudskippers (Chaiyara et al., 2013; Santoso et al., 2021). Microplastics can be vectors for heavy metals (e.g., iron, chromium, manganese, and nickel) and the interactions between microplastics and heavy metals increase the harmful effects of microplastics on organisms (Liu et al., 2022). A study by Hanun et al. (2023) revealed that structural changes in PET microplastics through sunlight exposure enhanced the capacity to adsorbing oxybenzone (UV filter). Moreover, Ammar et al. (2022) found that PE microplastics with 4-nonylphenol can damage the tissue of carp juveniles. Chemical additives in microplastics can leach out and cause toxicological effects on living organisms (Yuan et al., 2022). For instance, microplastics with a mixture of polycyclic aromatic hydrocarbons can reduce the viability of hemocytes of the blood clam (*Tegillarca granosa*) (Sun et al., 2021). In addition, polylactic acid (PLA) bioplastic debuted as an alternative plastic for eco-friendly utilization but also negatively affects living organisms. PLA residues in the environment can reduce the survival rate, impact development, and cause metabolic alterations in living organisms (Yadav et al., 2023).

10.4.2 IMPACTS ON HUMAN SUBSISTENCE

Since humans have exploited the rich biodiversity of the mangrove forest, this unique forest has been affected by urbanization and industrialization (John et al., 2022). Microplastic contamination in the mangrove forests originates from anthrophonic activities. Due to polluted mangrove forests, humans are going to lose another natural resource. Microplastics detected in commercial products from the mangrove forests, such as mudskippers (Kumkar et al., 2021), probably affect commercialization. Furthermore, contaminated mangrove biota, which is a delicacy for humans, can impact food safety (John et al., 2022). Normally, the digestive tracts of some animals, such as medium-sized and large fish, are removed in the preparation process before cooking. Only edible tissues of some animals are food for humans. However, small-size microplastics can penetrate from the digestive tracts to the internal organs of animals. Giacinto et al. (2023) found the presence of tiny microplastics (<10 µm) in the muscle of swordfish and bluefin tuna. Small-size microplastics demonstrate higher ability than large-size microplastics to do this. Moreover, small fish and mollusks are typically consumed as whole soft bodies. Microplastics contaminated in the tissues of such species can be transferred to humans. Nevertheless, the literature on microplastics contamination in humans remains limited. The toxicological effects of microplastics can harm human health. Microplastics ranging from 0.1 to 150 µm can possibly translocate to the human lymphatic system (Sangkham et al., 2022). Moreover, small-size microplastics (50 µm) can penetrate from the gastrointestinal tract to the liver and spleen and may cause an inflammatory response (Yuan et al., 2022). In addition, adsorbed pollutants or chemical additives in microplastics can have detrimental effects on human health (Yuan et al., 2022; Sangkham et al., 2022)

10.5 POSSIBLE SOLUTIONS TO CONTROL MICROPLASTIC POLLUTION

In order to reduce microplastic pollution, primary source management is important. As mentioned above, microplastics from three different pathways (terrestrial, atmospheric, and marine sources) may enter and accumulate in mangrove forests. Microplastics from terrestrial and marine sources should be managed and controlled. Moreover, plastic waste should be reused and recycled as much as possible at primary sources (Li et al., 2021). Single-use plastics have been prohibited in many countries. They have been banned in Bangladesh but dumping plastic waste on the forest floor in the Sundarbans mangrove forest remains a concern (Adyel and Macreadie, 2021). Wastewater treatment also needs to be improved to remove microplastics before release into the environment (Deng et al., 2021). For primary microplastics, microbeads in cosmetics have been banned in the United States, the United Kingdom, Canada, Australia, Ireland, New Zealand, Italy, South Korea, and Thailand (Watkins et al., 2019; Mitrano and Wohlleben, 2020; Thailand Pollution Control Department, 2021).

Nevertheless, the prevention of leakage of primary microplastics from industries and plastic transportation is required. According to studies on microplastic contamination in mangrove forests, secondary microplastics (fibers and fragments) with PE and PP as dominant polymer types are frequently detected in mangrove sediment, water, and biota in many countries (Tables 10.1, 10.2, and 10.3). PE and PP are the main plastic materials in global production (Yuan et al., 2022). Various types of plastic products are still necessary for human daily life. Thus, plastic waste management at primary sources, especially PE and PP plastic waste, should be strictly and efficiently controlled.

Mangrove forests are potential receptacles for terrestrial and marine waste (Deng et al., 2021). They are considered to be a destination for mismanaged plastic waste. A basic solution for plastic waste management in mangrove forests is cleanup activities. In Thailand, several cleaning campaigns in the Bang Pu mangrove forest have been carried out for a number of years (Foundation for Environmental Education for Sustainable Development, 2022). Coastal cleaning and improved river quality can decrease the accumulation of microplastics in mangrove forests (Cordova et al., 2021). Moreover, some nonprofit organizations have been established for mangrove cleanup as volunteering activities. This solution requires continuous action. It takes time and requires a lot of workers. It should be underlined that cleanup programs cannot remove existing microplastics from the mangrove forests. Nevertheless, the eradication of sources of future microplastics is beneficial. Moreover, raising awareness about plastic usage is important.

In order to prevent plastic waste driven by currents to mangrove forests, a barrier may be a viable solution. In Thailand, bamboo fences are used to reduce coastal erosion and trap floating debris. However, due to their short lifespan and detachment, they become floating debris and destroy mangrove trees (Pranchai et al., 2019). Moreover, they are obstacles to debris-cleaning programs. A technological solution may solve this problem. For example, the Great Bubble Barrier has been invented to collect plastic waste and microplastics (1–5 mm) in a river using a bubble wall (The Great Bubble Barrier, 2019). However, this invention is costly, starting at 350,000 euros. According to their results, it is eco-friendly and efficient to deal with mismanaged plastic waste and large-size microplastics. There are also other technologies for removing or trapping plastic waste from the environment. However, mangrove forests with unique root systems resist feasible inventions for plastic waste management. Developing technology to remove accumulated plastic waste and microplastics from mangrove forests has been challenging.

Biodegradable plastics have been promoted to solve severe pollution from conventional plastics. Conversely, the accumulation of biodegradable plastic waste in mangrove forests and other ecosystems may aggravate microplastic pollution. Wei et al. (2021) found that the occurrence of microplastics generated from biodegradable plastic poly(butylene adipate-co-terephthalate) was higher than from LDPE in aquatic environments. Moreover, biodegradable microplastics can be stronger vectors for pollutants and contaminants in the surrounding environment (Wang et al., 2021b). PLA bioplastic, one of the most popular plant-based plastics, leaves PLA residues after aerobic composting and impacts on the growth, development, and survival of animals (Yadav et al., 2023). However, the further development of biodegradable plastics is expected to solve plastic and microplastic pollution.

Importantly, although many countries have developed and announced policies for plastic waste management, the policy for the mangrove ecosystem has been scarcely reported. Thus, policies and regulations for plastic debris and microplastic pollution in mangrove forests should be developed.

10.6 CONCLUSIONS

As the potential sink of enormous volumes of waste from different pathways, mangrove forests require more scientific attention. Both urbanized and preserved mangrove forests have been invaded with the widespread presence of microplastics. Remote mangrove areas have not been exempt from intrusion by microplastics. Moreover, microplastic pollution has interrupted the high biodiversity

Microplastic Contamination in the Biota of Mangrove Forests

in mangrove forests and may cause detrimental effects on mangrove biota. Potential pathways of microplastic contamination in mangrove forests should be managed and controlled. Currently, there are very few studies reporting the contamination of microplastics in mangrove biota. Microplastic abundance in mangrove biota and the surrounding environment (water and sediment) varied with differences in regions, mangrove density, waste loads, and hydrodynamics. Additionally, a lack of standard methodology leads to different processes for investigating microplastics. Various methods can affect the results, and it is difficult to directly compare these with other studies, which may result in ineffective comparison among studies. Secondary microplastics (fragments and fibers) predominate in mangrove forests, with PE and PP being the main polymer types. Since some mangrove animals are food for humans, the contamination of microplastics, especially small-size microplastics, can be transferred to humans and may have harmful effects on human health. Besides, mangrove forests require solutions and policies to manage microplastic pollution and protect mangrove areas from the invasion of enormous amounts of waste.

ACKNOWLEDGMENTS

The authors acknowledge the Sirindhorn International Institute of Technology, Thailand, for providing scholarship to the first author.

REFERENCES

Abbasi, S. 2021. Microplastics washout from the atmosphere during a monsoon rain event. *Journal of Hazardous Materials Advances* 4 (December), 100035. https://doi.org/10.1016/j.hazadv.2021.100035

Abu El-Regal, M. A., and N. K. Ibrahim. 2014. Role of mangroves as a nursery ground for juvenile reef fishes in the southern Egyptian Red Sea. *The Egyptian Journal of Aquatic Research* 40, no. 1, 71–78. https://doi.org/10.1016/j.ejar.2014.01.001

Adyel, T. M., and P. I. Macreadie. 2021. World's largest mangrove forest becoming plastic cesspit. *Frontiers in Marine Science* 8, 766876. https://doi.org/10.3389/fmars.2021.766876

Aguirre-Sanchez, A., S. Purca, and A. G. Indacochea. 2022. Microplastic presence in the mangrove crab *Ucides Occidentalis* (Brachyura: Ocypodidae) (Ortmann, 1897) derived from local markets in Tumbes, Peru. *Air, Soil and Water Research* 15 (January–December), 117862212211245. https://doi.org/10.1177/11786221221124549

Alfonso, M. B., K. Takashima, S. Yamaguchi, M. Tanaka, and A. Isobe. 2021. Microplastics on plankton samples: Multiple digestion techniques assessment based on weight, size, and FTIR spectroscopy analyses. *Marine Pollution Bulletin* 173, no. part A (December), 113027. https://doi.org/10.1016/j.marpolbul.2021.113027

Aliabad, M. K., M. Nassiri, and K. Kor. 2019. Microplastics in the surface seawaters of chabahar bay, Gulf of Oman (Makran coasts). *Marine Pollution Bulletin* 143 (June), 125–133. https://doi.org/10.1016/j.marpolbul.2019.04.037

Ammar, E., M. S. Mohamed, and H. Sayed. 2022. Polyethylene microplastics increases the tissue damage caused by 4-nonylphenol in the common carp (*Cyprinus Carpio*) juvenile. *Frontiers in Marine Science* 9, 1041003. https://doi.org/10.3389/fmars.2022.1041003

Ansari, A. A., S. Trivedi, S. Saggu, and H. Rehman. 2014. Mudskipper: A biological Indicator for fnvironmental monitoring and assessment of coastal waters mudskipper. *Journal of Entomology and Zoology Studies* 2, no. 6, 22–33.

Arceo-Carranza, D., X. Chiappa-Carrara, R. C. López, and C. Y. Arenas. 2021. Mangroves as feeding and breeding grounds. In *Mangroves: Ecology, Biodiversity and Management*, eds. R. P. Rastogi, M. Phulwaria, and D. K. Gupta, 257–278. Singapore: Springer.

Brennecke, D., E. C. Ferreira, T. M. M. Costa, D. Appel, B. A. P. da Gama, and M. Lenz. 2015. Ingested microplastics (>100μm) are translocated to organs of the tropical fiddler crab *Uca Rapax*. *Marine Pollution Bulletin* 96, no. 1–2 (July), 491–495. https://doi.org/10.1016/j.marpolbul.2015.05.001

Capparelli, M. V., M. Martínez-Colón, O. Lucas-Solis, et al. 2022. Can the bioturbation activity of the fiddler crab *Minuca Rapax* modify the distribution of microplastics in sediments? *Marine Pollution Bulletin* 180 (July), 113798. https://doi.org/10.1016/j.marpolbul.2022.113798

Chaiyara, R., M. Ngoendee, and M. Kruatrachue. 2013. Accumulation of Cd, Cu, Pb, and Zn in water, sediments, and mangrove crabs (*Sesarma Mederi*) in the upper gulf of Thailand. *Science Asia* 39, no. 4: 376–383. https://doi.org/10.2306/scienceasia1513-1874.2013.39.376

Cordova, M. R., Y. I. Ulumuddin, T. Purbonegoro, and A. Shiomoto. 2021. Characterization of microplastics in mangrove sediment of Muara Angke wildlife reserve, Indonesia. *Marine Pollution Bulletin* 163 (February), 112012. https://doi.org/10.1016/j.marpolbul.2021.112012

Dehaut, A., A. L. Cassone, L. Frère, et al. 2016. Microplastics in seafood: Benchmark protocol for their extraction and characterization. *Environmental Pollution* 215 (August), 223–233. https://doi.org/10.1016/j.envpol.2016.05.018

Deng, H., J. He, D. Feng, et al. 2021. Microplastics pollution in mangrove ecosystems: A critical review of current knowledge and future directions. *Science of the Total Environment* 753 (January), 142041. https://doi.org/10.1016/j.scitotenv.2020.142041

Deng, J., P. Guo, X. Zhang, et al. 2020. Microplastics and accumulated heavy metals in restored mangrove wetland surface sediments at Jinjiang Estuary (Fujian, China). *Marine Pollution Bulletin* 159 (October), 111482. https://doi.org/10.1016/j.marpolbul.2020.111482

Duan, J., J. Han, S.G. Cheung, et al. 2021. How mangrove plants affect microplastic distribution in sediments of coastal wetlands: Case study in Shenzhen bay, South China. *Science of the Total Environment* 767 (May), 144695. https://doi.org/10.1016/j.scitotenv.2020.144695

Farrell, P., and K. Nelson. 2013. Trophic level transfer of microplastic: *Mytilus Edulis* (L.) to *Carcinus Maenas* (L.). *Environmental Pollution* 177 (June), 1–3. https://doi.org/10.1016/j.envpol.2013.01.046

Foundation for environmental education for sustainable development (FEED). 2022. The Bangpu nature education centre english annual report (January–December 2021). Report, FEED. www.feedthailand.org/wp-content/uploads/2022/10/B1AR2021_EN.pdf (accessed July 1, 2023).

Galicia, M. L. C., C. L. P. Flestado, J. R. S. Magalong, et al. 2023. Assessment of Microplastics in the surface water of Sitio Pulo, Navotas, Metro Manila. *Applied Environmental Research* 45, no. 1 (January–March). https://doi.org/10.35762/aer.2023001

Garcés-Ordóñez, O., V. A. Castillo-Olaya, A. F. Granados-Briceño, L. M. Blandón García, and L. F. Espinosa Díaz. 2019. Marine litter and microplastic pollution on mangrove soils of the Ciénaga Grande de Santa Marta, Colombian Caribbean. *Marine Pollution Bulletin* 145 (August), 455–462. https://doi.org/10.1016/j.marpolbul.2019.06.058

Garcés-Ordóñez, O., V. A. Castillo-Olaya, L. F. Espinosa-Díaz, and M. Canals. 2023. Seasonal variation in plastic litter pollution in mangroves from two remote tropical estuaries of the Colombian Pacific. *Marine Pollution Bulletin* 193 (August), 115210. https://doi.org/10.1016/j.marpolbul.2023.115210

Joint Group of Experts on the Scientific Aspects of Marine Environmental Protection (GESAMP). 2015. Sources, fate and effects of MP in the marine environment: A global assessment. *Journal Series GESAMP Reports and Studies, IMO*. www.gesamp.org/site/assets/files/1272/reports-and-studies-no-90-en.pdf (accessed July 2, 2023).

Giacinto, F. D., L. D. Renzo, G. Mascilongo, et al. 2023. Detection of microplastics, polymers and additives in edible muscle of swordfish (*Xiphias Gladius*) and bluefin tuna (*Thunnus Thynnus*) caught in the Mediterranean Sea. *Journal of Sea Research* 192 (April), 102359. https://doi.org/10.1016/j.seares.2023.102359

Govender, J., T. Naidoo, A. Rajkaran, S. Cebekhulu, A. Bhugeloo, and S. Sershen. 2020. Towards characterising microplastic abundance, typology and retention in mangrove-dominated estuaries. *Water* 12, no. 10 (October), 2802. https://doi.org/10.3390/w12102802

Hanun, J. N., F. Hassan, L. Theresia, et al. 2023. Weathering effect triggers the sorption enhancement of microplastics against oxybenzone. *Environmental Technology and Innovation* 30 (May), 103112. https://doi.org/10.1016/j.eti.2023.103112

Hassan, F., K. D. Prasetya, J. N. Hanun, et al. 2023. Microplastic contamination in sewage sludge: Abundance, characteristics, and impacts on the environment and human health. *Environmental Technology & Innovation* 31 (April), 103176. https://doi.org/10.1016/j.eti.2023.103176

Huang, J. S., J. B. Koongolla, H. X. Li, et al. 2020. Microplastic accumulation in fish from Zhanjiang mangrove wetland, South China. *Science of the Total Environment* 708 (March), 134839. https://doi.org/10.1016/j.scitotenv.2019.134839

John, J., A. R. Nandhini, P. V. Chellam, and M. Sillanpää. 2022. Microplastics in mangroves and coral reef ecosystems: A review. *Environmental Chemistry Letters* 20, no. 1 (February), 397–416. https://doi.org/10.1007/s10311-021-01326-4

Kannankai, M. P., R. K. Alex, V. V. Muralidharan, N. P. Nazeerkhan, A. Radhakrishnan, and S. P. Devipriya. 2022. Urban mangrove ecosystems are under severe threat from microplastic pollution: A case study from Mangalavanam, Kerala, India. *Environmental Science and Pollution Research* 29, no. 53 (November), 80568–80580. https://doi.org/10.1007/s11356-022-21530-1

Karami, A., A. Golieskardi, C. K. Choo, N. Romano, Y. B. Ho, and B. Salamatinia. 2017. A high-performance protocol for extraction of microplastics in fish. *Science of the Total Environment* 578 (February), 485–494. https://doi.org/10.1016/j.scitotenv.2016.10.213

Kedzierski, M., V. L. Tilly, G. César, O. Sire, and S. Bruzaud. 2017. Efficient microplastics extraction from sand. A cost effective methodology based on sodium iodide recycling. *Marine Pollution Bulletin* 115, no. 1–2 (February), 120–129. https://doi.org/10.1016/j.marpolbul.2016.12.002

Kristensen, E. 2008. Mangrove crabs as ecosystem engineers; with emphasis on sediment processes. *Journal of Sea Research* 59, no. 1–2 (February), 30–43. https://doi.org/10.1016/j.seares.2007.05.004

Kumkar, P., S. M. Gosavi, C. R. Verma, M. Pise, and L. Kalous. 2021. Big eyes can't see microplastics: Feeding selectivity and eco-morphological adaptations in oral cavity affect microplastic uptake in mud-dwelling amphibious mudskipper fish. *Science of the Total Environment* 786 (September), 147445. https://doi.org/10.1016/j.scitotenv.2021.147445

Li, D., L. Zhao, Z. Guo, et al. 2021. Marine debris in the Beilun estuary mangrove forest: Monitoring, assessment and implications. *International Journal of Environmental Research and Public Health* 18, no. 20 (October), 10826. https://doi.org/10.3390/ijerph182010826

Li, R., L. Yu, M. Chai, H. Wu, and X. Zhu. 2020b. The distribution, characteristics and ecological risks of microplastics in the mangroves of Southern China. *Science of the Total Environment* 708 (March), 135025. https://doi.org/10.1016/j.scitotenv.2019.135025

Li, R., L. Zhang, B. Xue, and Y. Wang. 2019. Abundance and characteristics of microplastics in the mangrove sediment of the semi-enclosed Maowei Sea of the south China sea: New implications for location, rhizosphere, and sediment compositions. *Environmental Pollution* 244 (January), 685–692. https://doi.org/10.1016/j.envpol.2018.10.089

Li, R., S. Zhang, L. Zhang, K. Yu, S. Wang, and Y. Wang. 2020a. Field study of the microplastic pollution in sea snails (*Ellobium chinense*) from mangrove forest and their relationships with microplastics in water/sediment located on the north of Beibu Gulf. *Environmental Pollution (Barking, Essex: 1987)* 263, no. part B (August), 114368. https://doi.org/10.1016/j.envpol.2020.114368

Li, W., B. Zu, L. Hu, L. Lan, Y. Zhang, and J. Li. 2022. Migration behaviors of microplastics in sediment-bearing turbulence: Aggregation, settlement, and resuspension. *Marine Pollution Bulletin* 180 (July), 113775. https://doi.org/10.1016/j.marpolbul.2022.113775

Liu, C., and C. Liu. 2023. Exploring plastic-management policy in China: Status, challenges and policy insights. *Sustainability* 15, no. 11 (June), 9087. https://doi.org/10.3390/su15119087

Liu, S., J. Huang, W. Zhang, et al. 2022. Microplastics as a vehicle of heavy metals in aquatic environments: A review of adsorption factors, mechanisms, and biological effects. *Journal of Environmental Management* 302, no. part A (January), 113995. https://doi.org/10.1016/j.jenvman.2021.113995

Lo, H. S., Y. K. Lee, B. H. K. Po, et al. 2020. Impacts of Typhoon Mangkhut in 2018 on the deposition of marine debris and microplastics on beaches in Hong Kong. *Science of the Total Environment* 716 (May), 137172. https://doi.org/10.1016/j.scitotenv.2020.137172

Luo, Y. Y., C. Not, and S. Cannicci. 2021. Mangroves as unique but understudied traps for anthropogenic marine debris: A review of present information and the way forward. *Environmental Pollution* 271 (February), 116291. https://doi.org/10.1016/j.envpol.2020.116291

Lusher, A. L., N. A. Welden, P. Sobral, and M. Cole. 2017. Sampling, isolating and identifying microplastics ingested by fish and invertebrates. *Analytical Methods*, no. 9: 1346–1360. https://doi.org/10.1039/c6ay02415g

Maghsodian, Z., A. M. Sanati, B. Ramavandi, A. Ghasemi, and G.A. Sorial. 2021. Microplastics accumulation in sediments and *Periophthalmus waltoni* fish, mangrove forests in southern Iran. *Chemosphere* 264, no. part 2 (February), 128543. https://doi.org/10.1016/j.chemosphere.2020.128543

Martin, C., H. Almahasheer, and C. M. Duarte. 2019. Mangrove forests as traps for marine litter. *Environmental Pollution* 247 (April), 499–508. https://doi.org/10.1016/j.envpol.2019.01.067

Martin, C., F. Baalkhuyur, L. Valluzzi, et al. 2020. Exponential increase of plastic burial in mangrove sediments as a major plastic sink. *Science Advances* 6, no. 44 (October), eaaz5593. https://doi.org/10.1126/sciadv.aaz5593

Mitrano, D. M., and W. Wohlleben. 2020. Microplastic regulation should be more precise to incentivize both innovation and environmental safety. *Nature Communications* 11, 5324. https://doi.org/10.1038/s41467-020-19069-1

Nabizadeh, R., M. Sajadi, N. Rastkari, and K. Yaghmaeian. 2019. Microplastic pollution on the Persian Gulf shoreline: A case study of Bandar Abbas city, Hormozgan Province, Iran. *Marine Pollution Bulletin* 145 (August), 536–546. https://doi.org/10.1016/j.marpolbul.2019.06.048

Nahian, S. A., R. Jahan, R. Kumar, S. M. B. Haider, P. Sharma, and A. M. Idris. 2023. Distribution, characteristics, and risk assessments analysis of microplastics in shore sediments and surface water of Moheshkhali Channel of Bay of Bengal, Bangladesh. *Science of the Total Environment* 855 (January), 158892. http://dx.doi.org/10.1016/j.scitotenv.2022.158892

Naidoo, T., Sershen, R. C. Thompson, and A. Rajkaran. 2020. Quantification and characterisation of microplastics ingested by selected juvenile fish species associated with mangroves in KwaZulu-Natal, South Africa. *Environmental Pollution* 257 (February), 113635. https://doi.org/10.1016/j.envpol.2019.113635

Navarro, C. K. P., C. G. L. A. Arcadio, K. M. Similatan, et al. 2022. Unraveling microplastic pollution in mangrove sediments of Butuan Bay, Philippines. *Sustainability* 14, no. 21 (November), 14469. https://doi.org/10.3390/su142114469

Nawar, N., M. Kurasaki, F. N. Chowdhury, et al. 2023. Characterization of microplastic pollution in the Pasur river of the Sundarbans ecosystem (Bangladesh) with emphasis on water, sediments, and fish. *Science of the Total Environment* 868 (April), 161704. https://doi.org/10.1016/j.scitotenv.2023.161704

Ni'am, A. C., F. Hassan, R. F. Shiu, and J. J. Jiang. 2022. Microplastics in sediments of east Surabaya, Indonesia: Regional characteristics and potential risks. *International Journal of Environmental Research and Public Health* 19, no. 19 (October), 12348. https://doi.org/10.3390/ijerph191912348

Not, C., C. Y. I. Lui, and S. Cannicci. 2020. Feeding behavior is the main driver for microparticle intake in mangrove crabs. *Limnology and Oceanography Letters* 5, no. 1 (February), 84–91. https://doi.org/10.1002/lol2.10143

Nuelle, M. T., J. H. Dekiff, D. Remy, and E. Fries. 2014. A new analytical approach for monitoring microplastics in marine sediments. *Environmental Pollution* 184 (January), 161–169. https://doi.org/10.1016/j.envpol.2013.07.027

Paes, E. D. S., T. V. Gloaguen, H. D. A. D. C. Silva, et al. 2022. Widespread microplastic pollution in mangrove soils of Todos os Santos Bay, northern Brazil. *Environmental Research* 210 (July), 112952. https://doi.org/10.1016/j.envres.2022.112952

Pan, Z., Y. Sun, Q. Liu, et al. 2020. Riverine microplastic pollution matters: A case study in the Zhangjiang River of Southeastern China. *Marine Pollution Bulletin* 159 (October), 111516. https://doi.org/10.1016/j.marpolbul.2020.111516

Patria, M. P., C. A. Santoso, and N. Tsabita. 2020. Microplastic Ingestion by Periwinkle Snail Littoraria scabra and Mangrove Crab *Metopograpsus quadridentata* in Pramuka Island, Jakarta Bay, Indonesia. *Sains Malaysiana* 49, no. 9: 2151–2158. https://doi.org/10.17576/jsm-2020-4909-13

Pete, A. J., P. J. Brahana, M. Bello, M. J. Benton, and B. Bharti. 2022. Biofilm formation influences the wettability and settling of microplastics. *Environmental Science & Technology Letters* 10, no. 2 (February), 159–164. https://doi.org/10.1021/acs.estlett.2c00728

Phuong, N. N., A. Zalouk-Vergnoux, L. Poirier, et al. 2016. Is there any consistency between the microplastics found in the field and those used in laboratory experiments? *Environmental Pollution* 211 (April), 111–123. https://doi.org/10.1016/j.envpol.2015.12.035

Plastics Europe. 2022. Plastics – The Facts 2022. Report, Plastics Europe. www.plasticseurope.org/knowledge-hub/plastics-the-facts-2022/ (accessed July 2, 2023).

Pranchai, A., M. Jenke, and U. Berger. 2019. Well-intentioned, but poorly implemented: Debris from coastal bamboo fences triggered mangrove decline in Thailand. *Marine Pollution Bulletin* 146 (September), 900–907. https://doi.org/10.1016/j.marpolbul.2019.07.055

Qian, J., S. Tang, P. Wang, at al. 2021. From source to sink: Review and prospects of microplastics in wetland ecosystems. *Science of the Total Environment* 758 (March), 143633. https://doi.org/10.1016/j.scitotenv.2020.143633

Rodrigues, M. O., A. M. M. Gonçalves, F. J. M. Gonçalves, and N. Abrantes. 2020. Improving cost-efficiency for MPs density separation by zinc chloride reuse. *MethodsX* 7 (September), 100785. https://doi.org/10.1016/j.mex.2020.100785

Rodríguez-Seijo, A., and R. Pereira. 2017. Morphological and physical characterization of microplastics. *Characterization and Analysis of Microplastics* 75, 49–66. https://doi.org/10.1016/bs.coac.2016.10.007

Rose, D., and M. Webber. 2019. Characterization of microplastics in the surface waters of Kingston Harbour. *Science of the Total Environment* 664 (May), 753–760. https://doi.org/10.1016/j.scitotenv.2019.01.319

Rose, P. K., M. Jain, N. Kataria, P. K. Sahoo, V. K. Garg, and A. Yadav. 2023b. Microplastics in multimedia environment: A systematic review on its fate, transport, quantification, health risk, and remedial measures. *Groundwater for Sustainable Development* 20 (February), 100889. https://doi.org/10.1016/j.gsd.2022.100889

Rose, P. K., S. Yadav, N. Kataria, and K. S. Khoo. 2023a. Microplastics and nanoplastics in the terrestrial food chain: Uptake, translocation, trophic transfer, ecotoxicology, and human health hisk. *TrAC Trends in Analytical Chemistry* 167 (October), 117249. https://doi.org/10.1016/j.trac.2023.117249

Rummel, C. D., A. Jahnke, E. Gorokhova, D. Kühnel, and M. Schmitt-Jansen. 2017. Impacts of biofilm formation on the fate and potential effects of microplastic in the aquatic environment. *Environmental Science & Technology Letters* 4, no. 7 (July), 258–267. https://doi.org/10.1021/acs.estlett.7b00164

Sangkham, S., O. Faikhaw, N. Munkong, et al. 2022. A review on microplastics and nanoplastics in the environment: Their occurrence, exposure routes, toxic studies, and potential effects on human health. *Marine Pollution Bulletin* 181 (August), 113832. https://doi.org/10.1016/j.marpolbul.2022.113832

Santana, M. F. M., F. T. Moreira, and A. Turra. 2017. Trophic transference of microplastics under a low exposure scenario: Insights on the likelihood of particle cascading along marine food-webs. *Marine Pollution Bulletin* 121, no. 1–2 (August), 154–159. https://doi.org/10.1016/j.marpolbul.2017.05.06

Santoso, H. B., and Hidayaturrahmah. 2021. Heavy metal concentrations in water, sediment and giant mudskipper (*Periophthalmodon schlosseri*) in the coastal wetlands of Kuala Lupak Estuary of the Barito River, Indonesia. *Repo Dosen ULM* 14, no. 5: 2878–2893. www.repo-dosen.ulm.ac.id//handle/123456789/27193

Sarker, S., A. N. M. S. Huda, Md. N. H. Niloy, and G. W. Chowdhury. 2022. Trophic transfer of microplastics in the aquatic ecosystem of Sundarbans mangrove forest, Bangladesh. *Science of the Total Environment* 838, no. part 2 (September), 155896. https://doi.org/10.1016/j.scitotenv.2022.155896

Seeruttun, L. D., P. Raghbor, and C. Appadoo. 2023. Mangrove and microplastic pollution: A case study from a small island (Mauritius). *Regional Studies in Marine Science* 62 (September), 102906. https://doi.org/10.1016/j.rsma.2023.102906

Shin, C. R., E. H. Choi, G. M. Kim, et al. 2021. Characterization of metapopulation of *Ellobium chinense* through Pleistocene expansions and four covariate COI guanine-hotspots linked to G-quadruplex conformation. *Scientific Reports* 11, 12239. https://doi.org/10.1038/s41598-021-91675-5

Su, L., H. Deng, B. Li, et al. 2019. The occurrence of microplastic in specific organs in commercially caught fishes from coast and estuary area of east China. *Journal of Hazardous Materials* 365 (March), 716–724. https://doi.org/10.1016/j.jhazmat.2018.11.024

Sun, S., W. Shi, Y. Tang, et al. 2021. The toxic impacts of microplastics (MPs) and polycyclic aromatic hydrocarbons (PAHs) on haematic parameters in a marine bivalve species and their potential mechanisms of action. *Science of the Total Environment* 783 (August), 147003. https://doi.org/10.1016/j.scitotenv.2021.147003

Swennen, C., N. Ruttanadakul, M. Haver, et al. 1995. The five sympatric mudskippers (Teleostei: Gobioidea) of Pattani area, Southern Thailand. *Natural History Bulletin of the Siam Society* 42, 109–129.

Ta, A. T., and S. Babel. 2020. Microplastics pollution with heavy metals in the aquaculture zone of the Chao Phraya River Estuary, Thailand. *Marine Pollution Bulletin* 161, no. part A (December), 111747. https://doi.org/10.1016/j.marpolbul.2020.111747

Tan, E., and N. B. M. Zanuri. 2023. Abundance and distribution of microplastics in tropical estuarine mangrove areas around Penang, Malaysia. *Frontiers in Marine Science* 10, 1148804. https://doi.org/10.3389/fmars.2023.1148804

Thailand Pollution Control Department. 2021. Action plan on plastic waste management phase I (2020–2022). Report, Thailand Pollution Control Department. www.pcd.go.th/publication/15038 (accessed July 1, 2023).

The Great Bubble Barrier. 2019. The great bubble barrier: A smart solution to plastic pollution. Press, The Great Bubble Barrier. www.thegreatbubblebarrier.com/wp-content/uploads/2019/01/One-pager-TGBB-2019-01-EN-1.pdf (accessed July 1, 2023).

Tirkey, A., and L. S. B. Upadhyay. 2021. Microplastics: An overview on separation, identification and characterization of microplastics. *Marine Pollution Bulletin* 170 (September), 112604. https://doi.org/10.1016/j.marpolbul.2021.112604

Tran, L. T., Y. T. N. Nguyen, T. T. K. Nguyen, and Q. M. Dinh. 2020. Burrow structure and utilization of *Periophthalmodon schlosseri* (Pallas, 1770) from Tran De coastal area, Soc Trang, Vietnam. *Egyptian Journal of Aquatic Biology and Fisheries* 24, no. 3 (May–June), 45–52. https://doi.org/10.21608/ejabf.2020.87819

Tran, L. X., Y. Maekawa, K. Soyano, and A. Ishimatsu. 2021. Morphological comparison of the feeding apparatus in herbivorous, omnivorous and carnivorous mudskippers (Gobiidae: Oxudercinae). *Zoomorphology* 140, no. 3 (September), 387–404. https://doi.org/10.1007/s00435-021-00530-8

Trindade, L. D. S., T. V. Gloaguen, T. D. S. F. Benevides, A. C. S. Valentim, M. R. Bomfim, and J. A. G. Santos. 2023. Microplastics in surface waters of tropical estuaries around a densely populated Brazilian bay. *Environmental Pollution* 323 (April), 121224. https://doi.org/10.1016/j.envpol.2023.121224

Vermeiren, P., K. Ikejima, Y. Uchida, and C. C. Muñoz. 2023. Microplastic distribution among estuarine sedimentary habitats utilized by intertidal crabs. *Science of the Total Environment* 866 (March), 161400. https://doi.org/10.1016/j.scitotenv.2023.161400

Villegas, L., M. Cabrera, G. M. Moulatlet, and M. Capparelli. 2022. The synergistic effect of microplastic and malathion exposure on fiddler crab *Minuca ecuadoriensis* microplastic bioaccumulation and survival. *Marine Pollution Bulletin* 175 (February), 113336. https://doi.org/10.1016/j.marpolbul.2022.113336

Wang, C., J. Yu, Y. Lu, D. Hua, X. Wang, and X. Zou. 2021b. Biodegradable microplastics (BMPs): A new cause for concern? *Environmental Science and Pollution Research* 28, no. 47 (December), 66511–66518. https://doi.org/10.1007/s11356-021-16435-4

Wang, T., M. Hu, G. Xu, H. Shi, J. Y. S. Leung, and Y. Wang. 2021a. Microplastic accumulation via trophic transfer: Can a predatory crab counter the adverse effects of microplastics by body defence? *Science of the Total Environment* 754 (February), 142099. https://doi.org/10.1016/j.scitotenv.2020.142099

Watkins, E., J. P. Schweitzer, E. Leinala, and P. Börkey. 2019. Policy approaches to incentivise sustainable plastic design. OECD environment working papers, OECD Publishing. www.oecd-ilibrary.org/environment/policy-approaches-to-incentivise-sustainable-plastic-design_233ac351-en (accessed July 3, 2023).

Wei, X. F., M. Bohlén, C. Lindblad, M. Hedenqvist, and A. Hakonen. 2021. Microplastics generated from a biodegradable plastic in freshwater and seawater. *Water Research* 198 (June), 117123. https://doi.org/10.1016/j.watres.2021.117123

Wicaksono, E. A., S. Werorilangi, T. S. Galloway, and A. Tahir. 2021. Distribution and seasonal variation of microplastics in Tallo River, Makassar, Eastern Indonesia. *Toxics* 9, no. 6 (June), 129. https://doi.org/10.3390/toxics9060129

Yadav, S., N. Kataria, P. Khyalia, et al. 2023. Recent analytical techniques, and potential eco-toxicological impacts of textile fibrous microplastics (FMPs) and associated contaminates: A review. *Chemosphere* 326 (June), 138495. https://doi.org/10.1016/j.chemosphere.2023.138495

Yeesin, P., and S. Bautip . 2021. Feeding behavior of Sesarmid Crab, *Episesarma Mederi* (H. Milne Edwards, 1853) (Decapoda; Sesarmidae) in the mangrove of Pattani Bay, Thailand. *KKU Science Journal* 49, no. 2 (April–June), 212–219. www.ph01.tci-thaijo.org/index.php/KKUSciJ/article/view/250270

Yuan, Z., R. Nag, and E. Cummins. 2022. Human health concerns regarding microplastics in the aquatic environment – From marine to food systems. *Science of the Total Environment* 823 (June), 153730. https://doi.org/10.1016/j.scitotenv.2022.153730

Zhang, L., S. Zhang, J. Guo, K. Yu, Y. Wang, and R. Li. 2020. Dynamic distribution of microplastics in mangrove sediments in Beibu Gulf, South China: Implications of tidal current velocity and tidal range. *Journal of Hazardous Materials* 399 (November), 122849. https://doi.org/10.1016/j.jhazmat.2020.122849

Zhou, Q., C. Tu, C. Fu, et al. 2020. Characteristics and distribution of microplastics in the coastal mangrove sediments of China. *Science of the Total Environment* 703 (February), 134807. https://doi.org/10.1016/j.scitotenv.2019.134807

11 Microplastics Pollution in the Aquatic Environments

Occurrences, Ecological Impacts, and Remediation Technologies

Sumarlin Shangdiar, Kassian T.T. Amesho,
Timoteus Kadhila, Chingakham Chinglenthoiba, Sioni Iikela,
Nastassia Thandiwe Sithole, and Mohd Nizam Lani

11.1 INTRODUCTION

Microplastics, defined as plastic particles smaller than 5 mm in size, have emerged as a pervasive environmental concern due to their widespread distribution and potential ecological impacts. These particles originate from various sources, including the breakdown of larger plastic debris, the fragmentation of synthetic textiles, and the direct release of microplastic-containing products into the environment (Lahens et al., 2018; Anderson et al., 2016). Microplastics exist in diverse forms, encompassing both primary microplastics, deliberately manufactured for industrial or consumer purposes, and secondary microplastics, formed through the degradation of larger plastic items (Yan et al., 2019; Alvim et al., 2020).

The significance and scope of the microplastic problem extend across terrestrial and aquatic ecosystems, with freshwater and marine environments serving as major reservoirs for microplastic accumulation (Alimi et al., 2018; Arias-Andres et al., 2018). Microplastics pose a multitude of threats to aquatic organisms, ranging from physical harm through ingestion and entanglement to chemical toxicity resulting from the release of hazardous additives and adsorbed pollutants (Auta et al., 2017; Bakir et al., 2020). Furthermore, microplastics have the potential to bioaccumulate and biomagnify within food webs, thereby posing risks to higher trophic levels, including humans, through the consumption of contaminated seafood (Baldwin et al., 2016; Gong et al., 2021). As shown in Figure 11.1, the keyword network analysis of articles related to "Microplastics" from 2018 to 2024 in the Web of Science (WoS) database highlights the main themes and connections between key topics in this field.

Given the pervasive nature and multifaceted impacts of microplastic pollution, there is a critical need to understand the sources, fate, and ecological consequences of microplastics in aquatic environments. This chapter aims to provide a comprehensive overview of microplastic pollution in aquatic ecosystems, focusing on occurrences, ecological impacts, and remediation strategies. By elucidating the intricacies of this complex environmental issue, we aim to underscore the urgency of concerted efforts to mitigate microplastic pollution and safeguard the health and integrity of aquatic ecosystems worldwide.

DOI: 10.1201/9781032706573-11

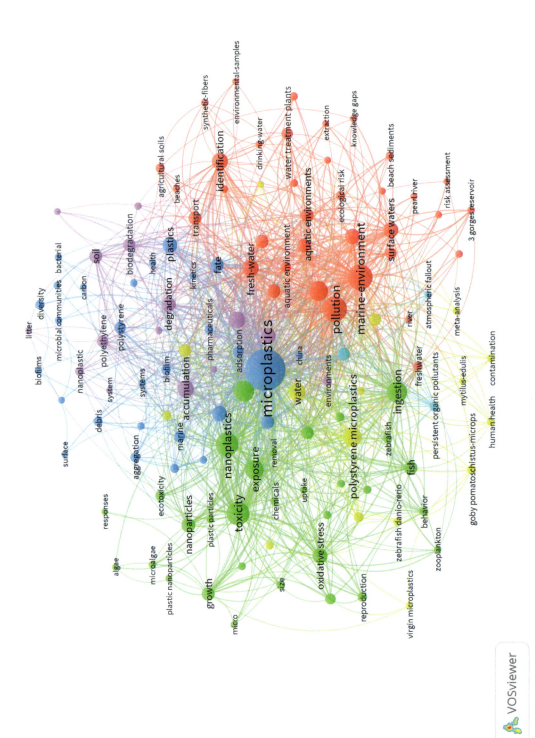

FIGURE 11.1 Keyword network analysis on "Microplastics" related articles from 2018 to 2024 found in Web of Science (WoS) database.

11.2 OCCURRENCES OF MICROPLASTICS IN AQUATIC ENVIRONMENTS

11.2.1 Sources of Microplastics in Aquatic Environments

Microplastics in aquatic environments originate from diverse sources, with primary and secondary pathways contributing to their abundance, as indicated in Figure 11.2. Primary microplastics are intentionally manufactured for various industrial and consumer applications, such as microbeads in personal care products and microfibers in textiles (Gewert et al., 2015; Gray et al., 2018). These primary microplastics may enter aquatic ecosystems directly through wastewater discharges or indirectly through atmospheric deposition and runoff from urban areas (He et al., 2021; Hermabessiere et al., 2018). Table 11.1 shows some sources of microplastics in aquatic environments.

Secondary microplastics, on the other hand, result from the fragmentation and degradation of larger plastic items, including plastic bags, bottles, and packaging materials (Horton et al., 2018; Huang et al., 2021a). Mechanical abrasion, UV radiation, and microbial degradation contribute to the breakdown of macroplastics into smaller particles, which subsequently enter aquatic environments through runoff, wind dispersion, and direct release (Huang et al., 2021b; Lares et al., 2018).

11.2.2 Distribution and Fate of Microplastics in Different Aquatic Environments

The distribution and fate of microplastics in aquatic environments (as indicated in Table 11.2) vary depending on factors such as hydrodynamics, sedimentation rates, and biological interactions. In freshwater systems, microplastics are commonly found in rivers, lakes, and reservoirs, where they accumulate in sediments, surface waters, and biota (Huang et al., 2021c; Hurley et al., 2018).

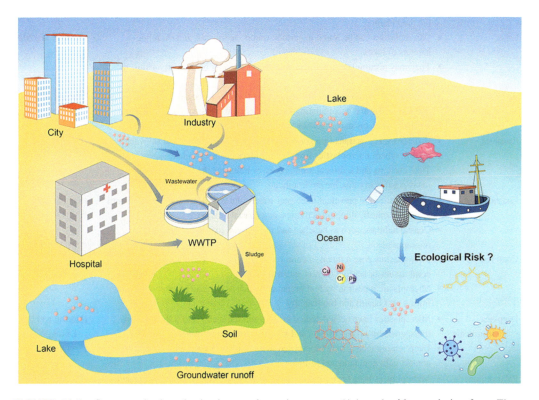

FIGURE 11.2 Sources of microplastics in aquatic environments. (Adapted with permission from Zhang et al., 2022, Copyright © 2022 Elsevier.)

TABLE 11.1
Sources of Microplastics in Aquatic Environments

Source	Description	References
Wastewater treatment plants	Effluent from sewage treatment plants	(Lares et al., 2018)
Urban runoff	Surface runoff from urban areas and roadways	(Huang et al., 2021d)
Atmospheric deposition	Deposition of airborne microplastics onto water bodies	(Horton et al., 2017)
Plastic debris	Fragmentation of larger plastic items	(Huang et al., 2021d)
Shipping and maritime activities	Discharge of cargo residues, fishing gear, and ship paints	(Hurley et al., 2018)
Recreational activities	Discarded plastic items from recreational boating, fishing, and beachgoers	(Jiang et al., 2020)
Agricultural runoff	Runoff from agricultural fields carrying plastic debris and microplastics	(Klein et al., 2015)
Industrial effluents	Discharge of microplastics from industrial processes	(Jiang et al., 2020)

TABLE 11.2
Distribution and Fate of Microplastics in Different Aquatic Environments

Aquatic Environment	Distribution and Fate of Microplastics	References
Rivers	• Microplastics can be transported downstream via river currents, accumulating in riverbeds, floodplains, and river mouths • Microplastics may also be ingested by aquatic organisms and transferred up the food chain	(Klein et al., 2015)
Lakes	• Microplastics can be found in surface waters, sediments, and shoreline habitats of lakes • They may accumulate in sediments or be ingested by benthic organisms, fish, and birds	(Kane et al., 2020)
Oceans	• Microplastics are widely distributed across the world's oceans, transported by ocean currents and accumulating in gyres, coastal areas, and on the seafloor • Marine organisms may ingest or become entangled in microplastics	(Kanhai et al., 2018)
Estuaries	• Estuaries serve as transition zones between freshwater rivers and saltwater oceans, where microplastics from both sources can accumulate • Tidal currents and sedimentation processes influence the distribution of microplastics in estuarine environments	(Kanhai et al., 2017)
Coastal areas	• Microplastics are prevalent in coastal waters and sediments due to inputs from terrestrial runoff, shipping activities, and coastal erosion • Coastal habitats, such as beaches and mangroves, may serve as sinks for microplastics	(Kanhai et al., 2018)
Coral reefs	• Microplastics pose a threat to coral reefs, with particles becoming trapped in coral tissues or settling on reef substrates • Microplastic ingestion by reef organisms, such as fish and invertebrates, can impact reef health and biodiversity	(Kane et al., 2020)
Deep-sea environments	• Microplastics have been detected in deep-sea sediments and water columns, transported by ocean currents and sinking from surface waters • Deep-sea organisms may be exposed to microplastics through ingestion or contact with contaminated sediments	(Horton and Dixon, 2018)
Polar regions	• Microplastics have been found in polar waters, transported by ocean currents and atmospheric deposition • Cold temperatures and ice formation can temporarily sequester microplastics in polar ice, but melting ice may release these particles back into the environment	(He et al., 2021)

Microplastics Pollution in the Aquatic Environments

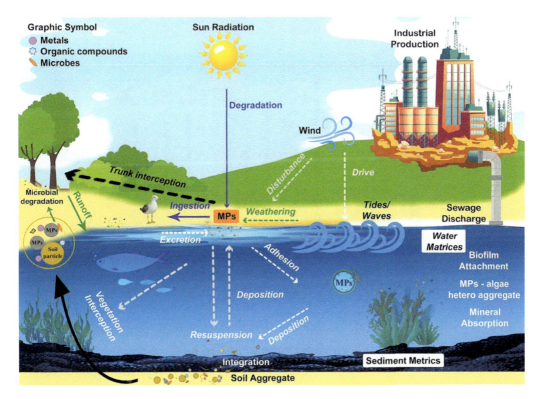

FIGURE 11.3 Schematic view of MPs' distribution and fate. (Obtained with permission from Chand and Suthar, 2024, Copyright (2024), Elsevier.)

Figure 11.3 shows a schematic view of MPs' distribution and fate. Urban and industrial activities, as well as agricultural runoff, contribute to the widespread distribution of microplastics in freshwater ecosystems (Kolandhasamy et al., 2018; Jiang et al., 2020).

In marine environments, microplastics are ubiquitously distributed across coastal regions, open oceans, and deep-sea sediments (Prata et al., 2019; Werbowski et al., 2021). Coastal areas adjacent to urban centers and shipping routes are particularly susceptible to microplastic contamination due to direct inputs from land-based sources and maritime activities (Kanhai et al., 2017). Once introduced into aquatic environments, microplastics undergo complex interactions with physical, chemical, and biological processes, influencing their transport, transformation, and ultimate fate (Klein et al., 2015; Kanhai et al., 2018).

11.2.3 Ecological Impacts of Microplastic Pollution

Microplastic pollution has become a significant concern in aquatic environments due to its adverse effects on various organisms and ecosystems. Understanding these ecological impacts is crucial for assessing the broader implications of microplastic contamination and formulating effective mitigation strategies. As illustrated in Figure 11.4, the processes related to the ecological impacts of microplastics in aquatic ecosystems demonstrate the various pathways and effects on marine life and water quality.

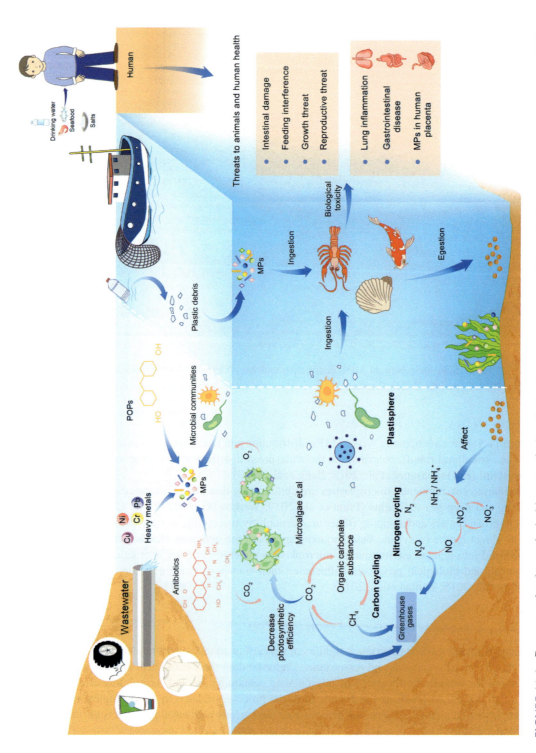

FIGURE 11.4 Processes related to ecological impacts of microplastics in aquatic ecosystems. (Adapted with permission from Zhang et al., 2022, Copyright © 2022 Elsevier.)

Microplastics Pollution in the Aquatic Environments

11.2.4 Effects of Microplastics on Different Organisms

Microplastics can have detrimental effects on a wide range of aquatic organisms, including fish, invertebrates, and marine mammals. These effects may vary depending on factors such as the size, shape, and composition of the microplastics, as well as the species and life stage of the organisms involved.

Fish: Studies have documented the ingestion of microplastics by fish species, resulting in detrimental effects on their health and physiology. Microplastics can accumulate in the gastrointestinal tract of fish, causing blockages, inflammation, and reduced nutrient absorption. Additionally, the leaching of chemical additives and adsorbed pollutants from microplastics can lead to toxicological effects, such as oxidative stress, DNA damage, and impaired immune function (Huang et al., 2021d; Horton et al., 2017).

Invertebrates: Benthic invertebrates are particularly susceptible to microplastic pollution, as they often inhabit sedimentary environments where microplastics tend to accumulate. Ingestion of microplastics by benthic organisms can disrupt feeding, reproduction, and energy allocation, ultimately affecting population dynamics and ecosystem functioning. Furthermore, the ingestion of microplastics can facilitate the transfer of associated contaminants into the food web, exacerbating ecological risks (Cole et al., 2013).

Marine mammals: Marine mammals, including cetaceans, pinnipeds, and sea otters, are vulnerable to microplastic ingestion and entanglement. Microplastics can be mistaken for prey items or ingested indirectly through the consumption of contaminated prey. Ingested microplastics can accumulate in the gastrointestinal tract, leading to digestive tract injuries, nutrient deficiencies, and physiological stress. Moreover, entanglement in plastic debris can impair mobility, feeding, and reproductive behaviors, ultimately affecting individual fitness and population viability (Han et al., 2020; Guo and Wang, 2021).

11.2.5 Mechanisms of Toxicological Effects of Microplastics

The toxicological effects of microplastics arise from various mechanisms, including physical abrasion, leaching of additives, and adsorption of contaminants. When ingested, microplastics can cause mechanical damage to the digestive tract, leading to tissue inflammation, ulceration, and perforation. Furthermore, microplastics can act as carriers for persistent organic pollutants (POPs) and other toxic compounds, facilitating their uptake and bioaccumulation in aquatic organisms. Once absorbed, these contaminants can elicit a range of adverse effects, including endocrine disruption, immunotoxicity, and reproductive impairment (Faure et al., 2015). As shown in Table 11.3, microplastics have diverse effects on different organisms, impacting their health, behavior, and overall ecosystem dynamics.

11.2.6 Potential Risks Associated with Human Consumption of Contaminated Seafood

The ingestion of microplastics by marine organisms raises concerns about human exposure through the consumption of contaminated seafood. Microplastics and associated contaminants can accumulate in edible tissues of fish and shellfish, posing potential risks to human health. While the extent of human exposure and the health impacts of microplastic ingestion are still being investigated, emerging evidence suggests the need for precautionary measures to mitigate the risks associated with seafood consumption (Di and Wang, 2018; Crew et al., 2020). Figure 11.5 illustrates the diverse potential long-term global impacts of accumulating and poorly reversible plastic pollution, as discussed by MacLeod et al. (2021).

TABLE 11.3

Effects of Microplastics on Different Organisms

Organism	Effects of Microplastics	References
Fish	Ingestion-induced blockages, inflammation, reduced nutrient absorption, oxidative stress, DNA damage, impaired immune function, bioaccumulation of contaminants	(Fu et al., 2020b)
Invertebrates	Disruption of feeding, reproduction, energy allocation, population dynamics, ecosystem functioning, transfer of associated contaminants into the food web	(Gray et al., 2018; Guo and Wang., 2021)
Marine mammals	Ingestion-induced injuries, nutrient deficiencies, physiological stress, entanglement-related injuries, impaired mobility, feeding, reproductive behaviors, bioaccumulation of contaminants, population viability concerns	(Han et al., 2020; He et al., 2021)
Birds	Ingestion-induced blockages, reduced nutrient absorption, impaired flight, reproductive impairment, entanglement-related injuries, bioaccumulation of contaminants	(Huang et al., 2021a; Lares et al., 2018)
Zooplankton	Ingestion-induced blockages, reduced feeding efficiency, reproductive impairment, transfer of associated contaminants into higher trophic levels	(Lares et al., 2018; Huang et al., 2021b)
Coral reefs	Mechanical damage, reduced growth rates, tissue inflammation, symbiotic algae expulsion, increased susceptibility to disease, compromised reproductive success	(Huang et al., 2021c; Huang et al., 2021d)
Sea turtles	Ingestion-induced blockages, digestive tract injuries, nutrient deficiencies, physiological stress, entanglement-related injuries, impaired mobility, reproductive impairment, bioaccumulation of contaminants, population viability concerns	(Hurley et al., 2018; Jiang et al., 2020)
Crustaceans	Ingestion-induced blockages, reduced feeding efficiency, reproductive impairment, altered behavior, transfer of associated contaminants into higher trophic levels	(Huang et al., 2021d; Jin et al., 2018)

11.3 REMEDIATION TECHNOLOGIES FOR MICROPLASTIC POLLUTION

Microplastic pollution in aquatic environments poses a significant threat to ecosystem health and human well-being. Various remediation technologies have been developed to address this pressing environmental challenge, each employing distinct approaches to remove or degrade microplastics. In this section, we provide an in-depth overview of physical, chemical, and biological methods used for microplastic remediation, along with their respective advantages, limitations, and recent advancements.

11.3.1 PHYSICAL METHODS FOR REMOVING MICROPLASTICS

Physical methods involve the mechanical removal of microplastics from the environment, targeting their physical properties such as size, shape, and density. Filtration and sedimentation are the two primary techniques employed in this approach.

Filtration is a commonly used method for removing microplastics from water bodies. It involves passing water through porous materials or membranes with specific pore sizes that can trap microplastics while allowing water to pass through. Various filtration systems, such as sand filters, membrane filters, and microfiltration systems, have been developed for this purpose. The efficiency

Microplastics Pollution in the Aquatic Environments

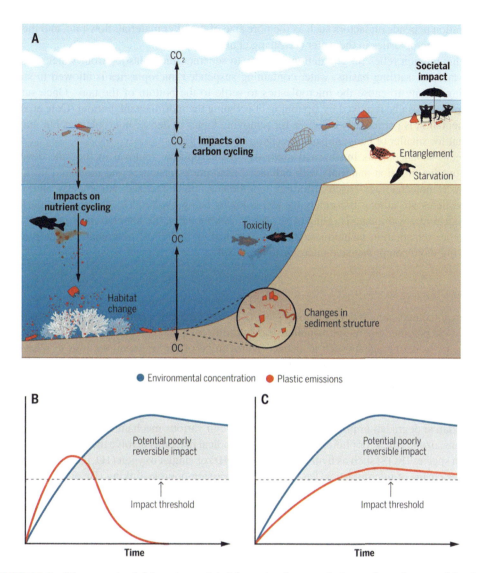

FIGURE 11.5 Diverse potential long-term global impacts of accumulating and poorly reversible plastic pollution. (Adapted with permission from MacLeod et al., 2021, Copyright © 2021 American Association for the Advancement of Science.)

(A) Potential impacts encompass geophysical effects on carbon and nutrient cycling, soil, and sediment habitats, alongside biological repercussions on endangered or keystone species and (eco)toxicity. Societal impacts arise from public perceptions of environmental quality and resultant policy shifts. (B and C) These diagrams illustrate how plastic pollution exceeding an impact threshold can lead to largely irreversible consequences. In (B), plastic pollution persists in the environment for extended periods, rendering concentrations resistant to emission reductions. In (C), while plastic pollution can be reversed through cleanup efforts or degradation, concentrations remain above the impact threshold due to ineffective emission control measures. CO_2 refers to carbon dioxide, while OC denotes organic carbon.

of filtration depends on factors such as the pore size of the filter material, flow rate, and the concentration of microplastics in the water (Z. Zhang et al., 2022).

Sedimentation relies on gravitational forces to separate microplastics from water. In sedimentation tanks or settling basins, water containing suspended microplastics is allowed to stand still, allowing gravity to cause the microplastics to settle to the bottom of the tank. Once settled, the microplastics can be removed by siphoning or scraping them off the tank bottom (Cole et al., 2014). Sedimentation is particularly effective for removing larger or denser microplastics but may be less efficient for smaller or buoyant particles (Ding et al., 2019).

Recent advancements in physical methods for microplastic removal include the development of innovative filtration materials with enhanced adsorption capacities and improved filtration efficiency. Nanotechnology-based filtration membranes, functionalized adsorbents, and magnetic nanoparticles are among the emerging technologies being explored to enhance the performance of physical removal methods (Ding et al., 2019). Additionally, the integration of physical methods with other remediation techniques, such as chemical or biological treatments, holds promise for achieving more comprehensive microplastic removal.

11.3.2 CHEMICAL METHODS FOR DEGRADING MICROPLASTICS

Chemical methods aim to degrade microplastics through oxidative or hydrolytic reactions, breaking down their polymer chains into smaller, less harmful compounds. Oxidative degradation involves the use of oxidizing agents such as ozone, hydrogen peroxide, or persulfate ions to react with the polymer chains of microplastics, leading to their fragmentation or depolymerization (Gies et al., 2018). Hydrolytic degradation, on the other hand, utilizes hydrolytic enzymes or strong acids to cleave the chemical bonds in microplastics, resulting in their breakdown into simpler molecules (Faure et al., 2015).

Oxidative degradation processes can occur through various mechanisms, including free radical reactions, ozone attack, or photodegradation. Free radical reactions involve the generation of reactive oxygen species (ROS) such as hydroxyl radicals (OH•) or singlet oxygen ($1O_2$), which react with the polymer chains of microplastics, leading to chain scission or cross-linking (He et al., 2021). Ozone attack involves the direct oxidation of polymer bonds by ozone molecules, while photodegradation utilizes ultraviolet (UV) or visible light to initiate chemical reactions that degrade microplastics (Dumichen et al., 2015).

Hydrolytic degradation processes involve the enzymatic or chemical cleavage of ester or amide bonds in the polymer chains of microplastics. Enzymatic degradation employs specific enzymes such as lipases, proteases, or esterases, which catalyze the hydrolysis of specific chemical bonds in microplastics. Chemical hydrolysis, on the other hand, utilizes strong acids or bases to break down polymer chains into smaller fragments (Gies et al., 2018).

Recent advancements in chemical methods for microplastic degradation focus on the development of novel catalysts, reaction conditions, and process optimization strategies to enhance degradation efficiency and minimize environmental impacts. Catalytic oxidation processes using supported metal catalysts, photocatalysts, or advanced oxidation processes (AOPs) have shown promise for accelerating microplastic degradation under mild reaction conditions. Additionally, the integration of chemical methods with physical or biological treatments offers synergistic effects, leading to more efficient microplastic removal and degradation (Dumichen et al., 2015).

11.3.3 BIOLOGICAL METHODS FOR DEGRADING MICROPLASTICS

Biological methods harness the activity of microorganisms or enzymes to degrade microplastics, utilizing natural biological processes to break down polymer chains into simpler compounds.

Microplastics Pollution in the Aquatic Environments

Enzymatic degradation and microbial biodegradation are the two primary approaches employed in this method. Enzymatic degradation involves the use of specific enzymes capable of cleaving chemical bonds in microplastics, leading to their fragmentation or depolymerization. Enzymes such as lipases, proteases, or esterases catalyze the hydrolysis of ester or amide bonds in the polymer chains of microplastics, breaking them down into smaller molecules that are more susceptible to microbial degradation (Gewert et al., 2015).

Microbial biodegradation relies on the metabolic activity of microorganisms to assimilate or mineralize microplastics as carbon sources. Various bacteria, fungi, and algae have been identified for their ability to degrade microplastics through enzymatic pathways or extracellular enzymatic activity. Microorganisms can colonize microplastic surfaces and produce extracellular enzymes that degrade polymer chains, leading to their degradation and mineralization into harmless by-products (Chia et al., 2020).

Recent advancements in biological methods for microplastic degradation focus on the isolation and characterization of novel microorganisms and enzymes with enhanced degradation capabilities. Metagenomic studies and high-throughput screening techniques have facilitated the discovery of microorganisms and enzymes capable of degrading a wide range of microplastic polymers under diverse environmental conditions. Genetic engineering and synthetic biology approaches offer potential solutions for enhancing the degradation efficiency and specificity of enzymes, leading to the development of tailored enzymatic systems for microplastic remediation (Bordos et al., 2019; Capolupo et al., 2020).

In summary, physical, chemical, and biological methods offer distinct approaches to addressing microplastic pollution in aquatic environments. Advances in each of these remediation technologies, coupled with interdisciplinary research and innovation, hold promise for mitigating the environmental impacts of microplastics and safeguarding aquatic ecosystems for future generations. As detailed in Table 11.4, various remediation technologies for microplastic pollution are compared, highlighting their effectiveness, feasibility, and potential limitations.

11.4 CURRENT STATE OF RESEARCH AND FUTURE DIRECTIONS

Microplastic pollution has emerged as a significant environmental concern, particularly in aquatic ecosystems, where it poses threats to ecosystem health, biodiversity, and human well-being. In recent years, there has been a growing body of research aimed at understanding the sources, distribution, fate, and impacts of microplastics in aquatic environments. This section provides an overview of the current state of research on microplastic pollution and outlines future directions for advancing our understanding and addressing the challenges associated with this global issue.

11.4.1 OVERVIEW OF THE CURRENT STATE OF RESEARCH

The current state of research on microplastic pollution in aquatic environments encompasses a wide range of topics, including the identification of sources, the characterization of microplastic particles, the assessment of environmental fate and transport, and the evaluation of ecological and human health impacts (Z. Zhang et al., 2022). Studies have identified multiple sources of microplastics in aquatic ecosystems, including primary sources such as plastic debris and microbeads, as well as secondary sources such as the fragmentation of larger plastic items and the degradation of synthetic textiles (Cai et al., 2018; Bharath et al., 2021).

Characterization studies have revealed the diverse nature of microplastic particles, which vary in size, shape, color, and polymer composition. Microplastics can be classified into primary microplastics, which are manufactured for specific purposes, and secondary microplastics, which result from the degradation of larger plastic items (Kappler et al., 2016). Common polymers found

TABLE 11.4
Comparison of Remediation Technologies for Microplastic Pollution

Method	Description	Advantages	Limitations	References
Physical	Mechanical removal of microplastics through filtration or sedimentation	• Simple and cost-effective • Versatile application	• Limited effectiveness for small or buoyant microplastics potential for system fouling	(Fok et al., 2020; Fu et al., 2020a)
Chemical	Degradation of microplastics using oxidative or hydrolytic reactions	Wide applicability; rapid degradation potential	Risk of generating harmful byproducts; energy-intensive processes	(Faure et al., 2015; Ding et al., 2019)
Biological	Utilization of microorganisms or enzymes to degrade microplastics	Environmentally friendly; potential for targeted degradation	Dependence on specific microbial strains or enzymes; variable degradation rates	(Crew et al., 2020; Danso et al., 2019)
Electrochemical	Utilizes electrical energy to degrade microplastics	High degradation efficiency; potential for in situ treatment	Energy-intensive; limited scalability for large-scale applications	(Chia et al., 2020; Danso et al., 2019)
Photocatalytic	Utilizes photocatalysts to degrade microplastics under light irradiation	Effective degradation under sunlight; potential for synergistic effects	Limited effectiveness in turbid or dark environments; catalyst stability issues	(Alvim et al., 2020; Alimi et al., 2018)
Sonication	Uses high-frequency sound waves to break down microplastics	Rapid and efficient degradation; noninvasive process	Limited penetration depth; energy consumption concerns	(Li et al., 2018b; Lin et al., 2018)
Nanotechnology	Utilizes engineered nanomaterials to capture or degrade microplastics	High surface area for adsorption; potential for tailored properties	Concerns regarding nanomaterial toxicity; environmental persistence	(Z. Zhang et al., 2022; Li et al., 2018a)

in microplastics include polyethylene (PE), polypropylene (PP), polyethylene terephthalate (PET), and polystyrene (PS), among others (Fu et al., 2020; Li et al., 2018b).

Environmental fate and transport studies have shown that microplastics can be transported over long distances through air and water currents, leading to their widespread distribution in aquatic environments. Microplastics can accumulate in sediments, surface waters, and biota, where they can persist for extended periods and interact with organisms at various trophic levels (Lin et al., 2018; Li et al., 2018a).

Ecological impact assessments have highlighted the adverse effects of microplastic pollution on aquatic organisms, including fish, invertebrates, and marine mammals (Andrady, 2011). Microplastics can cause physical harm, such as ingestion and entanglement, as well as chemical harm, such as the leaching of additives and the adsorption of contaminants. These impacts can disrupt biological processes, alter ecosystem structure and function, and ultimately threaten the sustainability of aquatic ecosystems (Alexander et al., 2016; Amaral-Zettler et al., 2020).

Human health studies have raised concerns about the potential risks associated with the ingestion of microplastics through the consumption of contaminated seafood and drinking water. While the health effects of microplastic exposure are still poorly understood, emerging evidence suggests that microplastics may act as vectors for the transfer of harmful contaminants and pathogens to humans (Anderson et al., 2017; Carr et al., 2016)

Microplastics Pollution in the Aquatic Environments

11.4.2 GAPS AND CHALLENGES IN OUR UNDERSTANDING

Despite significant progress in research on microplastic pollution, several gaps and challenges remain in our understanding of this complex issue. One key challenge is the lack of standardized methods for sampling, analysis, and quantification of microplastics in environmental matrices (Koelmans et al., 2019; Kokalj et al., 2018). Variability in sampling protocols, analytical techniques, and reporting standards makes it difficult to compare results across studies and assess the global extent of microplastic pollution (Chang, 2015; Cheung and Fok, 2017).

Another challenge is the limited knowledge of the ecological and human health impacts of microplastic exposure (Simon et al., 2018). While studies have documented adverse effects on individual organisms, the long-term consequences for ecosystem dynamics and human health are still poorly understood (Danso et al., 2019; Fok et al., 2020). Further research is needed to elucidate the mechanisms underlying these impacts and assess the cumulative effects of chronic microplastic exposure on aquatic ecosystems and human populations.

11.4.3 FUTURE DIRECTIONS FOR RESEARCH AND DEVELOPMENT

Moving forward, future research efforts should focus on addressing key knowledge gaps and advancing our understanding of microplastic pollution in aquatic environments. Priority areas for research include:

1. *Standardization of sampling and analysis methods:* Efforts should be made to develop standardized protocols for the collection, processing, and analysis of microplastic samples to improve data comparability and reliability.
2. *Assessment of ecological impacts:* Research is needed to assess the ecological impacts of microplastic pollution across different spatial and temporal scales, including ecosystem-level effects, community dynamics, and biodiversity loss.
3. *Evaluation of human health risks:* Studies should investigate the potential health risks associated with microplastic exposure through dietary intake, inhalation, and dermal contact, with a focus on vulnerable populations such as children and pregnant women.
4. *Development of remediation technologies:* Research and development efforts should continue to explore innovative remediation technologies for removing and degrading microplastics in aquatic environments, with an emphasis on cost-effectiveness, scalability, and environmental sustainability (Jin et al., 2018).
5. *Public awareness and policy interventions:* Initiatives to raise public awareness about the impacts of microplastic pollution and promote sustainable consumption and waste management practices are essential for mitigating the problem. Policy interventions such as bans on single-use plastics and extended producer responsibility schemes can help reduce the release of microplastics into the environment and promote the transition to a circular economy (Fu et al., 2020b; Fu and Wang, 2019).

In summary, addressing the complex issue of microplastic pollution in aquatic environments requires interdisciplinary collaboration, innovative research approaches, and concerted efforts from stakeholders across sectors. By advancing our understanding of microplastic sources, fate, and impacts, and developing effective remediation strategies and policy interventions, we can work toward mitigating the environmental and human health risks posed by microplastic pollution and safeguarding the health and integrity of aquatic ecosystems for future generations.

11.5 CONCLUSIONS

In summary, this chapter has provided a comprehensive overview of the issue of microplastic pollution in aquatic environments. We have explored the sources, distribution, fate, and impacts of microplastics, as well as the various remediation technologies available. Key points covered include the identification of sources such as plastic debris and microbeads, the characterization of microplastic particles, and the assessment of their ecological and human health impacts. We have discussed physical, chemical, and biological methods for removing or degrading microplastics, highlighting their respective advantages and limitations. The implications of this research for policy and management are significant. It is clear that microplastic pollution poses a serious threat to ecosystem health and human well-being, and urgent action is needed to address this issue. Policy interventions such as bans on single-use plastics and extended producer responsibility schemes can help reduce the release of microplastics into the environment. Furthermore, greater investment in research and development of remediation technologies is essential to effectively mitigate microplastic pollution in aquatic environments. By working collaboratively across disciplines and sectors, we can develop innovative solutions and implement evidence-based policies to safeguard the health and integrity of aquatic ecosystems for future generations.

REFERENCES

Alexander, J., Barregard, L., Bignami, M., Ceccatelli, S., Cottrill, B., Dinovi, M., et al., 2016. Presence of microplastics and nanoplastics in food, with particular focus on seafood. *EFSA J.* 14, 30.

Alimi, O.S., Budarz, J.F., Hernandez, L.M., Tufenkji, N., 2018. Microplastics and nanoplastics in aquatic environments: aggregation, deposition, and enhanced contaminant transport. *Environ. Sci. Technol.* 52, 1704–1724.

Alvim, C.B., Mendoza-Roca, J.A., Bes-Pia, A., 2020. Wastewater treatment plant as microplastics release source – quantification and identification techniques. *J. Environ. Manage.* 255, 109739.

Amaral-Zettler, L.A., Zettler, E.R., Mincer, T.J., 2020. Ecology of the plastisphere. *Nat. Rev. Microbiol.* 18, 139–151.

Anderson, J.C., Park, B.J., Palace, V.P., 2016. Microplastics in aquatic environments: implications for Canadian ecosystems. *Environ. Pollut.* 218, 269–280.

Anderson, P.J., Warrack, S., Langen, V., Challis, J.K., Hanson, M.L., Rennie, M.D., 2017. Microplastic contamination in lake winnipeg, *Canada. Environ. Pollut.* 225, 223–231.

Andrady, A.L., 2011. Microplastics in the marine environment. *Mar. Pollut. Bull.* 62, 1596–1605.

Arias-Andres, M., Kettner, M.T., Miki, T., Grossart, H.P., 2018. Microplastics: new substrates for heterotrophic activity contribute to altering organic matter cycles in aquatic ecosystems. *Sci. Total Environ.* 635, 1152–1159.

Auta, H.S., Emenike, C.U., Fauziah, S.H., 2017. Distribution and importance of microplastics in the marine environment: a review of the sources, fate, effects, and potential solutions. *Environ. Int.* 102, 165–176.

Bakir, A., Desender, M., Wilkinson, T., Van Hoytema, N., Amos, R., Airahui, S., et al., 2020. Occurrence and abundance of meso and microplastics in sediment, surface waters, and marine biota from the South Pacific region. *Mar. Pollut. Bull.* 160, 111572.

Baldwin, A.K., Corsi, S.R., Mason, S.A., 2016. Plastic debris in 29 great lakes tributaries: relations to watershed attributes and hydrology. *Environ. Sci. Technol.* 50, 10377–10385.

Bharath, K.M., Natesan, U., Vaikunth, R., Kumar, R.P., Ruthra, R., Srinivasalu, S., 2021. Spatial distribution of microplastic concentration around landfill sites and its potential risk on groundwater. *Chemosphere* 277, 130263.

Bordos, G., Urbanyi, B., Micsinai, A., Kriszt, B., Palotai, Z., Szabo, I., et al., 2019. Identification of microplastics in fish ponds and natural freshwater environments of the Carpathian basin, Europe. *Chemosphere* 216, 110–116.

Cai, M., He, H., Liu, M., Li, S., Tang, G., Wang, W., et al., 2018. Lost but can't be neglected: huge quantities of small microplastics hide in the South China Sea. *Sci. Total Environ.* 633, 1206–1216.

Capolupo, M., Sorensen, L., Jayasena, K.D., Booth, A.M., Fabbri, E., 2020. Chemical composition and ecotoxicity of plastic and car tire rubber leachates to aquatic organisms. *Water Res.* 169, 11.

Carr, S.A., Liu, J., Tesoro, A.G., 2016. Transport and fate of microplastic particles in wastewater treatment plants. *Water Res.* 91, 174–182.

Chang, M., 2015. Reducing microplastics from facial exfoliating cleansers in wastewater through treatment versus consumer product decisions. *Mar. Pollut. Bull.* 101, 330–333.

Cheung, P.K., Fok, L., 2017. Characterisation of plastic microbeads in facial scrubs and their estimated emissions in Mainland China. *Water Res.* 122, 53–61.

Chia, W.Y., Ying Tang, D.Y., Khoo, K.S., Kay Lup, A.N., Chew, K.W., 2020. Nature's fight against plastic pollution: Algae for plastic biodegradation and bioplastics production. *Environmental Science and Ecotechnology* 4, 100065.

Cole, M., Webb, H., Lindeque, P.K., Fileman, E.S., Halsband, C., Galloway, T.S., 2014. Isolation of microplastics in biota-rich seawater samples and marine organisms. *Sci. Rep.* 4, 8.

Crew, A., Gregory-Eaves, I., Ricciardi, A., 2020. Distribution, abundance, and diversity of microplastics in the upper St. Lawrence River. *Environ. Pollut.* 260, 113994.

Danso, D., Chow, J., Streit, W.R., 2019. Plastics: environmental and biotechnological perspectives on microbial degradation. *Appl. Environ. Microbiol.* 85(19), e01095-19.

Di, M., Wang, J., 2018. Microplastics in surface waters and sediments of the three Gorges reservoir, China. *Sci. Total Environ.* 616–617, 1620–1627.

Ding, L., Mao, R.F., Guo, X.T., Yang, X.M., Zhang, Q., Yang, C., 2019. Microplastics in surface waters and sediments of the Wei River, in the northwest of China. *Sci. Total Environ.* 667, 427–434.

Dumichen, E., Barthel, A.K., Braun, U., Bannick, C.G., Brand, K., Jekel, M., et al., 2015. Analysis of polyethylene microplastics in environmental samples, using a thermal decomposition method. *Water Res.* 85, 451–457.

Faure, F., Demars, C., Wieser, O., Kunz, M., de Alencastro, L.F., 2015. Plastic pollution in Swiss surface waters: nature and concentrations, interaction with pollutants. *Environ. Chem.* 12, 582–591.

Fok, L., Lam, T.W.L., Li, H.X., Xu, X.R., 2020. A meta-analysis of methodologies adopted by microplastic studies in China. *Sci. Total Environ.* 718, 135371.

Fu, W., Min, J., Jiang, W., Li, Y., Zhang, W., 2020a. Separation, characterization and identification of microplastics and nanoplastics in the environment. *Sci. Total Environ.* 721, 137561.

Fu, Z., Chen, G., Wang, W., Wang, J., 2020. Microplastic pollution research methodologies, abundance, characteristics and risk assessments for aquatic biota in China. *Environ. Pollut.* 266, 115098. https://doi.org/10.1016/j.envpol.2020.115098

Fu, Z.L., Chen, G.L., Wang, W.J., Wang, J., 2020b. Microplastic pollution research methodologies, abundance, characteristics and risk assessments for aquatic biota in China. *Environ. Pollut.* 266, 115098.

Fu, Z.L., Wang, J., 2019. Current practices and future perspectives of microplastic pollution in freshwater ecosystems in China. *Sci. Total Environ.* 691, 697–712.

Gewert, B., Plassmann, M.M., MacLeod, M., 2015. Pathways for degradation of plastic polymers floating in the marine environment. *Environ. Sci. Proc. Imp.* 17, 1513–1521.

Gies, E.A., LeNoble, J.L., Noel, M., Etemadifar, A., Bishay, F., Hall, E.R., et al., 2018. Retention of microplastics in a major secondary wastewater treatment plant in Vancouver, Canada. *Mar. Pollut. Bull.* 133, 553–561.

Gong, Y., Wang, Y.X., Chen, L., Li, Y.K., Chen, X.J., Liu, B.L., 2021. Microplastics in different tissues of a pelagic squid (Dosidicus gigas) in the northern Humboldt Current ecosystem. *Mar. Pollut. Bull.* 169, 112509.

Gray, A.D., Wertz, H., Leads, R.R., Weinstein, J.E., 2018. Microplastic in two South Carolina Estuaries: occurrence, distribution, and composition. *Mar. Pollut. Bull.* 128, 223–233.

Guo, X., Wang, J.L., 2021. Projecting the sorption capacity of heavy metal ions onto microplastics in global aquatic environments using artificial neural networks. *J. Hazard Mater.* 402, 123709.

Han, M., Niu, X.R., Tang, M., Zhang, B.T., Wang, G.Q., Yue, W.F., et al., 2020. Distribution of microplastics in surface water of the lower Yellow River near estuary. *Sci. Total Environ.* 707.

He, Z.W., Yang, W.J., Ren, Y.X., Jin, H.Y., Tang, C.C., Liu, W.Z., et al., 2021. Occurrence, effect, and fate of residual microplastics in anaerobic digestion of waste activated sludge: A state-of-the-art review, *Bioresource Technol.* 331,125035. https://doi.org/10.1016/j.biortech.2021.125035

Hermabessiere, L., Himber, C., Boricaud, B., Kazour, M., Amara, R., Cassone, A.L., et al., 2018. Optimization, performance, and application of a pyrolysis-GC/MS method for the identification of microplastics. *Anal. Bioanal. Chem.* 410, 6663–6676.

Horton, A.A., Dixon, S.J., 2018. Microplastics: an introduction to environmental transport processes. *Wires Water* 5, e1268.

Horton, A.A., Walton, A., Spurgeon, D.J., Lahive, E., Svendsen, C., 2017. Microplastics in freshwater and terrestrial environments: evaluating the current understanding to identify the knowledge gaps and future research priorities. *Sci. Total Environ.* 586, 127–141.

Huang, D.F., Li, X.Y., Ouyang, Z.Z., Zhao, X.N., Wu, R.R., Zhang, C.T., et al., 2021a. The occurrence and abundance of microplastics in surface water and sediment of the West River downstream, in the south of China. *Sci. Total Environ.* 756, 134857.

Huang, H., Sun, Z.H., Liu, S.C., Di, Y.N., Xu, J.Z., Liu, C.C., et al., 2021b. Underwater hyperspectral imaging for in situ underwater microplastic detection. *Sci. Total Environ.* 776, 145960.

Huang, J., Chen, H., Zheng, Y., Yang, Y., Zhang, Y., Gao, B., 2021c. Microplastic pollution in soils and groundwater: characteristics, analytical methods and impacts. *Chem. Eng. J.* 425, 131870.

Hurley, R.R., Lusher, A.L., Olsen, M., Nizzetto, L., 2018. Validation of a method for extracting microplastics from complex, organic-rich, environmental matrices. *Environ. Sci. Technol.* 52, 7409–7417.

Jin, Y.X., Xia, J.Z., Pan, Z.H., Yang, J.J., Wang, W.C., Fu, Z.W., 2018. Polystyrene microplastics induce microbiota dysbiosis and inflammation in the gut of adult zebrafish. *Environ. Pollut.* 235, 322–329.

Kane, I.A., Clare, M.A., Miramontes, E., Wogelius, R., Rothwell, J.J., Garreau, P., et al., 2020. Seafloor microplastic hotspots controlled by deep-sea circulation. *Science* 368, 1140.

Kanhai, L.K., Gardfeldt, K., Lyashevska, O., Hassellov, M., Thompson, R.C., O'Connor, I., 2018. Microplastics in sub-surface waters of the Arctic central basin. *Mar. Pollut. Bull.* 130, 8–18.

Kanhai, L.K., Officer, R., Lyashevska, O., Thompson, R.C., O'Connor, I., 2017. Microplastic abundance, distribution and composition along a latitudinal gradient in the Atlantic Ocean. *Mar. Pollut. Bull.* 115, 307–314.

Kappler, A., Fischer, D., Oberbeckmann, S., Schernewski, G., Labrenz, M., Eichhorn, K.J., et al., 2016. Analysis of environmental microplastics by vibrational microspectroscopy: FTIR, Raman or both? *Anal. Bioanal. Chem.* 408, 8377–8391.

Kolandhasamy, P., Su, L., Li, J.N., Qu, X.Y., Jabeen, K., Shi, H.H., 2018. Adherence of microplastics to soft tissue of mussels: a novel way to uptake microplastics beyond ingestion. *Sci. Total Environ.* 610, 635–640.

Klein, S., Worch, E., Knepper, T.P., 2015. Occurrence and spatial distribution of microplastics in river shore sediments of the Rhine-main area in Germany. *Environ. Sci. Technol.* 49, 6070–6076.

Koelmans, A.A., Nor, N.H.M., Hermsen, E., Kooi, M., Mintenig, S.M., De France, J., 2019. Microplastics in freshwaters and drinking water: critical review and assessment of data quality. *Water Res.* 155, 410–422.

Kokalj, A.J., Kunej, U., Skalar, T., 2018. Screening study of four environmentally relevant microplastic pollutants: uptake and effects on Daphnia magna and Artemia franciscana. *Chemosphere* 208, 522–529.

MacLeod, M., et al., 2021. The global threat from plastic pollution. *Science* 373, 61–65. https://doi.org/10.1126/science.abg5433

Lahens, L., Strady, E., Kieu Le, T.C., Dris, R., Kada, B., Rinnert, E., Gasperi, J., Tassin, B., 2018. Macroplastic and microplastic contamination assessment of a tropical river (Saigon River, Vietnam) transversed by a developing megacity. *Environ. Pollut.* 236, 661–671.

Lares, M., Ncibi, M.C., Sillanpaa, M., Sillanpaa, M., 2018. Occurrence, identification and removal of microplastic particles and fibers in conventional activated sludge process and advanced MBR technology. *Water Res.* 133, 236–246.

Li, H.X., Ma, L.S., Lin, L., Ni, Z.X., Xu, X.R., Shi, H.H., Yan, Y., Zheng, G.M., Rittschof, D., 2018a. Microplastics in oysters saccostrea cucullata along the Pearl River estuary, China. *Environ. Pollut.* 236, 619–625.

Li, X., Chen, L., Mei, Q., Dong, B., Dai, X., Ding, G., Zeng, E.Y., 2018b. Microplastics in sewage sludge from the wastewater treatment plants in China. *Water Res.* 142, 75–85.

Lin, L., Zuo, L.Z., Peng, J.P., Cai, L.Q., Fok, L., Yan, Y., Li, H.X., Xu, X.R., 2018. Occurrence and distribution of microplastics in an urban river: a case study in the Pearl River along Guangzhou City, China. *Sci. Total Environ.* 644, 375e381.

Prata, J. C., da Costa, J. P., Girão, A. V., Lopes, I., Duarte, A. C., Rocha-Santos, T., 2019. Identifying a quick and efficient method of removing organic matter without damaging microplastic samples. *Sci. Total Environ.* 686(71) 131–139.

Simon, M., van Alst, N., Vollertsen, J., 2018. Quantification of microplastic mass and removal rates at wastewater treatment plants applying focal plane array (FPA)-based Fourier transform infrared (FT-IR) imaging. *Water Res.* 142, 1–9.

Werbowski, L. M., Gilbreath, A. N., Munno, K., Zhu, X., Grbic, J., Wu, T., Sutton, R., Sedlak, M. D., Deshpande, A. D., Rochman, C. M., 2021. Urban stormwater runoff: A major pathway for anthropogenic particles, black rubbery fragments, and other types of microplastics to urban receiving waters. *ACS EST Water* 1, 1420–1428.

Yan, M., Nie, H., Xu, K., He, Y., Hu, Y., Huang, Y., Wang, J., 2019. Microplastic abundance, distribution and composition in the Pearl River along Guangzhou city and Pearl River estuary, China. *Chemosphere* 217, 879–886. https://doi.org/10.1016/j.chemosphere.2018.11.093

Zhang, Z., Gao, S.H., Luo, G., Kang, Y., Zhang, L., Pan, Y., Zhou, X., Fan, L., Liang, B., Wang, A., 2022. The contamination of microplastics in China's aquatic environment: occurrence, detection and implications for ecological risk. *Environ. Pollut.* 296, 118737. https://doi.org/10.1016/j.envpol.2021.118737.

12 Health and Environmental Impact of Microplastics
A Closer View

Niranjan Koirala, Arjun Sharma, Saru Gautam, Nabin Chaulagain, Proestos Charalampos, and Jian Bo Xiao

12.1 INTRODUCTION

Microplastics have had a significant impact on both health and the environment in recent years due to their widespread presence. The global issue of pollution from plastic is present currently. From the busiest beaches to isolated islands, large pieces of plastic have been found practically everywhere (Hasan Anik et al., 2021). A vital component of modern life, plastics are now widely used and they have different powerful qualities such as affordability as well as durability. In technical terms, rubbers and plastic are both polymeric materials. Chemical elements with high molar masses, whether synthetic or natural, are the primary building blocks of plastics (De Sousa, 2021). Even though their widespread manufacturing and use began in the 1950s, a world without plastics or organic polymers made from synthetic substances seems beyond comprehension to us today. Even though synthetic polymers like Bakelite initially existed at the beginning of the 20th century, it took until after World War II for plastics to be widely used in non-military applications (Geyer et al., 2017). The amount of plastic generated on a global scale rose over 200-fold between 1950 and 2015. Due to the COVID-19 outbreak, it was predicted that in 2020, there would be a worldwide consumption of plastics of 367 million tonnes (Mt), which represents a 0.3% decrease from the 368 Mt of 2019 (Uddin et al., 2022). This conduct generates a situation that is unsustainable for the natural ecosystem and, most critically, for the well-being of humans (Hossain et al., 2022). According to a report by Plastics Europe (2018), plastic output has increased dramatically worldwide and nearly reached 350 million tonnes in 2017. Thus, if the current rate of plastic pollution continues, scientists are currently issuing a warning that by 2050, there will be more plastic in the ocean than wildlife (Kumar et al., 2021). The UN estimates that there may be 51 trillion microplastic particles in the ocean, approximately 500 times more than there are stars in our galaxy (Nirmala et al., 2023). They can be either primary microplastics or secondary microplastics. The different sources and types of microplastic pollutants in the atmospheric environment, terrestrial environment, and aquatic environment are shown in Figure 12.1.

Microplastic release into the atmosphere directly or indirectly affects human health in various ways. After microplastic is released into the atmosphere it can enter the human body through food, air, drinking water sources, etc. Microplastics release different harmful chemicals, and can absorb toxic compounds that impact on the human digestive system, lungs, and respiratory system which can cause various chronic illnesses (Campanale et al., 2020). Microplastics (MPs) are generally found in different environmental areas, such as rivers, lakes, oceans, soil, and even in the atmosphere. However, the production of MPs, which are present throughout each environmental sector as well as in drinking water, is caused by industrial debris in addition to the decomposition of macroplastics and the dumping of consumer items (Biginagwa et al., 2016). WHO reported that

Health and Environmental Impact of Microplastics: A Closer View

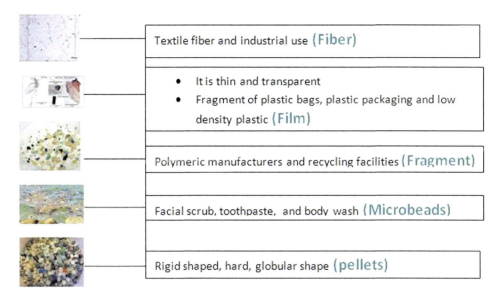

FIGURE 12.1 Sources and types of microplastic.

although MPs are present at low levels in treated tap and bottled water due to repeated consumption, this could have a long-term negative impact on people's health (Hasan Anik et al., 2021). Microplastics are taken as serious issues with a greater role in degrading ecosystems as well as causing various types of human health issues. The research conducted by Cashman et al. (2020) also shows around 8.3 billion tons of plastic products were produced in the early 1950s. As much as the use of plastic products has been increasing globally, the majority of the plastic products are released into oceans and ultimately cause ocean pollution which plays a direct role in degrading our natural ecosystem. Microplastic pollution has created an impact on several habitats including terrestrial ecosystems, marine ecosystems, and freshwater ecosystems, causing habitat destruction of several animals and chemical pollution. More than 359 million tons of plastic products are produced annually and a research report shows the rate will double in around the next 20 years (Napper & Thompson, 2020).

12.2 SOURCES OF MICROPLASTICS

Microplastics in aquatic settings come from two main sources: the initially identified source, referred to as primary microplastics, is when plastic gets created directly as microparticles, and the following source, known as secondary microplastics, is when larger plastic fragments are broken up.

12.2.1 Primary Microplastics

Primary microplastics are made to add to consumer and industrial goods such as beauty products, personal hygiene products, medications, laundry detergents, and pesticides (Osman et al., 2023; Martinho et al., 2022). They can enter the environment either as a byproduct of the breakdown of larger fragments of plastic or as a direct outcome of a primary source, including the inclusion of microbeads in PCCPs (personal care and cosmetic products) (Bashir et al., 2021). Two examples of microplastics are microbeads and microfibers. Microbeads can be produced in a variety of forms and from a variety of polymers, and they are typically between 0.1 and 1 mm in size. Polythene (PE) is the most widely used polymer, but it is also used in conjunction with polyethylene terephthalate (PET),

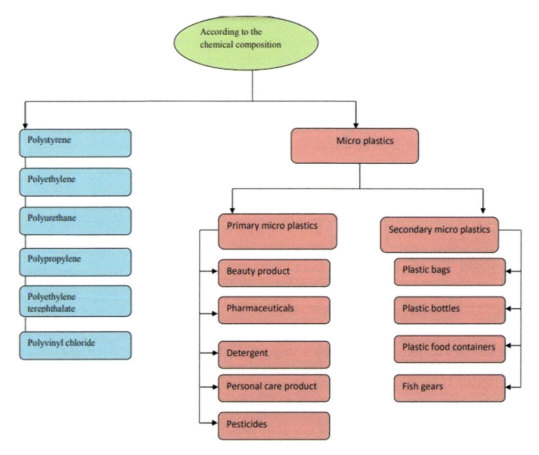

FIGURE 12.2 Types of microplastics according to their chemical composition.

polypropylene (PP), polymethyl methacrylate (PMMA), nylons (PA), polyester, and polyurethanes (Bashir et al., 2021). The most common types of microplastics discovered in the environment are microplastic fibers, commonly referred to as microfibers. When washing textile clothing at home, the wastewater and/or sludge releases a large quantity of microfibers from the clothing (Gaylarde et al., 2021). Additionally, a brief examination of the potential negative impacts of microfibers on aquatic and marine organisms as well as human health has been conducted. According to studies, numerous microfibers are discharged during washing of textile clothes (Acharya et al., 2021). According to their chemical compositions, microplastics are divided into primary and secondary microplastics, as shown in Figure 12.2.

12.2.2 Secondary Microplastics

At some point, secondary microplastics are produced as a result of the environment's degradation of big plastic. Figure 12.3 shows that all the microplastics are formed in an environment that may come from different sources such as UV radiation, wave action, temperature fluctuation, etc. Secondary microplastics are the fragments that result from the decomposition of larger plastic products including plastic bags, bottles, spoons, and jars in the marine environment (Lujan-Vega et al., 2021; Yuan et al., 2022a). Research has highlighted that UV light and low temperature are responsible for the breakdown of larger plastic products into smaller particles, which we generally refer to as secondary microplastics (Gola et al., 2021). Usually, these larger plastic pieces are improperly disposed

Health and Environmental Impact of Microplastics: A Closer View

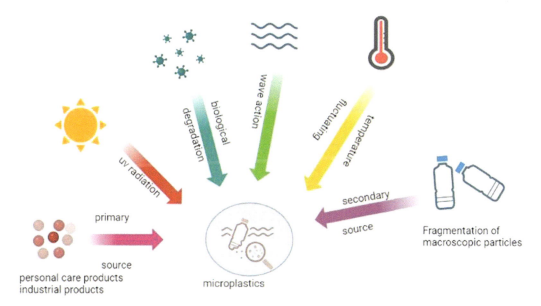

FIGURE 12.3 Weathering or aging process of microplastics.

of, such as plastic debris, including fishing gear, wrapping, or plastic from open dumps. According to the European Chemicals Agency, the abrasion and weathering of plastic items cause 176,000 tonnes of accidentally generated microplastics to be dispersed into European waterways each year (Wojnowska-Baryła et al., 2022). On an annual basis, the environment receives a further 42,000 tonnes of microplastics that have been purposefully included in products (European Environment Agency, 2021; Lofty et al., 2022).

12.2.3 Pathways of Microplastic Release

Enormous plastics are continuously degraded in the environment, creating a vast quantity of microplastics and nanoplastics (MNPLs) that are dispersed through the land, air, and oceans. Mainly microplastics are released from sources including around 37% from synthetic textiles, 35% from dust in cities, and around 28% from tires, which is displayed in Figure 12.4. The produced and released microplastics directly affect human health, and as a result they cause various health issues in humans through processes like inhalation, ingestion, and skin contact, which are three ways that humans are chronically exposed to MNPLs (Domenech & Marcos, 2021). According to a study by ENSSER (European Network of Scientists for Social and Environmental Responsibility), microplastics can enter the ecosystem via several different mechanisms, such as the discharge of wastewater, air accumulation, and the movement of plastic trash from land to sea through stormwater drainage (Prapanchan et al., 2023). Understanding these pathways is important for effectively addressing the release and distribution of different types of microplastics (Duis & Coors, 2016).

Direct release from manufacturing facilities: Microplastics can be released into the environment directly during the production and manufacturing procedure. Production of plastics generally generates byproducts, and these microplastics can be discharged into nearby water or released into the air/atmosphere, which directly impacts health and the environment.

Improper disposal and waste management practices: The main causes of this enormous mass of plastic garbage are a lack of technical expertise in handling toxic materials, inadequate construction of facilities for recycling and recovery, and, most importantly, a lack of knowledge of

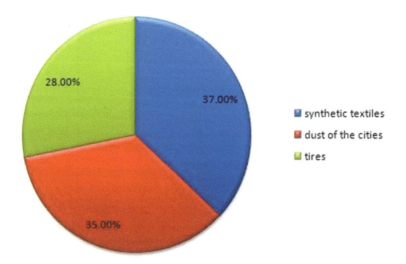

FIGURE 12.4 Formation of microplastic from different sources.

the laws and regulations. The extent of plastic pollution harms the ecology as a whole (Kibria et al., 2023). Inadequate knowledge and illiteracy regarding waste management, insufficient recycling products, and improper disposal can cause microplastic pollution.

Stormwater runoff: One of the pollutant sources that is expanding rapidly is stormwater runoff. Rainfall that falls on rooftops, parking lots, and highways as opposed to marshes, jungles, and grasslands usually empties into storm drains that have connections to rivers. This runoff which is actively and frequently loaded with microplastics is then dropped into the water and can cause contamination as well as pollution (Handbook of Water Purity and Quality, 2021; Piñon-Colin et al., 2020).

Atmospheric deposition: Recently, microplastics have been found in the atmosphere of cities, suburbs, as well as distant locations that are far from microplastic source areas, indicating the possibility of distant atmospheric transmission of microplastics. Microplastics can be transferred to the atmosphere by different mechanisms which can include wind erosion, emission from the industry during the manufacturing process, and even the transfer of microfibers from textiles during drying (Zhang et al., 2020).

12.2.4 Distribution of Microplastics

The roots and main distribution of microplastics are widespread but, usually, they are found in seawater, food and drinks, air, soil, etc. Microplastics are available in aquatic as well as terrestrial environments. In the marine environment, they are widely spread in every region, such as from the surface to the deep seas, creating microplastic pollution. Different parameters are responsible for the spreading of microplastics in the marine environment. The risks of microplastics are creating a threat to human health as well as the environment, and they have significantly impacted the growth, reproduction, and survival of humans as well as different flora and fauna. Oceans are considered huge reservoirs of microplastics as the microplastics there are accumulated from different sources. The main sources of microplastics are from domestic household activities and industrial sewage, waste treatment, air blasting, etc. as illustrated in Figure 12.5. The daily used cosmetic products as well as detergents produce high quantities of microplastics (Boyle & Örmeci, 2020). The research performed by Lebreton et al. (2017) shows that the estimated plastics production from the rivers to the oceans ranges from 1.15 to 2.41 million tons per year. The huge bulk of production of

Health and Environmental Impact of Microplastics: A Closer View

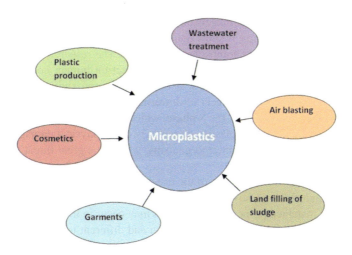

FIGURE 12.5 Sources of microplastics.

microplastics is in Asian countries. Economically unstable and poor countries have not adopted quality standards for waste management techniques.

Marine and aquatic organisms such as fish, seabirds, and other aquatic animals ingest microplastic particles from the water, which poses a significant risk to human health and also to the consumption of sea and marine foods. Freshwater sources such as rivers, ponds, and lakes are also responsible for the deposition of microplastics (Takarina et al., 2022). The microplastics generated from industrial procedures, urban and city areas, and wastewater deposition are the main sources of microplastic pollution in freshwater systems. Rivers are considered the main transporter system of microplastics to the seas and oceans (Koelmans et al., 2019).

Not only the water sources, but microplastics also are found and deposited in the terrestrial environment such as agricultural land and soil (Sajjad et al., 2022). Microplastics slowly degrade the quality of the soil and the fertilizers which are used in agricultural fields are also responsible for microplastic pollution (Lwanga et al., 2022). The research conducted by Yukioka et al. (2020) shows that more rubber-like microplastics are present in the dust samples taken from the roads than the dust from the general environment of the surveyed area in Kusatsu (Japan), Da Nang (Vietnam), and Kathmandu (Nepal). Urban cities produce a larger amount of plastic waste particles which results in the deposition of more microplastics. The accumulation of more plastic waste and a poor waste management system of urban cities creates a higher risk towards human health as well as to the natural ecosystem (Wojnowska-Baryła et al., 2022; Ng et al., 2023). Microplastics are also present in atmospheric areas. Different kinds of microplastic particles are transferred from the breakdown of larger plastic products, industrial waste, tires, and the different microfibers from the fiber industries (Torres-Agullo et al., 2021; Evangeliou et al., 2020). When the microplastics are released into the air they can move a greater distance and cause pollution to water sources, land, and ecosystems (Park & Park, 2021).

12.3 HEALTH

12.3.1 Health Impacts of Microplastics

Over the past 70 years, the production of plastic products has significantly increased and as a result microplastics have spread all over the world rapidly. These plastic synthetic materials create pollution in the environment which as a result spreads various chemicals and causes health impacts and this has raised greater concern due to the impacts on human health as well as wildlife. Different research works conducted on the health impacts show that microplastics enters through ingestion, inhalation, or contact with the environment (Thompson et al., 2009). The research shows that microplastics have had harmful effects on human health by their impact on human cells and different tissues causing cell death and allergic reactions, and also causing neurotoxicity and carcinogenic effects. These types of microplastics create effects by lowering the cell viability, mainly altering the cell cycles and arresting the cell cycle, i.e., the S phase, which is responsible for altering protein expression (Campanale et al., 2020). Microplastics do not only cause physical damage and issues but also spread different toxic chemicals and microorganisms, which is a great threat and risk towards the health of humans.

Modern people are more attracted toward the use of microplastic products and therefore production and packaging industries widely apply plastic materials. When the waste generated from the plastic sources is not managed properly health issues may occur in individuals (Evode et al., 2021). Plastic materials contain a high percentage of carbon and different toxic chemical substances. The used plastic products interact with the environment and cause adverse effects which may have primary as well as secondary adverse effects on human health including different genetic, reproductive, chronic, and mental health issues in healthy humans. The research conducted by Evode et al. (2021) also shows the development of different health problems due to microplastics is globally increasing with the development of communicable as well as non-communicable diseases such as cancer, birth defects, ophthalmological disorders, leukemia, etc. Figure 12.6 shows the impact of microplastics on human health, the human food chain, and environmental and factorial distribution that have huge effects on individuals as well as nature.

Health impact through ingestion: During the ingestion of food and water, microplastics may enter the human body including through different kinds of sea foods, water, and drinks that contain microplastics. The spread of microplastics can contaminate tap water, bottled water, and all other sources that directly interact with the gastrointestinal system of the human body after ingestion which can create an impact on the gut, including difficulties in nutrition absorption, and degrade the digestive health of the affected individuals (Pironti et al., 2021; Yuan et al., 2022a). The research conducted by Campanale et al. (2020) and Osman et al. (2023) shows that microplastics can physically impact the human body's digestive system, resulting in inflammation and changes to the function of the gut barrier. Microplastics release dangerous chemical

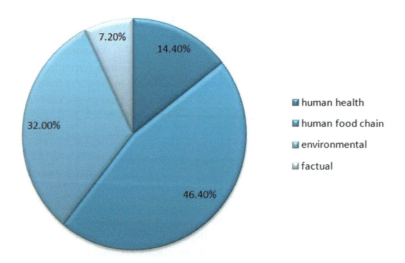

FIGURE 12.6 Percentages of sources of microplastic impacts.

pollutants from the environment such as heavy metals and persistent organic pollutants (POPs) can be adsorbed and accumulated by microplastics. These chemical pollutants may be released when consumed, creating different health hazards to humans. Microplastics are ingested from the consumed plastic particles. The research conducted by Yee et al. (2021) showed that with the observation and analysis of human stool samples that were analyzed, plastic particles were excreted, which provides evidence to show that plastic particles are consumed through food and water that are consumed. Therefore, the increased use of plastic products and their excessive use in daily life will surely increase the instances of consumption and impact on human health.

Health impact by inhalation: During the process of inhalation, microplastics in the air can be ingested, mostly due to the different kinds of pollution. Microplastics can be released into the atmosphere in several ways, such as the breaking down of large plastic products, such as when tire wear particles and textile microfibers are released. Microplastics can enter the pulmonary system by inhalation and have an impact on lung health (Kacprzak & Tijing, 2022). These microplastics have been found in human lungs and sputum, which plays a pivotal role in causing adverse effects on the lungs. The research done by Lu et al. (2022) and Chen et al. (2022) shows that microplastics have a significant impact and cause inflammation in the lung tissues and create breathing problems. There is insufficient research to demonstrate the evidence but those who are more exposed to work environments such as waste management, textile factories, and plastic production industries are more prone toward the development of lung problems due to increased contact with microplastics and synthetic plastic products (Prata, 2018).

Dermal exposure and skin health: Different products which humans use, such as personal care products that directly come into contact with the skin, also may cause an impact on the health of individuals. They may cause skin irritation, and they also serve as the carrier for other different types of pollution and cause different impacts on human health. The plastic content is not only caused by physical contact but also through the release of different toxic chemicals in the environment, such as endocrine-disrupting compounds (Manisalidis et al., 2020). Out of the many sources of microplastics, personal care and cosmetic products (PCCPs) contain abundant amounts of microplastics. A study conducted in Macao, China, found that PCCP products contain the latest microbeads (Khalid & Abdollahi, 2021; Bashir et al., 2021).

Endocrine disruption: Endocrine disruption is one of the most important global impacts of microplastics, which plays a significant role in altering human hormone levels, mainly affecting the reproductive and developmental tissues. The microplastic compounds which are responsible for the alternation and regulation of hormones are phthalates and bisphenol A (BPA). The research conducted by Hyland et al. (2019) also shows that exposure to phthalates has created adverse neurodevelopmental effects in and also affects cognitive development and behavioral dysfunction.

Development of cancer risk: Plastic particles like polyvinyl chloride (PVC) and other compounds have been found to be carcinogenic. PVC is the most common type of plastic, it is produced by different industries and poses a great risk of developing cancer. The research conducted by Brandt-Rauf et al. (2012) shows that more exposure to microplastic compounds like PVC results in a greater risk of cancer development by altering the DNA and causing mutations. The International Agency for Research on Cancer (IARC) has listed vinyl chloride, which is a compound of PVC, as causing cancer and a group 1 human carcinogen that impacts and stimulates the development of liver as well as lung cancers. The detection of microplastics in tumors is around 58%, whereas the percentage of microplastics in normal tissue is around 46% (Girardi et al., 2022).

Respiratory problems: Plastic products when heated release harmful chemicals such as volatile organic compounds and many other harmful types of chemicals when they are released into the atmosphere. When humans inhale these types of chemicals released from plastic particles it directly affects the respiratory system and can lead to the development of different types of respiratory problems (Lu et al., 2022). The over-exposure, inhalation, and consumption of products and goods with high microplastic contents results in a greater risk of developing lung diseases by reducing the lung-repairing capacity which leads to lung injuries. The development of allergens and toxins has a greater chance of entering the bloodstream which can initiate the development of asthma and chronic obstructive pulmonary disease (Lu et al., 2022).

The research by Campanale et al. (2020) shows that dermal exposure to microplastics has also caused skin irritation and inflammation, which results in the development of endocrine disruption. Some research also suggests overexposure to microplastics is responsible for a hormonal disbalance which impacts regular body metabolism and increasing the body weight of individuals who may go on to develop obesity (Ghosh et al., 2023).

12.3.2 MECHANISMS OF HEALTH IMPACTS

For the elimination of the possible risks posed by microplastics and the development of measures to prevent the health hazards to humans, it is necessary to understand the actual mechanism of microplastics' effects on human health. A serious impact of microplastics is the cause of physical damage to the different tissues and organs of the body which are in contact with the microplastics (Yuan et al., 2022b). Different health issues include ulceration of the gastrointestinal tract and inflammation of organs which results in different types of disorders. In the respiratory tract, the inhalation of microplastics causes lung problems and damages the tissues causing different types of respiratory problems (Campanale et al., 2020). Chemical exposure through microplastics causes different toxicological effects which impact the hormone level causing unbalancing, damaging the cellular functions in the body, and causing other chronic health issues such as mental and reproductive disorders (Jin et al., 2022).

Microplastics impact the immune system and cause different inflammations in the human body. Microplastics come in contact with the body's surface and different proinflammatory molecules are released from the body which alters the function of tissues and can lead to the development of chronic health issues including cardiovascular and autoimmune diseases (Yang et al., 2022). Microplastics impact the production of reactive oxygen species (ROS) in the body and as a result the antioxidant defense mechanism of the body is halted with the development of oxidative stress in the tissues and cells. They may lead to several problems through DNA damage which results in gene alterations and as a result there is a high chance of development of genetic disorders. For better nutrition absorption, metabolism, digestion, and immune function as well as for maintaining the mental well-being of the individual, the gut microbiota plays a significant role, however microplastics alter the balance of the gut microbiota (Das, 2023). The impact of microplastics also depends on their shape, size, composition, and the contamination exposure time.

Microplastics have several effects on human health due to their small size which means they have a higher surface area by volume. The particles which have greater surface area are highly cytotoxic for the cells and tissues and they may damage the DNA of the cells (Osman et al., 2023). Figure 12.7 also shows the impacts of different sources of microplastics which are used daily as well the food we consume which may contain particles of microplastics and cause different health effects.

12.3.3 PROTECTION AGAINST HEALTH IMPACTS OF MICROPLASTICS

For the minimization of the health impacts of microplastics, there are several steps to reduce the use of plastic products and minimize them spreading in the environment. The effective ways to minimize the possible health threat of microplastics to humans include by adopting individual action plans, industry and organizational actions, and policymakers; And also reducing the overuse

Health and Environmental Impact of Microplastics: A Closer View

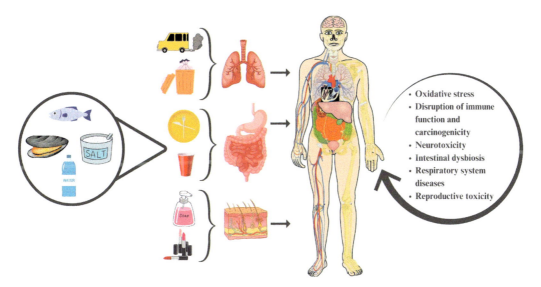

FIGURE 12.7 Impacts of different sources of microplastics on human health.

of microplastic sources such as preparing guidelines and regulation policies in the use of plastic products which are used for personal caring items, and more use of environment-friendly and sustainable products. Separating plastic waste from organic waste to reduce the release of plastic into the natural environment and adapting to different strategic plans for recycling products with the proper waste management strategies should also be done (Ray et al., 2022). Water and air treatment filtration systems can be utilized to reduce the amount of microplastics and remove them from the source by applying several steps and measures. Different governmental organizations and nongovernmental organizations should focus on developing policies and procedures to reduce the overuse of microplastics and implement actions for the protection and promotion of human health (Coffin et al., 2021; Osman et al., 2023).

According to the UN environment program, individuals can minimize their overuse of plastic products including plastic bags, plastic bottles, and utensils. Instead of using plastic products alternative and reusable products such as stainless steel, eco-friendly products and cloth bags should be utilized. While using personal care products, check the added ingredients in the products and try to avoid products that contain microbeads or any microplastic-related ingredients instead of applying different biodegradable alternative products (Habib et al., 2022; Bashir et al., 2021). Properly manage waste and dispose of it properly, which helps to minimize the spread of plastic materials and reduces the environmental impact. Properly dispose of plastic waste in the proper waste bins which helps in reducing the microplastic pollution in the environment and ultimately helps in reducing the health impacts for humans. Try to use and wear sustainable fashion wear such as clothing that is manufactured using natural fibers rather than synthetic fibers.

12.4 POLICY AND REGULATION

For the minimization of microplastics in the environment and to reduce the impact of microplastics on human health, governments should also prioritize and prepare effective legislation and policies by implementing different policies such as banning the use of plastic products, and restricting the use of microplastic ingredients in cosmetic and personal care products (Li, 2022). Governments should also focus on developing eco-friendly and alternative products to minimize the overuse of plastic items. Government bodies should also focus on the development of different waste management technologies for the collection of waste, sorting of the different waste products, and providing proper education to the public about the impact of microplastics on human health and other issues (Hopewell et al., 2009). For the minimization of plastic waste, governments should also develop innovative ideas and techniques with research and development for the promotion and protection of human health from the effects of microplastics. As microplastic pollution and its impact on humans is a global threat, every government should consult and cooperate with the different international bodies and apply the proper global standards of limiting the use of microplastics and help in the protection of the health of the population (Haque & Fan, 2022; Yuan et al., 2022a). Governments should also launch different educational and awareness programs to spread accurate knowledge about microplastics, and their impacts and health hazards with the best possible ways to reduce the use of microplastics (Coffin et al., 2021; Ray et al., 2022).

12.5 ENVIRONMENT

12.5.1 ENVIRONMENTAL IMPACTS OF MICROPLASTICS

A recent study (Yuan et al., 2022a) indicates that microplastic contamination of the environment creates a serious health risk to humans along with having a greater impact on wildlife which is a great threat to the ecosystem (Zolotova et al., 2022). Microplastics are a pollutant of which ecotoxicologists are mainly concerned with both human health and aquatic habitats. Particles of microplastics are pervasive in the atmosphere including the air, water, and land ecosystem (Akdogan & Guven, 2019). These microplastic particles have an ongoing detrimental effect on the ecosystem. The average lifespan of polymers is between 70 and 450 years, with a decomposition time of 10–20 years or 500–1000 years (Yuan et al., 2022b). It is important to comprehend the adverse environmental impacts of microplastics to counteract the urgent adverse outcomes. The environmental effects of microplastics are listed below.

Marine ecosystems: Microplastics are often made of nylon, polystyrene (PS), polyethylene terephthalate (PET), polyvinyl chloride (PVC), and other materials. As a result of inadequate management, the abundance of these microplastic particles is rising alarmingly, which has an impact on both the marine ecosystem and marine life (Gola et al., 2021). The disposition of microplastics in large water resources like oceans and seas directly impacts marine life, contributing to various levels of destruction when microplastics are transferred along the food chain. The smaller marine organisms, like zooplankton and filter-feeding marine species, when ingesting plastic residues, can have detrimental negative impacts including physical harm, impaired feeding, loss of nutritional value of the diet, reduced ability to reproduce, and increased death rates (Avio et al., 2017). They can severely impact the larger marine bodies including fish, marine mammals, and seabird species when microplastics move up the food chain. This eventually has an impact on human health. The occurrence of microplastics can be found in different sources such as seafood, which is a carrier of microplastics that can cause adverse health effects in humans (Curren et al., 2020).

Freshwater ecosystems: Freshwater ecosystems such as rivers, lakes, and streams are also susceptible to microplastics. Diverse aquatic creatures find a home in these environments. Although freshwater microplastic research has increased recently, the toxicity effects of microplastics in freshwater environments is less well understood than it is in marine environments (Talbot & Chang, 2022). Freshwater environments may become contaminated by microplastics, which can have a detrimental effect on the ecological diversity and health. Freshwater creatures that ingest them may have developmental, reproductive, and overall wellness problems (Castro-Castellon et al., 2022). In addition to this, microplastics can build up in sediments, changing the chemical makeup and altering the cycling of nutrients with the ability to affect the entire freshwater environment (Sarijan et al., 2021).

Terrestrial ecosystems: The contamination of microplastics has been detected in soils, implying that they are a potential threat to the terrestrial ecosystem. Studies have shown that these plastic residues can enter the soil through other environments, such as through atmospheric deposition or runoff from urban areas (Bertrim & Aherne, 2023). The detection of microplastics in soil has revealed its implications for soil health, nutrient cycling, and plant growth (Chia et al., 2022). Microplastics can also have an impact on soil-dwelling organisms, such as earthworms and other soil microorganisms, hence disrupting soil ecological function and their overall ecosystem. The global use and increase of microplastic pollution have covered a wider area of the environment which is posing a global challenge and threat to all terrestrial ecosystems as well as natural habitats throughout the globe (Souza Machado et al., 2018).

Ecological interactions: Microplastics have emerged as a severe environmental pollutant of major concern (Lamichhane et al., 2023). Environmental interactions and processes can be impacted by microplastics that can, for instance, alter how organisms behave, eat, and reproduce, which can alter population dynamics and the equilibrium of the ecosystem. Additionally, microplastics can serve as carriers of invading species and chemical contaminants, which may intensify the ecological implications (Mammo et al., 2020).

Habitat modification: When the deposition of microplastics occurs in natural habitats, they can result in severe physical changes and alter the habitat (U & M, 2017). Microplastics, for instance, may smother coral reefs and the seafloor by covering them, making it challenging for species to survive and thrive, and disrupting the function and structure of the natural habitat (Bednarz et al., 2021; Lincoln et al., 2022). In addition, sunlight penetration, cycling of nutrients, and ecological productivity can all be impacted by the existence of microplastics in water bodies and sediments. These modifications may affect food chains which could cause reductions in biodiversity, with cascading impacts on the entire ecological system (Entezari et al., 2022).

Chemical contamination: Microplastics can bind and retain chemical contaminants from their surroundings (Guzzetti et al., 2018). These substances include hydrophobic substances including heavy metals, polycyclic aromatic hydrocarbons (PAHs), and persistent organic pollutants (POPs) (Verla et al., 2019). Sadly, conventional water treatment facilities are unable to completely eradicate every trace of microplastics from water. When creatures consume microplastics, the pollutants that have collected inside them may be discharged into their digestive systems, which could have toxicological consequences and cause bioaccumulation in higher levels of trophic structure (Au et al., 2017). The researchers and scientists who are examining the effects of microplastics on the natural world and public health are concerned about this situation.

12.5.2 Mechanisms of Environmental Impacts

Numerous possibilities exist for microplastics to impact the natural world. The various ways they can impact ecosystems and living beings are crucial to understanding how they accomplish this. The primary ways that microplastics might damage the planet's ecosystem are listed below.

Ingestion: When organisms ingest microplastics inadvertently, it is one of the major ways they may damage the ecosystem. Fishes, turtles, seagulls, aquatic mammals, algae, and other living things might consider microplastics as food (Smith et al., 2018). There are several ways that feeding microplastics might harm these creatures. It could make it more difficult for their bodies to break down and absorb food, and it might even clog their GI tracts (Cheung et al., 2018). Additionally, hazardous compounds that are toxic to organisms when consumed can be found in microplastics.

Chemical transfer: Microplastics can absorb and acquire chemical pollutants from their immediate surroundings. These stored compounds may be released into creatures' digestive systems after intake, which might have toxicological consequences. Persistent organic pollutants (POPs), polycyclic aromatic hydrocarbons (PAHs), heavy metals, and other harmful compounds are among the chemicals linked to microplastics that have been revealed (Yuan et al., 2022a). The conveyance of chemicals that occurs within microplastics that are consumed by species at various trophic levels has the potential to cause bioaccumulation as well as biomagnification in the food chain (Au et al., 2017).

Disruption of feeding and reproduction: The widespread distribution of microplastics, according to current study findings, entails serious risks to the ingestion and reproduction patterns of numerous animal species, having a substantial influence on the dynamics of populations and the health of ecosystems. More specifically, filter-feeding creatures that rely on collecting tiny particles from water, such as aquatic organisms and some fish species, unintentionally consume microplastics via their feeding operations (Setälä et al., 2016). Because of the disruption in nutrient uptake caused by this inadvertent ingestion of microplastics, these creatures' general health and psychological well-being could potentially be at risk. Additionally, the penetration of tiny microplastic particles into an animal's organs of reproduction can severely impair its ability to reproduce, resulting in decreased offspring generation and demographic decreases (Moyo, 2022). Figure 12.8 illustrates the process of microplastics that impacts the environment which indicates that microplastic contamination of the environment creates a serious risk to human health.

Habitat alteration and ecological interactions: Microplastics have a significant and far-reaching impact on the ecosystem, expressing themselves in a variety of habitat changes and complex disturbances of biological connections. Such a complex network of relationships necessitates careful inspection and a deeper comprehension of the resembling interactions between these tiny fragments and the natural environment (Bhuyan, 2022).

Microplastics have several negative effects, one of which is their gradual buildup inside fragile habitats such as coral reefs, beds of seagrass, and estuaries, where they cause physical changes that choke species and seriously jeopardize their ability to develop and survive (Zheng et al., 2023).

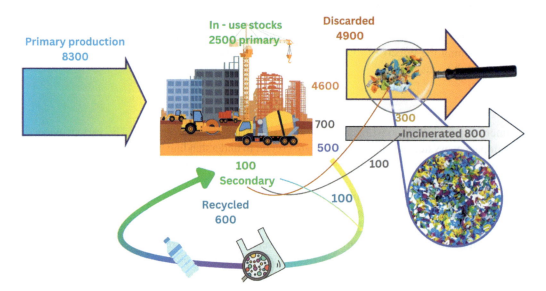

FIGURE 12.8 Process of microplastics impacting the environment in different ways.

Additionally, they probe the underwater world of marine and terrestrial microbial populations in their quest to affect and invade ecosystems (X. Zhang et al., 2021). This in turn triggers a chain reaction of effects that affect the health of ecosystems, cycling of nutrients, and degradative processes (Shen et al., 2022).

Various ideas about mitigating microplastics give rise to some optimism in the face of this complex environmental chaos. A comprehensive strategy that incorporates approaches to reduce the release of microplastics into the natural world, improve waste management procedures, promote environmentally friendly options, and establish strict rules and guidelines to prevent the flow of microplastic contamination must be diligently woven (Ng et al., 2023). The persistent commitment to continuous research and diligent monitoring is equally important for figuring out the evading processes behind environmental consequences and developing effective defenses against their harm.

12.5.3 PROTECTION AGAINST ENVIRONMENTAL IMPACTS OF MICROPLASTICS

To safeguard the planet from the effects of microplastics, an integrated approach incorporating several stakeholders, including governments, businesses, societies, and people, is required. The negative effects of microplastics on our planet can be mitigated by several important techniques and actions. Here are some of the crucial steps:

Source reduction: Implementing efficient actions targeted at reducing the generation and use of plastic materials is a scientifically reasonable strategy for protecting the natural world from the negative effects of microplastic contamination (Schuhen & Sturm, 2021). The use of plastics that are only used once must be restricted (Nanthini Devi et al., 2022) or gradually phased out, while also actively promoting environmentally friendly packaging substitutes and promoting the adoption of biodegradable or compostable plastics (Periyasamy, 2023). We may significantly reduce our global plastic footprint by diligently implementing such steps, which will eventually result in a significant decrease in the infiltration of microplastic particles into the fragile structure of our ecosystem.

Improved waste management: The development of robust and effective waste management systems must be given top priority if we are to save our stunning environments and stop the gradual intrusion of microplastics (Prata et al., 2019). This entails enhancing the facilities used for garbage collection, putting in place efficient recycling initiatives, and raising the awareness of appropriate and responsible disposal of waste (Garcia-Vazquez & Garcia-Ael, 2021). By following these actions, we can prevent plastic garbage from entering our valuable waterways, where it ultimately degrades into dangerous microplastics. These methods for controlling waste also aid in the fight against present-day sources of microplastics, such as the plastic garbage prevalent in rivers and coastal regions (Schmaltz et al., 2020).

Education and awareness: Creating a meaningful bond with the natural world as well as encouraging an improvement in consumer behavior requires the empowerment of individuals via public awareness. We can increase awareness of the causes, consequences, and impacts of microplastic contamination via educational initiatives that awaken each person to their role (Dowarah et al., 2022). People are empowered to make sensible decisions and take an active role in the fight against pollution caused by microplastic when they are informed about the need to reduce plastic usage, implement proper waste management methods, and embrace sustainable options (Garcia-Vazquez & Garcia-Ael, 2021). We can then save our precious natural environment.

Research and innovation: Identifying the actual effects of microplastic particles on our ecosystem and coming up with innovative solutions depend largely on maintaining our curiosity and supporting continuing investigation (Raddadi & Fava, 2019). We can learn more about the origins of microplastics, how they harm the environment, and the best ways to address the issue via research. We can stimulate invention in fields like materials science, waste management, and sustainable development by allowing the collaboration of professionals from business, the academic community, and authorities, paving the way for a cleaner and greener environment (Kalčíková & Žgajnar Gotvajn, 2019; Sarkar et al., 2022).

Policy and regulation: Governments have a critical role to play in protecting our ecosystem against the dangers of microplastic contamination by developing and implementing laws that are compassionate and long-sighted (Usman et al., 2022). The legislation could limit the number of microplastics that are permitted to be used in things that we use every day and enforce extended producer responsibility (EPR), which holds firms liable for the goods they manufacture (Leal Filho et al., 2019). Through sensible regulations, governments may also have a significant impact on how waste management methods are developed (Yu et al., 2023). Given the worldwide scope of this problem, international cooperation and coordination are crucial since the effects of microplastic contamination transcend national borders and have an influence on the whole planet (Stoll et al., 2022).

Collaboration and partnerships: We must work together to solve the complex problem of microplastic contamination. The government, businesses, non-profit organizations, academics, and the public at large must work together. Together they can successfully address microplastic contamination by exchanging information, gaining insight from each other, and developing methods together (Coffin et al., 2021). We can find more efficient solutions and develop comprehensive approaches to protect the natural world from the harmful effects of microplastics when everybody participates in this collaborative endeavor and brings their distinct ideas and skills to the table (Ghosh et al., 2023). We can alter things for the better so that the world and its residents can live in harmony.

12.5.4 Industry and Manufacturing Practices

Different industries and manufacturing companies must work to produce eco-friendly products and minimize the production of plastic materials. Production organizations should focus their consumers on the use of sustainable eco-friendly and natural products to minimize the environmental impact. From the industries, different kinds of waste are generated and larger amounts of plastic waste are generated so these manufacturing industries should focus on utilizing waste treatment methods and purification methods by applying the relevant waste management strategies and techniques (Alhazmi et al., 2021). Manufacturing companies and industries should prioritize the utilization of sustainable and reusable packaging materials instead of promoting plastic packaging, which ultimately helps in reducing the over-pollution of microplastic in the environment. These organizations and industries should focus on following the 5R principles of Refuse, Reduce, Reuse, Repurpose, and finally Recycle (Prata et al., 2019).

12.6 CONCLUSION

Microplastics have created the greatest challenge in the world. They are present in our daily food and drink items, water, and soil. These microplastics are tiny, measuring less than 5 mm, and are produced from various sources. Mainly these microplastics are separated from larger plastic materials. Different plastics products, cosmetic items, textiles, and paints are the main sources of production of microplastics. These microplastics are transferred a greater distance through the air and river water and stored in locations where they create microplastic pollution in the aquatic as well as terrestrial environment. In particular, microplastics are responsible for numerous issues and challenges toward human health and flora, and are found in the natural ecosystem. Microplastics create a risk to humans as well as animals through consuming products, ingestion, and inhalation. Microplastics create serious impacts on human health as well as on the environment. They can lead to mental health issues, cardiovascular problems, the development of cancer, and many types of chronic illnesses, while also affecting and altering the genes and DNA of cells and resulting in hormone imbalances, physical disabilities, etc. Moreover, microplastics impact the environment and disturb the overall natural ecosystem. The overall chain of the ecosystem is disrupted by this accumulation that causes microplastic pollution in the environment. For the solution and minimization of the health and environmental impacts of microplastics, several measures should be adopted by individuals, communities, and government bodies. For the minimization of microplastics in the ecosystem, the production of plastic materials should be reduced and their use should be minimal, with the utilization of products that can be recycled and reused, creating awareness to use sustainable, eco-friendly, and less carbon-emitting products. Government bodies should prepare effective policies and legislation regarding the use of plastic products and laws to prohibit the use of microplastic particles in cosmetics and sanitization products which are widely consumed and used. Along with that, following all the international standards and collaborating with the different international agencies, government bodies and international conservation agencies should work for the protection and promotion of human health and help to balance the ecosystem by reducing the overuse of plastic products and deposition of microplastics in the environment.

REFERENCES

Acharya, S., Rumi, S. S., Hu, Y., & Abidi, N. (2021). Microfibers from synthetic textiles as a major source of microplastics in the environment: A review. *Textile Research Journal*, *91*(17–18), 2136–2156. https://doi.org/10.1177/0040517521991244

Akdogan, Z., & Guven, B. (2019). Microplastics in the environment: A critical review of current understanding and identification of future research needs. *Environmental Pollution*, *254*, 113011. https://doi.org/10.1016/j.envpol.2019.113011

Alhazmi, H., Almansour, F. H., & Aldhafeeri, Z. (2021). Plastic waste management: A review of existing life cycle assessment studies. *Sustainability*, *13*(10), 5340. https://doi.org/10.3390/su13105340

Au, S. Y., Lee, C. M., Weinstein, J. E., Van Den Hurk, P., & Klaine, S. J. (2017). Trophic transfer of microplastics in aquatic ecosystems: Identifying critical research needs: Factors influencing microplastic trophic transfer. *Integrated Environmental Assessment and Management*, *13*(3), 505–509. https://doi.org/10.1002/ieam.1907

Avio, C. G., Gorbi, S., & Regoli, F. (2017). Plastics and microplastics in the oceans: From emerging pollutants to emerged threat. *Marine Environmental Research*, *128*, 2–11. https://doi.org/10.1016/j.marenvres.2016.05.012

Bashir, S. M., Kimiko, S., Mak, C.-W., Fang, J. K.-H., & Gonçalves, D. (2021). Personal care and cosmetic products as a potential source of environmental contamination by microplastics in a densely populated Asian City. *Frontiers in Marine Science*, *8*, 683482. https://doi.org/10.3389/fmars.2021.683482

Bednarz, V., Leal, M., Béraud, E., Ferreira Marques, J., & Ferrier-Pagès, C. (2021). The invisible threat: How microplastics endanger corals. *Frontiers for Young Minds*, *9*, 574637. https://doi.org/10.3389/frym.2021.574637

Bertrim, C., & Aherne, J. (2023). Moss bags as biomonitors of atmospheric microplastic deposition in urban environments. *Biology*, *12*(2), 149. https://doi.org/10.3390/biology12020149

Bhuyan, Md. S. (2022). Effects of microplastics on fish and in human health. *Frontiers in Environmental Science*, *10*, 827289. https://doi.org/10.3389/fenvs.2022.827289

Biginagwa, F. J., Mayoma, B. S., Shashoua, Y., Syberg, K., & Khan, F. R. (2016). First evidence of microplastics in the African Great Lakes: Recovery from Lake Victoria Nile perch and Nile tilapia. *Journal of Great Lakes Research*, *42*(1), 146–149. https://doi.org/10.1016/j.jglr.2015.10.012

Boyle, K., & Örmeci, B. (2020). Microplastics and nanoplastics in the freshwater and terrestrial environment: A review. *Water*, *12*(9), 2633. https://doi.org/10.3390/w12092633

Brandt-Rauf, P., Long, C., Kovvali, G., Li, Y., Monaco, R., & Marion, M.-J. (2012). Plastics and carcinogenesis: The example of vinyl chloride. *Journal of Carcinogenesis, 11*(1), 5. https://doi.org/10.4103/1477-3163.93700

Campsanale, C., Massarelli, C., Savino, I., Locaputo, V., & Uricchio, V. F. (2020). A detailed review study on potential effects of microplastics and additives of concern on human health. *International Journal of Environmental Research and Public Health, 17*(4), 1212. https://doi.org/10.3390/ijerph17041212

Cashman, M. A., Ho, K. T., Boving, T. B., Russo, S., Robinson, S., & Burgess, R. M. (2020). Comparison of microplastic isolation and extraction procedures from marine sediments. *Marine Pollution Bulletin, 159,* 111507. https://doi.org/10.1016/j.marpolbul.2020.111507

Castro-Castellon, A. T., Horton, A. A., Hughes, J. M. R., Rampley, C., Jeffers, E. S., Bussi, G., & Whitehead, P. (2022). Ecotoxicity of microplastics to freshwater biota: Considering exposure and hazard across trophic levels. *Science of The Total Environment, 816,* 151638. https://doi.org/10.1016/j.scitotenv.2021.151638

Chen, Q., Gao, J., Yu, H., Su, H., Yang, Y., Cao, Y., Zhang, Q., Ren, Y., Hollert, H., Shi, H., Chen, C., & Liu, H. (2022). An emerging role of microplastics in the etiology of lung ground glass nodules. *Environmental Sciences Europe, 34*(1), 25. https://doi.org/10.1186/s12302-022-00605-3

Cheung, L., Lui, C., & Fok, L. (2018). Microplastic contamination of wild and captive flathead grey mullet (Mugil cephalus). *International Journal of Environmental Research and Public Health, 15*(4), 597. https://doi.org/10.3390/ijerph15040597

Chia, R. W., Lee, J.-Y., Jang, J., Kim, H., & Kwon, K. D. (2022). Soil health and microplastics: A review of the impacts of microplastic contamination on soil properties. *Journal of Soils and Sediments, 22*(10), 2690–2705. https://doi.org/10.1007/s11368-022-03254-4

Coffin, S., Wyer, H., & Leapman, J. C. (2021). Addressing the environmental and health impacts of microplastics requires open collaboration between diverse sectors. *PLOS Biology, 19*(3), e3000932. https://doi.org/10.1371/journal.pbio.3000932

Curren, E., Leaw, C. P., Lim, P. T., & Leong, S. C. Y. (2020). Evidence of marine microplastics in commercially harvested seafood. *Frontiers in Bioengineering and Biotechnology, 8,* 562760. https://doi.org/10.3389/fbioe.2020.562760

Das, A. (2023). The emerging role of microplastics in systemic toxicity: Involvement of reactive oxygen species (ROS). *Science of The Total Environment, 895,* 165076. https://doi.org/10.1016/j.scitotenv.2023.165076

De Sousa, F. D. B. (2021). The role of plastic concerning the sustainable development goals: The literature point of view. *Cleaner and Responsible Consumption, 3,* 100020. https://doi.org/10.1016/j.clrc.2021.100020

Domenech, J., & Marcos, R. (2021). Pathways of human exposure to microplastics, and estimation of the total burden. *Current Opinion in Food Science, 39,* 144–151. https://doi.org/10.1016/j.cofs.2021.01.004

Dowarah, K., Duarah, H., & Devipriya, S. P. (2022). A preliminary survey to assess the awareness, attitudes/behaviours, and opinions pertaining to plastic and microplastic pollution among students in India. *Marine Policy, 144,* 105220. https://doi.org/10.1016/j.marpol.2022.105220

Duis, K., & Coors, A. (2016). Microplastics in the aquatic and terrestrial environment: Sources (with a specific focus on personal care products), fate and effects. *Environmental Sciences Europe, 28*(1), 2. https://doi.org/10.1186/s12302-015-0069-y

Entezari, S., Al, M. A., Mostashari, A., Ganjidoust, H., Ayati, B., & Yang, J. (2022). Microplastics in urban waters and its effects on microbial communities: A critical review. *Environmental Science and Pollution Research, 29*(59), 88410–88431. https://doi.org/10.1007/s11356-022-23810-2

European Environment Agency. (2021). *Microplastics from textiles: Towards a circular economy for textiles in Europe*. Publications Office. www.data.europa.eu/doi/10.2800/512375

Evangeliou, N., Grythe, H., Klimont, Z., Heyes, C., Eckhardt, S., Lopez-Aparicio, S., & Stohl, A. (2020). Atmospheric transport is a major pathway of microplastics to remote regions. *Nature Communications, 11*(1), 3381. https://doi.org/10.1038/s41467-020-17201-9

Evode, N., Qamar, S. A., Bilal, M., Barceló, D., & Iqbal, H. M. N. (2021). Plastic waste and its management strategies for environmental sustainability. *Case Studies in Chemical and Environmental Engineering, 4,* 100142. https://doi.org/10.1016/j.cscee.2021.100142

Garcia-Vazquez, E., & Garcia-Ael, C. (2021). The invisible enemy. Public knowledge of microplastics is needed to face the current microplastic crisis. *Sustainable Production and Consumption, 28,* 1076–1089. https://doi.org/10.1016/j.spc.2021.07.032

Gaylarde, C., Baptista-Neto, J. A., & Da Fonseca, E. M. (2021). Plastic microfibre pollution: How important is clothes laundering? *Heliyon, 7*(5), e07105. https://doi.org/10.1016/j.heliyon.2021.e07105

Geyer, R., Jambeck, J. R., & Law, K. L. (2017). Production, use, and fate of all plastics ever made. *Science Advances, 3*(7), e1700782. https://doi.org/10.1126/sciadv.1700782

Ghosh, S., Sinha, J. K., Ghosh, S., Vashisth, K., Han, S., & Bhaskar, R. (2023). Microplastics as an emerging threat to the global environment and human health. *Sustainability, 15*(14), 10821. https://doi.org/10.3390/su151410821

Girardi, P., Barbiero, F., Baccini, M., Comba, P., Pirastu, R., Mastrangelo, G., Ballarin, M. N., Biggeri, A., & Fedeli, U. (2022). Mortality for lung cancer among PVC baggers employed in the Vinyl chloride industry. *International Journal of Environmental Research and Public Health, 19*(10), 6246. https://doi.org/10.3390/ijerph19106246

Gola, D., Kumar Tyagi, P., Arya, A., Chauhan, N., Agarwal, M., Singh, S. K., & Gola, S. (2021). The impact of microplastics on the marine environment: A review. *Environmental Nanotechnology, Monitoring & Management, 16*, 100552. https://doi.org/10.1016/j.enmm.2021.100552

Guzzetti, E., Sureda, A., Tejada, S., & Faggio, C. (2018). Microplastic in marine organism: Environmental and toxicological effects. *Environmental Toxicology and Pharmacology, 64*, 164–171. https://doi.org/10.1016/j.etap.2018.10.009

Habib, R. Z., Aldhanhani, J. A. K., Ali, A. H., Ghebremedhin, F., Elkashlan, M., Mesfun, M., Kittaneh, W., Al Kindi, R., & Thiemann, T. (2022). Trends of microplastic abundance in personal care products in the United Arab Emirates over the period of 3 years (2018–2020). *Environmental Science and Pollution Research International, 29*(59), 89614–89624. https://doi.org/10.1007/s11356-022-21773-y

Haque, F., & Fan, C. (2022). The prospect of microplastic pollution control under the "New normal" concept beyond the COVID-19 pandemic. *Journal of Cleaner Production, 367*, 133027. https://doi.org/10.1016/j.jclepro.2022.133027

Hasan Anik, A., Hossain, S., Alam, M., Binte Sultan, M., Hasnine, Md. T., & Rahman, Md. M. (2021). Microplastics pollution: A comprehensive review on the sources, fates, effects, and potential remediation. *Environmental Nanotechnology, Monitoring & Management, 16*, 100530. https://doi.org/10.1016/j.enmm.2021.100530

Hopewell, J., Dvorak, R., & Kosior, E. (2009). Plastics recycling: Challenges and opportunities. *Philosophical Transactions of the Royal Society of London. Series B, Biological Sciences, 364*(1526), 2115–2126. https://doi.org/10.1098/rstb.2008.0311

Hossain, R., Islam, M. T., Ghose, A., & Sahajwalla, V. (2022). Full circle: Challenges and prospects for plastic waste management in Australia to achieve circular economy. *Journal of Cleaner Production, 368*, 133127. https://doi.org/10.1016/j.jclepro.2022.133127

Hyland, C., Mora, A. M., Kogut, K., Calafat, A. M., Harley, K., Deardorff, J., Holland, N., Eskenazi, B., & Sagiv, S. K. (2019). Prenatal exposure to phthalates and neurodevelopment in the CHAMACOS cohort. *Environmental Health Perspectives, 127*(10), 107010. https://doi.org/10.1289/EHP5165

Jin, H., Yan, M., Pan, C., Liu, Z., Sha, X., Jiang, C., Li, L., Pan, M., Li, D., Han, X., & Ding, J. (2022). Chronic exposure to polystyrene microplastics induced male reproductive toxicity and decreased testosterone levels via the LH-mediated LHR/cAMP/PKA/StAR pathway. *Particle and Fibre Toxicology, 19*(1), 13. https://doi.org/10.1186/s12989-022-00453-2

Kacprzak, S., & Tijing, L. D. (2022). Microplastics in indoor environment: Sources, mitigation and fate. *Journal of Environmental Chemical Engineering, 10*(2), 107359. https://doi.org/10.1016/j.jece.2022.107359

Kalčíková, G., & Žgajnar Gotvajn, A. (2019). Plastic Pollution in Slovenia: From Plastic Waste Management to Research on Microplastics. In F. Stock, G. Reifferscheid, N. Brennholt, & E. Kostianaia (Eds.), *Plastics in the Aquatic Environment—Part I* (Vol. 111, pp. 307–322). Springer International Publishing. https://doi.org/10.1007/698_2019_402

Khalid, M., & Abdollahi, M. (2021). Environmental distribution of personal care products and their effects on human health. *Iranian Journal of Pharmaceutical Research, 20*(1). https://doi.org/10.22037/ijpr.2021.114891.15088

Kibria, Md. G., Masuk, N. I., Safayet, R., Nguyen, H. Q., & Mourshed, M. (2023). Plastic waste: Challenges and opportunities to mitigate pollution and effective management. *International Journal of Environmental Research, 17*(1), 20. https://doi.org/10.1007/s41742-023-00507-z

Koelmans, A. A., Mohamed Nor, N. H., Hermsen, E., Kooi, M., Mintenig, S. M., & De France, J. (2019). Microplastics in freshwaters and drinking water: Critical review and assessment of data quality. *Water Research, 155*, 410–422. https://doi.org/10.1016/j.watres.2019.02.054

Kumar, R., Verma, A., Shome, A., Sinha, R., Sinha, S., Jha, P. K., Kumar, R., Kumar, P., Shubham, Das, S., Sharma, P., & Vara Prasad, P. V. (2021). Impacts of plastic pollution on ecosystem services, sustainable development goals, and need to focus on circular economy and policy interventions. *Sustainability*, *13*(17), 9963. https://doi.org/10.3390/su13179963

Lamichhane, G., Acharya, A., Marahatha, R., Modi, B., Paudel, R., Adhikari, A., Raut, B. K., Aryal, S., & Parajuli, N. (2023). Microplastics in the environment: Global concern, challenges, and control measures. *International Journal of Environmental Science and Technology*, *20*(4), 4673–4694. https://doi.org/10.1007/s13762-022-04261-1

Leal Filho, W., Saari, U., Fedoruk, M., Iital, A., Moora, H., Klöga, M., & Voronova, V. (2019). An overview of the problems posed by plastic products and the role of extended producer responsibility in Europe. *Journal of Cleaner Production*, *214*, 550–558. https://doi.org/10.1016/j.jclepro.2018.12.256

Lebreton, L. C. M., Van Der Zwet, J., Damsteeg, J.-W., Slat, B., Andrady, A., & Reisser, J. (2017). River plastic emissions to the world's oceans. *Nature Communications*, *8*(1), 15611. https://doi.org/10.1038/ncomms15611

Li, Y. (2022). Legislation and policy on pollution prevention and the control of marine microplastics. *Water*, *14*(18), 2790. https://doi.org/10.3390/w14182790

Lincoln, S., Andrews, B., Birchenough, S. N. R., Chowdhury, P., Engelhard, G. H., Harrod, O., Pinnegar, J. K., & Townhill, B. L. (2022). Marine litter and climate change: Inextricably connected threats to the world's oceans. *Science of The Total Environment*, *837*, 155709. https://doi.org/10.1016/j.scitotenv.2022.155709

Lofty, J., Muhawenimana, V., Wilson, C. A. M. E., & Ouro, P. (2022). Microplastics removal from a primary settler tank in a wastewater treatment plant and estimations of contamination onto European agricultural land via sewage sludge recycling. *Environmental Pollution*, *304*, 119198. https://doi.org/10.1016/j.envpol.2022.119198

Lu, K., Zhan, D., Fang, Y., Li, L., Chen, G., Chen, S., & Wang, L. (2022). Microplastics, potential threat to patients with lung diseases. *Frontiers in Toxicology*, *4*, 958414. https://doi.org/10.3389/ftox.2022.958414

Lujan-Vega, C., Ortega-Alfaro, J. L., Cossaboon, J., Acuña, S., & Teh, S. J. (2021). How are microplastics invading the world? *Frontiers for Young Minds*, *9*, 606974. https://doi.org/10.3389/frym.2021.606974

Lwanga, E. H., Beriot, N., Corradini, F., Silva, V., Yang, X., Baartman, J., Rezaei, M., Van Schaik, L., Riksen, M., & Geissen, V. (2022). Review of microplastic sources, transport pathways and correlations with other soil stressors: A journey from agricultural sites into the environment. *Chemical and Biological Technologies in Agriculture*, *9*(1), 20. https://doi.org/10.1186/s40538-021-00278-9

Mammo, F. K., Amoah, I. D., Gani, K. M., Pillay, L., Ratha, S. K., Bux, F., & Kumari, S. (2020). Microplastics in the environment: Interactions with microbes and chemical contaminants. *Science of The Total Environment*, *743*, 140518. https://doi.org/10.1016/j.scitotenv.2020.140518

Manisalidis, I., Stavropoulou, E., Stavropoulos, A., & Bezirtzoglou, E. (2020). Environmental and health impacts of air pollution: A review. *Frontiers in Public Health*, *8*, 14. https://doi.org/10.3389/fpubh.2020.00014

Martinho, S. D., Fernandes, V. C., Figueiredo, S. A., & Delerue-Matos, C. (2022). Microplastic pollution focused on sources, distribution, contaminant interactions, analytical methods, and wastewater removal strategies: A review. *International Journal of Environmental Research and Public Health*, *19*(9), 5610. https://doi.org/10.3390/ijerph19095610

Moyo, S. (2022). An enigma: A meta-analysis reveals the effect of ubiquitous microplastics on different taxa in aquatic systems. *Frontiers in Environmental Science*, *10*, 999349. https://doi.org/10.3389/fenvs.2022.999349

Nanthini Devi, K., Raju, P., Santhanam, P., & Perumal, P. (2022). Impacts of microplastics on marine organisms: Present perspectives and the way forward. *Egyptian Journal of Aquatic Research*, *48*(3), 205–209. https://doi.org/10.1016/j.ejar.2022.03.001

Napper, I. E., & Thompson, R. C. (2020). Plastic Debris in the marine environment: History and future challenges. *Global Challenges*, *4*(6), 1900081. https://doi.org/10.1002/gch2.201900081

Ng, C. H., Mistoh, M. A., Teo, S. H., Galassi, A., Ibrahim, A., Sipaut, C. S., Foo, J., Seay, J., Taufiq-Yap, Y. H., & Janaun, J. (2023). Plastic waste and microplastic issues in Southeast Asia. *Frontiers in Environmental Science*, *11*, 1142071. https://doi.org/10.3389/fenvs.2023.1142071

Nirmala, K., Rangasamy, G., Ramya, M., Shankar, V. U., & Rajesh, G. (2023). A critical review on recent research progress on microplastic pollutants in drinking water. *Environmental Research*, *222*, 115312. https://doi.org/10.1016/j.envres.2023.115312

Osman, A. I., Hosny, M., Eltaweil, A. S., Omar, S., Elgarahy, A. M., Farghali, M., Yap, P.-S., Wu, Y.-S., Nagandran, S., Batumalaie, K., Gopinath, S. C. B., John, O. D., Sekar, M., Saikia, T., Karunanithi,

P., Hatta, M. H. M., & Akinyede, K. A. (2023). Microplastic sources, formation, toxicity and remediation: A review. *Environmental Chemistry Letters*, *21*(4), 2129–2169. https://doi.org/10.1007/s10311-023-01593-3

Park, H., & Park, B. (2021). Review of microplastic distribution, Toxicity, analysis methods, and removal technologies. *Water*, *13*(19), 2736. https://doi.org/10.3390/w13192736

Periyasamy, A. P. (2023). Environmentally friendly approach to the reduction of microplastics during domestic washing: Prospects for machine vision in microplastics reduction. *Toxics*, *11*(7), 575. https://doi.org/10.3390/toxics11070575

Pinheiro, C., Oliveira, U., & Vieira, M. (2017). Occurrence and impacts of microplastics in freshwater fish. *Journal of Aquaculture & Marine Biology*, *5*(6). https://doi.org/10.15406/jamb.2017.05.00138

Piñon-Colin, T. de J., Rodriguez-Jimenez, R., Rogel-Hernandez, E., Alvarez-Andrade, A., & Wakida, F. T. (2020). Microplastics in stormwater runoff in a semiarid region, Tijuana, Mexico. *Science of the Total Environment*, *704*, 135411. https://doi.org/10.1016/j.scitotenv.2019.135411

Pironti, C., Ricciardi, M., Motta, O., Miele, Y., Proto, A., & Montano, L. (2021). Microplastics in the environment: Intake through the food web, human exposure and toxicological effects. *Toxics*, *9*(9), 224. https://doi.org/10.3390/toxics9090224

Prapanchan, V. N., Kumar, E., Subramani, T., Sathya, U., & Li, P. (2023). A global perspective on microplastic occurrence in sediments and water with a special focus on sources, analytical techniques, health risks, and remediation technologies. *Water*, *15*(11), 1987. https://doi.org/10.3390/w15111987

Prata, J. C. (2018). Airborne microplastics: Consequences to human health? *Environmental Pollution*, *234*, 115–126. https://doi.org/10.1016/j.envpol.2017.11.043

Prata, J. C., Silva, A. L. P., Da Costa, J. P., Mouneyrac, C., Walker, T. R., Duarte, A. C., & Rocha-Santos, T. (2019). Solutions and integrated strategies for the control and mitigation of plastic and microplastic pollution. *International Journal of Environmental Research and Public Health*, *16*(13), 2411. https://doi.org/10.3390/ijerph16132411

Raddadi, N., & Fava, F. (2019). Biodegradation of oil-based plastics in the environment: Existing knowledge and needs of research and innovation. *Science of The Total Environment*, *679*, 148–158. https://doi.org/10.1016/j.scitotenv.2019.04.419

Ray, S. S., Lee, H. K., Huyen, D. T. T., Chen, S.-S., & Kwon, Y.-N. (2022). Microplastics waste in environment: A perspective on recycling issues from PPE kits and face masks during the COVID-19 pandemic. *Environmental Technology & Innovation*, *26*, 102290. https://doi.org/10.1016/j.eti.2022.102290

Sajjad, M., Huang, Q., Khan, S., Khan, M. A., Liu, Y., Wang, J., Lian, F., Wang, Q., & Guo, G. (2022). Microplastics in the soil environment: A critical review. *Environmental Technology & Innovation*, *27*, 102408. https://doi.org/10.1016/j.eti.2022.102408

Sarijan, S., Azman, S., Said, M. I. M., & Jamal, M. H. (2021). Microplastics in freshwater ecosystems: A recent review of occurrence, analysis, potential impacts, and research needs. *Environmental Science and Pollution Research*, *28*(2), 1341–1356. https://doi.org/10.1007/s11356-020-11171-7

Sarkar, B., Dissanayake, P. D., Bolan, N. S., Dar, J. Y., Kumar, M., Haque, M. N., Mukhopadhyay, R., Ramanayaka, S., Biswas, J. K., Tsang, D. C. W., Rinklebe, J., & Ok, Y. S. (2022). Challenges and opportunities in sustainable management of microplastics and nanoplastics in the environment. *Environmental Research*, *207*, 112179. https://doi.org/10.1016/j.envres.2021.112179

Schmaltz, E., Melvin, E. C., Diana, Z., Gunady, E. F., Rittschof, D., Somarelli, J. A., Virdin, J., & Dunphy-Daly, M. M. (2020). Plastic pollution solutions: Emerging technologies to prevent and collect marine plastic pollution. *Environment International*, *144*, 106067. https://doi.org/10.1016/j.envint.2020.106067

Schuhen, K., & Sturm, M. T. (2021). Microplastic Pollution and Reduction Strategies. In T. Rocha-Santos, M. Costa, & C. Mouneyrac (Eds.), *Handbook of Microplastics in the Environment* (pp. 1–33). Springer International Publishing. https://doi.org/10.1007/978-3-030-10618-8_53-2

Setälä, O., Norkko, J., & Lehtiniemi, M. (2016). Feeding type affects microplastic ingestion in a coastal invertebrate community. *Marine Pollution Bulletin*, *102*(1), 95–101. https://doi.org/10.1016/j.marpolbul.2015.11.053

Shen, M., Song, B., Zhou, C., Almatrafi, E., Hu, T., Zeng, G., & Zhang, Y. (2022). Recent advances in impacts of microplastics on nitrogen cycling in the environment: A review. *Science of The Total Environment*, *815*, 152740. https://doi.org/10.1016/j.scitotenv.2021.152740

Smith, M., Love, D. C., Rochman, C. M., & Neff, R. A. (2018). Microplastics in seafood and the implications for human health. *Current Environmental Health Reports*, *5*(3), 375–386. https://doi.org/10.1007/s40 572-018-0206-z

Souza Machado, A. A., Kloas, W., Zarfl, C., Hempel, S., & Rillig, M. C. (2018). Microplastics as an emerging threat to terrestrial ecosystems. *Global Change Biology*, *24*(4), 1405–1416. https://doi.org/10.1111/gcb.14020

Stoll, T., Stoett, P., Vince, J., & Hardesty, B. D. (2022). Governance and Measures for the Prevention of Marine Debris. In T. Rocha-Santos, M. F. Costa, & C. Mouneyrac (Eds.), *Handbook of Microplastics in the Environment* (pp. 1129–1151). Springer International Publishing. https://doi.org/10.1007/978-3-030-39041-9_26

Takarina, N., Purwiyanto, A., Rasud, A., Arifin, A., & Suteja, Y. (2022). Microplastic abundance and distribution in surface water and sediment collected from the coastal area. *Global Journal of Environmental Science and Management*, *8*(2). https://doi.org/10.22034/GJESM.2022.02.03

Talbot, R., & Chang, H. (2022). Microplastics in freshwater: A global review of factors affecting spatial and temporal variations. *Environmental Pollution*, *292*, 118393. https://doi.org/10.1016/j.envpol.2021.118393

Thompson, R. C., Moore, C. J., vom Saal, F. S., & Swan, S. H. (2009). Plastics, the environment and human health: Current consensus and future trends. *Philosophical Transactions of the Royal Society of London. Series B, Biological Sciences*, *364*(1526), 2153–2166. https://doi.org/10.1098/rstb.2009.0053

Torres-Agullo, A., Karanasiou, A., Moreno, T., & Lacorte, S. (2021). Overview on the occurrence of microplastics in air and implications from the use of face masks during the COVID-19 pandemic. *Science of the Total Environment*, *800*, 149555. https://doi.org/10.1016/j.scitotenv.2021.149555

Uddin, M. A., Afroj, S., Hasan, T., Carr, C., Novoselov, K. S., & Karim, N. (2022). Environmental impacts of personal protective clothing used to combat COVID-19. *Advanced Sustainable Systems*, *6*(1), 2100176. https://doi.org/10.1002/adsu.202100176

Usman, S., Abdull Razis, A. F., Shaari, K., Azmai, M. N. A., Saad, M. Z., Mat Isa, N., & Nazarudin, M. F. (2022). The burden of microplastics pollution and contending policies and regulations. *International Journal of Environmental Research and Public Health*, *19*(11), 6773. https://doi.org/10.3390/ijerph1 9116773

Verla, A. W., Enyoh, C. E., Verla, E. N., & Nwarnorh, K. O. (2019). Microplastic–toxic chemical interaction: A review study on quantified levels, mechanism and implication. *SN Applied Sciences*, *1*(11), 1400. https://doi.org/10.1007/s42452-019-1352-0

Wojnowska-Baryła, I., Bernat, K., & Zaborowska, M. (2022). Plastic waste degradation in landfill conditions: The problem with microplastics, and their direct and indirect environmental effects. *International Journal of Environmental Research and Public Health*, *19*(20), 13223. https://doi.org/10.3390/ijerph192013223

Yang, W., Jannatun, N., Zeng, Y., Liu, T., Zhang, G., Chen, C., & Li, Y. (2022). Impacts of microplastics on immunity. *Frontiers in Toxicology*, *4*, 956885. https://doi.org/10.3389/ftox.2022.956885

Yee, M. S.-L., Hii, L.-W., Looi, C. K., Lim, W.-M., Wong, S.-F., Kok, Y.-Y., Tan, B.-K., Wong, C.-Y., & Leong, C.-O. (2021). Impact of microplastics and nanoplastics on human health. *Nanomaterials*, *11*(2), 496. https://doi.org/10.3390/nano11020496

Yu, A. J. G., Yap-Dejeto, L. G., Parilla, R. B., & Elizaga, N. B. (2023). Microplastics in Perna viridis and Venerupis species: Assessment and impacts of plastic pollution. *International Journal of Environmental Science and Technology*. https://doi.org/10.1007/s13762-023-04982-x

Yuan, Z., Nag, R., & Cummins, E. (2022a). Human health concerns regarding microplastics in the aquatic environment—From marine to food systems. *Science of The Total Environment*, *823*, 153730. https://doi.org/10.1016/j.scitotenv.2022.153730

Yuan, Z., Nag, R., & Cummins, E. (2022b). Ranking of potential hazards from microplastics polymers in the marine environment. *Journal of Hazardous Materials*, *429*, 128399. https://doi.org/10.1016/j.jhaz mat.2022.128399

Yukioka, S., Tanaka, S., Nabetani, Y., Suzuki, Y., Ushijima, T., Fujii, S., Takada, H., Van Tran, Q., & Singh, S. (2020). Occurrence and characteristics of microplastics in surface road dust in Kusatsu (Japan), Da Nang (Vietnam), and Kathmandu (Nepal). *Environmental Pollution*, *256*, 113447. https://doi.org/10.1016/j.envpol.2019.113447

Zhang, X., Li, Y., Ouyang, D., Lei, J., Tan, Q., Xie, L., Li, Z., Liu, T., Xiao, Y., Farooq, T. H., Wu, X., Chen, L., & Yan, W. (2021). Systematical review of interactions between microplastics and microorganisms in the soil environment. *Journal of Hazardous Materials, 418,* 126288. https://doi.org/10.1016/j.jhazmat.2021.126288

Zhang, Y., Kang, S., Allen, S., Allen, D., Gao, T., & Sillanpää, M. (2020). Atmospheric microplastics: A review on current status and perspectives. *Earth-Science Reviews, 203,* 103118. https://doi.org/10.1016/j.earscirev.2020.103118

Zheng, X., Sun, R., Dai, Z., He, L., & Li, C. (2023). Distribution and risk assessment of microplastics in typical ecosystems in the South China Sea. *Science of The Total Environment, 883,* 163678. https://doi.org/10.1016/j.scitotenv.2023.163678

Zolotova, N., Kosyreva, A., Dzhalilova, D., Fokichev, N., & Makarova, O. (2022). Harmful effects of the microplastic pollution on animal health: A literature review. *PeerJ, 10,* e13503. https://doi.org/10.7717/peerj.13503

13 Uptake, Accumulation, and Ecotoxicological Impacts of Microplastic on Plant Production and Soil Ecosystem

Sangita Yadav, Navish Kataria, Jun Wei Roy Chong, Pawan Kumar Rose, Seema Joshi, Pau Loke Show, and Kuan Shiong Khoo

13.1 INTRODUCTION

The daily usage of plastic is can be seen in almost every daily application; as a result, tonnes of plastic are generated annually. In 2019, the world produced 359 million metric tonnes of plastic, and by the year 2050, it is predicted that the world will produce up to 33 billion tonnes of plastic annually (Wu et al., 2020). Plastic debris has become a threat to both terrestrial and aquatic ecosystems due to its widespread usage and manufacture, especially micro- (less than 5 mm) and nano- (less than 100 nm) plastics (Rose et al., 2023a). Plastics are manufactured on land and subsequently transported to aquatic and terrestrial ecosystems. According to estimates, terrestrial ecosystems are exposed to ~4–23% more microplastics than marine ecosystems (Rose et al., 2023b). Household and industry waste account for almost 30% of all plastic waste that is discarded into the environment. The application of plastic mulching films, where residuals often end up in soils due to their resilience to degradation and the high cost of removal and recycling, is the main source of large plastic particles in agricultural fields. On the other hand, applying sludge, sewage plant wastewater, and domestic compost permits M/NPs produced from commercial abrasives, cosmetics, and textile fibres to infiltrate soils (Yadav et al., 2023).

At first, authorities and policymakers primarily concentrated on the management and disposal of large-sized plastic products, ignoring the impact of environmental deterioration. This process results in the development of smaller plastic particles known as microplastics, which are divided into four groups based on their size: MPs (<0.5 cm), mesoplastics (0.5–5 cm), macroplastics (5–50 cm), and megaplastics (>50 cm). MPs are finally less than 5 mm of plastic (Prakash et al., 2020). There are two types of MPs, according to the source: primary and secondary MPs. Primary MPs are manufactured plastic particles less than 5 mm in size, which are often present in clothing, pharmaceuticals, and personal care items. The term "secondary MPs" refers to plastic particles which are formed from the larger original plastics as a result of various physical, chemical, and biological processes and have a particle size of less than 5 mm (Jiang et al., 2019). Microplastic pollution has the potential to result in unforeseen ecological disasters because of its widespread dispersion and high abundance. Microplastics have been identified as an emerging issue because they change the physicochemical properties of soil. Organisms easily ingest them and can serve as a substratum for harmful organism consortia, raising concerns about possible adverse consequences on biodiversity and ecosystem function (Rillig & Lehmann, 2020). Microplastics are a type of global contaminant that are extremely persistent and can remain in the environment for decades. They can also intricately interact with the abiotic environment, have an impact on terrestrial animals either directly or indirectly, and combine with other pollutants to hasten the transport of those pollutants and have properties and behaviours that make them more likely to cause ecological shocks within terrestrial environments (Rillig & Lehmann, 2020).

Research on microplastics in the terrestrial environment is still in its early phases, despite a wealth of evidence pointing to terrestrial systems as the main sinks of microplastics. Furthermore, the potential harm to the biota have been investigated and ecotoxicological techniques have been used to find probable toxicity mechanisms. The long-term consequences of plastic materials on soil health and crop production are receiving more attention despite their short-term advantages (Dissanayake et al., 2022). Through complex interactions, the presence of plastics can have various detrimental consequences on soil organisms. Numerous soil invertebrates, such as earthworms, nematodes, isopods, collembolas, and snails, can be impacted by plastic pollution as well as due to the release of associated chemicals. MP-induced entanglement, asphyxia, and swallowing of organisms are well-known issues. Additionally, due to their small sizes and difficulties degrading naturally, MPs have a high likelihood of being consumed or inhaled by a variety of species, which will adversely affect them. Even worse, MPs have a strong propensity for other contaminants. For instance, MPs can be a vector for transferring contaminants in different ecosystems by adsorbing heavy metals and hydrophobic organic compounds on the plastic surface (Wu et al., 2020). Exposing microplastics to the biota results in oxidative stress, reproductive impairment, increased mortality, detrimental effects on growth and motility, and behavioural abnormalities in terrestrial species (such as earthworms, snails, and plants) (Huerta Lwanga et al., 2017). Plastic particles, particularly M/NPs, can accumulate on root surfaces and enter plants, where they can have a significant negative impact on plants, including germination and growth inhibition, alternation of antioxidant stress, and reduced rate of photosynthesis. Recent research that suggests microplastics may affect soil carbon and possibly nitrogen cycling is particularly alarming. The biophysical characteristics of soil, like soil aggregation, bulk density, and water-holding capacity of the soil, can be affected by MPs (Rillig et al., 2019). Consequently, the purpose of this chapter is to clarify the ecotoxicological effects of MPs on the soil ecosystem and plant production, to identify knowledge gaps in existing MP studies, and to offer pertinent suggestions for future research.

13.2 BIOACCUMULATION OF MICRO-/NANOPLASTICS

One of the most dramatic and pervasive changes to our planet's surface in recent years has been the buildup and fragmentation of plastics. Since the mass manufacture of plastic goods began in the 1950s, plastic garbage has quickly accumulated in terrestrial environments, the open ocean, on the shorelines of even the most remote islands, and in the deep water. This accumulation has taken place in just a few decades. M/NPs, which are new persistent pollutants, have been recognised as a significant environmental problem of growing interest. Different types of M/NPs exist in the soil, and their bioavailability is affected by soil factors such as particle size, particle density, abundance, co-occurrence, chemical properties, and specific receptor characteristics (plant or organism). MPs are directly ingested by a variety of organisms, and their bioavailability increases with their size. Planktivores can consume MPs directly as natural prey during normal feeding behaviour since they have a size fraction that is comparable to that of sediments and planktonic organisms. The chemical properties of the soil, including pH, the sorption and adsorption of heavy metals, and their redox potential, have a significant impact on the bioavailability of plastic in the soil (Azeem et al., 2021).

NPs have the ability to infiltrate plants and cause stress. Jiang et al. (2019) discovered that broad bean roots can deposit nanoplastics polystyrene fluorescence (100 nm) and subsequently plug cell wall pores, preventing the passage of nutrients. Additionally, rubber ash NPs appear to gather in the root cells of cucumber (Moghaddasi et al., 2015). Therefore, we can presume that crop productivity is seriously threatened by soil NP pollution, and the safety of organisms, including humans, is also impacted due to the use of these crops. Sun et al. (2020) directly demonstrated that plants may gather 100-nm NPs based on their surface charge. Another investigation revealed that cucumber plants' roots could acquire 100–700-nm polystyrene nanoplastics in a size-dependent manner (Li et al., 2021). In samples of plant tissue, polyethylene was found in three different diameters (0.2–1.2 mm, 0.05–1.5 mm, and 0.4–1.5 mm). Therefore, it is reasonable to assume that amendments/compost from waste are unsustainable. There is an unregulated buildup of plastic particles in soils and plants due to the use of additives that are likely to diminish soil productivity and increase food toxicity, respectively (Iqbal et al., 2021).

Microplastics' potential to contaminate food is a growing concern. Fruits and vegetables that have been exposed to microplastic are currently available worldwide. Recently, numerous fruits and vegetables in Catania's neighbourhood markets tested positive for plastic (Conti et al., 2020). Among the fruits and vegetables that were analysed, lettuce seemed to be the least infected, while apples and carrots appeared to be heavily contaminated. The tiniest microplastics were discovered in carrots (1.51 μm), whereas the largest plastic particles were found in lettuce (2.52 μm). Therefore, it is now necessary for us to set new standards for food security. This requires developing a precise system for food testing (Iqbal et al., 2021).

13.3 MICROPLASTIC-ASSOCIATED ALTERATIONS IN SOIL FUNCTIONS

Numerous ecosystem functions are supported by soil. Microplastics can endanger soil ecosystems in several ways, including biogeochemically, ecologically, and in terms of biodiversity due to their prolonged residence time. Microplastic buildup in soil profiles may impact the soil's physiochemical and biological functions. Microplastics considerably impact soil water repellency, water-holding capacity, bulk density, and porosity (Iqbal et al., 2021). Microplastics can also interact with the soil microbiome and change its microbial community and nutrient dynamics because of their small size and large surface area (Meng et al., 2022). Table 13.1 depicts how MPs have affected the soil ecosystem.

13.3.1 SOIL STRUCTURE AND WATER DYNAMICS

A soil structure is created in the soil by the accumulation and arrangement of solids made from organic compounds, mineral ions, and pores. The health of the soil can be determined using these aggregates. Well-managed soil in agricultural cultivation areas should have a continuous pore space network for free flow of air and water and root development (Chia et al., 2022). External factors, such as bioturbation and human activities, can quickly integrate microplastics that have been deposited on soil surfaces into the soil matrix. The texture and structure of the soil may alter due to this assimilation. Commercial polymers typically have a lower density than characteristic soil particles, which makes it possible to visually detect changes in soil structure and composition in heavily polluted soils. Soil environmental characteristics, as well as microplastic type, size, and shape, have been discovered to be factors in this effect (Dissanayake et al., 2022). MPs can modify soil parameters by altering their physical structures because of their unique characteristics. First, due to their affinity for the soil's organic and mineral components, MPs may change how soil aggregates. However, depending on the size, kind, and shape of the plastic particles, the impact is predicted to vary. Second, because plastics are typically less dense than a number of soil minerals, MPs can reduce

TABLE 13.1
Effects of Microplastics on Soil Functions

Type of Soil	Plastics Type and Size	MP Exposure Concentration	Impacts	References
Agricultural soil	PP, PVC	0.1%, 0.3%, and 1.0%, w/w	To some extent, PP and PVC increased soil enzyme activity, total organic carbon (TOC) content, and cation exchange capacity (CEC), showing that they can increase soil fertility	(Li et al., 2023)
Silt loam soil	PE; 25, 150, 550, and 1000 μm	1, 3, and 5% w/w	The saturated hydraulic conductivity and water-holding capacity of soil were shown to be dramatically reduced by MPs, along with effects on bulk density, water content, and soil particle composition	(Jing et al., 2023)
Saline-alkali soil	PE, PP; <5 mm	0.1, 1, and 5% w/w (in culture media)	Although MP treatment considerably improved the enzyme activity and useful nutrient content of the soil, it significantly decreased the electrical conductivity	(Yuan et al., 2023)
Agricultural soil	PE, PS, PP; 150 μm	0.01–20%, w/w (in culture media)	The MP concentrations caused an increase in the taxonomic diversity of *Patescibacteria*	(Sun et al., 2022)
Agricultural soils	PP; 0, 200, and 500 μm	0.5%, 1%, 2%, 4%, and 6% soil dry weight	The saturated hydraulic conductivity and water retention capacity were decreased by the addition of microplastic	(Guo et al., 2022a)
Agricultural soils	LDPE, bio-MPs; < 53–1000 μm	0.5%, 1.0%, 1.5%, 2.0%, and 2.5% (w/w, weight ratio of microplastics to air-dry soil)	Compared to LDPE, bio-MPs had a greater impact on soil carbon and nitrogen cycling dynamics	(Meng et al., 2022)
Farmland soil	PE; <2 mm	1% and 5% w/w of soil weight	*Arthrobacter* and *norank_f_Gemmatimonadaceae* were considerably altered by 1% (w/w) PE microplastics, whereas *Arthrobacter*, *Nocardia*, *Bacillus*, and *Blastococcus* were dramatically reduced by 5% (w/w) PE microplastics	(Ya et al., 2022)
Yellow-brown agricultural soils	LDPE, PS, PA, PU; 100–350 μm	0.5%, 1%, 2%, and 4% (w/w)	Alterations in the properties of yellow-brown soil and the impacts of MPs could lead to changes in the chemical speciation of heavy metals	(Wen et al., 2022)
Fluvo-aquic and latosol	LDPE; < 1 mm	10%, w/w	Microplastics decreased the mineralisation and breakdown of soil organic carbon (SOC) in the presence of straw, which led to an 18.9% drop in the amount of SOC that was microbially accessible	(Yu et al., 2021)
Soil	LDPE; 150–250 μm	2% and 7% w/w	Both 2% and 7% LDPE MPs have an effect on the bacterial network structure in soil and change functional groups engaged in soil nitrogen cycling processing	(Rong et al., 2021)

Abbreviations: Polyethylene (PE), polystyrene (PS), polypropylene (PP), low-density polyethylene (LDPE), biodegradable (Bio-MPs), polyamide (PA), polyurethane (PU), polyvinyl chloride (PVC).

the bulk density of soil and increase soil aeration, both of which may promote root penetration (Zhou et al., 2021a). According to Chen et al. (2022), MPs can interact with different soil elements, affecting the amount of dissolved organic carbon (DOC), aggregate formation, and contaminant buildup in soils. The inclusion of MPs decreased aggregate stability and macro-aggregate formation. De Souza Machado et al. (2018) showed that larger fragments (160–1200 μm) appeared to be bound only loosely within re-aggregated soil samples when 0.5–20 g of juvenile plastic per kg of dry soil was applied, whereas microbeads (15–20 μm) and fibres were more integrated into rebuilding macroaggregate structure. These well-defined occlusion dynamics have been accompanied by an inconsistent pattern of increased water-holding capacity (WHC) due to decompaction, decreased bulk density, and less water-stable aggregates (WSA). In summary, the majority of evidence points to the detrimental effects of soil microplastic on WSA and supports the hypothesis that seriously contaminated soils with microplastic could result in soil structure loss and increased erodibility (Büks, & Kaupenjohann, 2021).

MPs also affect the soil's porosity, and maintaining soil structure depends on controlling the distribution of pore size, which affects hydraulic conductivity and water retention (Wan et al., 2019). According to the linkage between soil bulk density and soil porosity, a rise in bulk density is associated with a decrease in macropores and an increase in meso- and micropores. Bulk density is another crucial indicator of soil health that promotes the passage of water and other substances, enhances the structure and aeration of the soil, and also controls soil compaction. The hydrological and agroecosystem processes of soil are closely related to soil bulk density (Chia et al., 2022). Overall bulk density is decreased by the addition of the less dense polymers, but there is no evident connection to WSA or WHC. An increase in mesopore space may be the reason for the greater WHC in some samples (Büks & Kaupenjohann, 2021). Polyester fibres (8 μm) improved water-holding capacity in terms of soil water dynamics, possibly keeping soils saturated for an extended time period (de Souza Machado et al., 2019), whereas 2-mm polyethene coatings increased soil water loss by causing more evaporation (Wan et al., 2019). The former was caused by fibres' capacity to group soil particles into clumps and entangle them at smaller spatial dimensions. Larger MPs, however, could have a deleterious impact on soil water retention and cause anoxia (Liu et al., 2014). Drought, which is anticipated to worsen owing to climate warming over the next few decades, could be mitigated or made worse by changes in soil water content (Lozano & Rillig, 2020). Overall data suggest that environmental factors such as vegetation, soil microbiota, and soil texture, together with microplastic type, shape, and concentration, impact the magnitude of WSA loss (Büks, & Kaupenjohann, 2021).

13.3.2 Impact of MPs on Nutrient/biogeochemical Cycling and GHG Emissions

In soils, microplastics can accumulate and significantly impact the biogeochemical cycle. The term "biogeochemical cycle" describes the cyclical exchange of vital chemical components between living organisms and their surroundings, which aids in synthesising and decomposing organic matter. The interaction between MPs and microbes in the natural ecosystem has substantially grown as a result of the widespread use of plastic products. Therefore, microbes will likely receive carbon sources, substrates, or cosubstrates from plastics and the intermediates of their decomposition. Trophic transmission involves the migration of microorganisms, MPs, and their degradation intermediates to produce biotic or abiotic processes within biogeochemical cycles. As a result, oxidation-reduction reactions are crucial to the destiny of MPs in the natural ecosystem (Wang et al., 2021a). In the trophic pyramid, microorganisms are essential because they enable saprotrophy, or the recycling of decaying organic matter, which releases and reuses chemical elements, primarily C, N, and P. As a result, MPs impact microbial oxidation-reduction reactions, which are a fundamental problem for the balance of the environment (de Almeida et al., 2023). The abundance, diversity, and population of microbes may change in the presence of MPs, which could affect how biogeochemical cycles are transformed. Increased microbial proliferation and community due to excess nutrients (carbon and nitrogen) affect the nutrient cycle in terrestrial soil (Kumar et al., 2022).

Microbes take part in a number of the most crucial carbon cycle processes, including methane metabolism, carbon fixation, and carbon degradation (the breakdown of organic matter). By secreting different enzymes to break down animal and plant waste, microbes can hasten the transformation and migration of carbon and are crucial in the process of carbon fixation (Shen et al., 2023). The CO_2 outflow rate was shown to be decreased by MPs on day 3 and then increased on day 15 of incubation, according to Xiao et al. (2022). The elevated hydrolase enzyme activities in MP-amended soil may be responsible for the increased CO_2 release. According to Cluzard et al. (2015), the addition of PE microbeads considerably raised the ammonium concentration and changed N-cycling, perhaps causing eutrophication. It was discovered that plastic particles could alter nitrification and denitrification processes, which are two important mechanisms for eliminating excess reactive nitrogen from the environment (Seeley et al., 2020). According to Li et al. (2020d), varying ammonia concentrations can result from adding 1000 particles L^{-1} of different MP types. Polyethylene, polystyrene, and polyethersulfone hindered the reaction, but PP and PVC increased the ammonia oxidation rate. Although MPs may have a significant impact on the nitrogen cycle, research into this topic is still in the pioneer stages. With the assistance of living organisms, nitrogen moves through the biogeochemical cycle, and this cycle involves a number of intricate procedures. The colonised microbial communities can be shaped by the MP type, size, surface topography, kind, and bioavailability, which can also affect nitrogen circulation.

The Xanthobacteraceae, Isosphaeraceae, and Rhizobiaceae groups of microorganisms, which are in charge of OM breakdown and nitrogen cycling, are impacted by MPs as polystyrene-contaminated soil (Jiang et al., 2020). MPs impact nematode (*Caenorhabditis elegans*) communities and interfere with soil biogeochemical cycles (Schöpfer et al., 2020). Changes in nutritional ratios (C/N) are also observed because MPs are absent from the soil, affecting the entire plant community (Qi et al., 2020). MPs' modifications to the nutrient cycle have unpredicted effects on GHG emissions, such as those of terrestrial CO_2, CH_4, and N_2O (Kumar et al., 2022).

13.3.3 MICROBIAL COMMUNITIES AND ACTIVITIES

Around 79% of the plastic garbage produced worldwide is built up in terrestrial areas. This can alter the microbial diversity and the function and structure of soils. MPs could affect soil organisms both directly and indirectly. On some soil animals, MPs may stick to the surface, preventing them from moving around freely. The majority of the indirect effects of MPs are caused by their adsorption of other pollutants, like heavy metals and organic contaminants, which would worsen soil contamination and increase the risks to soil organisms (Hüffer et al., 2019). One of the reliable indicators of the health of the soil is its principal decomposers, which are microorganisms. However, there are few studies currently available on how MPs affect soil microorganisms. Microbial activity in the soil is intimately correlated with soil characteristics. The transfer of MPs in soils can alter the soil microbial diversity (Li et al., 2020a). As the foundation for food production and climate regulation, soil microbes are crucial for biogeochemical cycling. Therefore, by comprehending how soil microorganisms react to MPs, we can forecast probable ecosystem-level effects of MP pollution (Zhou et al., 2021a). MPs can change the way soil aggregates, which may impact how bacteria evolve. They may also offer soil microbes a fresh ecological environment. MPs may also promote the development of microbial biofilms, which are a novel microbial niche known as the "plastisphere" and have an effect on the microbiota and function of microbes. This microbial niche creates hotspots for microbial activity, which causes the biogeochemical cycling in soil to be heterogeneous (Xiao et al., 2021). For instance, polyethylene fragments can alter or produce a large number of taxa, including viruses (pathogens) and bacteria that degrade plastic (Huang et al., 2019). Plenty of oligotrophic bacteria rose due to the presence of biodegradable MPs (PHBV), while copiotrophic species experienced a decline (Zhou et al., 2021b). In conclusion, MPs offer unique microbial ecological niches that encourage the growth of particular microbial populations, which can have unexpected effects on ecosystem processes. Gram-negative bacteria replaced Gram-positive bacteria in a wheat–soil system after adding polyvinyl chloride and polyethylene granules (10%). They reduced the activity of xylosidase and β-glucosidase by 16% to 43% (Zhang et al., 2021). The ability of nanoparticles (NPs) to penetrate and aggregate in soil organic detritus has an adverse biological impact on microbes, while changing soil qualities may not be as important to NPs. PS fibres reduce water-stable aggregates and have a significant detrimental impact on the health of plant soil. Therefore, water-stable aggregates are also essential for microbial activity. The impacts of MPs and fungus on soil–plant systems, however, are still not known (de Souza Machado et al., 2020). The polymers that most effectively prevented microbial activity were polypropylene fragments and polyethylene films. A crucial MP variable explaining the variation in microbial activity is polymer type (Lozano et al., 2021). Various MPs have diverse sorption capabilities, which results in different microbial habitats depending on their form. Alterations in the surface-to-volume ratios of MP particles cause varied effects on microbial populations according to their size. The connection between MPs and microbial communities is still not well understood mechanistically, however. To more accurately forecast the ecological effects of MPs in soils, this essential knowledge gap must be filled.

The fraction of microbial communities that MPs and NPs choose to inhabit can fluctuate as MP and NP concentrations in soil increase. Therefore, when mixed with other naturally occurring compounds and in conjunction with the plastisphere, freshly introduced MPs have an effect on environmental ecological functions. The relationship between soil and plants, the growth of fauna, and the recycling of nutrients may all be significantly impacted by the interaction between plastic and microbes, as previously discussed. Otherwise, a hotspot for microbial growth on the surface of MPs is recommended. Therefore, comprehension of M/NP environmental behaviours is required to better comprehend the response and function of microorganisms in the soil. It remains crucial for scientists to comprehend the mechanism of interaction between the microbial community and MPs in order to more properly evaluate the environmental impact of MPs on soil.

13.3.4 Impact on Soil Organic Matter (SOM)

The creation of "eco-corona" (EC) on plastic fragments or "biogenic aggregates" in the natural environment has generated new debates, challenges, and research opportunities for scientists about the interaction of M/NPs with natural organic matter (NOM). Allochthonous, autochthonous, and anthropogenic NOM can combine with M/NPs to create an EC coating or aggregation, which can alter their mobility, stability, bioreactivity, settling, and fate in the environment in addition to their physicochemical features (Ali et al., 2022). The soil bacteria and fauna depend on soil organic matter for a suitable habitat and food sources, which is essential to the long-term health of the soil. To date, however, there has not been much focus on how microplastics alter the pool of soil organic matter (SOM) (Meng et al., 2022). Microplastics can contribute to developing SOM by breaking large C materials. Additionally, under well-watered environments, microfibres have increased litter decomposition. An essential component of SOM is the pool of dissolved organic carbon (DOC). High levels of polypropylene exposure appeared to degrade DOC through increasing phenol-oxidase activity. Polyethylene, as opposed to DOC, seemed to have an impact on the relative functional groups of C. Reduced SOM in the rhizosphere was caused by polystyrene and polytetrafluorethylene. As a result, SOM turnover will change as a result of MP pollution (Iqbal et al., 2021).

The decomposition of organic matter by soil microorganisms controls the amount of organic matter in the soil (Xiao et al., 2022). As a result, after the soil is contaminated with microplastics, the number of soil microorganisms will decline, which will reduce the amount of organic matter that may be decomposed by microorganisms. For instance, microplastic pollution reduces the population of soil bacteria like *Chloroflexi* that fix carbon dioxide and organic materials (Liu et al., 2021). Additionally, Xiao et al. (2022) claim that the decline in soil organic matter following microplastic contamination is caused by modifications in microbial community succession and temporal turnover, which promote soil organic breakdown. For instance, soil microplastic pollution promotes the biomass growth of r-strategy bacteria (such as *Ruminiclostridium*-1), which in turn promotes the soil's ability to release carbon dioxide, which is necessary for the decomposition and reduction of soil organic matter.

13.4 IMPACTS ON TERRESTRIAL PLANTS

Ecotoxicologists have studied the toxicity and safety of MPs in relation to environmental health since they are present everywhere in the ecosystem, from the ground to the sea. The relationship between the transport routes and the pollutants, which is a severe environmental concern, can be linked using this methodology. The size of MPs, which makes them easier for organisms to transfer, and their surface, which functions as a vehicle for other organic pollutants through adsorption, are two main causes of their negative environmental effects. There are a variety of harmful impacts of these contaminants at the basic producer level, which forms the base of the food web and can occasionally harm the entire chain (Prakash et al., 2020). It is challenging to draw any broad generalisations from the few research studies on MPs' effects on terrestrial plants. The presence of microplastics may trigger different impacts on various plant organs. In general, the occurrence of MPs in the soil can modify its qualities, including its humidity, density, structure, and nutrient content. These changes may then affect the features of plant roots, their growth, and their ability to absorb nutrients (Dissanayake et al., 2022). MPs most significantly impact the roots of plants, then the leaves, shoots, and stems. This is because MPs can be easily absorbed by above-ground plants with roots from polluted soils and atmospheric deposition. MPs may postpone germination by preventing both above- and below-ground growth of wheat during both the vegetative and reproductive phases (Zhou et al., 2021a). Figure 13.1 and Table 13.2 present the toxicity of MPs to plants.

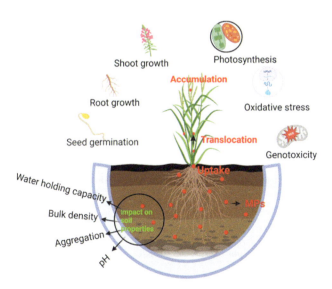

FIGURE 13.1 Uptake, translocation and toxicity of microplastics to the soil–plant ecosystem.

TABLE 13.2

Uptake and Impacts of Microplastics on Terrestrial Plants

Plant Species	Plastics Type and Size	MP Exposure Concentration	Impacts	References
Cucurbita pepo L.	PP, PE, PVC, and PET	< 0.5 mg/kg	MPs decreased leaf size, chlorophyll content, photosynthetic efficiency, and the composition of the micro- and macroelements	(Colzi et al., 2022)
Allium bulbs	PS; 80, 100, 200, 500, 1000, 2000, 4000, and 8000 nm	100 and 400 mg/L	Ps-MP particles showed cyto-genotoxicity by inducing many chromosomal and nuclear abnormalities in *A. cepa* root tip cells	(Kaur et al., 2022)
Senecio inaequidens and Centaurea cyanus	PVC; 250 μm	1% of PVC in 100 g of soil	Plant performance has been negatively impacted by PVC microparticles in the soil–plant system	(Gentili et al., 2022)
Trifolium repens, Orychophragmus violaceus, and Impatiens balsamina	PS; 2 μm and 80 nm	10, 50, 100, and 500 mg/L	Increased PS MP concentration decreases germination and root formation	(Guo et al., 2022b)
Tomato (Lycopersicon esculentum L.)	PS – 52.48 ± 20.93 μm, PE – 75.37 ± 17.55 μm, PP – 88.11 ± 28.53 μm,	10, 100, 500, and 1000 mg/L	The findings revealed that the three microplastics exhibited inhibitory effects on seed germination at concentrations lower than or equivalent to 500 mg/L, with an inhibition rate ranging from 10.1 to 23.6%	(Shi et al., 2022)
Rice (Oryza sativa L.)	PS, PVC	1.5 and 3.0 mg/L	Accumulation of PVC-MPs can be more detrimental to rice development and metabolism than PS-MPs. Repression in shoot and root weight as well as fresh and dried weight was observed	(Ma et al., 2022)
Soybean (Glycine max), mung bean (Vigna radiata)	PE; 6.5 μm and 13 μm	10, 50, 100, 200, and 500 mg/L	The phyto-toxicity of PE-MPs to soybean (*Glycine max*) was found to be higher than that of mung bean (*Vigna radiata*)	(Wang et al., 2021, b)
Rice (Oryza sativa L.)	PS; 100 nm and 1 μm	0.1, 1, and 10 mg/L	PS can induce oxidative stress. Rice's root length was reduced, and nutrient uptake was impeded	(Wu et al., 2021)

(Continued)

Impacts of Microplastic on Plants and Soil Ecosystem

TABLE 13.2 (Continued)
Uptake and Impacts of Microplastics on Terrestrial Plants

Plant Species	Plastics Type and Size	MP Exposure Concentration	Impacts	References
Raphanus sativus	ABS; 1–300 μm	10% w/w (in culture media)	Defects in the phenotype of the plants treated with polymer	(Tympa et al., 2021)
Common bean (Phaseolus vulgaris L.)	LDPE, PLA; <53–1000 μm	0.5%, 1.0%, 1.5%, 2.0%, 2.5% w/w dry soil weight	When compared to bio-MPs, notably at concentrations of 1.5%, 2.0%, and 2.5%, LDPE-MPs had no discernible effects on the shoot and root biomass	(Meng et al., 2021)

Abbreviations: Polypropylene (PP), polyethylene (PE), polyvinylchloride (PVC), polyethyleneterephthalate (PET), polystyrene (PS), low-density polyethylene (LDPE-MPs), polylactic acid (PLA), acrylonitrile butadiene styrene (ABS)

Jiang et al. (2019) investigated the effects of 5 μm and 100 nm with 10, 50, and 100 mg/L polystyrene fluorescent microplastics (PS MPs) for 48 h on *Vicia faba* root tips. The findings demonstrated that under 5 μm PS MPs, the biomass and catalase (CAT) enzyme of *V. faba* roots was reduced. In contrast, the levels of the enzymes peroxidase (POD) and superoxide dismutase (SOD) considerably increased. Only at the maximum concentration (100 mg L^{-1}) under the 100 nm PS MPs exposure was a noticeable growth reduction seen. However, genotoxic and oxidative damage to *V. faba* is caused by 100 nm PS MPs at a faster rate than by 5 μm PS MPs, according to the micronucleus (MN) test and antioxidative enzyme activity. Su et al. (2019) showed that MPs may be ingested by agricultural plants like wheat and subsequently move up the food chain and web. The tests conducted by Qi et al. (2018) revealed that MPs had a negative impact on the elemental composition of wheat growth tissue both during the vegetative and reproductive stages. Cucumber plants exposed to polystyrene nanoplastics showed a notable rise in their carbon, nitrogen, and biomass, according to Li et al. (2021). Nanoplastics may reduce the shoot-to-root biomass ratio in wheat seedlings. Nanoplastics were picked up and then transmitted to shoots from the top down, according to the images from a 3D laser confocal scanning microscope.

With regard to soil type (loamy, sandy soil, and clay obtained at various locations), exposure time, and microplastic concentrations, the experimental conditions used in these investigations varied greatly. Any cross-study comparisons are difficult due to the lack of uniformity. Conflicting results are frequently reported when numerous research works are compared because they share certain similar elements of the experimental setting. For instance, Boots et al. (2019) discovered that exposure to acrylic polymers and polyester at quantities of 0.1% w/w reduced the germination of perennial ryegrass. Conversely, Judy et al. (2019) discovered no adverse effects on the germination of ordinary wheat using identical quantities of polyester, polyethylene, and polyvinylchloride, indicating that species react differently.

Unexpectedly, different sizes of biodegradable plastic could be fatal. In wheat, biodegradable plastic mulch residue (1000–50 μm) appeared to be more detrimental than non-biodegradable plastic mulch in terms of decreasing leaf area, plant height, and biomass as well as delaying tillering and reducing seed setting (Qi et al., 2018). Perennial ryegrass seed germination and shoot height appeared to be greatly reduced when biodegradable polylactic acid (0.6–363 μm) was applied over high-density polyethylene and clothing microfibre (Boots et al., 2019). Additionally, maize biomass and chlorophyll content were decreased by a higher concentration (up to 10%) of biodegradable polylactic acid (100–154 µm) (Wang et al., 2020). It is now obvious that using biodegradable plastic won't help reduce the environment's growing plastic pollution. Agroecosystems may suffer irreparable harm as a result of their presence in agricultural soils. Therefore, the widespread promotion of it as an environmentally benign product needs to be carefully addressed. Positive or no significant effects have been reported, hence there are no consistent consequences of plastic exposure on soil organisms. Different plastic properties (such as kind, size, and presence of additives) and concentrations utilised in different research could be the cause of these conflicting results. The composition, size, form, and concentration of plastics have been found to be crucial factors in how they affect soil organisms.

13.4.1 UPTAKE AND TRANSLOCATION OF MPS BY PLANTS

M/NPs from soil and water were consumed and absorbed by roots. Some of these might enter through the fissures in the lateral roots. Roots of plants have a larger concentration of M/NPs compared to grass and above-ground parts of plants (Allouzi et al., 2021). Since MPs cannot pass through cellulose-rich plant cell walls due to their enormous size and high molecular weight, it is not envisaged that plants will be able to absorb them. However, MPs can cross biological membranes and penetrate plant cells when they are reduced to nanoparticles (< 0.1 µm), potentially infiltrating the food chain and web (Su et al., 2019). Crop health has been negatively impacted by PS NPs (0.2 µm) absorbed by vegetable roots (such as wheat and lettuce) and transmitted onto shoots because they disrupt intracellular metabolism (Li et al., 2020b). Additionally, the rate of uptake may differ depending on the MP type (size, shape, and stage of degradation), MP concentration in the soil, and the type of plant (Ebere et al., 2019). According to Li et al (2020c)'s observations, the 0.2-m PS luminescence signals were mostly observed in the vascular system and on the cell walls of the cortical tissue of the roots. This shows that the apoplastic transport system was used to move the beads through the intercellular space. The 0.2-µm PS beads were carried from the roots to the stems and leaves by the vascular system after entering the central cylinder, which followed the transpiration stream. Li et al. (2020c) also observed that the PS beads adhered to one another and methodically self-assembled into clusters within the intercellular space of the vascular tissue of the lettuce root and stem. PS beads were scattered throughout the leaf tissue, as opposed to the root and stem. In this study, the scientists demonstrated that higher plants can transport 0.2-micrometer-sized polymers from their roots to their shoots.

Aggregates of M/NPs can be absorbed or translocated by plants. The absorption and transport of plastic particles by plants are significantly influenced by transpiration pull. From the root to the shoot, the vascular system uses the transpiration stream to transfer plastic particles (Azeem et al., 2021). M/NPs can enter the vasculature responsible for transporting water by passing through small extracellular channels. The stem, leaves, and potentially fruits can swiftly absorb NPs supported by the water transportation system (Schwab et al., 2020). Foliar spray via the stomata is another manner in which plastic may enter plant leaves (Adeel et al., 2018). Stomatal uptake is one possible way for NPs to penetrate the leaves before moving to the vasculature, according to Sun et al. (2021). The maize plant's stems and roots both showed significant concentrations of micro-fluorescent PS NPs, and fluorescence microscope imaging research also shows that PS NPs are frequently found in and near the stem's vascular system. These results show that NPs are transported from the leaves to the stems and then from the stems to the roots via the vascular bundle. Similarly, a different study verified that PS NPs adhere to plant stomata, pierce through phloem, and enter the lettuce plant's roots (Lian et al., 2021). Currently, there is still very little MP and NP uptake and accumulation in plant shoots. The processes of MP and NP absorption and transportation in plant shoots need to be further studied. Some plant traits, such as root features, xylem characteristics, growth rate, water and lipid fractions, plasma membrane potential, cytoplasm pH, and pH vacuoles, might influence the uptake and absorption of NPs by plants (Allouzi et al., 2021).

Impacts of Microplastic on Plants and Soil Ecosystem

13.4.2 Toxicity Mechanism

In the literature, the phenomenon of particle adsorption on organism surface that results in the physical obstruction of light, gases, and nutrients, hence decreasing growth and photosynthetic activity, has received substantial study. According to recent studies, physical impacts are what largely cause toxicity since they restrict how often larger particles like MPs may interchange with the medium. Smaller particles, such as NPs, on the other hand, mostly have chemical impacts, such as a rise in the production of reactive oxygen species and a decrease in photosynthesis. The existence of additives developed during ageing, as well as the secondary metabolites produced as a result of the ageing processes of these compounds, can also be linked to the potential dangers of plastics (Larue et al., 2021).

A number of intricate elements likely cause MP phytotoxicity. We investigated the potential mechanisms of how MPs affect plant systems based on existing research and data, and they were as follows: (1) MPs modify the physical environment of the soil, the chemical fertility, the microbial activity, and the activity of enzymes, all of which have an indirect impact on plant performance; (2) toxicity of the MPs themselves (e.g., fillers, flame retardants, antioxidants, plasticisers, and colours); and (3) pollutants such as heavy metals, polycyclic aromatic hydrocarbons (PAHs), persistent organic pollutants (POPs), antibiotics, polychlorinated biphenyls (PCBs), and antibiotic resistance genes (ARGs), among others, can be absorbed by MPs (Zhang et al., 2022). It is hypothesised that with a reduction in particle size, effects on biota shift more toward chemical/toxic than physical. While it is not anticipated that microsized particles will penetrate the root, the situation is different with NP particles (Rillig et al., 2019). Phytotoxic chemicals that are already present in MPs when they enter the soil (for instance, when introduced during manufacturing) may do so along with these MP particles. By becoming adsorbed onto surfaces and within the particle "ecocorona" of MP particles in the soil, toxic substances may harm plant roots or their symbionts. They may also already be present in the particles. This could negatively impact the growth of plants (Rillig et al., 2019). NPs can interact with internal and extracellular biomolecules to facilitate the creation of protein-coronas or ecocoronas, which help move molecules through cell membranes (Junaid et al., 2023). Nanoplastic internalisation in terrestrial plants was reported by Sun et al. (2020). *Arabidopsis* is capable of absorbing and transporting NPs smaller than 200 nm. Positively charged nanoplastics accumulated in the root tips at relatively lower levels than negatively charged sulfonic-acid-modified nanoplastics. However, they produced more reactive oxygen species and significantly slowed plant growth.

13.5 CONCLUSION

The study of MPs in the terrestrial environment is still in its early stages, even though it has recently picked up momentum. Here, we present a summary of earlier research on MPs in terrestrial ecosystems, including soil and plants. Microplastics are one example of a contaminant that requires attention because of its prevalence and linkage to climate change. MPs and NPs negatively affect the physicochemical properties of soil, including its texture, porosity, and retention capacity. Microplastics have great potential for bioaccumulation because of their microscopic size. Microplastic contamination has a negative impact on the ecosystem in a number of ways, including weak economic services, molecular effects on organisms, and physiological consequences on people and animals. Different MP/NP sources and types can have an impact on the biogeochemical cycle by directly affecting soil physicochemical characteristics and soil organisms, as well as indirectly affecting soil biota through changes in soil material cycling. Plants may absorb and transmit microplastics, which could have an impact on plant growth. Some of the mechanisms will have a favourable impact on the growth of the roots and plants, while others will have detrimental effects. These effects will differ depending on the type of plant and are, therefore, likely to result in changes to the composition of plant communities and perhaps primary production. For instance, whereas onions are tolerant of soil microplastic pollution, wheat, lettuce, and broad beans are quite vulnerable. Microplastic characteristics, plant species, and environmental conditions all affected how microplastics affected higher plants in different ways. Direct and indirect pathways can be used to categorise the inhibitory effects of MPs on plant development. Direct ways include obstructing pores or light, harming roots mechanically, preventing genes from being expressed, and releasing chemicals. Examples of indirect mechanisms include modifying soil properties, altering soil bacteria or soil animals, and influencing the bioavailability of other pollutants.

REFERENCES

Adeel, M., Yang, Y. S., Wang, Y. Y., Song, X. M., Ahmad, M. A., & Rogers, H. J. (2018). Uptake and transformation of steroid estrogens as emerging contaminants influence plant development. *Environmental Pollution, 243*, 1487–1497. https://doi.org/10.1016/j.envpol.2018.09.016

Ali, I., Tan, X., Li, J., Peng, C., Naz, I., Duan, Z., & Ruan, Y. (2022). Interaction of microplastics and nanoplastics with natural organic matter (NOM) and the impact of NOM on the sorption behavior of anthropogenic contaminants–A critical review. *Journal of Cleaner Production*, 134314. https://doi.org/10.1016/j.jclepro.2022.134314

Allouzi, M. M. A., Tang, D. Y. Y., Chew, K. W., Rinklebe, J., Bolan, N., Allouzi, S. M. A., & Show, P. L. (2021). Micro (nano) plastic pollution: The ecological influence on soil-plant system and human health. *Science of the Total Environment, 788*, 147815. https://doi.org/10.1016/j.scitotenv.2021.147815

Azeem, I., Adeel, M., Ahmad, M. A., Shakoor, N., Jiangcuo, G. D., Azeem, K., & Rui, Y. (2021). Uptake and accumulation of nano/microplastics in plants: A critical review. *Nanomaterials, 11*(11), 2935. https://doi.org/10.3390/nano11112935

Boots, B., Russell, C. W., & Green, D. S. (2019). Effects of microplastics in soil ecosystems: Above and below ground. *Environmental Science & Technology, 53*(19), 11496– 11506. https://doi.org/10.1021/acs.est.9b03304

Büks, F., & Kaupenjohann, M. (2021). The impact of microplastic weathering on interactions with the soil environment: A review. *Soil.* https://doi.org/10.5194/soil-2021-67

Chen, L., Han, L., Feng, Y., He, J., & Xing, B. (2022). Soil structures and immobilization of typical contaminants in soils in response to diverse microplastics. *Journal of Hazardous Materials, 438*, 129555. https://doi.org/10.1016/j.jhazmat.2022.129555

Chia, R. W., Lee, J. Y., Jang, J., Kim, H., & Kwon, K. D. (2022). Soil health and microplastics: A review of the impacts of microplastic contamination on soil properties. *Journal of Soils and Sediments, 22*(10), 2690–2705. https://doi.org/10.1007/s11368-022-03254-4

Cluzard, M., Kazmiruk, T. N., Kazmiruk, V. D., & Bendell, L. I. (2015). Intertidal concentrations of microplastics and their influence on ammonium cycling as related to the shellfish industry. *Archives of Environmental Contamination and Toxicology, 69*, 310–319. https://doi.org/10.1007/s00244-015-0156-5

Colzi, I., Renna, L., Bianchi, E., Castellani, M. B., Coppi, A., Pignattelli, S., & Gonnelli, C. (2022). Impact of microplastics on growth, photosynthesis and essential elements in Cucurbita pepo L. *Journal of Hazardous Materials, 423*, 127238. https://doi.org/10.1016/j.jhazmat.2021.127238

Conti, G. O., Ferrante, M., Banni, M., Favara, C., Nicolosi, I., Cristaldi, A., & Zuccarello, P. (2020). Micro-and nano-plastics in edible fruit and vegetables. The first diet risks assessment for the general population. *Environmental Research, 187*, 109677. https://doi.org/10.1016/j.envres.2020.109677

de Almeida, M. P., Gaylarde, C. C., Baptista Neto, J. A., Delgado, J. D. F., Lima, L. D. S., Neves, C. V., & da Fonseca, E. M. (2023). The prevalence of microplastics on the earth and resulting increased imbalances in biogeochemical cycling. *Water Emerging Contaminants & Nanoplastics, 2*(2), 7. https://doi.org/10.20517/wecn.2022.20

de Souza Machado, A. A., Horton, A. A., Davis, T., & Maaß, S. (2020). Microplastics and their effects on soil function as a life-supporting system. In *Microplastics in terrestrial environments: Emerging contaminants and major challenges* (pp. 199–222). https://doi.org/10.1007/698_2020_450

de Souza Machado, A. A., Lau, C. W., Kloas, W., Bergmann, J., Bachelier, J. B., Faltin, E., & Rillig, M. C. (2019). Microplastics can change soil properties and affect plant performance. *Environmental Science & Technology, 53*(10), 6044–6052. https://doi.org/10.1021/acs.est.9b01339

de Souza Machado, A. A., Lau, C. W., Till, J., Kloas, W., Lehmann, A., Becker, R., & Rillig, M. C. (2018). Impacts of microplastics on the soil biophysical environment. *Environmental Science & Technology, 52*(17), 9656–9665. https://doi.org/10.1021/acs.est.8b02212

Dissanayake, P. D., Kim, S., Sarkar, B., Oleszczuk, P., Sang, M. K., Haque, M. N., & Ok, Y. S. (2022). Effects of microplastics on the terrestrial environment: A critical review. *Environmental Research, 209*, 112734. https://doi.org/10.1016/j.envres.2022.112734

Ebere, E. C., Wirnkor, V. A., & Ngozi, V. E. (2019). Uptake of microplastics by plant: A reason to worry or to be happy?. *World Scientific News, 131*, 256–267.

Gentili, R., Quaglini, L., Cardarelli, E., Caronni, S., Montagnani, C., & Citterio, S. (2022). Toxic impact of soil microplastics (PVC) on two weeds: Changes in growth, phenology and photosynthesis efficiency. *Agronomy, 12*(5), 1219. https://doi.org/10.3390/agronomy12051219

Guo, M., Zhao, F., Tian, L., Ni, K., Lu, Y., & Borah, P. (2022, b). Effects of polystyrene microplastics on the seed germination of herbaceous ornamental plants. *Science of the Total Environment, 809*, 151100. https://doi.org/10.1016/j.scitotenv.2021.151100

Guo, Z., Li, P., Yang, X., Wang, Z., Lu, B., Chen, W., & Xue, S. (2022, a). Soil texture is an important factor determining how microplastics affect soil hydraulic characteristics. *Environment International, 165*, 107293. https://doi.org/10.1016/j.envint.2022.107293

Huang, Y., Zhao, Y., Wang, J., Zhang, M., Jia, W., & Qin, X. (2019). LDPE microplastic films alter microbial community composition and enzymatic activities in soil. *Environmental Pollution, 254*, 112983. https://doi.org/10.1016/j.envpol.2019.112983

Huerta Lwanga, E., Mendoza Vega, J., Ku Quej, V., Chi, J. D. L. A., Sanchez del Cid, L., Chi, C., … & Geissen, V. (2017). Field evidence for transfer of plastic debris along a terrestrial food chain. *Scientific Reports, 7*(1), 1–7. https://doi.org/10.1038/s41598-017-14588-2

Hüffer, T., Metzelder, F., Sigmund, G., Slawek, S., Schmidt, T. C., & Hofmann, T. (2019). Polyethylene microplastics influence the transport of organic contaminants in soil. *Science of the Total Environment, 657*, 242–247. https://doi.org/10.1016/j.scitotenv.2018.12.047

Iqbal, S., Xu, J., Khan, S., Arif, M. S., Yasmeen, T., Nadir, S., & Schaefer, D. A. (2021). Deciphering microplastic ecotoxicology: Impacts on crops and soil ecosystem functions. *Circular Agricultural Systems, 1*(1), 1–7.

Jiang, X., Chang, Y., Zhang, T., Qiao, Y., Klobučar, G., & Li, M. (2020). Toxicological effects of polystyrene microplastics on earthworm (Eisenia fetida). *Environmental Pollution, 259*, 113896. https://doi.org/10.1016/j.envpol.2019.113896

Jiang, X., Chen, H., Liao, Y., Ye, Z., Li, M., & Klobučar, G. (2019). Ecotoxicity and genotoxicity of polystyrene microplastics on higher plant Vicia faba. *Environmental Pollution, 250*, 831–838. https://doi.org/10.1016/j.envpol.2019.04.055

Jing, X., Su, L., Wang, Y., Yu, M., & Xing, X. (2023). How do microplastics affect physical properties of silt loam soil under wetting–drying cycles?. *Agronomy, 13*(3), 844. https://doi.org/10.3390/agronomy13030844

Judy, J. D., Williams, M., Gregg, A., Oliver, D., Kumar, A., Kookana, R., & Kirby, J. K. (2019). Microplastics in municipal mixed-waste organic outputs induce minimal short to long-term toxicity in key terrestrial biota. *Environmental Pollution, 252*, 522– 531. https://doi.org/10.1016/j.envpol.2019.05.027

Junaid, M., Liu, S., Chen, G., Liao, H., & Wang, J. (2023). Transgenerational impacts of micro (nano) plastics in the aquatic and terrestrial environment. *Journal of Hazardous Materials, 443*, 130274. https://doi.org/10.1016/j.jhazmat.2022.130274

Kaur, M., Xu, M., & Wang, L. (2022). Cyto–genotoxic effect causing potential of polystyrene micro-plastics in terrestrial plants. *Nanomaterials, 12*(12), 2024. https://doi.org/10.3390/nano12122024

Kumar, A., Mishra, S., Pandey, R., Yu, Z. G., Kumar, M., Khoo, K. S., & Show, P. L. (2022). Microplastics in terrestrial ecosystems: Un-ignorable impacts on soil characterises, nutrient storage and its cycling. *TrAC Trends in Analytical Chemistry*, 116869. https://doi.org/10.1016/j.trac.2022.116869

Larue, C., Sarret, G., Castillo-Michel, H., & Pradas del Real, A. E. (2021). A critical review on the impacts of nanoplastics and microplastics on aquatic and terrestrial photosynthetic organisms. *Small, 17*(20), 2005834. https://doi.org/10.1002/smll.202005834

Li, J., Song, Y., & Cai, Y. (2020, a). Focus topics on microplastics in soil: Analytical methods, occurrence, transport, and ecological risks. *Environmental Pollution, 257*, 113570. https://doi.org/10.1016/j.envpol.2019.113570

Li, J., Yu, Y., Zhang, Z., & Cui, M. (2023). The positive effects of polypropylene and polyvinyl chloride microplastics on agricultural soil quality. *Journal of Soils and Sediments*, *23*(3), 1304–1314. https://doi.org/10.1007/s11368-022-03387-6

Li, L., Luo, Y., Li, R., Zhou, Q., Peijnenburg, W. J., Yin, N., & Zhang, Y. (2020, b). Effective uptake of submicrometre plastics by crop plants via a crack-entry mode. *Nature sustainability*, *3*(11), 929–937. https://doi.org/10.1038/s41893-020-0567-9

Li, L., Song, K., Yeerken, S., Geng, S., Liu, D., Dai, Z., & Wang, Q. (2020, d). Effect evaluation of microplastics on activated sludge nitrification and denitrification. *Science of the Total Environment*, *707*, 135953. https://doi.org/10.1016/j.scitotenv.2019.135953

Li, L., Yang, J., Zhou, Q., Peijnenburg, W. J., & Luo, Y. (2020, c). Uptake of microplastics and their effects on plants. *Microplastics in Terrestrial Environments: Emerging Contaminants and Major Challenges*, 279–298. https://doi.org/10.1007/698_2020_465

Li, Z., Li, Q., Li, R., Zhou, J., & Wang, G. (2021). The distribution and impact of polystyrene nanoplastics on cucumber plants. *Environmental Science and Pollution Research*, *28*, 16042–16053. https://doi.org/10.1007/s11356-020-11702-2

Lian, J., Liu, W., Meng, L., Wu, J., Chao, L., Zeb, A., & Sun, Y. (2021). Foliar-applied polystyrene nanoplastics (PSNPs) reduce the growth and nutritional quality of lettuce (Lactuca sativa L.). *Environmental Pollution*, *280*, 116978. https://doi.org/10.1016/j.envpol.2021.116978

Liu, E. K., He, W. Q., & Yan, C. R. (2014). 'White revolution' to 'white pollution'—agricultural plastic film mulch in China. *Environmental Research Letters*, *9*(9), 091001. 10.1088/1748-9326/9/9/091001

Liu, Y., Huang, Q., Hu, W., Qin, J., Zheng, Y., Wang, J., & Xu, L. (2021). Effects of plastic mulch film residues on soil-microbe-plant systems under different soil pH conditions. *Chemosphere*, *267*, 128901. https://doi.org/10.1016/j.chemosphere.2020.128901

Lozano, Y. M., & Rillig, M. C. (2020). Effects of microplastic fibers and drought on plant communities. *Environmental Science & Technology*, *54*(10), 6166–6173. https://doi.org/10.1021/acs.est.0c01051

Lozano, Y. M., Aguilar-Trigueros, C. A., Onandia, G., Maaß, S., Zhao, T., & Rillig, M. C. (2021). Effects of microplastics and drought on soil ecosystem functions and multifunctionality. *Journal of Applied Ecology*, *58*(5), 988–996. https://doi.org/10.1111/1365-2664.13839

Ma, J., Aqeel, M., Khalid, N., Nazir, A., Alzuaibr, F. M., Al-Mushhin, A. A., & Noman, A. (2022). Effects of microplastics on growth and metabolism of rice (Oryza sativa L.). *Chemosphere*, *307*, 135749. https://doi.org/10.1016/j.chemosphere.2022.135749

Meng, F., Yang, X., Riksen, M., & Geissen, V. (2022). Effect of different polymers of microplastics on soil organic carbon and nitrogen–A mesocosm experiment. *Environmental Research*, *204*, 111938. https://doi.org/10.1016/j.envres.2021.111938

Meng, F., Yang, X., Riksen, M., Xu, M., & Geissen, V. (2021). Response of common bean (Phaseolus vulgaris L.) growth to soil contaminated with microplastics. *Science of the Total Environment*, *755*, 142516. https://doi.org/10.1016/j.scitotenv.2020.142516

Moghaddasi, S., Khoshgoftarmanesh, A. H., Karimzadeh, F., & Chaney, R. (2015). Fate and effect of tire rubber ash nano-particles (RANPs) in cucumber. *Ecotoxicology and Environmental Safety*, *115*, 137–143. https://doi.org/10.1016/j.ecoenv.2015.02.020

Prakash, V., Dwivedi, S., Gautam, K., Seth, M., & Anbumani, S. (2020). Occurrence and ecotoxicological effects of microplastics on aquatic and terrestrial ecosystems. In *Microplastics in Terrestrial Environments: Emerging Contaminants and Major Challenges* (pp. 223–243). https://doi.org/10.1007/698_2020_456

Qi, Y., Ossowicki, A., Yang, X., Lwanga, E. H., Dini-Andreote, F., Geissen, V., & Garbeva, P. (2020). Effects of plastic mulch film residues on wheat rhizosphere and soil properties. *Journal of Hazardous Materials*, *387*, 121711. https://doi.org/10.1016/j.jhazmat.2019.121711

Qi, Y., Yang, X., Pelaez, A. M., Lwanga, E. H., Beriot, N., Gertsen, H., & Geissen, V. (2018). Macro-and micro-plastics in soil-plant system: Effects of plastic mulch film residues on wheat (Triticum aestivum) growth. *Science of the Total Environment*, *645*, 1048–1056. https://doi.org/10.1016/j.scitotenv.2018.07.229

Rillig, M. C., & Lehmann, A. (2020). Microplastic in terrestrial ecosystems. *Science*, *368*(6498), 1430–1431. https://doi.org/10.1126/science.abb5979

Rillig, M. C., Lehmann, A., de Souza Machado, A. A., & Yang, G. (2019). Microplastic effects on plants. *New Phytologist*, *223*(3), 1066–1070. https://doi.org/10.1111/nph.15794

Rong, L., Zhao, L., Zhao, L., Cheng, Z., Yao, Y., Yuan, C., & Sun, H. (2021). LDPE microplastics affect soil microbial communities and nitrogen cycling. *Science of the Total Environment*, *773*, 145640. https://doi.org/10.1016/j.scitotenv.2021.145640

Rose, P. K., Jain, M., Kataria, N., Sahoo, P. K., Garg, V. K., & Yadav, A. (2023, a). Microplastics in multimedia environment: A systematic review on its fate, transport, quantification, health risk, and remedial measures. *Groundwater for Sustainable Development*, *20*, 100889. https://doi.org/10.1016/j.gsd.2022.100889

Rose, P. K., Yadav, S., Kataria, N., & Khoo, K. S. (2023, b). Microplastics and nanoplastics in the terrestrial food chain: Uptake, translocation, trophic transfer, ecotoxicology, and human health risk. *TrAC Trends in Analytical Chemistry*, *167*, 117249. https://doi.org/10.1016/j.trac.2023.117249

Schöpfer, L., Menzel, R., Schnepf, U., Ruess, L., Marhan, S., Brümmer, F., & Kandeler, E. (2020). Microplastics effects on reproduction and body length of the soil-dwelling nematode Caenorhabditis elegans. *Frontiers in Environmental Science*, *8*, 41. https://doi.org/10.3389/fenvs.2020.00041

Schwab, F., Rothen-Rutishauser, B., & Petri-Fink, A. (2020). When plants and plastic interact. *Nature Nanotechnology*, *15*(9), 729–730. https://doi.org/10.1038/s41565-020-0762-x

Seeley, M. E., Song, B., Passie, R., & Hale, R. C. (2020). Microplastics affect sedimentary microbial communities and nitrogen cycling. *Nature Communications*, *11*(1), 2372. https://doi.org/10.1038/s41467-020-16235-3

Shen, M., Liu, S., Hu, T., Zheng, K., Wang, Y., & Long, H. (2023). Recent advances in the research on effects of micro/nanoplastics on carbon conversion and carbon cycle: A review. *Journal of Environmental Management*, *334*, 117529. https://doi.org/10.1016/j.jenvman.2023.117529

Shi, R., Liu, W., Lian, Y., Wang, Q., Zeb, A., & Tang, J. (2022). Phytotoxicity of polystyrene, polyethylene and polypropylene microplastics on tomato (Lycopersicon esculentum L.). *Journal of Environmental Management*, *317*, 115441. https://doi.org/10.1016/j.jenvman.2022.115441

Su, Y., Ashworth, V., Kim, C., Adeleye, A. S., Rolshausen, P., Roper, C., & Jassby, D. (2019). Delivery, uptake, fate, and transport of engineered nanoparticles in plants: A critical review and data analysis. *Environmental Science: Nano*, *6*(8), 2311–2331. https://doi.org/10.1039/C9EN00461K

Sun, H., Lei, C., Xu, J., & Li, R. (2021). Foliar uptake and leaf-to-root translocation of nanoplastics with different coating charge in maize plants *Journal of Hazardous Materials*, *416*125854.

Sun, X. D., Yuan, X. Z., Jia, Y., Feng, L. J., Zhu, F. P., Dong, S. S., & Xing, B. (2020). Differentially charged nanoplastics demonstrate distinct accumulation in Arabidopsis thaliana. *Nature Nanotechnology*, *15*(9), 755–760. https://doi.org/10.1038/s41565-020-0707-4

Sun, Y., Duan, C., Cao, N., Li, X., Li, X., Chen, Y., & Wang, J. (2022). Effects of microplastics on soil microbiome: The impacts of polymer type, shape, and concentration. *Science of the Total Environment*, *806*, 150516. https://doi.org/10.1016/j.scitotenv.2021.150516

Tympa, L. E., Katsara, K., Moschou, P. N., Kenanakis, G., & Papadakis, V. M. (2021). Do microplastics enter our food chain via root vegetables? A raman based spectroscopic study on Raphanus sativus. *Materials*, *14*(9), 2329. https://doi.org/10.3390/ma14092329

Wan, Y., Wu, C., Xue, Q., & Hui, X. (2019). Effects of plastic contamination on water evaporation and desiccation cracking in soil. *Science of the Total Environment*, *654*, 576–582. https://doi.org/10.1016/j.scitotenv.2018.11.123

Wang, F., Zhang, X., Zhang, S., Zhang, S., & Sun, Y. (2020). Interactions of microplastics and cadmium on plant growth and arbuscular mycorrhizal fungal communities in an agricultural soil. *Chemosphere*, *254*, 126791. https://doi.org/10.1016/j.chemosphere.2020.126791

Wang, J., Peng, C., Li, H., Zhang, P., & Liu, X. (2021, a). The impact of microplastic-microbe interactions on animal health and biogeochemical cycles: A mini-review. *Science of the Total Environment*, *773*, 145697. https://doi.org/10.1016/j.scitotenv.2021.145697

Wang, L., Liu, Y., Kaur, M., Yao, Z., Chen, T., & Xu, M. (2021, b). Phytotoxic effects of polyethylene microplastics on the growth of food crops soybean (Glycine max) and mung bean (Vigna radiata). *International Journal of Environmental Research and Public Health*, *18*(20), 10629. https://doi.org/10.3390/ijerph182010629

Wen, X., Yin, L., Zhou, Z., Kang, Z., Sun, Q., Zhang, Y., & Jiang, C. (2022). Microplastics can affect soil properties and chemical speciation of metals in yellow-brown soil. *Ecotoxicology and Environmental Safety*, *243*, 113958. https://doi.org/10.1016/j.ecoenv.2022.113958

Wu, J., Liu, W., Zeb, A., Lian, J., Sun, Y., & Sun, H. (2021). Polystyrene microplastic interaction with Oryza sativa: Toxicity and metabolic mechanism. *Environmental Science: Nano, 8*(12), 3699–3710. https://doi.org/10.1039/D1EN00636C

Wu, M., Yang, C., Du, C., & Liu, H. (2020). Microplastics in waters and soils: Occurrence, analytical methods and ecotoxicological effects. *Ecotoxicology and Environmental Safety, 202*, 110910. https://doi.org/10.1016/j.ecoenv.2020.110910

Xiao, M., Luo, Y., Zhang, H., Yu, Y., Yao, H., Zhu, Z., & Ge, T. (2022). Microplastics shape microbial communities affecting soil organic matter decomposition in paddy soil. *Journal of Hazardous Materials, 431*, 128589. https://doi.org/10.1016/j.jhazmat.2022.128589

Xiao, M., Shahbaz, M., Liang, Y., Yang, J., Wang, S., Chadwicka, D. R., & Ge, T. (2021). Effect of microplastics on organic matter decomposition in paddy soil amended with crop residues and labile C: A three-source-partitioning study. *Journal of Hazardous Materials, 416*, 126221. https://doi.org/10.1016/j.jhazmat.2021.126221

Ya, H., Xing, Y., Zhang, T., Lv, M., & Jiang, B. (2022). LDPE microplastics affect soil microbial community and form a unique plastisphere on microplastics. *Applied Soil Ecology, 180*, 104623. https://doi.org/10.1016/j.apsoil.2022.104623

Yadav, S., Kataria, N., Khyalia, P., Rose, P. K., Mukherjee, S., Sabherwal, H., & Khoo, K. S. (2023). Recent analytical techniques, and potential eco-toxicological impacts of textile fibrous microplastics (FMPs) and its associated contaminates: A review. *Chemosphere,* 138495. https://doi.org/10.1016/j.chemosphere.2023.138495

Yu, H., Zhang, Z., Zhang, Y., Song, Q., Fan, P., Xi, B., & Tan, W. (2021). Effects of microplastics on soil organic carbon and greenhouse gas emissions in the context of straw incorporation: A comparison with different types of soil. *Environmental Pollution, 288*, 117733. https://doi.org/10.1016/j.envpol.2021.117733

Yuan, Y., Zu, M., Li, R., Zuo, J., & Tao, J. (2023). Soil Properties, Microbial Diversity, and Changes in the Functionality of Saline-alkali Soil are Driven by Microplastics. *Journal of Hazardous Materials*, 130712. https://doi.org/10.1016/j.jhazmat.2022.130712

Zhang, S., Wang, J., Yan, P., Hao, X., Xu, B., Wang, W., & Aurangzeib, M. (2021). Non-biodegradable microplastics in soils: A brief review and challenge. *Journal of Hazardous Materials, 409*, 124525. https://doi.org/10.1016/j.jhazmat.2020.124525

Zhang, Z., Cui, Q., Chen, L., Zhu, X., Zhao, S., Duan, C., & Fang, L. (2022). A critical review of microplastics in the soil-plant system: Distribution, uptake, phytotoxicity and prevention. *Journal of Hazardous Materials, 424*, 127750. https://doi.org/10.1016/j.jhazmat.2021.127750

Zhou, J., Gui, H., Banfield, C. C., Wen, Y., Zang, H., Dippold, M. A., & Jones, D. L. (2021, b). The microplastisphere: Biodegradable microplastics addition alters soil microbial community structure and function. *Soil Biology and Biochemistry, 156*, 108211. https://doi.org/10.1016/j.soilbio.2021.108211

Zhou, J., Wen, Y., Marshall, M. R., Zhao, J., Gui, H., Yang, Y., & Zang, H. (2021, a). Microplastics as an emerging threat to plant and soil health in agroecosystems. *Science of the Total Environment, 787*, 147444. https://doi.org/10.1016/j.scitotenv.2021.147444

14 Ecotoxicological Impact of Microplastics in the Environment

Ajay Valiyaveettil Salimkumar, Mary Carolin Kurisingal Cleetus, Judith Osaretin Ehigie, Cyril Oziegbe Onogbosele, Dorcas Akua Essel, Ransford Parry, Bindhi S. Kumar, M.P. Prabhakaran, and V.J. Rejish Kumar

14.1 INTRODUCTION

Plastic, a valuable, beneficial, and convenient material that has become ubiquitous in our daily lives, has contributed significantly to the advancement of various industries and products. The mismanagement, mishandling, and irresponsible use of plastics have resulted in widespread pollution. The proliferation of plastic waste has emerged as a significant and escalating issue worldwide due to the continuous growth in plastic production and disposal practices. Over-reliance on single-use plastic has had significant environmental, social, economic, and health implications. Plastics, including microplastics (MPs), have become pervasive in our surroundings, they are integrating into the Earth's fossil record, serving as a distinctive marker of the Anthropocene, our present geological epoch. MPs are produced due to natural degradation and account for 99% of the total amount generated (Vo & Pham, 2021). MPs represent the enduring aftermath of plastic pollution. Being highly dispersive, their presence has been reported from the remote, pristine environments or locations of our planet that are largely untouched by human activities. MPs have several environmental concerns due to their physicochemical properties, and the risks associated with MPs depend on the exposure of these pollutants and the detrimental effects they induce at different exposure levels. The escalation of MP concentration in the environment heightens the likelihood of ecosystem exposure, resulting in augmented interactions, ingestions, and transfer across food webs. MPs, with their bioaccumulative potential, magnify within trophic levels of the food chain, intensifying detrimental impacts on organisms (Rose et al., 2023a). Despite the observed presence of MPs in various foods, such as bivalves, fish, fruits, vegetables, and table salts, a comprehensive understanding of the potential health risks to humans remains insufficient (Rose et al., 2023b; Yadav et al., 2023).

MPs constitute a relatively recent and underexplored research frontier marked by numerous uncertainties, yet there is a growing acknowledgment of the significant detrimental effects of MPs and the chemical pollutants they carry to aquatic and terrestrial organisms. This chapter systematically addresses key aspects of MPs, commencing with a precise definition of these minute plastic particles and their origin. It further probes into the environmental significance of MPs, their rising prevalence, ecological ramifications, chemical and physical properties, and their distribution across diverse ecosystems. The narrative extends to examining the toxicological effects of MPs, revealing

DOI: 10.1201/9781032706573-14

14.1.1 Definition of Microplastics

Plastics possess an essentially infinite shelf life, but various processes, such as thermal degradation, hydrolysis, mechanical degradation, thermo-oxidative degradation, photodegradation, and biodegradation, gradually diminish their structural integrity. As a result, plastics break down into smaller pieces, fragments, or even molecular components, perpetuating this degradation cycle and leaving a conspicuous presence in water, food, soil, and air (Mattsson et al., 2018). Based on the particle size, plastic debris can be categorized as macroplastics (>25 mm), mesoplastics (5–25 mm), microplastics or MPs (1 μm–5 mm), and nanoplastics or NPs (<1 μm) (Verschoor, 2015).

In 2018, Frias and Nash proposed the definition "MPs are any synthetic solid particle or polymeric matrix, with regular or irregular shape and with size ranging from 1 μm to 5 mm, of either primary or secondary manufacturing origin, which is insoluble in water" (Frias & Nash, 2019). The term MPs was coined by Thompson et al. (2004), while the lower size limit of MPs was fixed by the mesh size cutoff of the manta trawl net used for sampling MPs in the environment (Gigault et al., 2018), and the upper size limit was defined by Arthur et al. (2009). MPs are found to be diverse when it comes to their elemental composition, size, color, and shape.

14.1.2 Sources of Microplastics

The accumulation of MPs in the environment can be due to dumping directly or indirectly by humans. The potential sources of MPs are diverse. Generally, MPs in the environment come from two sources: primary and secondary. Primary MPs are intentionally manufactured to be of small size and include cosmetics, personal care products, industrial processing, textile applications or in, synthetic cloth production, air blasting, medicines as vectors, and aerospace, while secondary MPs originate from the weathering and breakdown of larger plastic items which are the main source of MPs (Kalčíková et al., 2017; An et al., 2020).

Plastic pellets, preproduction pellets, beads, or nurdles are granular plastics of 2–5 mm diameter that have applications in the production of various household appliances such as clothing, building chemicals, electrical, telecommunication automobile, medical equipment industry, and agriculture. Various personal care products and cosmetics contain MP beads. They enter the environment through the sewage network (An et al., 2020). Painting is another major cause of environmental plastic. This includes architectural coatings, marine coatings, automotive coatings, and road marking paints. Household laundry wastewater and washing plant wastewater release large amounts of plastic microfibers from various textiles into the environment. Effluents from sewage treatment plants are one of the largest sources of marine plastic in natural waters. Plastic running tracks, artificial turfs, rubber roads, vehicle tires, etc. release MPs into the environment due to wear and tear or weathering and aging (An et al., 2020).

Secondary sources of MPs are tiny fragments of plastics derived from larger plastic particles that have not been properly disposed of (Pettipas et al., 2016). The degradation of plastic structures occurs under the action of physical, biological, and chemical processes such as light irradiation aging, biological crushing, and mechanical grinding (Cole et al., 2011). These include plastic bags, plastic bottles, disposable plastic tableware, plastic packaging, fishing wastes (buoys, floating boxes, floating boxes, fishing rods, fish tanks, fishing nets, fishing lines, cables, feed bags, Styrofoam floats, boat paints, fishing nets, etc.), farming film, or film mulching (An et al., 2020).

14.1.3 Significance of Microplastics in the Environment

MPs can alter soil properties and directly or indirectly affect plants, animals, microorganisms, and ultimately humans. In an aquatic ecosystem, the density of MPs plays a crucial role in influencing their distribution. However, other factors such as color, shape, and abundance also impact the bio-availability of MPs in the aquatic environment (Wright et al., 2013; Crawford & Quinn, 2017). These factors collectively determine how MPs are dispersed and interact with aquatic organisms in the ecosystem. Their pervasive presence and similarity in appearance and size to plankton significantly increase the probability of MPs being ingested by aquatic fauna (Cole et al., 2011). The uptake of MP by aquatic organisms is unintentional due to their inability to distinguish them from natural prey items. After ingestion, MPs may initiate different toxico-physiological responses that will affect the health of aquatic organisms (Wright et al., 2013; De Sá et al., 2018). MPs have been found in the digestive tracts and tissues of organisms across various trophic levels in both freshwater and marine environments. This presence of MPs at different trophic levels indicates their transfer across the aquatic food web (Wright et al., 2013; Crawford & Quinn, 2017). The ingested MPs may accumulate in and even block the digestive tracts of aquatic animals, which thereby results in diminished feeding impetus due to false satiation (Cole et al., 2013; Watts et al., 2015; Welden & Cowie, 2016; De Sá et al., 2018), leading to reduced body weight, growth inhibition, impairment of the reproductive system (Lei et al., 2018), reduced mobility (Rehse et al., 2016), and death (Rist et al., 2016). MP intake can also cause physical damage to the digestive organs, oxidative stress, alterations in enzyme production and metabolism, and bioaccumulation in tissues (Welden & Cowie, 2016; Lei et al., 2018).

MPs have been observed to impede the growth of microalgae and have a detrimental impact on algal photosynthesis (Zhang et al., 2017). They can cause a wide range of physical damage and oxidative stress to algal cells and influence gene expression in specific metabolic pathways (Lagarde et al., 2016; Mao et al., 2018). The formation of aggregates between algal cells and MP particles has been observed to result in the release of extracellular sticky polysaccharides (Long et al., 2017).

The characteristics of MPs, such as their large surface area–volume ratio and hydrophobic nature, facilitate the accumulation of waterborne toxic contaminants like organic pollutants, antibiotics, and heavy metals (Holmes et al., 2012; Mao et al., 2018). MPs also serve as a habitat for diverse microbial communities, some of which can be pathogenic. Moreover, additives present in plastic may leach out, potentially causing toxic effects on aquatic biota. When organisms ingest these contaminated MPs, hazardous substances are introduced into the food web, posing risks to the aquatic ecosystem (Wang et al., 2019). As humans are the final consumers in the aquatic food web, the consumption of plastic-containing aquatic products can act as a major source of MPs in humans (Van & Janssen, 2014).

14.2 FATE AND DISTRIBUTION OF MICROPLASTICS IN THE ENVIRONMENT

The increase in the production and consumption of MPs has become a great concern over the past few decades. Because of their high disposability, limited recovery, durability, and inadvertent release, plastics pile up uncontrollably in the environment (Rocha-Santos & Duarte, 2015; Meng et al., 2020). There are two main sources of MPs: primary MPs, made specifically for a specific industrial or domestic use, such as cosmetics, toothpaste, and resin pellets used in the plastics industry, and secondary MPs, created when larger plastic items break down under the influence of ultraviolet radiation or mechanical abrasion (Auta et al., 2017), such as wastewater from plastic manufacturing industries. Due to their ubiquitous nature, they are present in the atmosphere (Wang et al., 2020), indoor air (Kacprzak & Tijing, 2022), soils or sediments (Huang et al., 2020), freshwater bodies like rivers and streams (Koutnik et al., 2021), and oceans and seas (Cau et al. 2020).

14.2.1 Physical and Chemical Properties of Microplastics

The fate and dispersion of MPs in the environment are significantly influenced by their physical and chemical characteristics, such as size, density, type, and surface characteristics. Due to their potential for easy dissemination in water or air currents, smaller particles typically move farther. Thus, smaller and lighter MPs could become airborne and travel to distant locations such as glacier zones and high mountains (Haque & Fan, 2023). Additionally, due to their smaller size, a vast variety of organisms, including plankton, fish, and other marine animals, can easily consume them by mistaking them for food (Thompson et al., 2009).

In aquatic systems, the density of MPs can influence the vertical dispersion and it has a significant impact on the distribution of MPs (Borges-Ramírez et al., 2020). Typical consumer plastic items have densities of 0.8–1.0 g/cm^3, however, polymers like polyvinyl chloride (PVC) and polyethylene terephthalate (PET) have densities that are higher than that of water (Yang et al., 2021). Thus, less dense polymers are buoyant while those with a density greater than that of seawater sink (Coyle et al., 2020). Since the density of plastic particles will influence their bioavailability in the water column, the sort of plastic that each organism consumes will differ (Wright et al., 2013). Wind can carry MPs that are less dense, which could lead to additional destruction of terrestrial and aquatic habitats (Wang et al., 2021).

MPs are composed of a variety of polymers, some of which include polyethylene (PE), polypropylene (PP), polystyrene (PS), and polyethylene terephthalate (PET). Due to their diverse chemical compositions, the persistence of different polymers in the environment may be influenced by their different rates of degradation and susceptibility to environmental conditions. According to these types, the surface properties of MPs can vary, including roughness, hydrophobicity, functional groups, and the presence of additives. These characteristics may have an impact on how other pollutants attach, which may have an impact on how those substances are transported, disposed of, and interact with living things. The stability of the structure, surface polarity, and hydrophobicity are determined by the kind of polymer and crystallinity, which in turn affects their transport mechanisms, environmental processes, and interactions with biota in the environment (Atugoda et al., 2022). Numerous characteristics of plastic, such as hardness, tensile strength, density, and oxygen permeability can be influenced by the crystallinity (Conradie et al., 2022), which can also influence their distribution in environmental compartments.

14.2.2 Environmental Fate of Microplastics

Numerous strategies have been implemented to reduce their uncontrollable presence in the environment. Due to their slow rate of disintegration, larger plastics have been placed in landfills, while other plastics have been processed through sewage treatment procedures comprising many elimination phases. While the average amount of MPs removed during primary treatment is 72%, secondary and tertiary wastewater treatment methods remove an average of 88% and 94% of MPs, respectively (Sooriyakumar et al., 2022). However, not all MPs are identified and eliminated from wastewater. It is then released back into the environment, where it is transported via several pathways such as domestic waste, industrial waste, and stormwater runoff (Horton & Dixon, 2018; Haque & Fan, 2023). The terrestrial ecosystems may be harmed by MP accumulation in the atmosphere (Wang et al., 2020; Hassan et al., 2023), landfills, soil (Rose et al., 2023b), and freshwater bodies found on land (Horton et al., 2017; Yadav et al., 2023).

According to Enyoh et al. (2019) and Wang et al. (2021), MPs found in soil have the potential to alter the bio-physiochemical properties of soil, which can affect microbial activity. The fate of MPs in the atmosphere can be affected by various factors such as the vertical gradient in pollution concentration, wind speed and direction, precipitation, and temperature. These MPs also act as carriers for various toxic pollutants such as heavy metals, pharmaceuticals, and personal care products (Wang et al., 2021). Due to their combination with other particles, MPs are now considered an emerging component of air pollution (Rose et al., 2023a). Plastic waste from land travels through different

Ecotoxicological Impact of Microplastics in the Environment

channels before reaching the ocean, which is the largest repository for plastic waste. They are spread through oceans and rivers, erosive processes such as storms and erosion, and mismanagement of marine or fishing waste (Horton & Dixon, 2018).

MPs are tiny particles that can sink to the ocean floor, mix with waterborne suspended matter, cling to organic particles, or be ingested by organisms (Rose et al., 2023a). These particles can also enter an organism through inhalation or dermal contact. Once in the environment, MPs can attach to heavy metals like zinc and mercury, which can make them very toxic when consumed (Hassan et al., 2023). In addition, MPs can be absorbed by other bioparticles, such as bacteria and nutrients, and can cause harm when ingested. Studies have found MPs in marine invertebrates like zooplankton and mollusks, as well as in vertebrates like fish and seabirds (Barboza et al., 2019). A review of emerging micropollutants suggests that crustaceans and amphibians are particularly sensitive to MP pollution, as they can inhibit their reproduction (Narwal et al., 2023). Other plastic pollutants, like plasticizers, can affect the hormone function of mollusks and amphibians.

14.2.3 Distribution of Microplastics in Different Environments

Marine environments are considered one of the major sinks of MPs. Once in the sea, MPs are carried by ocean currents around the world, where they persist and accumulate (Lusher, 2015). They can be found in a variety of marine habitats, such as sediments, deep ocean waters, and surface waters (Table 14.1). From zooplankton to larger carnivores, marine organisms at different levels of the food chain can

TABLE 14.1
Distribution of MPs in Different Environmental Compartments from Selected Recent Research Studies (2022–2023)

Country	Environment	Compartment	Polymer/Type	Source
Belgium	F	Surface water Sediment	PS, PP	Semmouri et al., 2023
China	F	Sediment	PET, PC, PP Fragments, fibers	Yan et al., 2022
China	F	Sediment	PP, PE	Zhao et al., 2023
China	F	Surface water Deep water	PS, PP, PE (surface water) PVC, PE, PET (deep water)	Liu et al., 2022
China	T	Soil	PE, PP Fragments and fibers	Cai et al., 2023
Denmark	O	Water column	PES, PP, PE, PVC Fragments and fibers	Gunaalan et al., 2023
India	F	Surface water	PET, cellulose Fibers, fragments, pellets, film, and foam	Warrier et al., 2022
India	M	Shore sediment	PE, PP Filaments	Dhineka et al., 2022
Italy	O	Surface and subsurface water	PE, PP, PS Fragments, sheets, fibers, filaments, foam, pellets	Sbrana et al., 2023
Italy	F	Water Sediment	PET, fibers (water) PVC, fragments (sediment)	Cera et al., 2022
South Africa	F	Surface water Sediment	PE, PP, PET Fibers, film	Apetogbor et al., 2023
Vietnam	W (P)	Sediment	PVC, PE, PP Fragments, film, fibers, foam	Nguyen et al., 2022

Abbreviations: F, freshwater; M, marine; O, ocean; T, terrestrial; W(P), wetlands (peatlands); PC, polycarbonate; PE, polyethylene; PES, polyester; PET, polyethylene terephthalate; PP, polypropylene; PS, polystyrene; PVC, polyvinyl chloride.

ingest MPs. MPs have been found in the top sediment layer of the deep seafloor, indicating that they have penetrated the marine ecosystem to such an extent that they appear to be present throughout the world's oceans and seas, including abyssal depths (Van Cauwenberghe et al., 2013). Once in the sediment, MPs can be consumed by deep-sea benthic organisms, thereby entering the food web. Numerous freshwater matrices, including rivers, lakes, ponds, lagoons, and estuaries, have been found to contain MPs. In each freshwater system, particles are dispersed widely over the benthic zone, coastline, water column, and surface water layers whereas surface water, the water column, the coastline, and marine sediment are all matrices of the ocean where MPs can be found (Atugoda et al., 2022). Urban streams and glaciers showed the highest levels of MPs of all the water bodies (Koutnik et al., 2021).

Compared to the aquatic environment, study of MPs in the terrestrial environment has received less attention and their distribution is quite a complex issue. MPs have been found in the soil, air, and even in places far from populated areas. They can contaminate the soil in several ways, such as when sewage sludge is used as fertilizer or when plastic waste degrades. Through air deposition, MP-containing items like synthetic textiles, and tire wear, they can get into terrestrial habitats. MPs in soil can have an impact on soil health and may get absorbed by plants and end up in the food chain. According to Azeem et al. (2021), plants absorb plastic from the soil through root absorption and transport from root to stem and stem transport to leaves and fruits.

MPs are now pervasive in the oceans all around the planet. They are carried by ocean currents strong dispersion patterns throughout the whole ocean, even to far-off places including the Arctic and Antarctic (Mishra et al., 2021). MPs have been found in these areas despite their remoteness and apparent pristine state, demonstrating the global reach of this concern. The prevalence of MP pollution even in the most pristine and well-protected regions is demonstrated by the discovery of MPs in Antarctic snow (Aves et al., 2022), the freshwater stream of an Antarctic specially protected area (González-Pleiter et al., 2020), and the abundance of MPs in Antarctic krill (Zhu et al., 2023) and Antarctic fish (Zhang et al., 2022). The Arctic has also shown increasing levels of MP pollution. Studies have discovered the presence of MPs in Arctic polar waters (Lusher et al., 2015), Arctic deep-sea sediments (Bergmann et al., 2017), and Arctic benthic organisms (Fang et al., 2018). Although compared to the Arctic ecosystem, Antarctica is still considered to be more untouched, over the years there have been several studies illustrating MP pollution there. Figure 14.1 represents the distribution routes of MPs in different environments.

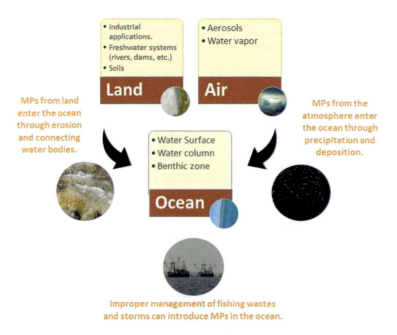

FIGURE 14.1 Distribution of MPs in different environments.

14.3 TOXICOLOGICAL EFFECTS OF MICROPLASTICS

14.3.1 Toxicological Mechanisms of Microplastics

The toxicological mechanisms of MPs in biological species depend on the species and the environmental compartment in which the exposure occurs. MPs adsorb organic contaminants and pathogens from microbial and invertebrate sources present in the environment. Following these, interactions such as synergy between adsorbed compounds and MPs cause toxicological effects (Alimba & Faggio, 2019).

Mainly, the induction of oxidative stress (Figure 14.2) is one of the established toxicological mechanisms of MPs, especially in aquatic organisms (Du et al., 2020). Oxidative stress is a condition that arises when there is an imbalance between the production and accumulation of reactive oxygen species (ROS) in cells and tissues and the body's ability to detoxify these reactive products. Organisms need adequate ROS (Du et al., 2018), and in turn, they play several physiological roles, including cell signaling. They are usually generated as by-products of oxygen metabolism (Pizzino et al., 2017).

However, environmental stressors such as MPs can increase ROS production, leading to an imbalance that causes cell and tissue damage (Pizzino et al., 2017). Oxidative stress occurs in different ways, such as generating free radicals that damage cell components, disruption of the endocrine system, immune-related responses, modified gene expressions, neurotoxicity, reproductive anomalies, and trans-generational effects. Cytotoxicity is another mechanism; although primarily found in terrestrial species like plants, and it is also a toxicological mechanism in aquatic organisms. Regarding cellular activities, cytotoxicity is reflected in apoptotic cell death, metabolic changes, and growth decrease (Du et al., 2020).

Depending on MPs age, size, and adsorption affinity, the degree of effects induced can vary in order of magnitude. Small-sized MPs readily translocate through the gastrointestinal membranes by endocytosis-like processes. As a result, MPs can be distributed into tissues and organs, leading to various effects on aquatic organisms (Alimba & Faggio, 2019). In the case of aged MPs, they have higher adsorptive affinities for organic contaminants, which thus enhances their effects.

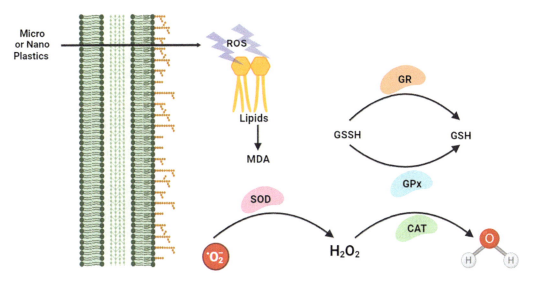

FIGURE 14.2 Oxidative stress due to micro-/nanoplastics (MNPs). CAT, catalase; GPx, glutathione peroxidase; GR, glutathione reductase; GSH, glutathione; GSSH, glutathione disulfide; ROS, reactive oxygen species; MDA, malonic dialdehyde; SOD, superoxide dismutases.

14.3.2 Effects on Bacteria and Phytoplankton

Biological species at all trophic levels, including bacteria and phytoplankton, contain MPs. The toxicity of MPs on bacteria can be determined using endpoints such as inhibition of growth, the efficiency of inorganic nitrogen conversion, the generation of ROS (an indication of oxidative stress), and chemical composition (Sun et al., 2018). MPs inhibit the growth of bacteria, for instance, *Halomonas alkaliphila* (Sun et al., 2018), and reduce the diversity of bacterial communities involved in denitrification and anammox reactions (Nie et al., 2022).

Phytoplankton play a crucial role in aquatic ecosystems, serving as critical components of the ecological community. They provide energy to food webs and contribute significantly to ecosystem functions such as carbon cycling. Generally, phytoplankton are relatively sensitive to MPs (Shiu et al., 2020). High concentrations of MPs can significantly alter the structure of the phytoplankton community. The alteration of the phytoplankton community structure due to high concentrations of MPs can have implications for carbon cycling and the overall health of aquatic ecosystems (Hitchcock, 2022). For instance, microalgae, which are primary producers in aquatic ecosystems, are susceptible to MP contamination, which can significantly impact aquatic food webs (Prata et al., 2019a). In the presence of PS on microalgae, chlorophyll concentration and photosynthetic activity decrease. This disrupts photosynthesis by interfering with the electron characters and producing ROS (Prata et al., 2019a; Jalaudin Basha et al., 2023).

MPs also actively modify the extracellular polymeric substances (EPS) chemical composition of phytoplankton, which respond to stress from pollution. These modifications serve as a coping mechanism for phytoplankton. The secretion of protein-rich EPS by phytoplankton facilitates the formation of aggregates and modifies the surface of plastic particles, influencing their fate and colonization (Shiu et al., 2020). Although induced toxicity is not expected from the current environmental concentrations of MPs, they can reduce the availability or absorption of nutrients or reduce the population of predator species (Prata et al., 2019a).

14.3.3 Effects on Zooplankton and Fish

Zooplankton and fish also endure their fair share of MP toxicity. Toxicity can be caused directly by MPs, when the MPs are ingested. However, there is also the occurrence of indirect toxicity resulting from the release of chemicals associated with the parent plastics, such as plasticizers (Issac & Kandasubramanian, 2021).

The ingestion of MPs has the potential to affect zooplankton fertility, feeding habits, and functioning (Malinowski et al., 2023). PS ingestion, for example, affects the health, feeding habits, and hatching of the copepods, *Calanus helgolandicus* (Cole et al., 2015). In some cases, it can result in death, especially when there is a problem of energy insufficiency (Cole et al., 2015) because there is a higher energy requirement for metabolic and physiological activities in the presence of MPs.

The common zooplankton model species, *Daphnia magna*, are sensitive to MPs. While MPs induce toxicity directly through ingestion by the organisms, they also modify the toxicity of other contaminants such as pesticides (Zocchi & Sommaruga, 2019).

In fish, the effects of MPs are well-known. MPs accumulate in body organs and tissues in fish, including the brain, gastrointestinal tract, dorsal muscle, and gills. MP ingestion by fish induces neurological toxicity which results from lipid oxidative damage (Barboza et al., 2020; Bao et al., 2023). In the brain, the acetylcholinesterase activity increases in the presence of MPs, while lipid peroxidation occurs in the gills, dorsal muscle, and also in the brain (Barboza et al., 2020; Choi et al., 2023). Additionally, MPs interfere with the folds of the intestines and genes related to the immune system of fish (Zhang et al., 2023a). MPs can further promote the bioaccumulation of adsorbed contaminants (e.g., bisphenol A) into the blood and body tissues such as the muscles and liver (Barboza et al., 2020; Hägg et al., 2023).

14.3.4 Effects on Mammals

The study of the effects of MPs in mammals faces difficulties arising from the variety of toxic behaviors of MPs, alongside physical and chemical properties, and also the technical and ethical challenges associated with mammalian studies. MPs can cause serious health problems in top predators like aquatic mammals. These health effects can reduce the population size and even cause extinction, especially in vulnerable populations (Nabi et al., 2022).

As mentioned above, one of the toxicity mechanisms of MPs is the induction of oxidative stress. PS MPs induced oxidative stress in the testicles and ovaries of mice (Meng et al., 2022; Wei et al., 2022). This oxidative stress can lead to a reduction in the number of viable epididymal sperm and spermatogenic cells in the testes, as well as a decrease in ovarian size and follicular quality. The MPs further reduced the rate of pregnancy and production of embryos, and this potentially affects the population (Wei et al., 2022).

Other endpoints that are considered regarding the effects of MPs on mammals include death rate, weight loss, kidney histological damage, and the alternation of biomarkers (Meng et al., 2022). More research is necessary, especially in validating the observed effects of MPs in mammals. Table 14.2 summarizes published literature highlighting the effects of MPs in different species.

14.4 POTENTIAL RISKS OF MICROPLASTICS FOR HUMAN HEALTH

14.4.1 Microplastics in Food

MPs are emerging contaminants of food. Ingestion of food contaminated with MPs is one of the major routes of exposure to humans (Figure 14.3). Human food is exposed to contamination with

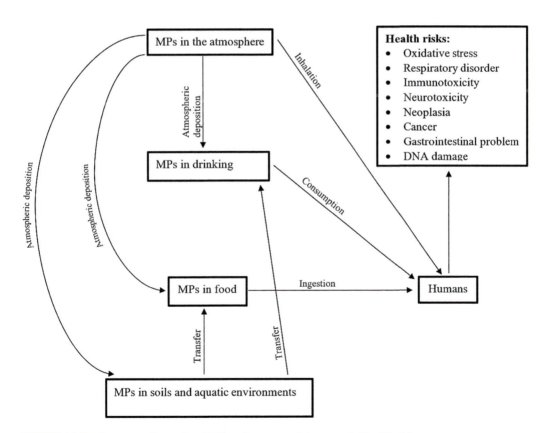

FIGURE 14.3 Sources and transfer of MPs to humans and the potential health risks.

TABLE 14.2

Summary of the Effects of Different Types of MPs on Different Species

MP Type	MP Size	Species	Effects	References
PS	80 nm and 8 μm	Largemouth bass, grass carp and Jian carp	Multiple intestinal abnormalities, upregulation of intestinal immune-related gene	Zhang et al., 2023b
Granular PLA and PVC	PLA (2.52 μm) and CMP-PVC (1.58 μm)	*Oreochromis mossambicus*	Disrupted the metabolic balance of the gut, oxidative stress in gut, damage to gut tissues and organelles, and elevated levels of gut microbiota dysbiosis	Bao et al., 2023
PE and PET	100–350 μm	*Astyanax altiparanae*	Increased mortality from dietary polymer accumulation in the gastrointestinal tract	Lourenço et al., 2023
PS	0.1 mm	Farmed tilapia	Impact on the antioxidant system and immune response	Zheng et al., 2023
Crumb rubber	3–5 mm	*Cyclopterus lumpus*	An uptake of para-phenylenediamines (PDs) into the blood	Hägg et al., 2023
Powder PA	4, 8, 16, 32, and 64 mg/L	*Carassius carassius*	Inhibition of acetylcholinesterase activity, increased stress indicators levels, and decreased immune responses	Choi et al., 2023
PE microspheres	27–32 μm	*Daphnia dentifera* and *Arctodiaptomus dorsalis*	Possibility of impacting the feeding habit of herbivorous zooplankton, hence promote agal growth	Malinowski et al., 2023
PA and PS	PA (38.34 μm) and PS (27.62 μm)	*Siniperca chuatsi*	Enhanced deformity of the musculoskeleton and juvenile mortality	Zhang et al., 2023b
PS beads	6.68 μm	*Acartia tonsa*	Reduction of body length and nauplii survival for exposure during oogenesis	Shore et al., 2021
PE and PA-nylon 6	PE (10–30 μm) and PA 6 (5–20 μm)	*Tigriopus japonicus*	24 h-EC50 of 57.6 and 58.9 mg/L for ingestion rates of females on exposure to PE and PA 6, respectively	Yu et al., 2020
Micro-PS	2 μm	*Tisochrysis lutea, Heterocapsa triquetra, Chaetoceros neogracile*	Potential impact on the distribution and bioavailability in the water column and experimental systems	Long et al., 2017
Micro-PS	2 μm	*Chaetoceros neogracile* and *Rhodomonas salina*	Great decrease in the sinking rates of *C. neogracile* aggregates	Long et al., 2015

Abbreviations: CMP, conventional microplastic; PA, polyamide; PE, polyethylene; PET, polyethylene terephthalate; PS, polystyrene; PVC, polyvinyl chloride

Ecotoxicological Impact of Microplastics in the Environment

MPs in different environmental matrices. Studies have shown that MPs occur in food derived from both plant and animal products (Table 14.3) and may contaminate them together with associated sorbed contaminants. MPs have been detected in foods such as cereals, potatoes, vegetables, fruits, fish and other seafood, meat, dairy products, sugar, honey, and beverages (Pironti et al., 2021).

MPs in food resources arise from the pollution of soils, aquatic environments, and the atmosphere. Food crops and other forms of plants used as food by humans may be contaminated with MPs through the uptake of MPs from soils or the incorporation of MPs in their bodies through atmospheric deposition. Seafood and other aquatic food resources are exposed to contamination with MPs due to the trophic transfer of MPs in food webs. Due to the abundance of MPs in the marine environment, seafood may be reservoirs of MPs. Humans, being at the top of the food web, ultimately ingest MPs through these foods (Vázquez-Rowe et al., 2021).

Species of shellfish (aquatic crustaceans and mollusks), fish, reptiles, birds, and mammals are common items in the food of many people of different cultural backgrounds. Food products from these animals are potentially exposed to contamination with MPs and their affiliated pollutants in the environment. The use of plastics as food containers and packages is a potential source of MPs in human food. When plastic food containers are subjected to high temperatures, they degrade slowly

TABLE 14.3
MPs in Food, Drinking Water, and Air

| Resource/Medium | Microplastics | | | | References |
	Particle	Shape	Size	Polymer	
Food	33–80 MP/kg	Fibers,	<500–5000	PE, PP	Makhdoumi
Sugar		fragments	μm		et al., 2023
Fish	0.72–4.68 MP/fish	Fibers, fragments, foams	1490–4950 μm	PE, PET, PP, PVC, PMMA, PU	Ndibe et al., 2023
Meat (chicken)	4–1190 MP/kg	Fibers	–	–	Prata & Dias-Pereira, 2023
Vegetables (lettuce, broccoli, carrots)	26,375–201,750 MP/g	–	1.36–2.95 μm	–	Garrido & Costanzo, 2022
Fruits (apples, pears)	52,600–307,750 MP/g	–	1.56–3.19 μm	–	Garrido & Costanzo, 2022
Drinking water (raw water, tap water, treated water and bottled water)	0–10,390 MP/L	Fibers, fragments, films	1–1349 μm	PP, PA, PE,PET, PES, PVC, PS, PAM, PMMA, PTT, PPTA	Garrido & Costanzo, 2022
Air	<1–1583 MP/m³	Fibers,	2181–	PET, PE, PP, PES, PS,	O'Brien
Indoor air	< 1 to >1000 MP/	fragments	5000 μm	PU, PVC	et al., 2023
Outdoor air	m³	Fibers, fragments, films, foams, granules	<50–100 μm	PET, PE, PP, PES, PS, PU, PVC, PMMA	O'Brien et al., 2023

Abbreviations: MP, microplastics; PA, polyamide; PAM, polyacrylamide; PE, polyethylene; PES, polyester; PET, Polyethylene terephthalate; PMMA, polymethyl methacrylate; PP, polypropylene; PPTA, p-phenylene terephthalamide; PS, polystyrene; PTT, polytrimethylene; PU, polyurethane; PVC, polyvinyl chloride.

and generate MPs in food (Sewwandi et al., 2023). Based on shape, the common MPs detected in food include pellets, fragments, and fibers. The MP polymers shown to contaminate food include PE, PP, PS, PVC, and PET (Cverenkárová et al., 2021).

14.4.2 MICROPLASTICS IN DRINKING WATER

Drinking water for humans comes from different sources: rainwater, underground water, surface water (rivers, streams, lakes, and reservoirs), water treatment plants (WTPs), and other sources. Drinking water may be obtained directly from raw or untreated sources and treated sources such as tap water, bottled water, and sachet water (in sealed plastic bags). Irrespective of the source of drinking water and packaging material, studies have shown that drinking water is potentially exposed to contamination with MPs. MPs in soils, aquatic environments, and the atmosphere (Figure 14.3) directly or indirectly contribute to the occurrence of MPs in drinking water (Semmouri et al., 2022; Nirmala et al., 2023).

WTPs are not 100% efficient in the removal of MPs from water for drinking purposes. MPs have been detected in both the influents and effluents of WTPs. In developing countries or human communities lacking access to treated or potable water, rainwater contaminated with MPs from the atmosphere may be used for drinking. In addition, surface run-off containing MPs can release MPs, particularly in surface waters where drinking water is obtained. Plastic bottles used in packaging of drinking water have been reported to be one of the sources of MPs in bottled drinking water. The concentration of MPs in bottled drinking water can increase due to the abrasion of plastic bottles (Nizamali et al., 2023).

MPs of different types and concentrations have been detected in drinking water (Table 14.3). The common particle shapes of MPs in drinking water are fragments and fibers. MP polymers which frequently contaminate drinking water are PE, PP, PS, PVC, and PET. Other polymers that have been detected in drinking water include high-density polyethylene (HDPE) and low-density polyethylene (LDPE) (Koelmans et al., 2019; Muhib et al., 2023). Also, the occurrence of MPs with adsorbed pollutants such as polycyclic aromatic hydrocarbons (PAHs), polychlorinated biphenyls (PCBs), metals, and pharmaceuticals in drinking water is a potential threat to human health.

MPs in drinking water and food consumed by humans may be transported to various organ systems of the body, including the digestive, circulatory, respiratory, and nervous systems, and accumulate in organs or tissues of the body.

14.4.3 INHALATION OF MICROPLASTICS

Atmospheric or airborne MPs exist and expose humans to contamination with MPs. Inhalation of MPs is a pathway of exposure of humans to MPs in the atmosphere (Figure 14.3). Different types of MPs of varying concentrations have been detected in indoor and outdoor air and dust (Table 14.3). The composition of atmospheric MPs may include fragments and fibers. In addition, air samples have been shown to contain MP polymers such as PE, PP, PS, PVC, polyester (PES) and polybutadiene (PB) (Rahman et al., 2021; Purwiyanto et al., 2022). The size of MPs influences the distribution of MPs in the air. Small-size and low-density MPs are more likely to be airborne than larger particles. Polymers <1 μm to < 2.5 μm have been detected in air samples. In the atmosphere, air transport of MPs over long distances is possible (Rahman et al., 2021).

Humans may inhale airborne MPs indoors or outdoors and transport them to the respiratory tract. Inhalation of MPs may be followed by deposition and accumulation in the lungs (Jenner et al., 2022). The inhalation of MPs has implications for human health. Oxidative stress in humans caused by inhalation of MPs has been observed.

14.5 MITIGATION STRATEGIES

14.5.1 CURRENT STRATEGIES FOR MICROPLASTIC MITIGATION

MPs in the environment can be mitigated at the source of discharge where plastics directly enter the environment through cosmetic products and detergents or a secondary source where plastics break down into MPs through good waste management practices. Various stakeholders have tackled this problem with different approaches including policy tools, good waste management practices, and incentivizing the use of alternatives. However, due to the global nature of plastic pollution, intergovernmental bodies have a role to play by creating legal frameworks which address the transboundary nature of plastic waste and are binding on its members. To this effect, the Basel Convention of Control of Transboundary Movements of Hazardous Wastes and their Disposal was adopted in March 1989. The objective of the convention was to protect environmental and human health against the adverse impacts of hazardous waste. The Basel Convention is binding on 186 parties and amended its position on plastic waste to include all plastic mixtures with lead and organo-halogenated compounds as hazardous waste (Raubenheimer & McIlgorm, 2018). At the regional level, the EU has adopted REACH, the Marine Strategic Framework Directive (MSFD), and Water Framework Directives (WFD) to regulate the use of plastics and mitigate plastic pollution. These regulations set a minimum target for EU member states. Some of the attempts at plastic mitigation by the EU include a ban on the 10 most widely used single-use plastic products and the target to reduce single-use plastic bags in 1997 (Stuart Braun, 2021). By 2029, the SUP directive has targeted that 90% of PET bottles should be recycled. The EU has set a target to reduce the amount of single-use plastics per person to 40 by 2025 (Directive (EU) 2015/720, n.d.).

At the national level, many countries worldwide have started tackling the problem of plastic waste using various policy tools at their disposal with a focus on single-use plastics and Styrofoam. A few countries, mostly in the Global North, have made efforts to ban microbeads wholly or partially in products (Anagnosti et al., 2021). Policy tools adopted to tackle plastic waste are regulatory tools, economic tools, or a combination of both. Regulatory tools to mitigate plastic waste can be top-down approaches like bans or cooperative approaches like public–private agreements. Economic tools states use to mitigate plastic waste include levies on suppliers, retailers, and consumers, which influence the willingness of the parties affected to use or buy single-use plastic products. These tools have had varying levels of success at the national level. Some of the states with successful plastic mitigation policies are Austria, Switzerland, Sweden, Spain, and Luxembourg. These states have used a combination of public–private partnership and economic levies to reduce single-use plastic bags by 40–80% (UNEP, 2018). Bans instituted at the national level in countries like Cameroon, Gambia, Guinea-Bissau, and Tanzania did not significantly reduce single-use plastic and Styrofoam products due to a lack of enforcement, lack of cheap alternatives, and a lack of awareness of the citizens. The impact of bans at the local level of governance had mixed results, with successes in Himachal Pradesh in India and 23 cities in Indonesia, and failures in Bai Lin (China), Bengal, and Karnataka in India (UNEP, 2018).

Besides policy interventions at the national level, good waste management practices can mitigate plastic waste. Good waste management practice based on the principles of the circular economy includes reducing pollution at source and recycling. Some companies and individuals mitigate plastic waste at source through strategies like installing meshes to prevent the discharge of plastic waste into the environment. Recycling plastics is a complex method which involves sorting plastics from non-plastics, separating recyclables from non-recyclables, sorting by polymer type and color, and extrusion into recycled pellets (Prata et al., 2019b). Recycled plastics are often used in textiles, bottles, and construction materials.

Alternatives like bio-based and biodegradable plastics reduce plastic pollution. Bio-based plastics are plastics wholly or partially made from biomass that do not necessarily disintegrate in the environment, while bio-degradable plastics are plastics that can be broken down by microorganisms in the right conditions (Das et al., 2018). Bioplastics and bio-degradable plastics have advantages over conventional plastics because they are renewable, compostable, and environmentally friendly. Bioplastics and bio-degradable plastics are heavily used in the biomedical industry for bone reconstruction, scaffolds for tissues, and in dentistry (Sidek et al., 2019).

In India, efforts to mitigate plastic pollution have mostly been state led. In 2022, the government banned the use of single-use plastics. The Environment Ministry in India mandates manufacturers of plastic packaging materials to collect their entire production by 2024. They must ensure that a minimum percentage of the plastic produced is recycled and incorporated into the supply chain. The Ministry has introduced a system allowing both manufacturers and users of plastic packaging to acquire Extended Producer Responsibility (EPR) certificates for trading with non-recyclable plastic being eligible for end-of-life disposal (Kapur-Bakshi et al., 2021).

14.5.2 FUTURE RESEARCH NEEDS

The majority of the efforts to mitigate plastic pollution are policy tools. Despite the popularity of policy tools to mitigate plastic pollution, it has its limits. It is of no surprise that plastic waste mitigation efforts were successful in the Global North. For policy tools to be effective, one needs strong state capacity. It can be argued that a country's state capacity determines the type of policy tools to use. In developing countries with large but segmented informal economies, public–private agreements with large supermarket chains, which was typical in Global North countries, are not possible. Research needs to be directed to understand the linkages between stakeholders and the public and how to shape those factors to ensure all parties understand each other and work efficiently to solve the problem of plastic litter and also to improve the state capacity of developing countries to address the challenges of plastic waste.

The problem of plastic pollution demands a multidisciplinary approach and technological solutions that could help to reduce the reliance on policy tools. However, current technological approaches can be costly when scaled up or too time consuming. Bioremediation of plastics has emerged as a promising strategy because microbes have been shown to break down conventional plastics like PET and PE (Wei & Zimmermann, 2017). However, the rate of plastic degradation is low compared to the rate of accumulation. Research should be directed toward engineering novel types of microbes which can break down plastic more efficiently.

Good waste management practices based on the circular economy mitigate plastic pollution. Research should be carried out to improve the efficiency of waste management across all levels from the discharge point to recycling. The bottleneck for recycling is the sorting process, which is complicated and time consuming (Prata et al., 2019b). New approaches in sorting which use artificial intelligence (AI)-based image processing could potentially reduce the amount of time it takes.

14.6 CONCLUSION

MPs cause a variety of toxicity effects on biological species via different toxicity mechanisms. The most common toxicity mechanism is the induction of oxidative stress, which is reflected in various body functioning of organisms. With no known exceptions, MPs have been found in all environmental compartments, and in every trophic level, including bacteria, phytoplankton, zooplankton, fish, and mammals. In bacteria, the inhibition of growth and the alteration of communities are some of the known impacts of MPs, while in phytoplankton, they cause reduced photosynthetic activity, which may further impact the whole population.

It is significant to acknowledge that different habitats can have varying MP distributions depending on their geographic location, human activity, and pollution levels. Understanding the concentrations of MPs through space and time in diverse environments around the globe is essential for comprehending their global distribution, as well as their degradation mechanisms and how they affect the fate and dispersal of MPs in the environment.

ACKNOWLEDGMENT

The first author acknowledges Erasmus Mundus Joint Master's Scholarship for support.

REFERENCES

Alimba, C. G., & Faggio, C. (2019). Microplastics in the marine environment: Current trends in environmental pollution and mechanisms of toxicological profile. *Environmental Toxicology and Pharmacology*, *68*, 61–74.

An, L., Liu, Q., Deng, Y., Wu, W., Gao, Y., & Ling, W. (2020). Sources of microplastic in the environment. In He, D., Luo, Y. (eds) *Microplastics in terrestrial environments: Emerging contaminants and major challenges* (vol 95, pp. 143–159). Cham: Springer. https://link.springer.com/chapter/10.1007/698_2 020_449#Sec13

Anagnosti, L., Varvaresou, A., Pavlou, P., Protopapa, E., & Carayanni, V. (2021). Worldwide actions against plastic pollution from microbeads and microplastics in cosmetics focusing on European policies. Has the issue been handled effectively? *Marine Pollution Bulletin*, *162*, 111883.

Apetogbor, K., Pereao, O., Sparks, C., & Opeolu, B. (2023). Spatio-temporal distribution of microplastics in water and sediment samples of the Plankenburg river, Western Cape, South Africa. *Environmental Pollution*, *323*, 121303.

Arthur, C., Baker, J. E., & Bamford, H. A. (2009). Proceedings of the International Research Workshop on the Occurrence, Effects, and Fate of Microplastic Marine Debris, September 9–11, 2008. Tacoma, WA, USA: University of Washington Tacoma.

Atugoda, T., Piyumali, H., Liyanage, S., Mahatantila, K., & Vithanage, M. (2022). Fate and behavior of microplastics in freshwater systems. In *Handbook of Microplastics in the Environment* (pp. 781–811). Cham: Springer International Publishing.

Auta, H., Emenike, C., & Fauziah, S. (2017). Distribution and importance of microplastics in the marine environment: A review of the sources, fate, effects, and potential solutions. *Environment International*, *102*, 165–176.

Aves, A. R., Revell, L. E., Gaw, S., Ruffell, H., Schuddeboom, A., Wotherspoon, N. E., & McDonald, A. J. (2022). First evidence of microplastics in Antarctic snow. *The Cryosphere*, *16*(6), 2127–2145.

Azeem, I., Adeel, M., Ahmad, M. A., Shakoor, N., Jiangcuo, G. D., Azeem, K., & Rui, Y. (2021). Uptake and accumulation of nano/microplastics in plants: A critical review. *Nanomaterials*, *11*(11), 2935.

Bao, R., Cheng, Z., Peng, L., Mehmood, T., Gao, L., Zhuo, S., … & Su, Y. (2023). Effects of biodegradable and conventional microplastics on the intestine, intestinal community composition, and metabolic levels in tilapia (*Oreochromis mossambicus*). *Aquatic Toxicology*, *265*, 106745.

Barboza, L. G. A., Cózar, A., Gimenez, B. C., Barros, T. L., Kershaw, P. J., & Guilhermino, L. (2019). Macroplastics pollution in the marine environment. In Sheppard, C (ed), *World seas: An environmental evaluation*, 2nd ed (pp. 305–328). Academic Press. ISBN 9780128050521, https://doi.org/10.1016/B978-0-12-805052-1.00019-X.

Barboza, L. G. A., Lopes, C., Oliveira, P., Bessa, F., Otero, V., Henriques, B., & Guilhermino, L. (2020). Microplastics in wild fish from North East Atlantic Ocean and its potential for causing neurotoxic effects, lipid oxidative damage, and human health risks associated with ingestion exposure. *Science of the Total Environment*, *717*, 134625.

Bergmann, M., Wirzberger, V., Krumpen, T., Lorenz, C., Primpke, S., Tekman, M. B., & Gerdts, G. (2017). High quantities of microplastic in Arctic deep-sea sediments from the HAUSGARTEN observatory. *Environmental Science & Technology*, *51*(19), 11000–11010.

Borges-Ramírez, M. M., Mendoza-Franco, E. F., Escalona-Segura, G., & Rendón-von Osten, J. (2020). Plastic density as a key factor in the presence of microplastic in the gastrointestinal tract of commercial fishes from Campeche Bay, Mexico. *Environmental Pollution*, *267*, 115659.

Cai, L., Zhao, X., Liu, Z., & Han, J. (2023). The abundance, characteristics and distribution of microplastics (MPs) in farmland soil—Based on research in China. *Science of the Total Environment*, *876*, 162782.

Cau, A., Avio, C. G., Dessì, C., Moccia, D., Pusceddu, A., Regoli, F., & Follesa, M. C. (2020). Benthic crustacean digestion can modulate the environmental fate of microplastics in the deep sea. *Environmental Science & Technology*, *54*(8), 4886–4892.

Cera, A., Pierdomenico, M., Sodo, A., & Scalici, M. (2022). Spatial distribution of microplastics in volcanic lake water and sediments: Relationships with depth and sediment grain size. *Science of the Total Environment, 829*, 154659.

Choi, J. H., Lee, J. H., Jo, A. H., Choi, Y. J., Choi, C. Y., Kang, J. C., & Kim, J. H. (2023). Microplastic polyamide toxicity: Neurotoxicity, stress indicators and immune responses in crucian carp, Carassius carassius. *Ecotoxicology and Environmental Safety, 265*, 115469.

Cole, M., Lindeque, P., Fileman, E., Halsband, C., & Galloway, T. S. (2015). The impact of polystyrene microplastics on feeding, function and fecundity in the marine copepod Calanus helgolandicus. *Environmental Science & Technology, 49*(2), 1130–1137.

Cole, M., Lindeque, P., Fileman, E., Halsband, C., Goodhead, R., Moger, J., & Galloway, T. S. (2013). Microplastic ingestion by zooplankton. *Environmental Science & Technology, 47*(12), 6646–6655.

Cole, M., Lindeque, P., Halsband, C., & Galloway, T. S. (2011). Microplastics as contaminants in the marine environment: A review. *Marine Pollution Bulletin, 62*(12), 2588–2597.

Conradie, W., Dorfling, C., Chimphango, A., Booth, A. M., Sørensen, L., & Akdogan, G. (2022). Investigating the physicochemical property changes of plastic packaging exposed to UV irradiation and different aqueous environments. *Microplastics, 1*(3), 456–476.

Coyle, R., Hardiman, G., & O'Driscoll, K. (2020). Microplastics in the marine environment: A review of their sources, distribution processes, uptake and exchange in ecosystems. *Case Studies in Chemical and Environmental Engineering, 2*, 100010.

Crawford, C. B., & Quinn, B. (2017). The biological impacts and effects of contaminated microplastics. In Crawford, C. B. & Quinn, B (eds), *Microplastic Pollutants* (pp. 159–178). Elsevier. ISBN 9780128094068, https://doi.org/10.1016/B978-0-12-809406-8.00007-4.

Cverenkárová, K., Valachovičová, M., Mackuľak, T., Žemlička, L., & Bírošová, L. (2021). Microplastics in the food chain. *Life, 11*(12), 1349.

Das, S. K., Sathish, A., & Stanley, J. (2018). Production of biofuel and bioplastic from Chlorella pyrenoidosa. *Materials today: Proceedings, 5*(8), 16774–16781.

De Sá, L. C., Oliveira, M., Ribeiro, F., Rocha, T. L., & Futter, M. N. (2018). Studies of the effects of microplastics on aquatic organisms: What do we know and where should we focus our efforts in the future?. *Science of the Total Environment, 645*, 1029–1039.

Dhineka, K., Sambandam, M., Sivadas, S. K., Kaviarasan, T., Pradhan, U., Begum, M., & Murthy, M. R. (2022). Characterization and seasonal distribution of microplastics in the nearshore sediments of the south-east coast of India, Bay of Bengal. *Frontiers of Environmental Science & Engineering, 16*, 1–11.

DIRECTIVE (EU) 2015/720 OF THE EUROPEAN PARLIAMENT AND OF THE COUNCIL of 29 April 2015 amending Directive 94/62/EC as regards reducing the consumption of lightweight plastic carrier bags (Text with EEA relevance). (n.d.). www.eur-lex.europa.eu/legal-content/EN/TXT/PDF/?uri=CELEX:32015L0720&rid=1 (accessed July 15 2023).

Du, J., Tang, J., Xu, S., Ge, J., Dong, Y., Li, H., & Jin, M. (2018). A review on silver nanoparticles-induced ecotoxicity and the underlying toxicity mechanisms. *Regulatory Toxicology and Pharmacology, 98*, 231–239.

Du, J., Xu, S., Zhou, Q., Li, H., Fu, L., Tang, J., & Du, X. (2020). A review of microplastics in the aquatic environmental: Distribution, transport, ecotoxicology, and toxicological mechanisms. *Environmental Science and Pollution Research, 27*, 11494–11505.

Enyoh, C. E., Verla, A. W., Verla, E. N., Ibe, F. C., & Amaobi, C. E. (2019). Airborne microplastics: A review study on method for analysis, occurrence, movement and risks. *Environmental Monitoring and Assessment, 191*, 1–17.

Fang, C., Zheng, R., Zhang, Y., Hong, F., Mu, J., Chen, M., & Bo, J. (2018). Microplastic contamination in benthic organisms from the Arctic and sub-Arctic regions. *Chemosphere, 209*, 298–306.

Frias, J. P., & Nash, R. (2019). Microplastics: Finding a consensus on the definition. *Marine Pollution Bulletin, 138*, 145–147.

Garrido Gamarro, E., & Constanzo, V. (2022). Microplastics in food commodities: A food safety review on human exposure through dietary sources. Série Sécurité sanitaire et qualité des aliments; FAO: Rome, Italy. ISBN 978-92-5-136982-1.

Gigault, J., Ter Halle, A., Baudrimont, M., Pascal, P. Y., Gauffre, F., Phi, T. L., & Reynaud, S. (2018). Current opinion: What is a nanoplastic?. *Environmental Pollution, 235*, 1030–1034.

González-Pleiter, M., Edo, C., Velázquez, D., Casero-Chamorro, M. C., Leganés, F., Quesada, A., … & Rosal, R. (2020). First detection of microplastics in the freshwater of an Antarctic Specially Protected Area. *Marine Pollution Bulletin, 161*, 111811.

Gunaalan, K., Almeda, R., Lorenz, C., Vianello, A., Iordachescu, L., Papacharalampos, K. & Nielsen, T. G. (2023). Abundance and distribution of microplastics in surface waters of the Kattegat/Skagerrak (Denmark). *Environmental Pollution, 318*, 120853.

Hägg, F., Herzke, D., Nikiforov, V. A., Booth, A. M., Sperre, K. H., Sørensen, L., Creese, M. E., & Halsband, C. (2023). Ingestion of car tire crumb rubber and uptake of associated chemicals by lumpfish (*Cyclopterus lumpus*). *Frontiers in Environmental Science, 11*, 1219248.

Haque, F., & Fan, C. (2023). Fate and impacts of microplastics in the environment: Hydrosphere, pedosphere, and atmosphere. *Environments, 10*(5), 70.

Hassan, F., Prasetya, K. D., Hanun, J. N., Bui, H. M., Rajendran, S., Kataria, N., & Jiang, J. J. (2023). Microplastic contamination in sewage sludge: Abundance, characteristics, and impacts on the environment and human health. *Environmental Technology & Innovation*, 103176.

Hitchcock, J. N. (2022). Microplastics can alter phytoplankton community composition. *Science of The Total Environment, 819*, 153074.

Holmes, L. A., Turner, A., & Thompson, R. C. (2012). Adsorption of trace metals to plastic resin pellets in the marine environment. *Environmental Pollution, 160*, 42–48.

Horton, A. A., & Dixon, S. J. (2018). Microplastics: An introduction to environmental transport processes. *Wiley Interdisciplinary Reviews: Water, 5*(2), e1268.

Horton, A. A., Walton, A., Spurgeon, D. J., Lahive, E., & Svendsen, C. (2017). Microplastics in freshwater and terrestrial environments: Evaluating the current understanding to identify the knowledge gaps and future research priorities. *Science of the Total Environment, 586*, 127–141.

Huang, Y., Liu, Q., Jia, W., Yan, C., & Wang, J. (2020). Agricultural plastic mulching as a source of microplastics in the terrestrial environment. *Environmental Pollution, 260*, 114096.

Issac, M. N., & Kandasubramanian, B. (2021). Effect of microplastics in water and aquatic systems. *Environmental Science and Pollution Research, 28*, 19544–19562.

Jalaudin Basha, N. N., Adzuan Hafiz, N. B., Osman, M. S., & Abu Bakar, N. F. (2023). Unveiling the noxious effect of polystyrene microplastics in aquatic ecosystems and their toxicological behavior on fishes and microalgae. *Frontiers in Toxicology, 5*, 1135081.

Jenner, L. C., Rotchell, J. M., Bennett, R. T., Cowen, M., Tentzeris, V., & Sadofsky, L. R. (2022). Detection of microplastics in human lung tissue using µFTIR spectroscopy. *Science of The Total Environment, 831*, 154907.

Kacprzak, S., & Tijing, L. D. (2022). Microplastics in indoor environment: Sources, mitigation and fate. *Journal of Environmental Chemical Engineering, 10*(2), 107359.

Kalčíková, G., Alič, B., Skalar, T., Bundschuh, M., & Gotvajn, A. Ž. (2017). Wastewater treatment plant effluents as source of cosmetic polyethylene microbeads to freshwater. *Chemosphere, 188*, 25–31.

Kapur-Bakshi, S., Kaur, M., & Gautam, S. (2021). Circular economy for plastics in India: A roadmap. 85. Retrieved from www.teriin.org/sites/default/files/2021-12/Circular-Economy-Plastics-India-Road map.pdf

Koelmans, A. A., Nor, N. H. M., Hermsen, E., Kooi, M., Mintenig, S. M., & De France, J. (2019). Microplastics in freshwaters and drinking water: Critical review and assessment of data quality. *Water Research, 155*, 410–422.

Koutnik, V. S., Leonard, J., Alkidim, S., DePrima, F. J., Ravi, S., Hoek, E. M., & Mohanty, S. K. (2021). Distribution of microplastics in soil and freshwater environments: Global analysis and framework for transport modeling. *Environmental Pollution, 274*, 116552.

Lagarde, F., Olivier, O., Zanella, M., Daniel, P., Hiard, S., & Caruso, A. (2016). Microplastic interactions with freshwater microalgae: Hetero-aggregation and changes in plastic density appear strongly dependent on polymer type. *Environmental Pollution, 215*, 331–339.

Lei, L., Wu, S., Lu, S., Liu, M., Song, Y., Fu, Z., & He, D. (2018). Microplastic particles cause intestinal damage and other adverse effects in zebrafish Danio rerio and nematode Caenorhabditis elegans. *Science of the Total Environment, 619*, 1–8.

Liu, Y., Cao, W., Hu, Y., Zhang, J., & Shen, W. (2022). Horizontal and vertical distribution of microplastics in dam reservoir after impoundment. *Science of the Total Environment, 832*, 154962.

Long, M., Moriceau, B., Gallinari, M., Lambert, C., Huvet, A., Raffray, J., & Soudant, P. (2015). Interactions between microplastics and phytoplankton aggregates: Impact on their respective fates. *Marine Chemistry*, *175*, 39–46.

Long, M., Paul-Pont, I., Hegaret, H., Moriceau, B., Lambert, C., Huvet, A., & Soudant, P. (2017). Interactions between polystyrene microplastics and marine phytoplankton lead to species-specific hetero-aggregation. *Environmental Pollution*, *228*, 454–463.

Lourenço, A. L. A., Olivatto, G. P., de Souza, A. J., & Tornisielo, V. L. (2023). Effects Caused by the Ingestion of Microplastics: First Evidence in the Lambari Rosa (Astyanax altiparanae). *Animals*, *13*(21), 3363.

Lusher, A. (2015). Microplastics in the marine environment: Distribution, interactions and effects. In *Marine Anthropogenic Litter*, 245–307. Cham: pringer International Publishing.

Lusher, A. L., Tirelli, V., O'Connor, I., & Officer, R. (2015). Microplastics in Arctic polar waters: The first reported values of particles in surface and sub-surface samples. *Scientific Reports*, *5*(1), 14947.

Makhdoumi, P., Pirsaheb, M., Amin, A. A., Kianpour, S., & Hossini, H. (2023). Microplastic pollution in table salt and sugar: Occurrence, qualification and quantification and risk assessment. *Journal of Food Composition and Analysis*, *119*, 105261.

Malinowski, C. R., Searle, C. L., Schaber, J., & Höök, T. O. (2023). Microplastics impact simple aquatic food web dynamics through reduced zooplankton feeding and potentially releasing algae from consumer control. *Science of The Total Environment*, *904*, 166691.

Mao, Y., Ai, H., Chen, Y., Zhang, Z., Zeng, P., Kang, L., & Li, H. (2018). Phytoplankton response to polystyrene microplastics: Perspective from an entire growth period. *Chemosphere*, *208*, 59–68.

Mattsson, K., Jocic, S., Doverbratt, I., & Hansson, L. A. (2018). Nanoplastics in the aquatic environment. *Microplastic Contamination in Aquatic Environments*, *17*, 379–399.

Meng, F., Fan, T., Yang, X., Riksen, M., Xu, M., & Geissen, V. (2020). Effects of plastic mulching on the accumulation and distribution of macro and micro plastics in soils of two farming systems in Northwest China. *PeerJ*, *8*, e10375.

Meng, X., Zhang, J., Wang, W., Gonzalez-Gil, G., Vrouwenvelder, J. S., & Li, Z. (2022). Effects of nano- and microplastics on kidney: Physicochemical properties, bioaccumulation, oxidative stress and immunoreaction. *Chemosphere*, *288*, 132631.

Mishra, A. K., Singh, J., & Mishra, P. P. (2021). Microplastics in polar regions: An early warning to the world's pristine ecosystem. *Science of the Total Environment*, *784*, 147149.

Muhib, M. I., Uddin, M. K., Rahman, M. M., & Malafaia, G. (2023). Occurrence of microplastics in tap and bottled water, and food packaging: A narrative review on current knowledge. *Science of The Total Environment*, *865*, 161274.

Nabi, G., Ahmad, S., Ullah, S., Zada, S., Sarfraz, M., Guo, X., & Wanghe, K. (2022). The adverse health effects of increasing microplastic pollution on aquatic mammals. *Journal of King Saud University-Science*, *34*(4), 102006.

Narwal, N., Katyal, D., Kataria, N., Rose, P. K., Warkar, S. G., Pugazhendhi, A., & Khoo, K. S. (2023). Emerging micropollutants in aquatic ecosystems and nanotechnology-based removal alternatives: A review. *Chemosphere*, *341*, 139945. https://doi.org/10.1016/j.chemosphere.2023.139945

Ndibe, L., Ndibe, G., & Patrick, O. (2023). Abundance and seasonal variation of microplastics detected in edible fish sold in Lagos State, Nigeria. *African Journal of Environment and Natural Science Research*, *6*, 158–168.

Nguyen, M. K., Lin, C., Hung, N. T. Q., Vo, D. V. N., Nguyen, K. N., Thuy, B. T. P., & Tran, H. T. (2022). Occurrence and distribution of microplastics in peatland areas: A case study in long An province of the Mekong Delta, Vietnam. *Science of The Total Environment*, *844*, 157066.

Nie, Z., Wang, L., Lin, Y., Xiao, N., Zhao, J., Wan, X., & Hu, J. (2022). Effects of polylactic acid (PLA) and polybutylene adipate-co-terephthalate (PBAT) biodegradable microplastics on the abundance and diversity of denitrifying and anammox bacteria in freshwater sediment. *Environmental Pollution*, *315*, 120343.

Nirmala, K., Rangasamy, G., Ramya, M., Shankar, V. U., & Rajesh, G. (2023). A critical review on recent research progress on microplastic pollutants in drinking water. *Environmental Research*, 115312.

Nizamali, J., Mintenig, S. M., & Koelmans, A. A. (2023). Assessing microplastic characteristics in bottled drinking water and air deposition samples using laser direct infrared imaging. *Journal of Hazardous Materials*, *441*, 129942.

O'Brien, S., Rauert, C., Ribeiro, F., Okoffo, E. D., Burrows, S. D., O'Brien, J. W., & Thomas, K. V. (2023). There's something in the air: A review of sources, prevalence and behaviour of microplastics in the atmosphere. *Science of the Total Environment*, *874*, 162193.

Pettipas, S., Bernier, M., & Walker, T. R. (2016). A Canadian policy framework to mitigate plastic marine pollution. *Marine Policy*, *68*, 117–122.

Pironti, C., Ricciardi, M., Motta, O., Miele, Y., Proto, A., & Montano, L. (2021). Microplastics in the environment: Intake through the food web, human exposure and toxicological effects. *Toxics*, *9*(9), 224.

Pizzino, G., Irrera, N., Cucinotta, M., Pallio, G., Mannino, F., Arcoraci, V., & Bitto, A. (2017). Oxidative stress: Harms and benefits for human health. *Oxidative Medicine and Cellular Longevity*, *2017*(1), 8416763. https://doi.org/10.1155/2017/8416763

Prata, J. C., & Dias-Pereira, P. (2023). Microplastics in terrestrial domestic animals and human health: Implications for food security and food safety and their role as sentinels. *Animals*, *13*(4), 661.

Prata, J. C., da Costa, J. P., Lopes, I., Duarte, A. C., & Rocha-Santos, T. (2019a). Effects of microplastics on microalgae populations: A critical review. *Science of the Total Environment*, *665*, 400–405.

Prata, J. C., Silva, A. L. P., Da Costa, J. P., Mouneyrac, C., Walker, T. R., Duarte, A. C., & Rocha-Santos, T. (2019b). Solutions and integrated strategies for the control and mitigation of plastic and microplastic pollution. *International Journal of Environmental Research and Public Health*, *16*(13), 2411.

Purwiyanto, A. I. S., Prartono, T., Riani, E., Naulita, Y., Cordova, M. R., & Koropitan, A. F. (2022). The deposition of atmospheric microplastics in Jakarta-Indonesia: The coastal urban area. *Marine Pollution Bulletin*, *174*, 113195.

Rahman, L., Mallach, G., Kulka, R., & Halappanavar, S. (2021). Microplastics and nanoplastics science: Collecting and characterizing airborne microplastics in fine particulate matter. *Nanotoxicology*, *15*(9), 1253–1278.

Raubenheimer, K., & McIlgorm, A. (2018). Can the Basel and Stockholm Conventions provide a global framework to reduce the impact of marine plastic litter?. *Marine Policy*, *96*, 285–290.

Rehse, S., Kloas, W., & Zarfl, C. (2016). Short-term exposure with high concentrations of pristine microplastic particles leads to immobilisation of Daphnia magna. *Chemosphere*, *153*, 91–99.

Rist, S. E., Assidqi, K., Zamani, N. P., Appel, D., Perschke, M., Huhn, M., & Lenz, M. (2016). Suspended micro-sized PVC particles impair the performance and decrease survival in the Asian green mussel Perna viridis. *Marine Pollution Bulletin*, *111*(1–2), 213–220.

Rocha-Santos, T., & Duarte, A. C. (2015). A critical overview of the analytical approaches to the occurrence, the fate and the behavior of microplastics in the environment. *TrAC Trends in Analytical Chemistry*, *65*, 47–53.

Rose, P. K., Jain, M., Kataria, N., Sahoo, P. K., Garg, V. K., & Yadav, A. (2023a). Microplastics in multimedia environment: A systematic review on its fate, transport, quantification, health risk, and remedial measures. *Groundwater for Sustainable Development*, *20*, 100889. https://doi.org/10.1016/j.gsd.2022.100889

Rose, P. K., Yadav, S., Kataria, N., & Khoo, K. S. (2023b). Microplastics and nanoplastics in the terrestrial food chain: Uptake, translocation, trophic transfer, ecotoxicology, and human health risk. *TrAC Trends in Analytical Chemistry*, *167*, 117249. https://doi.org/10.1016/j.trac.2023.117249

Sbrana, A., Valente, T., Bianchi, J., Franceschini, S., Piermarini, R., Saccomandi, F., & Silvestri, C. (2023). From inshore to offshore: Distribution of microplastics in three Italian seawaters. *Environmental Science and Pollution Research*, *30*(8), 21277–21287.

Semmouri, I., Vercauteren, M., Van Acker, E., Pequeur, E., Asselman, J., & Janssen, C. (2023). Distribution of microplastics in freshwater systems in an urbanized region: A case study in Flanders (Belgium). *Science of the Total Environment*, *872*, 162192.

Semmouri, I., Vercauteren, M., Van Acker, E., Pequeur, E., Asselman, J., & Janssen, C. (2022). Presence of microplastics in drinking water from different freshwater sources in Flanders (Belgium), an urbanized region in Europe. *International Journal of Food Contamination*, *9*(1), 1–11.

Sewwandi, M., Wijesekara, H., Rajapaksha, A. U., Soysa, S., & Vithanage, M. (2023). Microplastics and plastics-associated contaminants in food and beverages; Global trends, concentrations, and human exposure. *Environmental Pollution*, *317*, 120747.

Shiu, R. F., Vazquez, C. I., Chiang, C. Y., Chiu, M. H., Chen, C. S., Ni, C. W., & Chin, W. C. (2020). Nano-and microplastics trigger secretion of protein-rich extracellular polymeric substances from phytoplankton. *Science of the Total Environment*, *748*, 141469.

Shore, E. A., DeMayo, J. A., & Pespeni, M. H. (2021). Microplastics reduce net population growth and fecal pellet sinking rates for the marine copepod, Acartia tonsa. *Environmental Pollution, 284*, 117379.

Sidek, I. S., Draman, S. F. S., Abdullah, S. R. S., & Anuar, N. (2019). Current development on bioplastics and its future prospects: An introductory review. *INWASCON Technology Magazine, 1*, 3–8.

Sooriyakumar, P., Bolan, N., Kumar, M., Singh, L., Yu, Y., Li, Y., & Siddique, K. H. (2022). Biofilm formation and its implications on the properties and fate of microplastics in aquatic environments: A review. *Journal of Hazardous Materials Advances, 6*, 100077.

Stuart Braun. (2021). 5 things to know about the EU plastics ban. *n.d. Dw. Com* www.dw.com/en/5-things-to-know-about-the-eu-single-use-plastics-ban/a-58109909 (accessed July 16, 2023).

Sun, X., Chen, B., Li, Q., Liu, N., Xia, B., Zhu, L., & Qu, K. (2018). Toxicities of polystyrene nano-and microplastics toward marine bacterium Halomonas alkaliphila. *Science of the Total Environment, 642*, 1378–1385.

Thompson, R. C., Moore, C. J., Vom Saal, F. S., & Swan, S. H. (2009). Plastics, the environment and human health: Current consensus and future trends. *Philosophical Transactions of the Royal Society B: Biological Sciences, 364*(1526), 2153–2166.

Thompson, R. C., Olsen, Y., Mitchell, R. P., Davis, A., Rowland, S. J., John, A. W., & Russell, A. E. (2004). Lost at sea: Where is all the plastic?. *Science, 304*(5672), 838–838.

UNEP. (2018). Single-Use Plastics: A Roadmap for Sustainability. *UN Environment.* www.unep.org/resources/report/single-use-plastics-roadmap-sustainability (accessed July 16,2023)

Van Cauwenberghe, L., & Janssen, C. R. (2014). Microplastics in bivalves cultured for human consumption. *Environmental Pollution, 193*, 65–70.

Van Cauwenberghe, L., Vanreusel, A., Mees, J., & Janssen, C. R. (2013). Microplastic pollution in deep-sea sediments. *Environmental Pollution, 182*, 495–499.

Vázquez-Rowe, I., Ita-Nagy, D., & Kahhat, R. (2021). Microplastics in fisheries and aquaculture: Implications to food sustainability and safety. *Current Opinion in Green and Sustainable Chemistry, 29*, 100464.

Verschoor, A. J. (2015). Towards a definition of microplastics: Considerations for the specification of physico-chemical properties. National Institute for Public Health and the Environment: Bilthoven, The Netherlands, 2015; p. 42

Vo, H. C., & Pham, M. H. (2021). Ecotoxicological effects of microplastics on aquatic organisms: A review. *Environmental Science and Pollution Research, 28*, 44716–44725.

Wang, C., Zhao, J., & Xing, B. (2021). Environmental source, fate, and toxicity of microplastics. *Journal of Hazardous Materials, 407*, 124357.

Wang, W., Gao, H., Jin, S., Li, R., & Na, G. (2019). The ecotoxicological effects of microplastics on aquatic food web, from primary producer to human: A review. *Ecotoxicology and Environmental Safety, 173*, 110–117.

Wang, X., Li, C., Liu, K., Zhu, L., Song, Z., & Li, D. (2020). Atmospheric microplastic over the South China Sea and East Indian Ocean: Abundance, distribution and source. *Journal of Hazardous Materials, 389*, 121846.

Warrier, A. K., Kulkarni, B., Amrutha, K., Jayaram, D., Valsan, G., & Agarwal, P. (2022). Seasonal variations in the abundance and distribution of microplastic particles in the surface waters of a Southern Indian Lake. *Chemosphere, 300*, 134556.

Watts, A. J., Urbina, M. A., Corr, S., Lewis, C., & Galloway, T. S. (2015). Ingestion of plastic microfibers by the crab Carcinus maenas and its effect on food consumption and energy balance. *Environmental Science & Technology, 49*(24), 14597–14604.

Wei, R., & Zimmermann, W. (2017). Biocatalysis as a green route for recycling the recalcitrant plastic polyethylene terephthalate. *Microbial Biotechnology, 10*(6), 1302.

Wei, Z., Wang, Y., Wang, S., Xie, J., Han, Q., & Chen, M. (2022). Comparing the effects of polystyrene microplastics exposure on reproduction and fertility in male and female mice. *Toxicology, 465*, 153059.

Welden, N. A., & Cowie, P. R. (2016). Long-term microplastic retention causes reduced body condition in the langoustine, Nephrops norvegicus. *Environmental Pollution, 218*, 895–900.

Wright, S. L., Thompson, R. C., & Galloway, T. S. (2013). The physical impacts of microplastics on marine organisms: A review. *Environmental Pollution, 178*, 483–492.

Yadav, S., Kataria, N., Khyalia, P., Rose, P. K., Mukherjee, S., Sabherwal, H., & Khoo, K. S. (2023). Recent analytical techniques, and potential eco-toxicological impacts of textile fibrous microplastics (FMPs) and its associated contaminates: A review. *Chemosphere*, 326, 138495. https://doi.org/10.1016/j.chemosphere.2023.138495

Yan, M., Yang, J., Sun, H., Liu, C., & Wang, L. (2022). Occurrence and distribution of microplastics in sediments of a man-made lake receiving reclaimed water. *Science of The Total Environment*, *813*, 152430.

Yang, L., Zhang, Y., Kang, S., Wang, Z., & Wu, C. (2021). Microplastics in freshwater sediment: A review on methods, occurrence, and sources. *Science of the Total Environment*, *754*, 141948.

Yu, J., Tian, J. Y., Xu, R., Zhang, Z. Y., Yang, G. P., Wang, X. D., & Chen, R. (2020). Effects of microplastics exposure on ingestion, fecundity, development, and dimethylsulfide production in Tigriopus japonicus (Harpacticoida, copepod). *Environmental Pollution*, *267*, 115429.

Zhang, C., Chen, X., Wang, J., & Tan, L. (2017). Toxic effects of microplastic on marine microalgae Skeletonema costatum: Interactions between microplastic and algae. *Environmental Pollution*, *220*, 1282–1288.

Zhang, C., Wang, F., Wang, Q., Zou, J., & Zhu, J. (2023b). Species-specific effects of microplastics on juvenile fishes. *Frontiers in Physiology*, *14*. 1256005. https://doi.org/10.3389/fphys.2023.1256005

Zhang, M., Liu, S., Bo, J., Zheng, R., Hong, F., Gao, F., & Fang, C. (2022). First evidence of microplastic contamination in Antarctic fish (Actinopterygii, Perciformes). *Water*, *14*(19), 3070.

Zhang, X., Shi, J., Yuan, P., Li, T., Cao, Z., & Zou, W. (2023a). Differential developmental and proinflammatory responses of zebrafish embryo to repetitive exposure of biodigested polyamide and polystyrene microplastics. *Journal of Hazardous Materials*, *460*, 132472.

Zhao, X., Liu, Z., Cai, L., & Han, J. (2023). Occurrence and distribution of microplastics in surface sediments of a typical river with a highly eroded catchment, a case of the Yan River, a tributary of the Yellow River. *Science of The Total Environment*, *863*, 160932.

Zheng, Y., Addotey, T. N. A., Chen, J., & Xu, G. (2023). Effect of polystyrene microplastics on the antioxidant system and immune response in GIFT (Oreochromis niloticus). *Biology*, *12*(11), 1430.

Zhu, W., Liu, W., Chen, Y., Liao, K., Yu, W., & Jin, H. (2023). Microplastics in Antarctic krill (Euphausia superba) from Antarctic region. *Science of the Total Environment*, *870*, 161880.

Zocchi, M., & Sommaruga, R. (2019). Microplastics modify the toxicity of glyphosate on Daphnia magna. *Science of the Total Environment*, *697*, 134194.

15 Microbial Degradation of Plastic Polymers

Organism Diversity, Mechanism, and Influencing Factors Perspective

Pawan Kumar Rose, Nishita Narwal, Rakesh Kumar, Navish Kataria, Sangita Yadav, and Kuan Shiong Khoo

15.1 INTRODUCTION

Plastics have become an essential component in several aspects of our everyday lives owing to their remarkable attributes, including their affordability, stability, and durability, which are inherent to their polymeric nature (Yadav et al., 2023). The synthetic plastic polymers may be classified into two distinct types, namely the C–C polymers (polypropylene, polyethylene, polyvinyl chloride, and polystyrene), which together account for over 77% of the worldwide market, and the C–O polymers (polyethylene terephthalate and polyurethane), which cover more than 18% of the global market share (Ali et al., 2021). Nevertheless, the inherent characteristics of plastic make it resistant to biodegradation, leading to its build-up rather than breakdown in landfills and other ecosystems, including seas and coastlines (Narwal et al., 2023). The rising presence of plastic garbage in ecosystems poses a significant environmental contamination issue, resulting in detrimental effects on both animal and human populations (Rose et al., 2023a). Consequently, researchers and environmental advocates are actively pursuing an effective and environmentally sustainable strategy to mitigate plastic pollution. The utilisation of microorganisms for the remediation of plastic is considered an environmentally sustainable approach. Microbial enzymes, such as oxidoreductase, laccase, and peroxidase, can break down plastic polymers. The plastic polymers are enzymatically degraded into shorter chains of monomers, dimers, and oligomers. These smaller molecules can readily traverse the cellular membrane and serve as a carbon and energy substrate for microbial organisms. Furthermore, the application of contemporary biotechnological methodologies, including genetic engineering, systems biology, and the creation of synthetic microbial consortia, has been utilised to address the constraints associated with conventional management strategies of plastic pollution. The utilisation of genetic engineering techniques has proven advantageous in manipulating the genetic composition of microbes, hence augmenting their capacity to degrade plastic pollutants (Jaiswal et al., 2020). Therefore, this chapter discusses the contemporary scientific literature on microorganisms, including bacteria, fungi, algae, and invertebrates, as well as natural and manufactured enzymes for their ability to modify and biodegrade various plastic polymers. Additionally, the chapter explores the mechanisms involved in the biodegradation of plastics. Additionally, various factors that influence biodegradation processes and the limitations associated with the biodegradability of plastic polymers are presented.

266 DOI: 10.1201/9781032706573-15

15.2 ORGANISM DIVERSITY AND PLASTIC DEGRADATION

The microbial degradation of microplastics in different environmental matrices is an intricate phenomenon that is affected by various physicochemical factors. Plastics function as carbon and energy substrates for the growth and proliferation of microorganisms. The examination of microorganisms in their axenic condition, namely bacteria and fungi, as well as microbial communities, is often conducted within the framework of plastic degradation. The evaluation of plastic/microplastic degradation can be categorised into three primary classifications: (1) techniques that primarily focus on the elimination of small molecules; (2) techniques that primarily analyse chemical modifications, such as alterations in hydrophobicity and functional groups, within the polymer structure; and (3) techniques that primarily investigate physical changes, such as variations in tensile strength, surface morphology, and crystallinity, related to material properties. Mass loss, carbon dioxide evolution, and gel permeation chromatography techniques have garnered significant attention in academic studies due to their use in assessing degradation processes via the analysis of chemical bond cleavage. Biodegradation is assessed by monitoring changes in chemical functionality using techniques such as nuclear magnetic resonance, infrared spectroscopies, and contact angle measurements. The evaluation of biodegradation may be carried out by many analytical methods, such as dynamic mechanical analysis, thermal analysis, and surface analysis, utilising scanning electron microscopy and atomic force microscopy. These approaches assess changes in material properties as indications of biodegradation (Rose et al., 2023b; Chamas et al., 2020).

15.2.1 BACTERIA

Bacteria are single-celled organisms that exist independently and may be found in diverse habitats around the planet (Urbanek et al., 2018). These organisms possess dimensions on the order of micrometres and are classified under the taxonomic group known as Kingdom Protista. The capacity of some bacterial species to digest plastic/microplastic is attributed to their insolubility in water. The first step of the degradation process involves the attachment of bacterial species to the surface of the polymer (Yeom et al., 2022; Asiandu et al., 2021). Bacteria use plastic as a carbon source, leading to a slow degradation of its structure via changes in its physical and chemical properties (Yuan et al., 2020). The influence of biofilm growth on the bacterial breakdown of plastics has been seen to be significant, as it promotes the adhesion of bacterial colonies to the surface of the plastic and improves their long-term viability (Lobelle and Cunliffe, 2011). According to Dong et al. (2023), a range of bacterial species, such as *Rhodococcus*, *Pseudomonas*, *Acinetobacter*, *Paracoccus*, and *Bacillus*, have shown the ability to break down plastics into smaller sizes (microplastic). Prior studies have shown that the capability of bacteria to degrade plastic/microplastic is based on their intrinsic aptitude to break down long-chain fatty acids. Therefore, it is unsurprising that the breakdown of plastic polymers has attracted considerable interest and inquiry in connection to the bacterial species *Pseudomonas*. Hou et al. (2022) reported that *Pseudomonas aeruginosa* RD1-3 and *Pseudomonas knackmussii* N1-2 could degrade polyethylene microplastics. Several studies have documented the degradation of microplastic polymers by various bacterial species, such as low-density polyethylene by *Pseudomonas putida* and *Pseudomonas syringae* (Kyaw et al., 2012), polypropylene degradation by *Enterobacter* and *Pseudomonas* (Skariyachan et al., 2021), by *Rhodococcus* sp. ADL36 and *Pseudomonas* sp. ADL15 (Habib et al., 2020) and spherical polyphenylene sulphide by *Pseudomonas* sp. (Dong et al., 2023). The degradation of polystyrene and polycarbonate (i.e., thermoplastics) has been observed in the presence of various bacterial species, such as *Pseudomonas aeruginosa*, *Bacillus megaterium*, *Rhodococcus ruber*, *Serratia marcescens*, *Staphylococcus aureus*, and *Streptococcus pyogenes* (Arefian et al., 2020). Previous research has shown that several thermoset polymers, including polyurethane, may undergo degradation when exposed to bacteria from different taxonomic families, such as *Bacillus*, *Pseudomonas*, and *Micrococcus* (Espinosa et al., 2020). The predominant emphasis of existing research has been centred on investigating the biodegradability of individual bacterial strains. However, it is crucial to acknowledge that bacteria in their natural habitats often demonstrate synergistic interactions within consortia. This phenomenon has been seen and recorded in several empirical investigations. The examination of the microbial-facilitated breakdown of microplastics is usually carried out by using bacteria inside a controlled laboratory environment. The aforementioned methodology facilitates the expedient analysis of metabolic pathways, the evaluation of the impact of environmental conditions, and the observation of alterations transpiring in microplastics during the process of degradation (Rose et al., 2023b). The microbial consortia either directly contributed to the process of biodegradation or facilitated the removal of hazardous intermediates, hence enhancing the overall efficiency of biodegradation. Microbial consortia might potentially enhance biodegradation via mechanisms such as metabolic cross-feeding and the production of metabolites that stimulate co-metabolic degradation (Yuan et al., 2020). Mehmood et al. (2016) demonstrated that the degradation rate of polyethylene by certain bacteria, such as *Pseudomonas aeruginosa*, *Burkholderia seminalis*, and *Stenotrophomonas pavanii*, was enhanced when particular additives were introduced. The aforementioned enhancements included the integration of nanoparticles sensitised with food-grade colour and starch (Amobonye et al., 2021).

15.2.2 Fungi

Fungi are a group of eukaryotic organisms that include yeasts and moulds (Carlile et al., 2001). Fungi possess enzymes, such as peroxides and lactase, crucial in breaking down intricate polymers (Rao et al., 2023). These enzymes enable certain species of fungi to engage in plastic biodegradation. Recent studies indicate that the *Aspergillus* species has emerged as the predominant fungus group involved in the biodegradation of artificial polymers. Several studies have reported the ability of certain *Aspergillus* species, namely *Aspergillus clavatus*, *Aspergillus fumigatus*, and *Aspergillus niger*, to degrade polyethylene, polyurethane, and polypropylene. These species were isolated from diverse terrestrial environments (Amobonye et al., 2021). After that, *Penicillium* species are widely recognised as the most prominent and productive strains for the biodegradation of microplastics. Several fungal species have been identified for their significant ability to degrade plastic, including *Fusarium solani*, *Alternaria solani*, *Spicaria* spp., *Geomyces pannorum*, *Phoma* sp., *Penicillium* sp., *Trichoderma viride*, *Zalerion maritimum*, *Eupenicillium hirayamae*, *Phialophora alba*, and *Paecilomyces variotii* (Rose et al., 2023b; Thakur et al., 2023; Zhang et al., 2020). Lignin has similarities to plastic in terms of hydrophobicity and chemical composition, characterised by the presence of non-phenolic aromatic rings, ether linkages, and a carbon skeleton that undergoes oxidation throughout the process of lignin degradation (Ali et al., 2021). According to Jeyakumar et al. (2013), the structural similarity between lignin, polyethylene, and polypropylene allows laccase and manganese peroxidase enzymes to break down these materials effectively. Nevertheless, there is a scarcity of studies examining the fungal-mediated degradation of microplastics, highlighting the challenges associated with identifying fungal strains that possess effective microplastics-degrading capabilities by ectopic screening methods (Rose et al., 2023b; Yuan et al., 2020). Moreover, in contrast to the predominant emphasis on the potential of individual cultures, several studies have shown the collective action of fungal consortia in the efficient degradation of several types of plastics, such as polyethylene (Sowmya et al., 2015a). The research has emphasised fungal enzymes' significance, particularly depolymerase, in several biological processes. In addition, the enzymes possess a wide range of specificity, enabling them to degrade various polymers, a characteristic of considerable importance. According to a recent study conducted by Ekanayaka et al. (2022), it has been shown that Ascomycetes and Basidiomycetes have a significant capacity for plastic degradation when subjected to controlled

Microbial Degradation of Plastic Polymers

laboratory settings. According to Temporiti et al. (2022), it has been noted that fungus peroxidases and laccases possess the ability to break down polyvinyl chloride and polyethylene, whilst lipases and cutinases can degrade polyurethane and polyethylene terephthalate. Several fungal species, such as *Bjerkandera adusta*, *Phanerochaete chrysosporium*, and *Rhizopus oryzae*, are capable of degrading polyethylene plastic polymers (Srikanth et al., 2022). The degradation of high-density polyethylene has been investigated using *Aspergillus flavus* PEDX3 as a catalyst (Zhang et al., 2020). Previous research has shown the biodegradation of polyvinyl chloride by several organisms, including *Phanerochaete chrysosporium* (Nowak et al., 2021), *Cochliobolus* (Amobonye et al., 2021), and *Penicillium* (Pardo-Rodríguez et al., 2021). Olakanmi et al. (2023) reported the degradation of polystyrene by three fungal species, namely *Penicillium*, *Phanerochaete chrysosporium*, and *Pleurotus ostreatus*. The importance of fungal hyphae's dispersion and penetrative capacity has been recognised as a crucial element in their initial colonisation before subsequent depolymerisation and their ability to release hydrophobins to increase hyphal adhesion to hydrophobic surfaces. The pretreatment of various substrates has been shown to enhance the fungal biodegradation of plastics.

15.2.3 ALGAE

Algae are a kind of eukaryotic organism that possess the ability to carry out photosynthesis and contain chlorophyll pigment (Li et al., 2019). However, in contrast to other categories of microorganisms, little research has been conducted to explore the capabilities of algae in the degradation of plastic polymers. Various species of algae, such as *Anabaena*, *Chlorella*, *Spirogyra*, *Nostoc*, *Oscillatoria*, and *Spirulina*, have been seen to inhabit diverse plastic surfaces within terrestrial environments. However, no empirical data demonstrate their ability to metabolise these polymers (Sarmah and Rout, 2018). A study by Kumar et al. (2017) showed that *Scenedesmus dimorphus*, *Anabaena spiroides*, and *Navicula pupula* can degrade high- and low-density polyethylene. Notably, *Anabaena spiroides*, a kind of blue-green algae, exhibited the highest potential in this regard, as it achieved a degradation rate of 8.18% for low-density polyethylene over 30 days. Moreover, Khoironi and Anggoro (2019) reported that *Spirulina* sp. could biodegrade polyethylene terephthalate and polypropylene materials. However, it is worth noting that the degradation rate recorded over 112 days was much lower when compared to bacterial and fungal cells. The results align with expectations since algae, unlike bacteria, rely on atmospheric carbon dioxide as their principal carbon source and use sunlight as their primary energy source (Dineshbabu et al., 2020). Therefore, although these organisms can inhabit plastic surfaces and incorporate microplastics, their metabolic pathways do not possess a natural inclination to degrade them. This is a significant cause for concern, as this incomplete degradation process has been identified as a pathway for plastic accumulation in biological systems and subsequent entry into the food chain (Hoffmann et al., 2020). Nevertheless, a recent investigation has effectively used the notable capabilities of *Phaeodactylum tricornutum* as a genetic host, together with its cost-effective growing settings, to biodegrade polyethylene terephthalate. Moog et al. (2019) successfully introduced the gene responsible for the PETase enzyme, which is highly favoured, into the photosynthetic diatom *Ideonella sakaiensis*. Sarmah and Rout (2019) documented the breakdown of polyethylene polymers by *Oscillatoria subbrevis* and *Phormidium lucidum*. The potential of *Chlorella vulgaris* and *Chlorella fusca* var. *vacuolata*, two species of microalgae, to degrade bisphenol A has been reported by Haiping and Fanping (2023). Larue et al. (2021) observed comparable colonisation outcomes in the case of *Raphidocelis subcapitata* on plastic microbeads. According to Taipale et al. (2023), it has been demonstrated that *Cryptomonas* sp. colonies on polyethylene may attract a diverse range of species that use different plastic polymers as carbon sources. This phenomenon not only facilitates the development of algae but also expedites the decomposition of plastic materials.

15.2.4 INVERTEBRATES

Invertebrates are organisms that lack a vertebral column. Almost 90% of the overall animal species are classified as invertebrates (Imbs et al., 2021). Recently, there has been a notable increase in the focus on the biodegradation of plastic/microplastics, particularly polystyrene, inside the digestive systems of invertebrates. According to Rose et al. (2023b), the larvae of *Tenebrio molitor* (commonly known as yellow mealworms), *Zophobas atratus* (referred to as superworms), *Plodia interpunctella* (known as Indian mealmoths), *Galleria mellonella* (often called bigger waxworms), and *Achroia grisella* (referred to as lesser waxworms) were seen to consume plastic/microplastics and then undergo biodegradation inside their gastrointestinal tracts. According to the findings of Yang et al. (2023), the breakdown of plastic by insect vertebrates may be described as a sequential process consisting of five distinct steps: (1) The ingestion of plastics by insects and subsequent passage through the digestive system. (2) Microorganisms residing inside the gastrointestinal tract of insects can adhere to plastic surfaces and induce the gradual degradation of plastic materials. (3) Plastic undergoes degradation into smaller polymer units by enzymatic processes, including hydrolysis and oxidation. (4) The host provides bio-emulsifying substances to enhance the attacking capability of enzymes. (5) Fatty acids are generated by the cleavage of chemical bonds and the degradation of polymeric polymers due to insect metabolism. The larvae of *Tenebrio molitor* and *Zophobas atratus*, species of darkling beetle, have shown the ability to biodegrade polystyrene and low-density polyethylene within a concise timeframe of a few hours. The larvae of *Tenebrio molitor* can break down many materials, including polyvinyl chloride, polypropylene, and the hydrolysable bioplastic known as polylactic acid. A study by Yang et al. (2015) showed that polystyrene underwent fast biodegradation inside the gastrointestinal tract of *Tenebrio molitor Linnaeus* larvae. Several scientific studies have documented the degradation of plastic polymers by various insect species. For instance, the degradation of polyether-polyurethane foam by *Tenebrio molitor* (Liu et al., 2022), polystyrene by *Zophobas atratus* (Kim et al. 2020), low-density polyethylene and polystyrene by *Galleria mellonella* (Lou et al., 2020), polyethylene by *Plodia interpunctella* (Mahmoud et al., 2021), polystyrene by *Tribolium castaneum* (Wang et al., 2020), polystyrene and polyethylene by *Tribolium confusum* (Bilal et al., 2021), high-density polyethylene by *Achroia grisella* (Kundungal et al., 2019), polyvinyl chloride by *Spodoptera frugiperda* (Zhang et al., 2022), low-density polyethylene by *Corcyra cephalonica* (Kesti and Thimmappa, 2019), and polystyrene by *Stegobium paniceum* (Agrafioti et al., 2023). Peng et al. (2019) revealed that *Tenebrio obscurus* had a greater capacity for polystyrene breakdown throughout the gastrointestinal tract (26.03%) compared to *Tenebrio molitor* (11.67%). According to Yang et al. (2021), it has been proposed that the intestinal digestive system of *Tenebrio obscurus* has the ability to depolymerise low-density polyethylene. This process is facilitated by certain bacterial families, including Enterobacteriaceae, Enterococcaceae, and Streptococcaceae, as well as specific genera such as *Spiroplasma* sp. and *Enterococcus* sp. In a study conducted by Song et al. (2020), the researchers investigated the capacity of *Achatina fulica*, often known as land snails, to break down polystyrene. The findings of the study revealed that the process of polystyrene biodegradation was linked to certain gut microbes, including the families Enterobacteriaceae, Sphingobacteriaceae, and Aeromonadaceae (Rose et al., 2023b).

15.3 MECHANISM OF PLASTIC BIODEGRADATION

Plastic biodegradation refers to transforming persistent waste materials into less harmful, smaller molecular components that may be reintegrated into the biogeochemical cycle. The process of plastic polymer biodegradation may be discerned by examining the modifications in the physical characteristics of the polymers, particularly the decrease in molecular weight, decline in mechanical strength, and alteration of plastic surface properties (Ho et al., 2018). The process of polymer biodegradation begins with the first adhesion of microorganisms and the subsequent development of microbial biofilms on the plastic surface, which is referred to as the plastisphere. The plastisphere is known to support unique microbial communities and serve as a habitat for prospective microorganisms capable of decomposing plastic materials (Kirstein et al., 2019). According to Lobelle and Cunliffe (2011), the formation of microbial biofilms on plastic surfaces occurs quickly, leading to a notable reduction in plastic buoyancy and hydrophobicity. In the study conducted by Sivan et al. (2006), it was seen that *Rhodococcus ruber* successfully colonised and developed three-dimensional structures resembling mushrooms inside the mature biofilm. Plastic biodegradation encompasses many distinct processes, including biodeterioration, biofragmentation, assimilation, and mineralisation. These processes are facilitated by various enzyme activity and bond breakage mechanisms (Pathak, 2017).

15.3.1 BIODETERIORATION

Biodeterioration refers to the process by which the chemical, mechanical, and physical characteristics of polymers are altered due to the chemical and physical activities of microorganisms and other biological agents. These agents are affected by external factors like light, temperature, and environmental pollutants (Anjana et al., 2020). The adhesion and colonisation of microorganisms, as well as the production of microbial biofilms on the plastic surface (also known as a "plastisphere"), are the first steps in the process of polymer biodeterioration. Plastics lose their buoyancy and hydrophobicity when microbes form biofilms on them (Nauendorf et al., 2016). Polyethylene and other plastics with a high surface hydrophobicity need biofilm development to facilitate bacterial contact with the polymeric surface (Schwibbert et al., 2019). *Pseudomonas* and other biofilm-forming bacteria are known to attach more firmly to and destroy low-density polyethylene in comparison to other bacteria in the planktonic mode (Tribedi et al., 2015). On the fully developed biofilm, *Rhodococcus ruber* colonised and developed "mushroom-like" three-dimensional structures. Biofilms formed by bacteria provide a buffer against environmental stresses, allowing the microbial population to thrive in a wide range of environments. Microorganisms, using the endoenzymes and exoenzymes they release, may break down the plastic polymers. Microbe-secreted exopolysaccharides aid in biofilm's ability to stick to plastic. Polymer biodegradation is heavily influenced by exopolysaccharides and enzymes. *Rhodococcus ruber*, a polyethylene-degrading bacterium, has exopolysaccharides in its biofilm that were up to 2.5 times greater than the protein level (Sivan et al., 2006). The attachment and development of fungus on polymer solids causes localised swelling, leaving a polymer with markedly diminished mechanical characteristics (Amobonye et al., 2021). Fungus can grow on almost any surface in nature. White-rot and brown-rot fungi in particular are superior to bacteria at breaking down microplastics because their mycelia can penetrate deeply into the surface of polymeric substances and they secrete large amounts of extracellular enzymes (like lignin peroxidase, manganese peroxidase, versatile peroxidase, and multi-copper oxidase laccase) that break down the polymers into their oligomers, dimers, and monomers (Rose et al., 2023b; Ali et al., 2021). These polymers outside of cells operate as surfactants, helping bacteria penetrate more easily by promoting interactions between hydrophilic and hydrophobic phases (Lucas et al., 2008).

15.3.2 BIOFRAGMENTATION

Biofragmentation is the second phase in the degradation process, whereby biodeteriorated plastic polymers are depolymerised via catalytic cleavage. This process is facilitated by extracellular enzymes and free radicals produced by microorganisms (Jenkins et al., 2019). Polymers with a carbon–carbon backbone often provide difficulties during their degradation mechanism due to their high molecular weight, strong covalent bonds, and scarcity of reactive functional groups. Polymers such as polyethylene, polypropylene, and polyvinyl chloride undergo hydroxylation, synthesising primary alcohols. These alcohols transform into aldehydes and undergo further oxidation to provide carboxylic acids (Ru et al., 2020). Concurrently, the intermediates have the potential to undergo catalysis by subterminal oxidative monooxygenases, leading to the production of secondary alcohols. The dehydrogenases catalyse the oxidation of these secondary alcohols, resulting in the formation of ketones. The process of ketones undergoing subsequent transformation into esters, followed by the breakage is facilitated by esterases, keratinases, and lipases (Gao et al., 2022). Hence, the biofragmentation process is postulated to include two primary processes: the decrease in the polymer's molecular weight and the oxidation of the resulting lower-weight molecules. The aforementioned reactions play a crucial role in promoting the following functioning of microbial enzymatic systems, which often possess the capability to degrade smaller molecules (Restrepo-Flórez et al., 2014). The process of enzymatic depolymerisation of polymers results in the liberation of monomers that microbes may use as a carbon source, hence promoting an augmentation in microbial biomass (Degli-Innocenti, 2014). There exists a considerable body of data indicating the participation of enzymes, such as amidases, oxidases, laccases, and peroxidases, in the process of polymer breakdown (Gomez-Mendez et al., 2018).

15.3.3 Assimilation and Mineralisation

During the assimilation phase of plastic degradation, the low-molecular-weight molecules (oligomers with 10–50 carbon atoms) produced by depolymerisation are taken up by the cells of plastic-degrading organisms before the last step of plastic degradation, mineralisation. Although the mechanisms by which plastic molecules are assimilated across various microbial membranes remain unclear, it is hypothesised that, like hydrocarbons, they require both active and passive transport. According to Shahnawaz et al. (2019), octadecane, a byproduct of plastic polymer degradation, has been observed to be internalised by *Pseudomonas* sp. DG17 through facilitated passive transport mechanisms when present in higher concentrations. Conversely, it is assimilated at lower concentrations through energy-dependent active transport processes (Hua et al., 2013). Various membrane transport systems have been demonstrated to aid in translocating these substances into the cytoplasm for subsequent processing, for instance, terephthalic acid, a hydrolytic product of polyethylene terephthalate by a specific transporter (Hosaka et al., 2013), polyethylene glycol, a plastic degradative product by porins (Duret and Delcour, 2010), and polyethylene oligomeric intermediates by ATP binding cassette family of proteins (Amobonye et al., 2021). Once inside the cell, these tiny molecules are mineralised into water and carbon dioxide by aerobic metabolism or methane and carbon dioxide via anaerobic metabolism (Lim and Thian, 2022). One hypothesised mechanism for polyethylene breakdown involves the creation of acetic acid, which either acetyl-CoA generation or lipid synthesis may metabolise, ultimately entering the Krebs cycle (Wilkes and Aristilde, 2017). Tribedi and Sil (2014) showed that succinate, an intermediate in the Krebs cycle, may be produced by the breakdown of polyethersulphones by the esterase enzyme in *Pseudomonas* sp. AKS2. Furthermore, terephthalic acid undergoes metabolic processes resulting in the production of protocatechuic acid (PCA), which then undergoes a series of enzymatic reactions leading to the formation of 2-pyrone-4,6-dicarboxylic acid. This compound is subsequently directed into the Kreb's cycle as pyruvate and oxaloacetate, which finally undergoes complete mineralisation to carbon dioxide and water (Yoshida et al., 2016). Ali et al. (2023) compared the carbon dioxide generation in small waxworm larvae that were deprived and those that were given low-density polyethylene. The findings of the study indicate that, on the 15th day of the trial, larvae that were provided with low-density polyethylene exhibited a carbon dioxide production of 1.95 g, which was 85.7% more than the carbon dioxide production of larvae that did not get any low-density polyethylene supply, measuring at 1.05 g. The researchers determined that the overall efficiency of converting low-density polyethylene carbon into minerals in the larvae reached a maximum of 97.2%. This conversion process resulted in 62.3% of the carbon being transformed into biomass, 32.3% into carbon dioxide, and 2.6% into biomass. The process of mineralisation may occur under

Microbial Degradation of Plastic Polymers

either aerobic or anaerobic conditions. In both scenarios, the enzymatic activity of esterases, lipases, cutinases, peroxidases, and laccases is necessary for the mineralisation stage (Wu et al., 2023).

15.4 NATURAL AND ENGINEERED ENZYMES IN PLASTIC DEGRADATION

Enzymes are catalysts with biological activity that modulate the rate of a chemical process by particular substrate binding and subsequent production of a distinct product. The heightened interest among scientists in understanding the process by which plastic materials' increasing deterioration and disintegration has spurred microbes to break down plastics. The underlying process may be attributed to the existence of enzymes inside microbes. Enzymes may be categorised as either natural or artificial. Various organisms, such as fungi, algae, invertebrates, and bacteria, possess distinct enzymatic mechanisms for decomposing plastic materials (Figure 15.1). Likewise, experimental production of enzymes is possible. Similarly, enzymes can be synthesised in the laboratory.

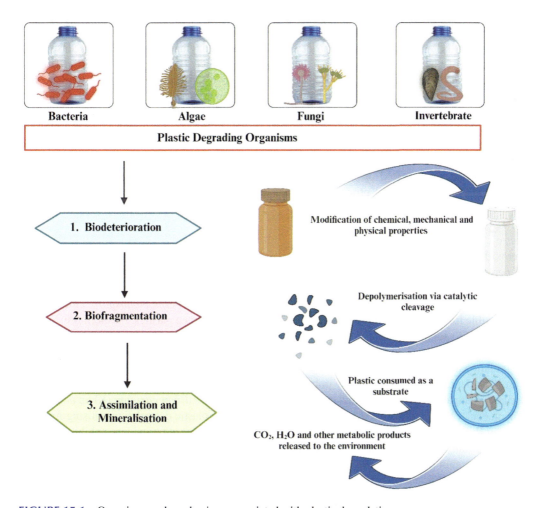

FIGURE 15.1 Organisms and mechanisms associated with plastic degradation.

15.4.1 Natural Enzymes

Natural enzymes are recognised for their remarkable efficacy and adaptability as biocatalysts, fulfilling pivotal functions in several processes. Proteins constitute the bulk of enzymes, serving as complex, sizable, chiral organic structures that often include ions and water molecules to maintain their enzymatic capabilities and structural integrity.

15.4.1.1 Oxidoreductases

The enzymes responsible for facilitating substances' oxidation and reduction reactions are referred to as oxidoreductases. Catalysis occurs via the transfer of electrons from the oxidant to the reductant. Oxidoreductases may be classified into two distinct types, namely dehydrogenases and oxidases. The enzymes with the most tremendous significance in plastic breakdown, categorised by their respective classes, are enumerated as follows.

Laccase

Laccases are classified as multi-copper enzymes since they possess spectroscopically different copper ions and function as phenol oxidases. Laccases are well acknowledged for their capacity to enzymatically degrade phenols, lignin, and diamines that have aryl moieties. The laccase enzyme exhibited a notable response when exposed to low-density polyethylene. Laccase is an enzyme belonging to the AlkB family (Othman et al., 2021). The enzymatic influence of laccase on the degradation of phenolic compounds has been examined since 1883, after its isolation from the *Rhus vernicifera* tree native to Japan. Laccase has been well documented as the enzyme most often connected with high-density polyethylene breakdown. Laccase, an enzyme belonging to the oxidase group, has been seen to facilitate the depolymerisation of polymers by oxidatively cleaving the amorphous area of high-density polyethylene. This process creates a carbonyl region inside the polymer chain, which is readily accessible (Kang et al., 2019). A scanning electron microscopy study showed that after 90 days of incubation in the presence of the laccase enzyme, pits and fractures appeared on the high-density polyethylene surface (Kang et al. 2019). Santo et al. (2013) documented the process of polyethylene biodegradation using laccases derived from *Rhodococcus ruber*.

Peroxidase

Peroxidases are a group of enzymes that play a role in facilitating redox reactions via the cleavage of peroxides. The active redox centres of the system consist of chemical moieties, including cysteine heme, selenium, manganese, thiols, and other chemical groups. Du et al. (2022) reported that dye-decolorising peroxidases derived from *Nostocaceae* and *Thermomonospora curvata* were responsible for the oxidative destruction of pre-oxidised polystyrene plastics. Furthermore, Sowmya et al. (2015b) reported that a consortium consisting of *Curvularia lunata*, *Alternaria alternata*, *Penicillium simplicissimum*, and *Fusarium* sp. exhibited the ability to degrade polyethylene. This degradation was attributed to the specific activity of the manganese peroxidase enzyme and laccase. The degradation of low-density polyethylene occurred due to the enzymatic activity of lipase, laccase, manganese peroxidase, and esterase derived from *Penicillium citrinum* (Khan et al. 2023). Enyoh et al. (2022) conducted a comparative investigation on plastic degradation utilising Autodock Vina in PyRx software. The study findings indicated that manganese peroxidase exhibited superior binding affinity, enhancing plastic degradation efficiency. Extracellular enzymes, manganese peroxidase, and lignin peroxidase, improve the hydrophilicity of microplastics by catalysing the conversion of their functional groups into alcohol or carbonyl groups. The transition from a hydrophobic to a somewhat hydrophilic nature facilitates the subsequent microbial metabolism (Chamas et al., 2020). The successful completion of this stage is of utmost importance in facilitating the transition towards the absorption of digested wastes by microbes. During the assimilation phase, the polymeric

Microbial Degradation of Plastic Polymers

particles traverse the semi-permeable membrane of the bacterium and get localised inside the cytoplasm, where they undergo further breakdown by intracellular enzymes (Kumar et al., 2022).

Hydrolase

Hydrolases are a class of enzymes that facilitate the hydrolysis of molecules, often functioning as biocatalysts and utilising water as a source of hydroxyl groups during the breakdown of substrates. The primary mechanism by which plastics degrade in the environment is the action of hydrolases (Müller et al., 2005). The enzymatic hydrolysis by a hydrolase enzyme enables substrate degradation in the presence of water. The process of cleaving extended carbon chains is a consecutive two-step mechanism. It is well-recognised that all polymers found in the natural environment exhibit hydrophobic properties. The first stage of enzyme–polymer interaction entails the occurrence of hydrophobic contacts between extracellular enzymes synthesised by microorganisms and the surface of the plastic material. Many hydrolases include a hydrophobic gap at the active site to accommodate polymer hydrophobic groups. This structural feature enhances the accessibility of the enzyme to the polymer. During the subsequent phase, the active site of the enzyme facilitates the hydrolysis of extensive polymer chains into monomers or dimers, which may be used by the microbial cell as a source of carbon. Certain hydrolases can generate energy upon activation. Hydrolases, particularly proteases, are attached to biological membranes by either a single transmembrane helix or peripheral membrane proteins. Other proteins have a multi-span transmembrane topology. Hydrolases are often referred to as "substrate hydrolase", however, a majority of them are instead designated as "substratease". In a study conducted by Müller et al. (2005), the authors documented the degradation of polyethylene terephthalate through the action of a hydrolase enzyme derived from *Thermobifida fusca*.

LIPASE

Lipase is an enzyme catalyst that facilitates the hydrolysis of ester bonds within triglycerides, transforming these compounds into glycerol and free fatty acids. Various fungal species, including *Thermomyces lanuginosus*, *Candida rugosa*, *Rhizopus delemer*, and *Candida antarctica*, have been recognised as lipase producers and have been implicated in the process of plastic breakdown (Vertommen et al., 2005; Eberl et al., 2009). *Rhizopus delemer* lipase degrades polyester-type polyurethane film by approximately 53% in a 24-hour reaction period (Tokiwa and Calabia, 2009). The effectiveness of Lipase B from *Candida antarctica* in hydrolysing polyethylene terephthalate to terephthalic acid was demonstrated by Carniel et al. (2017). Furthermore, Furukawa et al. (2019) investigated the enzymatic decomposition using lipase FE-01 from *Thermomyces languinosus*, which facilitates the breakdown of electrospun polycaprolactone fibre, suggesting its potential application in the field of biodegradable polymers. Further research in this area is warranted to explore the full potential of lipases in polymer degradation and to optimise their enzymatic activity for industrial applications.

PETASE

Polyethylene terephthalate is a widely used plastic that may be degraded into its monomeric form, mono-2-hydroxyethyl terephthalate, via hydrolysis. Enzymes known as PETases, which fall under the esterase class, have shown great potential as catalysts for this degradation process. The crystallographic analysis of PETase reveals its structural resemblance to both cutinase and lipase enzymes. The enzyme PETase has a structural shape typical of a/b hydrolases and demonstrates a higher level of accessibility to its active site than cutinase. The enzyme PETase has a physically accessible active site crevice that can accommodate 4-methylcatechol (4MHET) moieties. The catalytic triad comprises Ser160, Asp206, and His237, which have structural resemblances to other hydrolases. PETase can break the polymeric chain by both endo-fashion and exo-fashion pathways, in addition to its surface activity. Sevilla et al. (2023) observed the enzymatic breakdown of polyethylene terephthalate by the bacterium *Ideonella sakaiensis*. PETase, which exhibits similarities to cutinase, has been recognised as a potential depolymerisation enzyme for the polymer into its elemental components. Zhu et al. (2022) introduced a novel biocatalyst named BIND-PETase. This biocatalyst exhibits considerable potential for the biocatalytic degradation of polyethylene terephthalate plastics. The researchers achieved this by utilising genetic engineering techniques to integrate a functional PETase enzyme into the Curli of an *Escherichia coli* cell.

MHETase

The enzyme MHETase facilitates the hydrolysis of the ester bond found in MHET, producing terephthalic acid and ethylene glycol. The enzyme under consideration is synthesised concurrently with strain *I. sakaiensis* during the polyethylene terephthalate breakdown process. The enzyme MHETase is characterised by the presence of five disulphide bonds, one of which has a side chain including a catalytic triad composed of the amino acids S225, D492, and H528. This particular disulphide bond also contains an oxyanion hole, which is characterised by an amide nitrogen backbone formed by the amino acids G132 and E226. The existence of these architectural components enhances the process of MHET degradation by MHETase. The precise classification of MHETase among the enzyme categories of feruloyl esterase or tannase is still unclear, however its structural conformation demonstrates an apparent affinity for the para-carboxy group of the substrate.

15.4.2 Engineered Enzymes

Engineered enzymes refer to novel biocatalysts artificially synthesised by modifying the amino acid sequences that make up their structure. These modifications are made to enhance the enzymes' characteristics and functionality. The use of metagenomics and proteomic methodologies becomes advantageous in building enzymes capable of degrading plastic materials (Figure 15.2). Protein

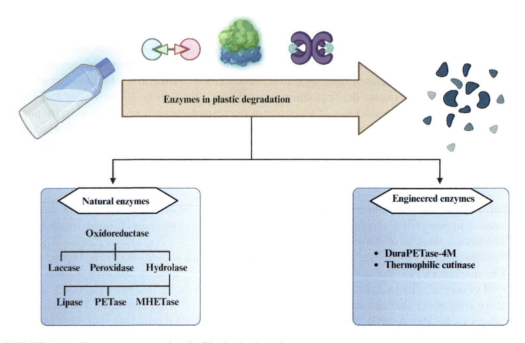

FIGURE 15.2 The enzymes associated with plastic degradation.

Microbial Degradation of Plastic Polymers

engineering has the potential to enhance the efficiency of these enzymes. By use of this process, the thermostability, performance, and active site may be improved (Perera et al., 2023). Directed evolution, semirational design, and rational design are prominent computational methods extensively employed in the field of enzyme engineering. Various changes may be performed depending on the structure of the binding site. These alterations include strengthening the contact between the enzyme and substrate, enhancing the thermal stability of the enzyme by electrostatic and chemical modifications, and introducing a disulphide link. Genetic engineering is a widely used technique that enables the production of significant amounts of degradative enzymes by engineered microbes via the efficient expression of imported, functioning foreign genes. An example of a thermophilic anaerobic bacterium, *Clostridium thermocellum*, has been genetically modified to synthesise thermophilic cutinase. It has been demonstrated that this bacterium can degrade 62% of a polyethylene terephthalate film (250 μm in thickness, 2 × 0.8 cm) into soluble monomer feedstocks within 14 days at a temperature of 60°C (Yan et al., 2021). Liu et al. (2022c) successfully produced a modified enzyme called "DuraPETase-4M" through expression in the *Escherichia coli* host JM109. This engineered enzyme exhibited remarkable thermal stability and biodegradability towards polyethylene terephthalate by enhancing its overall rigidity (N233C/S282C), modifying the flexibility of crucial regions (H214S), and optimising the protein's surface electrostatic charge (S245R). The altered enzyme, DuraPETase-4M, exhibited notable improvements in thermal stability and biodegradability. It demonstrated the ability to destroy amorphous polyethylene terephthalate powder and polyethylene terephthalate pre-formed film at a greatly accelerated pace, about 3.2 and 5.4 times quicker, respectively, compared to the unmodified DuraPETase enzyme (Wu et al., 2023).

15.5 FACTORS AFFECTING PLASTIC DEGRADATION

The degradation process is influenced by two main factors: intrinsic characteristics and external environmental circumstances that impact its degradation behaviour. The primary inherent characteristics of (micro)plastics include their composition, substance, structure, and form of existence. External environmental conditions include various factors such as temperature, pH level, humidity, diversity of microbes, hydrolysis, catalysts, enzymes, and other relevant variables (Figure 15.3).

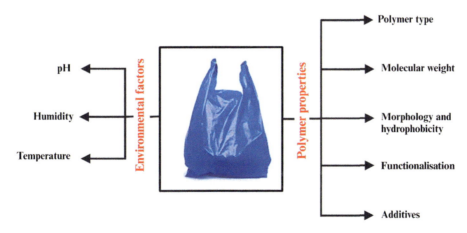

FIGURE 15.3 Factor affecting plastic degradation.

15.5.1 External Environment Conditions

The primary determinants of the external environment include catalysts, acidity, temperature, light, oxygen, dynamic elements such as shear stress, and microbes within the surrounding milieu. The pH level significantly influences the deterioration process of plastic materials. The pH value influences the degradation of microplastic, as it regulates the survival and activity of microorganisms. This, in turn, affects the structure of microbial populations, enzyme activity, and degradation rates (Auta et al., 2018). The optimisation of pH settings positively impacts the proliferation and activity of microorganisms significantly, hence maximising the breakdown of plastic/microplastic. As shown by Auta et al. (2018), the *Rhodococcus* sp. strain 36 exhibited enhanced growth rates and reached its peak microplastic degradation rates at pH 8.96. Furthermore, Xu et al. (2011) documented that microorganisms can efficiently break down polylactic acid in alkaline conditions with a pH of 8.0. During the microbial degradation process, some metabolites generated by microorganisms have the potential to influence pH levels, hence causing disruption to the structure of microplastics and facilitating their destruction (Shabbir et al., 2020). According to Tamnou et al. (2021), the presence of an acidic pH in the surrounding environment has been seen to impede the biodegradation of polyethylene polymers by the bacteria *Pseudomonas aeruginosa*. Temperatures could also affect plastic degradation efficiency. According to Gewert et al. (2015), the fragmentation of plastic at a temperature of 0°C resulted in the production of smaller particles with a greater surface area-to-volume ratio than the original plastic/microplastic. The microplastic undergoes fracturing when exposed to a temperature of 0°C (Ariza-Tarazona et al., 2020). This fracturing phenomenon leads to an increase in the surface area-to-volume ratio of the microplastic. The effect of temperature in the process of plastic deterioration is considered to be auxiliary rather than primary. The biodegradation process is influenced indirectly by its impact on biological activity, while the degradation rate is also indirectly influenced by the physical and chemical qualities of materials or the activation energy involved in the reaction process during photocatalytic degradation. Hence, the impact of elevated or reduced temperatures on the deterioration of plastics is not universally conclusive but rather contingent upon the underlying reaction mechanism governing the process of plastic degradation. Therefore, the impact of reduced temperature on plastic crushing is readily apparent, resulting in an increase in the contact area for the reaction. Hence, the occurrence of degradation is facilitated by low temperature in the presence of a catalyst. Moreover, without a catalyst, biodegradation facilitates the breakdown of plastics in conditions conducive to microorganisms' development and reproduction. The study conducted by Chen et al. (2020) revealed that plastic decomposition rates were greatly enhanced using ultra-high-temperature composting compared to high-temperature composting. This improvement resulted in an estimated 6.6-fold increase in degradation efficiency. The primary reason for this phenomenon is the rapid oxidation of the –C–C– link in the molecular composition of plastics under highly thermophilic circumstances, forming –C=O– and –C–O– bonds. This process leads to enhanced degradation rates and increased hydrophobicity. In addition, this mechanism facilitates the development of biofilms consisting of functional microorganisms on the surface of microplastics. In a study conducted by Abrusci et al. (2011), it was shown that *Brevibacillus borstelensis* exhibited polyethylene degradation rates ranging from 0.7% to 1.2% over 90 days at a temperature of 30°C. Interestingly, when the temperature was elevated to 45°C, the degradation rate significantly escalated to 11.5% (Lin et al., 2022). Moisture is a critical determinant in the proliferation of microbes, hence assuming a pivotal function in plastic breakdown. Adequate moisture is a prerequisite for the activity of microorganisms. The hydrolytic activity of microorganisms exhibits a positive correlation with higher levels of moisture content. The acceleration of degradation caused by higher temperatures and humidity may be significant (Urbanek et al., 2018). In regions characterised by warm climates, the breakdown of polymers by microbes is observed to escalate when the relative humidity surpasses the threshold of 70% (Singh and Sharma, 2008).

Microbial Degradation of Plastic Polymers 279

15.5.2 INTRINSIC PROPERTIES

Molecular weight is one of the factors determining the biodegradation of plastics. Low molecular weight favours biodegradation (Kale et al., 2015). Low size offers a large specific surface area with a faster degradation rate (Gewert et al., 2015). Their chemical composition and molecular structure affect plastic degradability, which can impact factors such as enzyme accessibility. Plastics polymers that incorporate ester linkages, such as polyesters, polyurethanes, and polyethylene terephthalate, exhibit a higher degree of biodegradability than those lacking ester connections (Barlow et al., 2020). Because of their high molecular weight, hydrophobicity, and highly stable $-C-C-$ bonds in their backbone, common plastics like polypropylene, polyethylene, polyvinyl chloride, and polystyrene have very low biodegradability in the natural environment. Furthermore, no enzymes have been identified that can effectively cleave their $-C-C-$ bonds (Chamas et al., 2020; Liu et. al., 2022).

15.6 LIMITATIONS AND RECOMMENDATIONS

Several critical limitations of existing research and prospective avenues for future study are as follows (Jain et al., 2023; Rose et al., 2023b):

1. Plastic degradation by different microorganisms and enzymes has been well studied, however, the precise mechanisms behind this breakdown process have yet to be fully elucidated. Additional research is required to comprehensively understand the mechanisms behind plastic degradation and their potential implications for practical use.
2. The practical employment of microbes and enzymes in waste management and bioremediation is limited due to the relatively slow rate of plastic decomposition. Subsequent investigations must prioritise the advancement of methodologies to augment the breakdown rates of microorganisms and enzymes responsible for plastic degradation.
3. Understanding the environmental implications associated with the plastic-degrading capabilities of microbes remains limited. Further investigation is necessary to assess the possible ecological consequences of introducing plastic-degrading microorganisms into the natural environment.
4. The intricate nature of plastic waste and the constraints associated with current bioprocessing techniques provide a substantial obstacle in expanding laboratory findings for industrial application. Future research should prioritise the development of plastic waste management solutions that are both efficient and cost-effective.
5. The information on microbes and enzymes capable of degrading plastic is now constrained to a few extensively investigated species. Future studies should investigate the microbial variety in various habitats and ascertain the presence of new microorganisms and enzymes that exhibit the capability to degrade a broader spectrum of plastics.
6. The introduction of plastic-degrading bacteria into the environment may lead to the release of harmful compounds or the disturbance of natural ecosystems. Hence, it is essential to engage in a thorough evaluation of the potential advantages and disadvantages associated with the use of microbes capable of decomposing plastic materials.

15.7 CONCLUSION

Wastes made from synthetic plastics pose many environmental and human health problems. Researchers have shown the crucial role played by various microorganisms and invertebrate species in the plastic polymer biodegradation process. Specifically, no yeast species have been found as plastic biodegraders, proving that the vast microbial diversity in various natural environments has not been fully explored. Due to their sluggish disintegration rate, synthetic polymer biodegradation has not been implemented on a commercial scale yet, although there is still a lot of study being done in the sector. Scientists recommend using engineered strains or enzymes with enhanced selectivity for synthetic polymers to combat plastic waste. Despite the widespread discovery of plastic-degrading enzymes, little is known about their biochemical or structural features. These details are essential for comprehending the processes involved in the biodegradation of rigid polymers. The study concluded that bacteria capable of breaking down lignin, algae, insects that feed on plastic, and other invertebrate species might all play significant roles in the biodegradation of plastic polymers. Finally, further research is required to create new technologies for plastic waste degradation since no efficient, ecofriendly, economical, socially acceptable plastic-degrading process is available.

REFERENCES

Abrusci, C., Pablos, J.L., Corrales, T., López-Marín, J., Marín, I., and Catalina, F. 2011. Biodegradation of photo-degraded mulching films based on polyethylenes and stearates of calcium and iron as pro-oxidant additives. *International Biodeterioration & Biodegradation* 65(3):451–459.

Agrafioti, P., Müller-Blenkle, C., Adler, C., and Athanassiou, C.G. 2023. Evaluation of zeolite dusts as grain protectants against Lepinotus reticulatus, Liposcelis decolor, Acarus siro and Stegobium paniceum. *Journal of Plant Diseases and Protection* 130(2):393–399.

Ali, S.S., Elsamahy, T., Al-Tohamy, R., Zhu, D., Mahmoud, Y.A.G., Koutra, E., Metwally, M.A., Kornaros, M., and Sun, J. 2021. Plastic wastes biodegradation: Mechanisms, challenges and future prospects. *Science of the Total Environment* 780:146590.

Ali, S.S., Elsamahy, T., Zhu, D., and Sun, J. 2023. Biodegradability of polyethylene by efficient bacteria from the guts of plastic-eating waxworms and investigation of its degradation mechanism. *Journal of Hazardous Materials* 443:130287.

Amobonye, A., Bhagwat, P., Singh, S., and Pillai, S. 2021. Plastic biodegradation: Frontline microbes and their enzymes. *Science of the Total Environment* 759:143536.

Anjana, K., Hinduja, M., Sujitha, K., and Dharani, G. 2020. Review on plastic wastes in marine environment–Biodegradation and biotechnological solutions. *Marine Pollution Bulletin* 150:110733.

Arefian, M., Tahmourespour, A., and Zia, M. 2020. Polycarbonate biodegradation by newly isolated Bacillus strains. *Archives of Environmental Protection* 46(1):14–20.

Ariza-Tarazona, M.C., Villarreal-Chiu, J.F., Hernández-López, J.M., De la Rosa, J.R., Barbieri, V., Siligardi, C., and Cedillo-González, E.I. 2020. Microplastic pollution reduction by a carbon and nitrogen-doped TiO2: Effect of pH and temperature in the photocatalytic degradation process. *Journal of Hazardous Materials* 395:122632.

Asiandu, A.P., Wahyudi, A., and Sari, S.W. 2021. A review: Plastics waste biodegradation using plastics-degrading bacteria. *Journal of Environmental Treatment Techniques* 9(1):148–157.

Auta, H.S., Emenike, C.U., Jayanthi, B., and Fauziah, S.H. 2018. Growth kinetics and biodeterioration of polypropylene microplastics by Bacillus sp. and Rhodococcus sp. isolated from mangrove sediment. *Marine Pollution Bulletin* 127:15–21.

Barlow, D.E., Biffinger, J.C., Estrella, L., Lu, Q., Hung, C.S., Nadeau, L.J., Crouch, A.L., Russell Jr, J.N., and Crookes-Goodson, W.J. 2020. Edge-localized biodeterioration and secondary microplastic formation by Papiliotrema laurentii unsaturated biofilm cells on polyurethane films. *Langmuir* 36(6):1596–1607.

Bilal, H., Raza, H., Bibi, H., and Bibi, T. 2021. Plastic biodegradation through insects and their symbionts microbes: A review. *Journal of Bioresource Management* 8(4):7.

Carlile, M.J., Watkinson, S.C., and Gooday, G.W. 2001. *The fungi*. Gulf Professional Publishing.

Carniel, A., Valoni, É., Junior, J.N., da Conceição Gomes, A., and de Castro, A.M. 2017. Lipase from Candida antarctica (CALB) and cutinase from Humicola insolens act synergistically for PET hydrolysis to terephthalic acid. *Process Biochemistry* 59:84–90.

Chamas, A., Moon, H., Zheng, J., Qiu, Y., Tabassum, T., Jang, J.H., Abu-Omar, M., Scott, S.L., and Suh, S. 2020. Degradation rates of plastics in the environment. *ACS Sustainable Chemistry & Engineering* 8(9):3494–3511.

Chen, Z., Zhao, W., Xing, R., Xie, S., Yang, X., Cui, P., Lü, J., Liao, H., Yu, Z., Wang, S., and Zhou, S. 2020. Enhanced in situ biodegradation of microplastics in sewage sludge using hyperthermophilic composting technology. *Journal of Hazardous Materials* 384:121271.

Degli-Innocenti, F. 2014. Biodegradation of plastics and ecotoxicity testing: When should it be done. *Frontiers in Microbiology* 5:475.

Dineshbabu, G., Uma, V.S., Mathimani, T., Prabaharan, D., and Uma, L. 2020. Elevated CO_2 impact on growth and lipid of marine cyanobacterium Phormidium valderianum BDU 20041–towards microalgal carbon sequestration. *Biocatalysis and Agricultural Biotechnology* 25:101606.

Dong, X., Zhu, L., He, Y., Li, C., and Li, D. 2023. Salinity significantly reduces plastic-degrading bacteria from rivers to oceans. *Journal of Hazardous Materials* 451:131125.

Du, Y., Yao, C., Dou, M., Wu, J., Su, L., and Xia, W. 2022. Oxidative degradation of pre-oxidated polystyrene plastics by dye decolorising peroxidases from Thermomonospora curvata and Nostocaceae. *Journal of Hazardous Materials* 436:129265.

Duret, G., and Delcour, A.H. 2010. Size and dynamics of the Vibrio cholerae porins OmpU and OmpT probed by polymer exclusion. *Biophysical Journal* 98(9):1820–1829.

Eberl, A., Heumann, S., Brückner, T., Araujo, R., Cavaco-Paulo, A., Kaufmann, F., and Guebitz, G.M. 2009. Enzymatic surface hydrolysis of poly (ethylene terephthalate) and bis (benzoyloxyethyl) terephthalate by lipase and cutinase in the presence of surface active molecules. *Journal of Biotechnology* 143(3):207–212.

Ekanayaka, A.H., Tibpromma, S., Dai, D., Xu, R., Suwannarach, N., Stephenson, S.L., and Karunarathna, S.C. 2022. A review of the fungi that degrade plastic. *Journal of Fungi* 8(8):772.

Enyoh, C.E., Maduka, T.O., Duru, C.E., Osigwe, S.C., Ikpa, C.B., and Wang, Q. 2022. In sillico binding affinity studies of microbial enzymatic degradation of plastics. *Journal of Hazardous Materials Advances* 6:100076.

Espinosa, M.J.C., Blanco, A.C., Schmidgall, T., Atanasoff-Kardjalieff, A.K., Kappelmeyer, U., Tischler, D., Pieper, D.H., Heipieper, H.J., and Eberlein, C. 2020. Toward biorecycling: Isolation of a soil bacterium that grows on a polyurethane oligomer and monomer. *Frontiers in Microbiology* 11:404.

Furukawa, M., Kawakami, N., Tomizawa, A., and Miyamoto, K. 2019. Efficient degradation of poly (ethylene terephthalate) with Thermobifida fusca cutinase exhibiting improved catalytic activity generated using mutagenesis and additive-based approaches. *Scientific Reports* 9(1):16038.

Gao, R., Liu, R., and Sun, C. 2022. A marine fungus Alternaria alternata FB1 efficiently degrades polyethylene. *Journal of Hazardous Materials* 431:128617.

Gewert, B., Plassmann, M.M., and MacLeod, M. 2015. Pathways for degradation of plastic polymers floating in the marine environment. *Environmental Science: Processes & Impacts* 17(9):1513–1521.

Gómez-Méndez, L.D., Moreno-Bayona, D.A., Poutou-Pinales, R.A., Salcedo-Reyes, J.C., Pedroza-Rodríguez, A.M., Vargas, A., and Bogoya, J.M. 2018. Biodeterioration of plasma pretreated LDPE sheets by Pleurotus ostreatus. *PLoS One* 13(9):e0203786.

Habib, S., Iruthayam, A., Abd Shukor, M.Y., Alias, S.A., Smykla, J., and Yasid, N.A. 2020. Biodeterioration of untreated polypropylene microplastic particles by Antarctic bacteria. *Polymers* 12(11):2616.

Haiping, L., and Fanping, M. 2023. Efficiency, mechanism, influencing factors, and integrated technology of biodegradation for aromatic compounds by microalgae: A review. *Environmental Pollution* 335:122248.

Ho, B.T., Roberts, T.K., and Lucas, S. 2018. An overview on biodegradation of polystyrene and modified polystyrene: The microbial approach. *Critical Reviews in Biotechnology* 38(2):308–320.

Hoffmann, L., Eggers, S.L., Allhusen, E., Katlein, C., and Peeken, I. 2020. Interactions between the ice algae Fragillariopsis cylindrus and microplastics in sea ice. *Environment International* 139:105697.

Hosaka, M., Kamimura, N., Toribami, S., Mori, K., Kasai, D., Fukuda, M., and Masai, E. 2013. Novel tripartite aromatic acid transporter essential for terephthalate uptake in Comamonas sp. strain E6. *Applied and Environmental Microbiology* 79(19):6148–6155.

Hou, L., Xi, J., Liu, J., Wang, P., Xu, T., Liu, T., and Lin, Y.B. 2022. Biodegradability of polyethylene mulching film by two Pseudomonas bacteria and their potential degradation mechanism. *Chemosphere* 286:131758.

Hua, F., Wang, H.Q., Li, Y., and Zhao, Y.C. 2013. Trans-membrane transport of n-octadecane by Pseudomonas sp. DG17. *Journal of Microbiology* 51:791–799.

Imbs, A.B., Ermolenko, E.V., Grigorchuk, V.P., Sikorskaya, T.V., and Velansky, P.V. (2021). Current progress in lipidomics of marine invertebrates. *Marine Drugs* 19(12):660.

Jain, R., Gaur, A., Suravajhala, R., Chauhan, U., Pant, M., Tripathi, V., and Pant, G. 2023. Microplastic pollution: Understanding microbial degradation and strategies for pollutant reduction. *Science of The Total Environment* 9:167098.

Jaiswal, S., Sharma, B., and Shukla, P. 2020. Integrated approaches in microbial degradation of plastics. *Environmental Technology & Innovation 17:*100567.

Jenkins, S., Quer, A.M.I., Fonseca, C., and Varrone, C. 2019. Microbial degradation of plastics: new plastic degraders, mixed cultures and engineering strategies. In N. Jamil, P. Kumar, and R. Batool (eds), *Soil Microenvironment for Bioremediation and Polymer Production:*213–238.

Jeyakumar, D., Chirsteen, J., and Doble, M. 2013. Synergistic effects of pretreatment and blending on fungi mediated biodegradation of polypropylenes. *Bioresource Technology 148:*78–85.

Kale, S.K., Deshmukh, A.G., Dudhare, M.S., and Patil, V.B. 2015. Microbial degradation of plastic: A review. *Journal of Biochemical Technology 6*(2):952–61.

Kang, B.R., Kim, S.B., Song, H.A., and Lee, T.K. 2019. Accelerating the biodegradation of high-density polyethylene (HDPE) using Bjerkandera adusta TBB-03 and lignocellulose substrates. *Microorganisms 7*(9):304.

Kesti, S.S.K., and Thimmappa, S.C.T. 2019. First report on biodegradation of low density polyethylene by rice moth larvae, Corcyra cephalonica (Stainton). *The Holistic Approach to Environment 9*(4):79–83.

Khan, S., Ali, S.A., and Ali, A.S. 2023. Biodegradation of low density polyethylene (LDPE) by mesophilic fungus 'Penicillium citrinum' isolated from soils of plastic waste dump yard, Bhopal, India. *Environmental Technology 44*(15):2300–2314.

Khoironi, A., Anggoro, S., and Sudarno, 2019. Evaluation of the interaction among microalgae Spirulina sp, plastics polyethylene terephthalate and polypropylene in freshwater environment. *Journal of Ecological Engineering 20*(6):161–173.

Kim, H.R., Lee, H.M., Yu, H.C., Jeon, E., Lee, S., Li, J., and Kim, D.H. 2020. Biodegradation of polystyrene by Pseudomonas sp. isolated from the gut of superworms (larvae of Zophobas atratus). *Environmental Science & Technology 54*(11):6987–6996.

Kirstein, I.V., Wichels, A., Gullans, E., Krohne, G., and Gerdts, G. 2019. The plastisphere–uncovering tightly attached plastic "specific" microorganisms. *PLoS One 14*(4):e0215859.

Kumar, R.V., Kanna, G.R., and Elumalai, S.,2017. Biodegradation of polyethylene by green photosynthetic microalgae. *Journal of the Bioremediation Biodegradation 8*(381):2.

Kumar, V., Sharma, N., Duhan, L., Pasrija, R., Thomas, J., Umesh, M., Lakkaboyana, S.K., Andler, R., Vangnai, A.S., Vithanage, M., and Awasthi, M.K. 2022. Microbial engineering strategies for synthetic microplastics clean up: A review on recent approaches. *Environmental Toxicology and Pharmacology,* 98:104045.

Kundungal, H., Gangarapu, M., Sarangapani, S., Patchaiyappan, A., and Devipriya, S.P. 2019. Efficient biodegradation of polyethylene (HDPE) waste by the plastic-eating lesser waxworm (Achroia grisella). *Environmental Science and Pollution Research 26:*18509–18519.

Kyaw, B.M., Champakalakshmi, R., Sakharkar, M.K., Lim, C.S., and Sakharkar, K.R. 2012. Biodegradation of low density polythene (LDPE) by Pseudomonas species. *Indian Journal of Microbiology 52:*411–419.

Larue, C., Sarret, G., Castillo-Michel, H., and Pradas del Real, A.E. 2021. A critical review on the impacts of nanoplastics and microplastics on aquatic and terrestrial photosynthetic organisms. *Small 17*(20):2005834.

Li, X., Patena, W., Fauser, F., Jinkerson, R.E., Saroussi, S., Meyer, M.T., and Jonikas, M.C. 2019. A genome-wide algal mutant library and functional screen identifies genes required for eukaryotic photosynthesis. *Nature Genetics 51*(4):627–635.

Lim, B.K.H., and San Thian, E. 2022. Biodegradation of polymers in managing plastic waste—A review. *Science of The Total Environment 813:*151880.

Lin, Z., Jin, T., Zou, T., Xu, L., Xi, B., Xu, D., He, J., Xiong, L., Tang, C., Peng, J., and Zhou, Y. 2022. Current progress on plastic/microplastic degradation: Fact influences and mechanism. *Environmental Pollution 304:*119159.

Liu, J., Liu, J., Xu, B., Xu, A., Cao, S., Wei, R., and Dong, W. 2022. Biodegradation of polyether-polyurethane foam in yellow mealworms (Tenebrio molitor) and effects on the gut microbiome. *Chemosphere 304:*135263.

Liu, Y., Liu, Z., Guo, Z., Yan, T., Jin, C., and Wu, J. 2022c. Enhancement of the degradation capacity of IsPETase for PET plastic degradation by protein engineering. *Science of The Total Environment 834:*154947.

Lobelle, D., and Cunliffe, M. 2011. Early microbial biofilm formation on marine plastic debris. *Marine Pollution Bulletin 62*(1):197–200.

Lou, Y., Ekaterina, P., Yang, S.S., Lu, B., Liu, B., Ren, N., and Xing, D. 2020. Biodegradation of polyethylene and polystyrene by greater wax moth larvae (Galleria mellonella L.) and the effect of co-diet supplementation on the core gut microbiome. *Environmental Science & Technology 54*(5):2821–2831.

Lucas, N., Bienaime, C., Belloy, C., Queneudec, M., Silvestre, F., and Nava-Saucedo, J.E. 2008. Polymer biodegradation: Mechanisms and estimation techniques–A review. *Chemosphere 73*(4):429–442.

Mahmoud, E.A., Al-Hagar, O.E., and El-Aziz, M.F.A. 2021. Gamma radiation effect on the midgut bacteria of Plodia interpunctella and its role in organic wastes biodegradation. *International Journal of Tropical Insect Science 41:*261–272.

Mehmood, C.T., Qazi, I.A., Hashmi, I., Bhargava, S., and Deepa, S. 2016. Biodegradation of low density polyethylene (LDPE) modified with dye sensitized titania and starch blend using Stenotrophomonas pavanii. *International Biodeterioration & Biodegradation 113:*276–286.

Moog, D., Schmitt, J., Senger, J., Zarzycki, J., Rexer, K.H., Linne, U., Erb, T., and Maier, U.G. 2019. Using a marine microalga as a chassis for polyethylene terephthalate (PET) degradation. *Microbial Cell Factories 18*(1):1–15.

Müller, R.J., Schrader, H., Profe, J., Dresler, K., and Deckwer, W.D. 2005. Enzymatic degradation of poly (ethylene terephthalate): Rapid hydrolyse using a hydrolase from T. fusca. *Macromolecular Rapid Communications 26*(17):1400–1405.

Narwal, N., Katyal, D., Kataria, N., Rose, P.K., Warkar, S.G., Pugazhendhi, A., Ghotekar, S., and Khoo, K.S. 2023. Emerging micropollutants in aquatic ecosystems and nanotechnology-based removal alternatives: A review. *Chemosphere 341*:139945.

Nauendorf, A., Krause, S., Bigalke, N.K., Gorb, E.V., Gorb, S.N., Haeckel, M., Wahl, M., and Treude, T. 2016. Microbial colonization and degradation of polyethylene and biodegradable plastic bags in temperate fine-grained organic-rich marine sediments. *Marine Pollution Bulletin 103*(1–2):168–178.

Nowak, B., Rusinowski, S., Korytkowska-Wałach, A., and Chmielnicki, B. 2021. The composition of poly (vinyl chloride) with polylactide/poly (butylene terephthalate-co-butylene sebacate) and its biodegradation by Phanerochaete chrysosporium. *International Biodeterioration & Biodegradation 157:*105153.

Olakanmi, G.B., Lateef, S.A., and Ogunjobi, A.A. 2023. Utilisation of disposable face masks as substrate for Pleurotus Ostreatus mushroom production. *Available at SSRN 4325859.*

Othman, A.R., Hasan, H.A., Muhamad, M.H., Ismail, N.I., and Abdullah, S.R.S. 2021. Microbial degradation of microplastics by enzymatic processes: A review. *Environmental Chemistry Letters 19:*3057–3073.

Pardo-Rodríguez, M.L., and Zorro-Mateus, P.J.P. 2021. Biodegradation of polyvinyl chloride by Mucor sp and Penicillium sp isolated from soil. *Revista de Investigación, Desarrollo e Innovación 11*(2):387–399.

Pathak, V.M. 2017. Review on the current status of polymer degradation: A microbial approach. *Bioresources and Bioprocessing 4*(1):1–31.

Peng, B.Y., Su, Y., Chen, Z., Chen, J., Zhou, X., Benbow, M.E., Criddle, C.S., Wu, W.M., and Zhang, Y. 2019. Biodegradation of polystyrene by dark (Tenebrio obscurus) and yellow (Tenebrio molitor) mealworms (Coleoptera: Tenebrionidae). *Environmental Science & Technology 53*(9):5256–5265.

Perera, I.C., Abeywickrama, T.D., and Rahman, F.A. 2023. Role of genetically engineered yeast in plastic degradation. In A. Daverey, K. Dutta, S. Joshi, T. Gea (eds), *Advances in Yeast Biotechnology for Biofuels and Sustainability* (pp. 567–584). Elsevier.

Rao, A.S., Nair, A., Salu, H.A., Pooja, K.R., Nandyal, N.A., Joshi, V.S., and More, S.S. 2023. Carbohydrases: A class of all-pervasive industrial biocatalysts. In G. Brahmachari (ed), *Biotechnology of Microbial Enzymes* (pp. 497–523). Academic Press.

Restrepo-Flórez, J.M., Bassi, A., and Thompson, M.R. 2014. Microbial degradation and deterioration of polyethylene–A review. *International Biodeterioration & Biodegradation 88*:83–90.

Rose, P.K., Jain, M., Kataria, N., Sahoo, P.K., Garg, V.K., and Yadav, A. 2023b. Microplastics in multimedia environment: A systematic review on its fate, transport, quantification, health risk, and remedial measures. *Groundwater for Sustainable Development 20:*100889.

Rose, P.K., Yadav, S., Kataria, N., and Khoo, K.S. 2023a. Microplastics and nanoplastics in the terrestrial food chain: Uptake, translocation, trophic transfer, ecotoxicology, and human health risk. *TrAC Trends in Analytical Chemistry 167*:117249.

Ru, J., Huo, Y., and Yang, Y. 2020. Microbial degradation and valorization of plastic wastes. *Frontiers in Microbiology 11:*442.

Santo, M., Weitsman, R., and Sivan, A. 2013. The role of the copper-binding enzyme–laccase–in the biodegradation of polyethylene by the actinomycete Rhodococcus ruber. *International Biodeterioration & Biodegradation 84:*204–210.

Sarmah, P., and Rout, J. 2018. Algal colonization on polythene carry bags in a domestic solid waste dumping site of Silchar town in Assam. *Phykos 48*(67):e77.

Sarmah, P., and Rout, J. 2019. Cyanobacterial degradation of low-density polyethylene (LDPE) by Nostoc carneum isolated from submerged polyethylene surface in domestic sewage water. *Energy, Ecology and Environment 4:*240–252.

Schwibbert, K., Menzel, F., Epperlein, N., Bonse, J., and Krüger, J. 2019. Bacterial adhesion on femtosecond laser-modified polyethylene. *Materials 12*(19):3107.

Sevilla, M.E., Garcia, M.D., Perez-Castillo, Y., Armijos-Jaramillo, V., Casado, S., Vizuete, K., and Cerda-Mejía, L. (2023). Degradation of PET Bottles by an Engineered Ideonella sakaiensis PETase. *Polymers 15*(7):1779.

Shabbir, S., Faheem, M., Ali, N., Kerr, P.G., Wang, L.F., Kuppusamy, S., and Li, Y., 2020. Periphytic biofilm: An innovative approach for biodegradation of microplastics. *Science of the Total Environment 717:*137064.

Shahnawaz, M., Sangale, M.K., Ade, A.B., Shahnawaz, M., Sangale, M.K., and Ade, A.B. 2019. Analysis of the plastic degradation products. In M. Shahnawaz, M.K. Sangale, A.B. Ade (eds), *Bioremediation Technology for Plastic Waste:*93–101.

Singh, B., and Sharma, N. 2008. Mechanistic implications of plastic degradation. *Polymer Degradation and Stability 93*(3):561–84.

Sivan, A., Szanto, M., and Pavlov, V. 2006. Biofilm development of the polyethylene-degrading bacterium Rhodococcus ruber. *Applied Microbiology and Biotechnology 72:*346–352.

Skariyachan, S., Taskeen, N., Kishore, A.P., Krishna, B.V., and Naidu, G. 2021. Novel consortia of Enterobacter and Pseudomonas formulated from cow dung exhibited enhanced biodegradation of polyethylene and polypropylene. *Journal of Environmental Management 284:*112030.

Song, Y., Qiu, R., Hu, J., Li, X., Zhang, X., Chen, Y., Wu, W.M., and He, D. 2020. Biodegradation and disintegration of expanded polystyrene by land snails Achatina fulica. *Science of the Total Environment 746:*141289.

Sowmya, H.V., Ramalingappa, B., Nayanashree, G., Thippeswamy, B., and Krishnappa, M. 2015b. Polyethylene degradation by fungal consortium. *International Journal of Environmental Research 9*(3):823–830.

Sowmya, H.V., Ramalingappa, Krishnappa, M., and Thippeswamy, B. 2015a. Degradation of polyethylene by Penicillium simplicissimum isolated from local dumpsite of Shivamogga district. *Environment, Development and Sustainability 17:*731–745.

Srikanth, M., Sandeep, T.S.R.S., Sucharitha, K., and Godi, S. 2022. Biodegradation of plastic polymers by fungi: A brief review. *Bioresources and Bioprocessing 9*(1):42.

Taipale, S.J., Rigaud, C., Calderini, M.L., Kainz, M.J., Pilecky, M., Uusi-Heikkilä, S., Vesamäki, J.S., Vuorio, K., and Tiirola, M. 2023. The second life of terrestrial and plastic carbon as nutritionally valuable food for aquatic consumers. *Ecology Letters 26:*1336–1347.

Tamnou, E.B.M., Arfao, A.T., Nougang, M.E., Metsopkeng, C.S., Ewoti, O.V.N., Moungang, L.M., Nana, P.A., Takang-Etta, L.R.A., Perrière, F., Sime-Ngando, T., and Nola, M. 2021. Biodegradation of polyethylene by the bacterium Pseudomonas aeruginosa in acidic aquatic microcosm and effect of the environmental temperature. *Environmental Challenges 3:*100056.

Temporiti, M.E.E., Nicola, L., Nielsen, E., and Tosi, S. 2022. Fungal enzymes involved in plastics biodegradation. *Microorganisms 10*(6):1180.

Thakur, B., Singh, J., Singh, J., Angmo, D., and Vig, A.P. 2023. Biodegradation of different types of microplastics: Molecular mechanism and degradation efficiency. *Science of The Total Environment 877:*162912.

Tokiwa, Y., Calabia, B.P., Ugwu, C.U., and Aiba, S. 2009. Biodegradability of plastics. *International Journal of Molecular Sciences 10*(9):3722–3742.

Tribedi, P., and Sil, A.K. 2014. Cell surface hydrophobicity: A key component in the degradation of polyethylene succinate by Pseudomonas sp. AKS2. *Journal of Applied Microbiology 116*(2):295–303.

Tribedi, P., Gupta, A.D., and Sil, A.K. 2015. Adaptation of Pseudomonas sp. AKS2 in biofilm on low-density polyethylene surface: An effective strategy for efficient survival and polymer degradation. *Bioresources and Bioprocessing 2:*1–10.

Urbanek, A.K., Rymowicz, W., and Mirończuk, A.M. 2018. Degradation of plastics and plastic-degrading bacteria in cold marine habitats. *Applied Microbiology and Biotechnology 102:*7669–7678.

Vertommen, M.A.M.E., Nierstrasz, V.A., Van Der Veer, M., and Warmoeskerken, M.M.C.G. 2005. Enzymatic surface modification of poly (ethylene terephthalate). *Journal of Biotechnology 120*(4):376–386.

Wang, Z., Xin, X., Shi, X., and Zhang, Y. 2020. A polystyrene-degrading Acinetobacter bacterium isolated from the larvae of Tribolium castaneum. *Science of the Total Environment 726:*138564.

Wilkes, R.A., and Aristilde, L. 2017. Degradation and metabolism of synthetic plastics and associated products by Pseudomonas sp.: Capabilities and challenges. *Journal of Applied Microbiology 123*(3):582–593.

Wu, Z., Shi, W., Valencak, T.G., Zhang, Y., Liu, G., and Ren, D. 2023. Biodegradation of conventional plastics: Candidate organisms and potential mechanisms. *Science of The Total Environment 885:*163908.

Xu, L., Crawford, K., and Gorman, C.B. 2011. Effects of temperature and pH on the degradation of poly (lactic acid) brushes. *Macromolecules 44*(12):4777–4782.

Yadav, S., Kataria, N., Khyalia, P., Rose, P.K., Mukherjee, S., Sabherwal, H., Chai, W.S., Rajendran, S., Jiang, J.J., and Khoo, K.S. 2023. Recent analytical techniques, and potential eco-toxicological impacts of textile fibrous microplastics (FMPs) and its associated contaminates: A review. *Chemosphere 326:*138495.

Yan, F., Wei, R., Cui, Q., Bornscheuer, U.T., and Liu, Y.J. 2021. Thermophilic whole-cell degradation of polyethylene terephthalate using engineered Clostridium thermocellum. *Microbial Biotechnology 14*(2):374–385.

Yang, S.S., Ding, M.Q., Zhang, Z.R., Ding, J., Bai, S.W., Cao, G.L., Zhao, L., Pang, J.W., Xing, D.F., Ren, N.Q., and Wu, W.M. 2021. Confirmation of biodegradation of low-density polyethylene in dark-versus yellow-mealworms (larvae of Tenebrio obscurus versus Tenebrio molitor) via. gut microbe-independent depolymerization. *Science of the Total Environment 789:*147915.

Yang, X.G., Wen, P.P., Yang, Y.F., Jia, P.P., Li, W.G., and Pei, D.S. 2023. Plastic biodegradation by *in vitro*environmental microorganisms and *in vivo* gut microorganisms of insects. *Frontiers in Microbiology 13:*1001750.

Yang, Y., Yang, J., Wu, W.M., Zhao, J., Song, Y., Gao, L., Yang, R., and Jiang, L. 2015. Biodegradation and mineralization of polystyrene by plastic-eating mealworms: Part 1. Chemical and physical characterization and isotopic tests. *Environmental Science & Technology 49*(20):12080–12086.

Yeom, S.J., Le, T.K., and Yun, C.H. 2022. P450-driven plastic-degrading synthetic bacteria. *Trends in Biotechnology 40*(2):166–179.

Yoshida, S., Hiraga, K., Takehana, T., Taniguchi, I., Yamaji, H., Maeda, Y., Toyohara, K., Miyamoto, K., Kimura, Y., and Oda, K. 2016. A bacterium that degrades and assimilates poly (ethylene terephthalate). *Science 351*(6278):1196–1199.

Yuan, J., Ma, J., Sun, Y., Zhou, T., Zhao, Y., and Yu, F. 2020. Microbial degradation and other environmental aspects of microplastics/plastics. *Science of the Total Environment 715:*136968.

Zhang, J., Gao, D., Li, Q., Zhao, Y., Li, L., Lin, H., and Zhao, Y. 2020. Biodegradation of polyethylene microplastic particles by the fungus Aspergillus flavus from the guts of wax moth Galleria mellonella. *Science of the Total Environment 704:*135931.

Zhang, Z., Peng, H., Yang, D., Zhang, G., Zhang, J., and Ju, F. 2022. Polyvinyl chloride degradation by a bacterium isolated from the gut of insect larvae. *Nature Communications 13*(1):5360.

Zhu, B., Ye, Q., Seo, Y., and Wei, N. 2022. Enzymatic degradation of polyethylene terephthalate plastics by bacterial Curli display PETase. *Environmental Science & Technology Letters 9*(7):650–657.

16 Advanced Technologies for the Production of Bioplastics

Kassian T.T. Amesho, E.I. Edoun, Sumarlin Shangdiar, Abner Kukeyinge Shopati, Timoteus Kadhila, Nastassia Thandiwe Sithole, Chandra Mohan, Sioni Iikela, and Manoj Chandra Garg

16.1 INTRODUCTION

Bioplastics, also known as bio-based plastics, represent a promising avenue toward mitigating the environmental impact of conventional petroleum-based plastics. Defined as polymers derived from renewable resources such as plant-based materials, bioplastics exhibit properties similar to traditional plastics while offering enhanced sustainability and reduced ecological footprint (Rosenboom et al., 2022; Koch and Mihalyi, 2018). These materials are characterized by their biodegradability, compostability, and reduced carbon footprint, making them attractive alternatives in the quest for more environmentally friendly packaging and materials (Sicotte, 2020).

The imperative for sustainable alternatives to conventional plastics arises from the significant environmental challenges posed by the widespread use and disposal of petroleum-based plastics. These challenges include pollution of terrestrial and aquatic ecosystems, depletion of finite fossil fuel resources, and contribution to greenhouse gas emissions and climate change (Nakajima et al., 2017). As the global demand for plastics continues to rise, driven by population growth, urbanization, and industrialization, the need for more sustainable alternatives becomes increasingly urgent. As depicted in Figure 16.1, the keyword network analysis of "Bioplastic"-related articles from 2018 to 2024 in the Web of Science (WoS) database reveals the primary themes and interconnections within this research area.

In response to this imperative, advanced technologies have emerged for the production of bioplastics (as indicated in Figure 16.2), offering innovative approaches to harnessing renewable resources and optimizing the properties and performance of bioplastic materials. These technologies encompass a range of methodologies, including biocatalysis, metabolic engineering, synthetic biology, and nanotechnology, each contributing to the development of novel bioplastic formulations with tailored properties and applications (Meereboer et al., 2020).

Biocatalysis represents one of the key technologies driving advancements in bioplastic production. By harnessing the catalytic activity of enzymes, biocatalysis enables the conversion of renewable feedstocks such as starch, cellulose, and lignin into monomers suitable for polymerization (Moshood, 2021). Enzymatic processes offer several advantages, including high specificity, mild reaction conditions, and reduced environmental impact, making them attractive for sustainable bioplastic synthesis (Hobbs et al., 2019; Jaiswal et al., 2020).

286
DOI: 10.1201/9781032706573-16

Advanced Technologies for the Production of Bioplastics 287

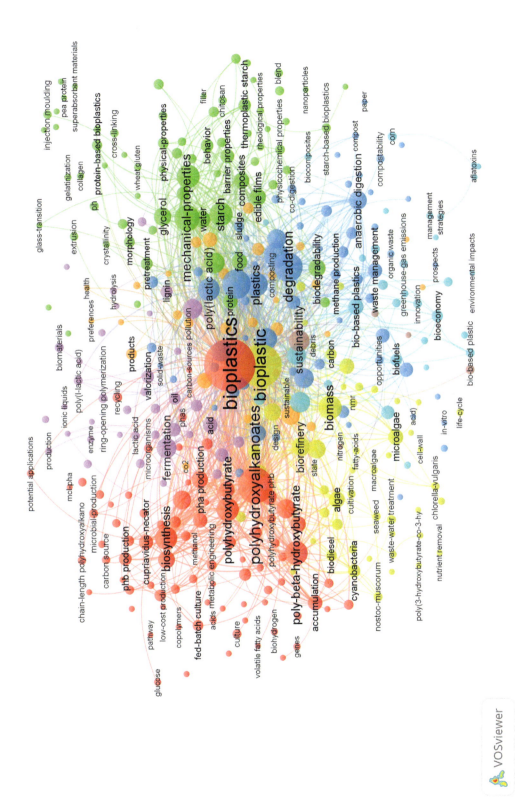

FIGURE 16.1 Keyword network analysis on "Bioplastic"-related articles from 2018 to 2024 found in Web of Science (WoS) database.

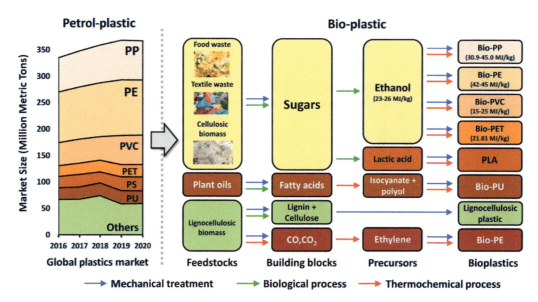

FIGURE 16.2 Production process of bioplastics, which incorporates mechanical, biological, and thermochemical methods. These methods have the potential to replace petroleum-based plastics in high-demand applications. The numbers in brackets indicate the energy content associated with each step of the process. (Obtained with permission from Patria et al., 2024, Copyright © 2024 Elseiver.)

Metabolic engineering, another prominent technology in bioplastic production, involves the modification of microbial metabolic pathways to enhance the production of desired biopolymers. Through genetic manipulation and strain optimization, metabolic engineering enables the development of microbial hosts capable of efficiently converting renewable substrates into bioplastic precursors, such as polyhydroxyalkanoates (PHA) and polylactic acid (PLA) (Espinosa, 2020; Chen et al., 2020). This approach offers opportunities for tailored bioplastic synthesis and optimization of production processes.

Synthetic biology, a rapidly evolving discipline, has also revolutionized bioplastic production by enabling the design and construction of biological systems for specific applications. Through the rational design of genetic circuits and cellular pathways, synthetic biologists can engineer microbial hosts with enhanced capabilities for bioplastic synthesis, as well as novel properties such as biodegradation and responsiveness to environmental stimuli (Wei et al., 2019; Lee and Shan, 2021). Synthetic biology holds promise for the development of next-generation bioplastics with advanced functionalities and environmental compatibility.

Nanotechnology, at the forefront of materials science, has further expanded the horizons of bioplastic production by offering innovative approaches to reinforcement, modification, and functionalization of biopolymer matrices. By incorporating nanoscale additives such as nanoparticles, nanofibers, and nanocomposites, researchers can enhance the mechanical strength, barrier properties, and thermal stability of bioplastic materials, opening up new opportunities for diverse applications ranging from packaging to biomedical devices (Geyer et al., 2017; Karan et al., 2019).

In summary, the introduction of advanced technologies for bioplastic production marks a significant step toward realizing the vision of a more sustainable and circular economy. By harnessing renewable resources, optimizing production processes, and enhancing material properties, these

16.2 BIOCATALYSIS FOR BIOPLASTICS PRODUCTION

Biocatalysis plays a pivotal role in the production of bioplastics, offering sustainable and environmentally friendly pathways for the synthesis of biopolymer precursors and the polymerization of bioplastics. This section provides an in-depth exploration of biocatalysis in bioplastics production, encompassing enzymatic breakdown of plant-based materials into monomers, enzymatic polymerization of biopolymers, and an analysis of the advantages and limitations associated with biocatalysis for bioplastics production.

Enzymatic breakdown of plant-based materials into monomers serves as a fundamental step in the biocatalytic route towards bioplastics production. Plant-based feedstocks such as starch, cellulose, and lignin are abundant renewable resources that can be enzymatically hydrolyzed into their constituent monomers, such as glucose, xylose, and lactic acid (Pellis et al., 2021; Walker and Rothman, 2020). This enzymatic hydrolysis process involves the action of specific enzymes, such as amylases, cellulases, and ligninases, which catalyze the cleavage of glycosidic bonds in polysaccharides and lignocellulosic materials, releasing monomeric sugars for subsequent biopolymer synthesis (Henry et al., 2019; Straub et al., 2017).

Following the enzymatic breakdown of plant-based materials, enzymatic polymerization represents a key enzymatic route for the production of biopolymers from monomeric precursors. Enzymes such as polymerases, polycondensases, and transglutaminases catalyze the polymerization of monomers into biopolymers through various mechanisms, including chain extension, condensation, and cross-linking (Sheldon et al., 2018; Ioannidou et al., 2020). This enzymatic polymerization process offers several advantages, including high specificity, mild reaction conditions, and precise control over polymer structure and properties, enabling the synthesis of tailored bioplastics with desired functionalities.

16.2.1 ADVANTAGES AND LIMITATIONS OF BIOCATALYSIS FOR BIOPLASTICS PRODUCTION

Biocatalysis offers numerous advantages for bioplastics production, including:

1. *Sustainability*: Biocatalytic processes utilize renewable feedstocks and biodegradable enzymes, reducing reliance on fossil resources and minimizing environmental impact (Ioannidou et al., 2020).
2. *Specificity*: Enzymes exhibit high substrate specificity, enabling selective synthesis of biopolymers with desired properties and functionalities.
3. *Mild reaction conditions*: Enzymatic reactions typically occur under mild temperature and pH conditions, reducing energy consumption and process costs (Ahmed et al., 2018).
4. *Green chemistry*: Biocatalysis conforms to principles of green chemistry, producing minimal waste and avoiding the use of hazardous chemicals.
5. *Versatility*: Biocatalytic processes can be tailored to accommodate a wide range of feedstocks and polymerization reactions, offering flexibility and versatility in bioplastics production.

Despite these advantages, biocatalysis for bioplastics production also presents certain limitations, including:

1. *Limited substrate compatibility:* Enzyme activity may be restricted to specific substrates, limiting the range of feedstocks that can be utilized for bioplastic synthesis.

2. *Sensitivity to reaction conditions:* Enzymatic reactions may be sensitive to variations in temperature, pH, and substrate concentrations, necessitating careful control and optimization.
3. *Enzyme stability:* Enzymes may exhibit reduced stability under certain process conditions, requiring supplementation or immobilization strategies to enhance enzyme longevity and activity (Al-Tohamy et al., 2023a).
4. *Process scale-up:* Scaling up biocatalytic processes for industrial production may pose challenges related to enzyme cost, process efficiency, and reactor design.

In summary, biocatalysis offers a promising approach for the production of bioplastics, leveraging the catalytic prowess of enzymes to drive sustainable and efficient synthesis routes (Ali et al., 2023). By capitalizing on the unique advantages of biocatalytic processes while addressing inherent limitations, researchers and industry stakeholders can advance the development and commercialization of bioplastics toward a more sustainable future.

16.3 METABOLIC ENGINEERING FOR BIOPLASTICS PRODUCTION

Metabolic engineering involves the modification of microbial metabolic pathways to enhance the production of desired biopolymers for bioplastics. This section explores the application of metabolic engineering in bioplastics production, focusing on the modification of metabolic pathways in microorganisms.

16.3.1 Modification of Metabolic Pathways in Microorganisms for Biopolymer Production

Metabolic engineering strategies aim to redirect the flux of precursor metabolites toward the synthesis of biopolymers, such as polyhydroxyalkanoates (PHAs) or polylactic acid (PLA), in microorganisms. This process typically involves the manipulation of key enzymes or regulatory elements within metabolic pathways to optimize substrate utilization and product formation (Aragosa et al., 2021). Microorganisms commonly employed for bioplastics production include bacteria (e.g., *Escherichia coli*) and yeast (e.g., *Saccharomyces cerevisiae*). Through genetic manipulation techniques such as gene knockout, overexpression, or heterologous gene expression, metabolic engineers can tailor microbial hosts to efficiently convert renewable feedstocks into biopolymers (Ali et al., 2021). Table 16.1 shows different types of biological wastes used for the production of biodegradable plastics, their source feedstock, and applications.

16.3.2 Advantages and Limitations of Metabolic Engineering for Bioplastics Production

Metabolic engineering offers several advantages for bioplastics production, including:

1. *Tailored biopolymer synthesis*: Allows for the engineering of microorganisms to produce specific biopolymers with desired properties, such as molecular weight and thermal stability (George et al., 2021; Al-Tohamy et al., 2023b).
2. *Renewable feedstock utilization*: Enables the utilization of renewable carbon sources, including sugars, lignocellulosic biomass, and industrial waste streams, for biopolymer production (Koller and Braunegg, 2018).
3. *Process efficiency*: Enhances biopolymer production yields and rates through targeted genetic modifications, optimizing metabolic flux toward desired pathways (Kopecka et al., 2020).

Advanced Technologies for the Production of Bioplastics

TABLE 16.1

Different Types of Biological Wastes Used for the Production of Biodegradable Plastics, Their Source Feedstock, and Applications

Type of Bioplastic	Source Feedstock	Application of Synthesized Bioplastic
Polyhydroxyalkanoates (PHA)	Microbial fermentation of organic waste (e.g., agricultural residues, food waste, wastewater)	• Packaging materials • Disposable cutlery • Agricultural mulch films • Medical implants
Polylactic acid (PLA)	Fermentation of starch-rich crops (e.g., corn, sugarcane)	• Food packaging • Disposable tableware • Textiles • Biomedical implants
Polybutylene succinate (PBS)	Fermentation of lignocellulosic biomass (e.g., forestry residues, agricultural waste)	• Agricultural films • Compostable bags • Disposable utensils
Polyethylene furanoate (PEF)	Fermentation of plant-based sugars (e.g., glucose, fructose)	• Bottles for beverages • Packaging films • Textile fibers
Polyhydroxyurethane (PHU)	Microbial fermentation of waste glycerol from biodiesel production	• Coatings • Adhesives • Biomedical applications

However, metabolic engineering also presents certain limitations, including:

1. *Complexity of genetic manipulation*: Genetic modification of microbial hosts may require extensive knowledge of metabolic pathways and cellular physiology, as well as sophisticated molecular biology techniques (Israni et al., 2020).
2. *Metabolic burden*: Overexpression of heterologous genes or introduction of exogenous pathways may impose metabolic burden on host organisms, affecting cell viability and growth (Kabir et al., 2020).
3. *Regulatory hurdles*: Regulatory approval for genetically modified organisms (GMOs) and bioprocesses may pose challenges, particularly in the context of commercial-scale bioplastics production (Lee et al., 2021).

16.4 SYNTHETIC BIOLOGY FOR BIOPLASTICS PRODUCTION

Synthetic biology represents a cutting-edge approach that harnesses the principles of engineering and biology to design and construct biological systems for specific applications, including the production of bioplastics (Taufik et al., 2020). This innovative field offers unprecedented opportunities to engineer microbial hosts to produce biopolymers with precisely tailored properties, such as biodegradability, mechanical strength, and thermal stability (Changwichan et al., 2018).

At the heart of synthetic biology lies the ability to rationally design and assemble genetic components, pathways, and regulatory elements to create novel biological functions or enhance existing ones (Ali et al., 2022). In the context of bioplastics production, synthetic biology enables the customization of microbial hosts to efficiently synthesize target biopolymers through metabolic engineering approaches (Akgül et al., 2022).

One of the key aspects of synthetic biology in bioplastics production is the design and construction of biological systems optimized for biopolymer synthesis. This involves several essential steps, including the selection of suitable microbial hosts, identification of target biopolymers, and engineering of metabolic pathways to enhance precursor flux toward biopolymer production (Adamcova et al., 2017). By precisely manipulating genetic elements and metabolic networks, synthetic biologists can tailor microbial hosts to produce bioplastics with desired properties, such as composition, molecular weight, and mechanical properties (Accinelli et al., 2012). Moreover, synthetic biology enables the implementation of modular design principles, allowing for the rapid prototyping and optimization of bioproduction processes. This modular approach facilitates the assembly of genetic elements into functional units, streamlining the design–build–test cycle and accelerating the development of bioplastic-producing microbial strains (Bishop et al., 2021).

Overall, synthetic biology offers a powerful toolkit for engineering microbial hosts to produce bioplastics with tailored properties, paving the way for the development of sustainable and environmentally friendly alternatives to conventional petroleum-based plastics. By combining advanced genetic engineering techniques with metabolic engineering strategies, synthetic biology holds great promise for revolutionizing the bioplastics industry and contributing to a more sustainable future (Jiang, et al., 2020)

16.4.1 DESIGN AND CONSTRUCTION OF BIOLOGICAL SYSTEMS FOR BIOPOLYMER PRODUCTION

Synthetic biology entails the rational design and assembly of biological components, such as genes, pathways, and regulatory elements, to engineer microorganisms capable of producing bioplastics. This process involves several key steps:

1. *Genetic design*: Identification of target biopolymers and selection of appropriate biosynthetic pathways for their production.
2. *Genome engineering*: Introduction of genetic modifications, such as gene knockout, overexpression, or insertion of exogenous genes, to optimize metabolic flux toward biopolymer synthesis.
3. *Pathway optimization*: Fine-tuning of metabolic pathways and regulatory networks to maximize biopolymer yields and enhance product purity and quality.
4. *Host selection*: Selection of suitable microbial hosts based on their natural metabolic capabilities, genetic tractability, and compatibility with the desired biopolymer synthesis pathways.

16.4.2 ADVANTAGES AND LIMITATIONS OF SYNTHETIC BIOLOGY FOR BIOPLASTICS PRODUCTION

Synthetic biology offers several advantages for bioplastics production:

1. *Customized biopolymer synthesis*: Enables the design of microbial hosts tailored for the production of specific biopolymers with desired properties, such as biodegradability, thermal stability, and mechanical strength (Questell-Santiago et al., 2020).
2. *Modular design*: Facilitates the modular assembly of genetic components and pathways, allowing for rapid prototyping and optimization of bioproduction processes (Pleissner et al., 2016).
3. *Scalability and sustainability*: Offers scalable and sustainable routes for bioplastics production using renewable feedstocks and environmentally friendly bioprocesses (Brizga et al., 2020).

However, synthetic biology also presents certain limitations:

1. *Complexity and uncertainty*: Designing and engineering complex biological systems for bioplastics production may pose challenges due to limited understanding of cellular physiology and metabolic regulation (van den Oever et al., 2017).

Advanced Technologies for the Production of Bioplastics

2. *Genetic stability*: Genetic instability and off-target effects in engineered microbial hosts may affect biopolymer yields and compromise product quality (Coates et al., 2020).
3. *Regulatory hurdles*: Regulatory approval for synthetic biology applications in bioplastics production may be hindered by concerns regarding biosafety, biocontainment, and environmental impact (Fadeeva et al., 2021).

16.5 NANOTECHNOLOGY FOR BIOPLASTICS PRODUCTION

Nanotechnology has emerged as a promising approach for enhancing the properties of bioplastics through the use of nanoparticles for reinforcement and modification. This section explores the application of nanotechnology in bioplastics production, focusing on the utilization of nanoparticles to improve the performance and functionality of biopolymer materials.

Nanoparticles, typically ranging in size from 1 to 100 nanometers, exhibit unique physical and chemical properties compared to bulk materials due to their high surface area-to-volume ratio and quantum effects (Hundertmark et al., 2018). In the context of bioplastics, nanoparticles can be incorporated into polymer matrices to impart desirable characteristics such as increased mechanical strength, enhanced thermal stability, improved barrier properties, and controlled biodegradability (Zheng and Suh, 2019).

One of the primary applications of nanotechnology in bioplastics production is the reinforcement of polymer matrices with nanoparticles to enhance mechanical properties. For example, the incorporation of nanocellulose, derived from renewable sources such as wood pulp or agricultural residues, has been shown to significantly improve the tensile strength, modulus, and toughness of biopolymer composites (Klemeš et al., 2020). Similarly, the addition of nanoclays, such as montmorillonite or halloysite, can enhance the stiffness, thermal stability, and gas barrier properties of bioplastics (Vanapalli et al., 2021).

In addition to reinforcement, nanotechnology enables the modification of biopolymer surfaces and interfaces to achieve specific functionalities. Surface functionalization of nanoparticles allows for the introduction of desired properties such as antimicrobial activity, UV protection, flame retardancy, and hydrophobicity/hydrophilicity (European Commission, 2018). These tailored surface properties can significantly expand the application range of bioplastics in various sectors, including packaging, biomedical devices, electronics, and automotive industries.

Despite the numerous advantages offered by nanotechnology in bioplastics production, several challenges and limitations must be addressed to realize its full potential. One of the key challenges is ensuring the uniform dispersion and compatibility of nanoparticles within biopolymer matrices to prevent aggregation, phase separation, and loss of mechanical properties (WEF et al., 2019). Achieving optimal nanoparticle loading levels and distribution patterns is crucial to maximizing the performance benefits while minimizing cost and processing complexities.

Furthermore, concerns regarding the potential environmental and health impacts of nanoparticle release during bioplastic degradation or recycling processes warrant careful consideration. The release of nanoparticles into the environment could pose risks to ecosystems and human health, highlighting the need for comprehensive risk assessment and mitigation strategies.

In conclusion, nanotechnology offers exciting opportunities for enhancing the performance and functionality of bioplastics through the incorporation of nanoparticles. By leveraging the unique properties of nanomaterials, bioplastics can be tailored to meet the diverse requirements of modern applications while promoting sustainability and environmental stewardship. However, addressing the technical, environmental, and regulatory challenges associated with nanotechnology in bioplastics production is essential to ensure safe and sustainable implementation.

16.6 CHALLENGES AND OPPORTUNITIES FOR ADVANCED TECHNOLOGIES IN BIOPLASTICS PRODUCTION

The advancement of technologies in bioplastics production holds promise for addressing environmental concerns associated with conventional petroleum-based plastics. However, along with opportunities, several challenges must be addressed to realize the full potential of these technologies. One of the primary challenges in bioplastics production lies in assessing and mitigating its environmental impact. While bioplastics offer the advantage of being derived from renewable resources, their production processes may still entail energy-intensive operations and generate greenhouse gas emissions (Zhu et al., 2016). Additionally, concerns exist regarding the land and water usage associated with cultivating feedstock crops for bioplastics production, which may lead to deforestation, habitat loss, and water scarcity (Chinthapalli et al., 2019). Moreover, the scalability of bioplastics production remains a critical consideration, as large-scale adoption of these materials would necessitate substantial infrastructure investments and technological advancements to meet global demand while minimizing the environmental footprint (UNIDO, 2019).

Another significant challenge pertains to the economic feasibility of bioplastics production compared to conventional plastics. Despite advancements in technology, bioplastics often incur higher production costs due to factors such as raw material procurement, processing, and manufacturing (Bucknall, 2020). Moreover, uncertainties regarding market demand, pricing dynamics, and regulatory frameworks further complicate the economic viability of bioplastics as a mainstream alternative to traditional plastics (WEF, 2016). Additionally, the lack of standardized metrics for assessing the life cycle costs and benefits of bioplastics production poses challenges in evaluating their competitiveness in the global market (WEF, 2016).

Public awareness and acceptance play a crucial role in shaping the adoption and utilization of bioplastics. While bioplastics offer environmental benefits, misconceptions regarding their biodegradability, compostability, and recycling capabilities abound among consumers (Narancic and O'Connor, 2017). Moreover, inadequate infrastructure for separate collection, recycling, and composting of bioplastics in many regions hinders their end-of-life management and contributes to contamination of conventional plastic waste streams (Gewert et al., 2015). Therefore, enhancing public education and engagement initiatives, along with investments in waste management infrastructure, are essential for promoting responsible use and disposal practices of bioplastics (Yang et al., 2014). While advanced technologies hold immense potential for revolutionizing bioplastics production and contributing to a more sustainable future, addressing the challenges of environmental impact, economic feasibility, and public awareness remains imperative. Collaborative efforts involving policymakers, industry stakeholders, researchers, and consumers are essential for overcoming these challenges and unlocking the full benefits of advanced bioplastics technologies.

16.7 LIFE CYCLE ASSESSMENT (LCA) AND TECHNO-ECONOMIC ANALYSIS (TEA) OF BIOPLASTICS PRODUCTION

The growing concern over environmental degradation and the need to reduce reliance on fossil fuels have spurred significant interest in bioplastics as an alternative to conventional plastics. Bioplastics, derived from renewable biomass sources, offer the promise of reduced carbon footprint and enhanced sustainability (Terzopoulou et al., 2016). However, the production of bioplastics involves complex processes that require careful consideration of environmental impacts, economic feasibility, and technological advancements. In this chapter, we explore advanced technologies employed in the production of bioplastics, focusing on life cycle assessment (LCA) and techno-economic analysis (TEA) as key methodologies for evaluating their environmental and economic sustainability.

16.7.1 Life Cycle Assessment (LCA) of Bioplastics Production

Life cycle assessment (LCA) is a systematic methodology used to evaluate the environmental impacts associated with all stages of a product's life cycle, from raw material extraction to end-of-life disposal. In the context of bioplastics production, LCA plays a crucial role in assessing the overall environmental performance of different manufacturing processes and feedstock sources. Advanced technologies in bioplastics production aim to minimize environmental impacts by optimizing resource utilization, reducing energy consumption, and mitigating emissions (Stagner, 2016). One advanced technology in bioplastics production is the utilization of novel feedstock sources, such as algae, agricultural residues, and organic waste streams (Gajendiran et al., 2016; Harding et al., 2007). LCA studies have demonstrated that certain feedstock sources, such as algae, have the potential to significantly reduce greenhouse gas emissions and land use compared to traditional feedstocks like corn or sugarcane. Furthermore, advances in bioprocessing technologies, such as fermentation and enzymatic conversion, have enabled the efficient conversion of biomass into bioplastic precursors, further enhancing the environmental sustainability of bioplastics production.

16.7.2 Techno-economic Analysis (TEA) of Bioplastics Production

Techno-economic analysis (TEA) plays a pivotal role in the evaluation of economic feasibility and competitiveness within the realm of bioplastics production. It serves as a systematic methodology employed to scrutinize the viability of different technologies or processes. Specifically tailored to the bioplastics sector, TEA serves as a vital tool in assessing the cost-effectiveness of cutting-edge technologies while pinpointing crucial cost determinants influencing overall production expenses.

Within the landscape of bioplastics production, TEA serves multifaceted functions essential for strategic decision-making and resource allocation. By conducting comprehensive analyses, TEA enables stakeholders to gain insights into the intricate economic dynamics governing bioplastics production. Through meticulous examination of capital investment, operating costs, and revenue streams, TEA facilitates a comprehensive understanding of the financial landscape, enabling informed decision-making and strategic planning (Yamada and Fukumoto, 2021). Furthermore, TEA acts as a driving force behind the advancement of bioplastics technologies. By systematically evaluating process efficiency and cost-effectiveness, TEA incentivizes the development and adoption of innovative technologies aimed at enhancing production efficiency and reducing overall costs. This iterative process of technological refinement fosters continuous improvement within the bioplastics industry, driving innovation and fostering competitiveness in the global market.

The pursuit of enhanced market competitiveness underscores the significance of TEA in the bioplastics domain. As industries seek sustainable alternatives to traditional plastics, TEA provides a roadmap for enhancing market competitiveness through strategic cost optimization and value proposition refinement. By identifying key cost drivers and market opportunities, TEA empowers stakeholders to develop tailored strategies aimed at maximizing profitability and market penetration (Lowentha, 2020). In a nutshell, TEA serves as a cornerstone in the realm of bioplastics production, offering invaluable insights into economic viability and market competitiveness. Through meticulous analysis of costs, revenues, and market dynamics, TEA enables stakeholders to make informed decisions, driving innovation and fostering sustainability within the bioplastics industry (Karan et al., 2019). As advancements in bioplastics technologies continue to evolve, TEA remains an indispensable tool for navigating the complex economic landscape and shaping the future of sustainable plastics production. Figure 16.3 illustrates the non-renewable energy consumption and global warming potential for different types of bioplastics and petroleum-based plastics, comparing (a) production alone (cradle-to-gate) and (b) production throughout its lifespan (cradle-to-grave).

One key aspect of TEA in bioplastics production is the consideration of economies of scale and process integration. By scaling up production facilities and integrating bioplastic production with existing industrial processes, significant cost savings can be achieved through economies of scale

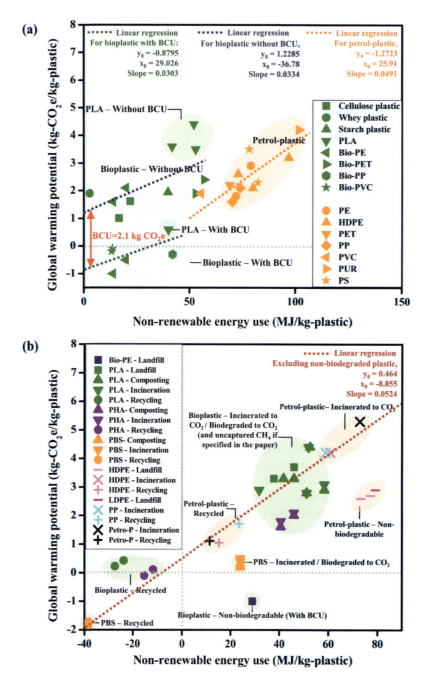

FIGURE 16.3 The non-renewable energy consumption and global warming potential for (a) production alone (cradle-to-gate) and (b) production throughout its lifespan (cradle-to-grave) for different types of bioplastics and petroleum-based plastics. (Obtained with permission from Patria et al., 2024, Copyright © 2024 Elseiver.)

and synergistic resource utilization. Additionally, advances in process optimization, catalyst development, and downstream processing techniques have contributed to cost reductions and improved product quality in bioplastics production. Figure 16.4 highlights two key aspects: (a) the essential elements for standardizing TEA and LCA in sustainable plastic management, and (b) a comparative

Advanced Technologies for the Production of Bioplastics 297

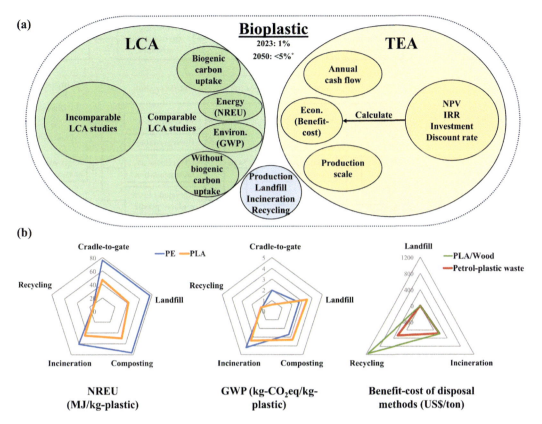

FIGURE 16.4 (a) Essential elements for standardizing TEA/LCA in sustainable plastic management. (b) Comparative analysis of ³E indicator parameters between petroleum-based and bioplastic materials. (*The percentage of bioplastic in total plastic production is estimated based on an overall assessment of various reports.) (Obtained with permission from Patria et al., 2024, Copyright © 2024 Elseiver.)

analysis of 3E indicator parameters between petroleum-based and bioplastic materials, with the bioplastic percentage in total plastic production estimated from various reports.

Several case studies and real-world applications demonstrate the practical implementation of advanced technologies for bioplastics production. For example, companies like NatureWorks LLC and Braskem have successfully commercialized bioplastics derived from renewable feedstocks like corn and sugarcane, demonstrating the feasibility of large-scale bioplastic production. Figure 16.5 illustrates the study's scope and methodology for comparing bioplastics and petroleum-based plastics, including standardization of LCA and TEA studies. Furthermore, research initiatives and collaborative projects, such as the European Bioplastics Association and the Bioplastics Feedstock Alliance, are actively exploring innovative technologies and sustainable feedstock sources to further advance the bioplastics industry.

To sum up, advanced technologies play a crucial role in advancing the production of bioplastics, offering solutions to environmental challenges and economic constraints associated with traditional plastics. LCA and TEA provide valuable insights into the environmental and economic sustainability of bioplastics production, guiding decision-making processes and driving innovation in the bioplastics industry (George et al., 2021). As technological advancements continue to evolve and new opportunities emerge, the future of bioplastics holds great promise in contributing to a more sustainable and circular economy. Table 16.2 shows an overview of the distinct characteristics and objectives of LCA and TEA in the context of bioplastics production.

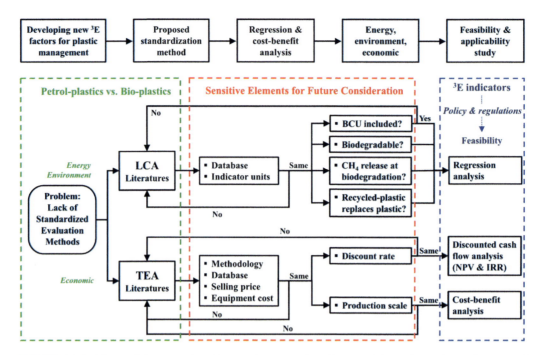

FIGURE 16.5 The study's scope, focusing on the comparison between bioplastics and petroleum-based plastics. This comparison spans from raw material acquisition through plastic production to end-of-life treatment. Additionally, the figure outlines the methodology employed for standardizing independent LCA and TEA studies, facilitating the comparison of ^{3}E (Environmental, Economic, and Energy) indicators. (Obtained with permission from Patria et al., 2024, Copyright © 2024 Elseiver.)

TABLE 16.2
Overview of the Distinct Characteristics and Objectives of Life Cycle Assessment (LCA) and Techno-Economic Analysis (TEA) in the Context of Bioplastic Production

Aspect	Life Cycle Assessment (LCA)	Techno-economic Analysis (TEA)
Purpose	Evaluate environmental impacts of bioplastic production	Assess economic feasibility and competitiveness of production
Scope	Considers entire life cycle of bioplastic, from cradle to grave	Focuses on economic aspects of production process
Methodology	Systematic assessment of environmental impacts using standardized methods and indicators	Economic analysis of production costs, revenues, and profitability
Parameters considered	Resource consumption, energy use, emissions, waste generation, land use, etc.	Capital investment, operating costs, revenue streams, market demand, etc.
Output	Environmental impact indicators (e.g., carbon footprint, ecological footprint)	Financial indicators (e.g., net present value, internal rate of return, payback period)
Decision support	Helps identify environmental hotspots and opportunities for improvement	Guides investment decisions, process optimization, and technology selection
Data requirements	Comprehensive data on raw materials, energy use, emissions, transportation, etc.	Detailed cost data, market prices, production yields, discount rates, etc.
Applicability	Used for comparing different production processes, feedstock sources, and end-of-life scenarios	Applied to assess project feasibility, determine project economics, and inform business strategies

16.8 CONCLUSIONS

In conclusion, the exploration of advanced technologies for bioplastics production offers a promising pathway toward addressing environmental concerns associated with conventional petroleum-based plastics. Throughout this chapter, we have examined various innovative approaches, including biocatalysis, metabolic engineering, synthetic biology, and nanotechnology, each contributing to the development of sustainable alternatives to traditional plastics. A summary of key points reveals the multifaceted nature of bioplastics production, from enzymatic breakdown of plant-based materials to the design of biological systems for tailored biopolymer synthesis. Biocatalysis enables the efficient conversion of renewable feedstocks into bioplastic monomers, while metabolic engineering facilitates the optimization of microbial pathways for enhanced biopolymer production. Synthetic biology provides a platform for designing custom biological systems capable of producing bioplastics with desired properties, and nanotechnology offers opportunities for reinforcement and modification of bioplastics through the use of nanoparticles.

As we reflect on these advancements, it is evident that continued research and development are essential for overcoming existing challenges and maximizing the potential of advanced bioplastics technologies. Environmental impact assessment and scalability studies are critical for ensuring the sustainability of bioplastics production processes, while economic feasibility analyses are necessary for assessing their competitiveness in the global market. Additionally, efforts to raise public awareness and engagement in the use and disposal of bioplastics are essential for fostering responsible consumption behaviors and promoting a circular economy.

In light of these considerations, a call to action is warranted for stakeholders across academia, industry, and government to collaborate in advancing the field of bioplastics production. By investing in research, innovation, and education, we can accelerate the transition toward a more sustainable materials economy, where bioplastics play a central role in mitigating environmental pollution and promoting resource conservation.

In closing, the journey toward sustainable bioplastics production is ongoing, and it requires collective commitment and collaboration to realize its full potential. Let us seize this opportunity to drive positive change and shape a future where bioplastics contribute to a cleaner, healthier planet for generations to come.

REFERENCES

Accinelli, C., Sacca, M. L., Mencarelli, M., & Vicari, A. 2012. Deterioration of bioplastic carrier bags in the environment and assessment of a new recycling alternative. *Chemosphere* 89 (2), 136–143. https://doi.org/10.1016/j.chemosphere.2012.05.028.

Adamcova, D., Elbl, J., Jan, Z., Vaverkova, M. D., Kintl, A., Juricka, D., Jan, H., & Martin, B. 2017. Study on the (Bio)Degradation process of bioplastic materials under industrial composting conditions. *Acta Univ. Agric. Silvic. Mendelianae Brunensis* 65 (3), 791–798. https://doi.org/10.11118/actaun201765030791.

Ahmed, T., Shahid, M., Azeem, F., Rasul, I., Shah, A. A., Noman, M., Hameed, A., Manzoor, N., Manzoor, I., & Muhammad, S. 2018. Biodegradation of plastics: current scenario and future prospects for environmental safety. *Environ. Sci. Pollut. Control Ser.* 25 (8), 7287–7298. https://doi.org/10.1007/s11356-018-1234-9

Akgül, A., Palmeiro-Sanchez, T., Lange, H., Magalhaes, D., Moore, S., Paiva, A., Kazanç, F., & Anna, T. 2022. Characterization of tars from recycling of PHA bioplastic and synthetic plastics using fast pyrolysis. *J. Hazard Mater.* 439, 129696. https://doi.org/10.1016/j.jhazmat.2022.129696

Al-Tohamy, R., Ali, S. S., Zhang, M., Sameh, M., Zahoor, Y. A. G. M., Waleed, N., Okasha, K. M., Sun, & S. J., Sun. 2023a. Can wood-feeding termites solve the environmental bottleneck caused by plastics? A critical state-of-the-art review. *J. Environ. Manag.* 326, 116606 https://doi.org/10.1016/j.jenvman.2022.116606.

Al-Tohamy, R., Samir Ali, S., Zhang, M., Elsamahy, T., Abdelkarim, E. A., Jiao, H., Sun, S., & Sun, J. 2023b. Environmental and human health impact of disposable face masks during the COVID-19 pandemic: woodfeeding termites as a model for plastic biodegradation. *Appl. Biochem. Biotechnol.* 195 (3), 2093–2113. https://doi.org/10.1007/s12010-022-04216-9

Ali, S. S., Elsamahy, T., Abdelkarim, E. A., Al-Tohamy, R., Kornaros, M., Ruiz, H. A., Tong, Z., Li, F., & Sun, J. 2022. Biowastes for biodegradable bioplastics production and end-of-life scenarios in circular bioeconomy and biorefinery concept. *Bioresour. Technol.* 363, 127869 https://doi.org/10.1016/j.biortech.2022.127869

Ali, S. S., Abdelkarim, E. A., Elsamahy, T., Al-Tohamy, R., Li, F., Kornaros, M., Zuorro, A., Zhu, D., & Sun, J. 2023. Bioplastic production in terms of life cycle assessment: a state-of-the-art review. *Environ. Sci. Ecotechnol.* 15, 100254 https://doi.org/10.1016/j. ese.2023.100254.

Amiri, S., Zeydi, M. M., & Amiri, N. 2021. Bacillus cereus Saba.Zh, a novel bacterial strain for the production of bioplastic (polyhydroxybutyrate). *Braz. J. Microbiol.* 52 (4), 2117–2128. https://doi.org/10.1007/s42770-021-00599-9.

Aragosa, A., Specchia, V., & Frigione, M. 2021. Isolation of two bacterial species from argan soil in Morocco associated with polyhydroxybutyrate (PHB) accumulation: current potential and future prospects for the bio-based polymer production. *Polymers* 13 (11), 1870. https://doi.org/10.3390/polym13111870

Bishop, G., Styles, D., & Lens, P. N. L. 2021. Environmental performance comparison of bioplastics and petrochemical plastics: a review of life cycle assessment (LCA) methodological decisions. *Resour. Conserv. Recycl.* 168, 105451.

Brizga, J., Hubacek, K., & Feng, K. 2020. The unintended side effects of bioplastics: carbon, land, and water footprints. *One Earth* 3, 45–53.

Bucknall, D. G. 2020. Plastics as a materials system in a circular economy. *Philos. Trans. R. Soc. A* 378, 20190268.

Changwichan, K., Silalertruksa, T., & Gheewala, S. 2018. Eco-efficiency assessment of bioplastics production systems and end-of-life options. *Sustainability* 10 (4), 952. https://doi.org/10.3390/su10040952.

Chen, C., Dai, L., Ma, L., & Guo, R. 2020. Enzymatic degradation of plant biomass and synthetic polymers. *Nat. Rev. Chem.* 4, 114–126.

Chinthapalli, R. et al. 2019. Biobased building blocks and polymers — global capacities, production and trends, 2018–2023. *Ind. Biotechnol.* 15, 237–241.

Coates, G. W., & Getzler, Y. D. Y. 2020. Chemical recycling to monomer for an ideal, circular polymer economy. *Nat. Rev. Mater.* 5, 501–516.

Espinosa, M. J. C. et al. 2020. Toward biorecycling: isolation of a soil bacterium that grows on a polyurethane oligomer and monomer. *Front. Microbiol.* 11, 404.

European Commission. 2018. A European strategy for plastics in a circular economy. *European Commission.* www.europarc.org/wp-content/uploads/2018/01/Eu-plastics-strategy-brochure.pdf

Fadeeva, Z., & Berkel, R. 2021. Van Unlocking circular economy for prevention of marine plastic pollution: an exploration of G20 policy and initiatives. *J. Environ. Manag.* 277, 111457.

Gajendiran, A., Krishnamoorthy, S., & Abraham, J. 2016. Microbial degradation of low-density polyethylene (LDPE) by Aspergillus clavatus strain JASK1 isolated from landfill soil. *3 Biotech* 6, 52.

Gandini, A., Lacerda, T. M., Carvalho, A. J. F., & Trovatti, E. 2016. Progress of polymers from renewable resources: furans, vegetable oils, and polysaccharides. *Chem. Rev.* 116, 1637–1669.

George, N., Debroy, A., Bhat, S., Bindal, S., & Singh, S. 2021. Biowaste to bioplastics: an ecofriendly approach for a sustainable future. *J. Appl. Biotechnol. Rep.* 8 (3). https://doi.org/10.30491/jabr.2021.259403.1318.

Gewert, B., Plassmann, M. M., & Macleod, M. 2015. Pathways for degradation of plastic polymers floating in the marine environment. *Environ. Sci. Process. Impacts* 17, 1513–1521.

Geyer, R., Jambeck, J. R., & Law, K. L. 2017. Production, use, and fate of all plastics ever made. *Sci. Adv.* 3, 25–29.

Harding, K. G., Dennis, J. S., von Blottnitz, H., & Harrison, S. T. L. 2007. Environmental analysis of plastic production processes: comparing petroleum-based polypropylene and polyethylene with biologically-based poly-β-hydroxybutyric acid using life cycle analysis. *J. Biotechnol.* 130, 57–66.

Henry, B., Laitala, K., & Klepp, I. G. 2019. Microfibres from apparel and home textiles: prospects for including microplastics in environmental sustainability assessment. *Sci. Total Environ.* 652, 483–494.

Hobbs, S. R., Parameswaran, P., Astmann, B., Devkota, J. P., & Landis, A. E. 2019. Anaerobic codigestion of food waste and polylactic acid: effect of pretreatment on methane yield and solid reduction. *Adv. Mater. Sci. Eng.* 2019, 4715904.

Hundertmark, T., McNally, C., Simons, T. J., & Vanthournout, H. *No Time to Waste: What Plastics Recycling Could Offer. McKinsey on Chemicals* (McKinsey & Company, 2018).

Advanced Technologies for the Production of Bioplastics

Ioannidou, S. M. et al. 2020. Sustainable production of bio-based chemicals and polymers via integrated biomass refining and bioprocessing in a circular bioeconomy context. *Bioresour. Technol.* 307, 123093.

Israni, N., Venkatachalam, P., Gajaraj, B., Varalakshmi, K. N., & Shivakumar, S. 2020. Whey valorization for sustainable polyhydroxyalkanoate production by Bacillus megaterium: production, characterization and in vitro biocompatibility evaluation. *J. Environ. Manag.* 255, 109884 https://doi.org/10.1016/j.jenvman.2019.109884.

Jaiswal, S., Sharma, B., & Shukla, P. 2020. Integrated approaches in microbial degradation of plastics. *Environ. Technol. Innov.* 17, 100567.

Jiang, L. et al. 2020. PEF plastic synthesized from industrial carbon dioxide and biowaste. *Nat. Sustain.* 3, 761–767.

Kabir, E., Kaur, R., Lee, J., Kim, K.-H., Eilhann, E. K. 2020. Prospects of biopolymer technology as an alternative option for non-degradable plastics and sustainable management of plastic wastes. *J. Clean. Prod.* 258, 120536. https://doi.org/10.1016/j.jclepro.2020.120536.

Karan, H., Funk, C., Grabert, M., Oey, M., & Hankamer, B. 2019. Green bioplastics as part of a circular bioeconomy. *Trends Plant Sci.* 24, 237–249.

Klemeš, J. J., Fan, Y., Van, Tan, R. R., & Jiang, P. 2020. Minimising the present and future plastic waste, energy and environmental footprints related to COVID-19. *Renew. Sustain. Energy Rev.* 127, 109883

Koch, D., & Mihalyi, B. 2018. Assessing the change in environmental impact categories when replacing conventional plastic with bioplastic in chosen application fields. *Chem. Eng. Trans.* 70, 853–858.

Koller, M., & Braunegg, G. 2018. Advanced approaches to produce polyhydroxyalkanoate (PHA) biopolyesters in a sustainable and economic fashion. *The EuroBiotech J* 2 (2), 89–103. https://doi.org/10.2478/ebtj-2018-0013.

Kopecka, R., Kubínova, I., Katerina, S., Mravcova, L., Vítez, T., & Vítezova, M. 2022. Microbial degradation of virgin polyethylene by bacteria isolated from a landfill site. *SN Appl. Sci.* 4 (11), 302. https://doi.org/10.1007/s42452-022-05182-x.

Lee, A., & ShanLiew, M. 2021. Tertiary recycling of plastics waste: an analysis of feedstock, chemical and biological degradation methods. *J. Mater. Cycles Waste Manag.* 23, 32–43.

Meereboer, K. W., Misra, M., & Mohanty, A. K. 2020. Review of recent advances in the biodegradability of polyhydroxyalkanoate (PHA) bioplastics and their composites. *Green Chem.* 22, 5519–5558.

Moshood, T. D. 2021. Expanding policy for biodegradable plastic products and market dynamics of bio-based plastics: challenges and opportunities. *Sustainability* 13, 6170.

Nakajima, H., Dijkstra, P., & Loos, K. 2017. The recent developments in biobased polymers toward general and engineering applications: polymers that are upgraded from biodegradable polymers, analogous to petroleum-derived polymers, and newly developed. *Polymers* 9, 523.

Narancic, T., & O'Connor, K.E. 2017. Microbial biotechnology addressing the plastic waste disaster. *Microb. Biotechnol.* 10, 1232–1235.

Lowenthal, A. S. 2020. *Break Free From Plastic Pollution Act. 116th US Congress* 1–127 (US Congress, 2020).

Patria, R. D., Rehma, S., Yuen, C. B., Lee, D. J., Vuppaladadiyam, A. K., & Leu, S.Y. 2024. Energy-environment-economic (3E) hub for sustainable plastic management – Upgraded recycling, chemical valorization, and bioplastics. *Appl. Energy* 357, 122543. https://doi.org/10.1016/j.apenergy.2023.122543

Pellis, A., Malinconico, M., Guarneri, A., & Gardossi, L. 2021. Renewable polymers and plastics: performance beyond the green. *New Biotechnol.* 60, 146–158.

Pleissner, D. et al. 2016. Valorization of organic residues for the production of added value chemicals: a contribution to the bio-based economy. *Biochem. Eng. J.* 116, 3–16.

Rosenboom, JG., Langer, R., & Traverso, G. 2022. Bioplastics for a circular economy. *Nat Rev Mater* 7, 117–137. https://doi.org/10.1038/s41578-021-00407-8

Sheldon, R. A. 2018. Chemicals from renewable biomass: a renaissance in carbohydrate chemistry. *Curr. Opin. Green Sustain. Chem.* 14, 89–95.

Sicotte, D. M. 2020. From cheap ethane to a plastic planet: regulating an industrial global production network. *Energy Res. Soc. Sci.* 66, 101479.

Stagner, J. 2016. Methane generation from anaerobic digestion of biodegradable plastics–a review. *Int. J. Environ. Stud.* 73, 462–468.

Straub, S., Hirsch, P. E., & Burkhardt-Holm, P. 2017. Biodegradable and petroleum-based microplastics do not differ in their ingestion and excretion but in their biological effects in a freshwater invertebrate *Gammarus fossarum. Int. J. Environ. Res. Public Health* 14, 774.

Taufik, D., Reinders, M. J., Molenveld, K., Onwezen, M. C. 2020. The paradox between the environmental appeal of bio-based plastic packaging for consumers and their disposal behaviour. *Sci. Total Environ.* 705, 135820 https://doi.org/10.1016/j.scitotenv.2019.135820.

Terzopoulou, Z. et al. 2016. Biobased poly(ethylene furanoate-co-ethylene succinate) copolyesters: solid state structure, melting point depression and biodegradability. *RSC Adv.* 6, 84003–84015

Questell-Santiago, Y. M., Galkin, M. V., Barta, K., & Luterbacher, J. S. 2020. Stabilization strategies in biomass depolymerization using chemical functionalization. *Nat. Rev. Chem.* 4, 311–330.

United Nations Industrial Development Organization (UNIDO). 2019. Addressing the challenge of marine plastic litter using circular economy methods: relevant considerations. *UNIDO* www.unido.org/sites/default/files/files/2019-06/UNIDO_Addressing_the_challenge_of_Marine_Plastic_Litter_Using_Circular_Economy.pdf.

Vanapalli, K. R. et al. 2021. Challenges and strategies for effective plastic waste management during and post COVID-19 pandemic. *Sci. Total Environ.* 750, 141514.

van den Oever, M., Molenveld, K., van der Zee, M., & Bos, H. 2017. Bio-based and biodegradable plastics – facts and figures. Focus on food packaging in the Netherlands. Report 1722. *Wageningen Food & Biobased Research* www.edepot.wur.nl/408350

Venkata Mohan, S., Modestra, J. A., Amulya, K., Butti, S. K., & Velvizhi, G. 2016. A circular bioeconomy with biobased products from CO_2 sequestration. *Trends Biotechnol.* 34, 506–519.

Walker, S., & Rothman, R. 2020. Life cycle assessment of bio-based and fossil-based plastic: a review. *J. Clean. Prod.* 261, 121158.

Wei, R. et al. 2019. Biocatalytic degradation efficiency of postconsumer polyethylene terephthalate packaging determined by their polymer microstructures. *Adv. Sci.* 6, 1900491.

World Economic Forum (WEF). *Top 10 Emerging Technologies* 4–15 (WEF, 2019).

Yamada, S., & Fukumoto, Y. 2021. China aims to go as big in bioplastics as it did in solar panels. *Nikkei Asia.* www.asia.nikkei.com/Spotlight/Environment/China-aims-to-go-as-big-in-bioplastics-as-it-did-in-solar-panels

Yang, J., Yang, Y., Wu, W., Zhao, J., & Jiang, L. 2014. Evidence of polyethylene biodegradation by bacterial strains from the guts of plastic-eating waxworms. *Environ. Sci. Technol.* 48, 13776–13784.

Zheng, J., & Suh, S. 2019. Strategies to reduce the global carbon footprint of plastics. *Nat. Clim. Change* 9, 374–378.

Zhu, Y., Romain, C., & Williams, C. K. 2016. Sustainable polymers from renewable resources. *Nature* 540, 354–364.

17 Role of Nanotechnology in Plastic and Microplastic Management

Manisha Gulati, Khushboo Singhal, Malya, Piyush Kumar Gupta, Deepansh Sharma, and Sunny Dholpuria

17.1 INTRODUCTION

Plastic pollution has escalated into a global environmental crisis with dire consequences for terrestrial and marine ecosystems, including humans. Plastics play a crucial role in our daily lives, as they are strong, lightweight, and adaptable, but due to their long-standing nature, they persist in the environment for a very long time resulting in the contamination of nearly all natural resources (Thompson R C et al., 2009). The rising plastic pollution is an alarming situation and according to a report of 2023 published by the United Nations, more than 400 million tonnes of plastic is produced every year worldwide, 50% of which can be used only once and less than 10% is recycled. Plastics not only pose a threat to marine life through ingestion and entanglement, but they also have the potential to disrupt entire ecosystems, ranging from the tiniest microorganisms to large marine mammals. It has been estimated that by 2050, plastic waste will accumulate in landfills or the environment at a staggering 12 billion metric tons. Globally, 259 million metric tonnes of waste were generated in 2018, illustrating a disconcerting trajectory in waste production and management (Rose P K et al., 2023). The problems are not just associated with plastic waste, in recent years, microplastics (MPs), plastic particles having a size less than 5 mm, have become a major issue (Ziani K et al., 2023) and these have not only been discovered in the extreme locations of the world like the Arctic and Antarctic oceans (Hale R C et al., 2020) but have been found in the blood of animals including humans (Lin Y D., 2023). These tiny plastic particles arise from a variety of sources, including tiny beads found in personal care items, tiny fibers found in clothes, and may be produced from the breakdown of larger plastic trash (Duis K et al., 2016). The existence of microplastic has been observed in air also and this has become one of the causes of metabolic disorders, neurotoxicity, and enhanced cancer susceptibility (Campanale C et al., 2020). The impacts of microplastics have also been observed on different organs (Table 17.1) as well as on the developing embryo (Figure 17.1).

Traditional waste management techniques like recycling, incineration, and landfilling have their own limitations, such as landfilling can lead to the leaching of hazardous substances into the soil and water, and incinerating waste produces air pollution and greenhouse gas emissions. Additionally, these techniques are ineffective in addressing the problem of MPs, as it is challenging to collect and remove them from the environment. NT (nanotechnology), being a groundbreaking technology, has the potential to address a wide range of environmental and other problems at the nanoscale level due to their unique physical, chemical, and biological properties (Bayda S et al., 2019). Nanomaterials have shown great potential in a wide range of applications related to plastic and microplastic management, such as degrading plastics, enhancing their properties, and filtering MPs (Goh P S et al., 2022). NT-based approaches, such as photocatalysis, nanofilters (NFs), magnetic separation, and

DOI: 10.1201/9781032706573-17

TABLE 17.1
Effect of Microplastics on Various Organs and Glands of the Human Body

Organ	Effect on the Organs
Liver	Liver inflammation and fibrosis, disrupted metabolic function, carcinogenicity, hepatic steatosis, altered amino acid metabolism
Gastrointestinal tract	Hinder digestion and absorption, alter microbial flora, metabolic stress, obesity, diabetes
Placenta	Growth restriction, pregnancy toxemia, less vascularization
Respiratory tract	Asthma, allergic alveolitis, chronic obstructive pulmonary disease, respiratory irritation
Blood	Immune disorder, hemolysis, cardiovascular diseases
Skin	Cellular degeneration and toxicity, induce inflammation, epithelial oxidative stress
Thyroid gland	PDBE (polybrominated diphenyl ethers) affects thyroid homeostasis
Kidney	Accumulation of phthalate causes glomerular atrophy, glomerular necrosis
Reproductive organs	Complications in pregnancy, antiandrogen action, bisphenol A affects spermatogenesis, decreased testosterone, reduced LH and FHS in testes, ovarian fibrosis, less growth of follicles in ovary
Heart	Circulatory system stresses, disruption in endocrine system, cardiovascular problems, myocardium vascular congestion
Spleen	Decreased leukocytes, altered immune response
Brain	Cognitive impairment, reduced AChE activity, oxidative stress

Source: Based on data from Hofstede L T et al. (2023).

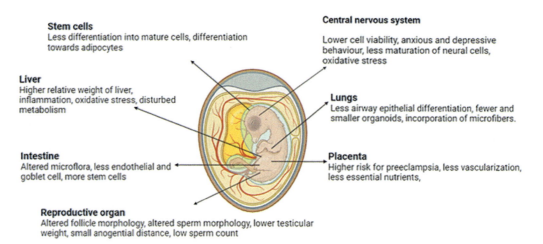

FIGURE 17.1 Impact of microplastics on various organs of a developing fetus. (Based on data from Hofstede L T et al., 2023.)

bionanomaterial-based degradation, have shown potential in accelerating the degradation of plastics including polyethylene, polypropylene, and polystyrene (Yaqoob A A et al., 2020). This chapter highlights the mechanisms of various nanotechnological approaches used to deal with the problems of plastic and microplastic pollution.

17.2 NANOTECHNOLOGY-ASSISTED REMEDIATION OF PLASTIC POLLUTION

Nanotechnology has proven to be a valuable asset in the recycling of plastic waste and offers the potential to create next-generation valuable raw materials. Unlike the conventional methods

Role of Nanotechnology in Plastic and MP Management 305

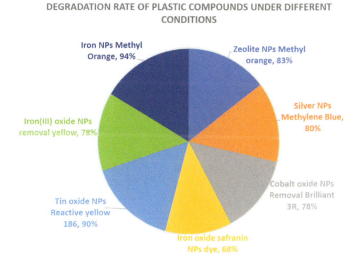

FIGURE 17.2 Degradation of plastic compound from the enviornment by using different NPs. (Based on data from Gondal A H et al., 2022.)

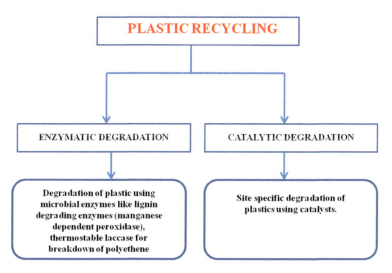

FIGURE 17.3 Methods of plastic recycling: enzymatic polymerization and catalytic polymerization.

including melting and reforming plastics, NT-based approaches seek to maintain the properties of the recycled material and enhance its performance (Figure 17.2) (Gondal A H et al., 2022). In one of the approaches involving the use of multiwalled carbon nanotubes (MWCNTs), nickel zinc ferrite NPs (nanoparticles) along with recycled polystyrene lead to the production of composite materials with enhanced thermal and physical properties (Khan W S et al., 2014).

The recycling of plastics can be achieved by two methods namely enzymatic depolymerization and catalytic depolymerization (Figure 17.3).

17.2.1 Enzymatic Depolymerization

Enzymes are predominantly found in all types of living organisms and as a biological catalyst, they play an important role in regulating the metabolic processes in a highly specific manner.

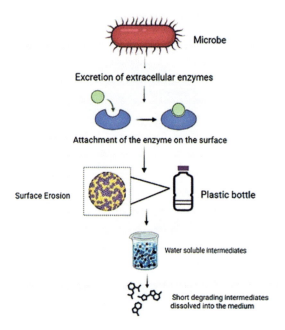

FIGURE 17.4 Enzymatic degradation of plastics.

A generalized mechanism of enzyme-based degradation of plastics can be seen in Figure 17.4. There are a number of enzymes isolated from microbial sources that can cause the degradation of plastics such as PETase (polyethylene terephthalate hydrolase) found in the bacteria *Ideonella sakaiensis* (Carr C M et al., 2020) and esterase enzyme found in the yeast *Pseudozyma antarctica,* which are mainly used for the degradation of polyethylene and plastic mulch films, respectively (Sato S et al., 2017). The applications of nanotechnology greatly enhance the activity of such enzymes and the NPs act as a scaffold for these enzymes and make them more stable and more efficient (Ellis G A et al., 2020), for example, the immobilization of lipase and cutinase enzymes over NPs of magnetite (Fe_3O_4) and silicon dioxide (SiO_2) increased the stability and activity of these enzymes for up to 144 hours (Krakor et al., 2021). A similar approach has been used in the degradation of plastics by immobilizing PETase enzyme on cobaltous phosphate nanoflowers at 37°C and pH 6–10, and this enhanced the consistency and activity of the enzyme several fold. When the enzyme encounters the PET plastic surface, it forms noncovalent interactions (particularly), hydrogen bonds and electrostatic interactions which facilitate the hydrolysis of the ester bond (Zurier H S et al., 2022). This led to the depolymerization of the polymer chain of PET into monomers like mono-(2-hydroxyethyl) terephthalate (MHET), bis-(2-hydroxyethyl) terephthalate (BHET), and ethylene glycol (EG) (Yin Q et al., 2022). MHET so formed can further be hydrolyzed by enzyme MHETase into terephthalic acid and ethylene glycol (Palm G J et al., 2019). The application of nanoflowers increase the enzyme recyclability, reusability, and contact efficiency between PET and PETase enzymes, thereby facilitating 3.5 times higher degradation of plastics (Jia Y. et al., 2021). Another approach involves the use of lipases for the degradation of polylactic acid by reinforcing it with nanocellulose crystals, which results in the enhancement of the degradation process by providing a suitable substrate. This process involves the adsorption of lipases on the surface of the nanocomposite followed by hydrolyses of ester bond in PLA (polylactic acid), and forms by-products such as lactic acid under optimum conditions of pH 8.6 and 37°C (Lee S H et al., 2011). However, this process is associated with many challenges including degradation of enzymes below or above optimal conditions, limited affinity toward a wide range of substrates, slow rate of degradation, and the availability of enzymes (Zdarta J et al., 2018). NPs help in overcoming these issues by increasing the porosity and availability of enzymes in the deeper areas of plastics. Another challenge is the specificity of an enzyme and because the different kinds of plastics vary in the composition of their polymers and this increases the requirement for different enzymes for efficient degradation (Neupane S et al., 2019). The recent advances in genetic engineering help in overcoming this issue through the production of engineered enzymes that show specificity against a wide range of plastic polymers and a cocktail of enzymes can also be used along with different NPs. For example, the active site of cutinase Tfu_0883 from *Thermobifida fusca* was modified by site-directed mutagenesis to increase the affinity of cutinase to polyethylene terephthalate (PET) and the ability to hydrolyze it. This mutation at a particular site of the enzyme (amino acid 218) resulted in the replacement of isoleucine (I) with alanine (A), which created space and a second double mutation Q132A/T101A possessing glutamine (Q) and tyrosine (T) was replaced with alanine (A), which also created space and increased hydrophobicity. All these changes cause considerably greater hydrolysis efficiency toward PET fibers—a double mutant exhibited 1.6-fold increased hydrolysis activity (Urbanek A K et al., 2021). In a similar context, *Escherichia coli* has been genetically engineered in such a manner that it acts on plastic bottles resulting in the formation of vanillin as an end-product through the enzyme-catalyzed hydrolysis pathway (Gondal A H et al., 2022).

17.2.2 CATALYTIC DEPOLYMERIZATION

This process utilizes catalysts for the breakdown of plastic into its constituent monomers. These catalysts are chemical compounds which increase the speed of a chemical reaction resulting in the degradation of plastics. Due to their highly specific nature, they break down plastics by a number of methods such as pyrolysis, gasification, and hydrocracking, and the application of NPs greatly enhances the activity of catalysts. Ruthenium nanomaterial supported on carbon is one of the most commonly used catalysts for depolymerizing plastics like polypropylene and polyolefins into n-alkanes under conditions of 200–250°C temperature, 20–50 bar pressure, and hydrogen gas (Rorrer E J et al., 2020). In this process, hydrogenolysis occurs which involves the breakdown of the C–C bond by introducing hydrogen molecules into the polypropylene chain converting the substrate (i.e., plastic) in the combination of linear or branched liquid alkanes like iso-alkanes (5–32 carbons chain), n-decane, n-nonane, n-octane, n-heptane, n-hexane, and others (Zichittella G et al., 2022). The main advantage associated with this process is that the catalyst can be reused a number of times for subsequent depolymerization (Thiyagarajan S et al., 2022). Another process involves the utilization of ionic liquids and ZnO (zinc oxide) NPs as catalysts for thermoplastic polycarbonate depolymerization, in which the polycarbonate is initially dissolved in tetrabutylammonium chloride (ionic solution) causing the breakdown of intermolecular bonds in the polymer chain. This is followed by the adsorption of the subsequent material on the ZnO surface facilitating a transesterification reaction forming an intermediate biphenyl A which further gets broken down into monomers such as cyclic carbonate 4-methyl-1,3-dioxolan-2-one, phenol, and esters under different conditions (Li J P H et al., 2014). The optimum temperature condition that results in the complete degradation of polycarbonate is 100°C for 7 hours. ZnO nanocatalysts can be recycled several times without considerable loss of their activity. The mild operational conditions, low cost, and recyclability of catalysts make this procedure quite effective (Iannone F et al., 2017).

17.3 ALTERATION OF PLASTICS

Another way of overcoming the problem of plastic pollution is by increasing its shelf life so that it can be used for a much longer time. The incorporation of NPs in plastics changes their physical and chemical properties, and makes them more durable, UV resistant, and heat resistant. The altered plastic has an increased shelf life, which causes a reduction in the frequent requirement leading to minimization of plastic waste. In this context, silica NPs have been used quite effectively for enhancing the properties of plastics. These nanoparticles are mainly used in the form of supporting fillers in plastics like polyethylene and polypropylene, and act as a barrier that reduces the tendency of plastics to become soft and deform under the effect of external forces like high temperature and pressure (Gondal A H et al., 2022). Similarly, the addition of clay NPs to polyethylene enhances its mechanical properties by making it more resistant to fracture and deformation. NT can also be used to modulate the self-healing properties of plastic materials in order to minimize their use. In this approach, plastics can restore themselves to their normal condition without any kind of human intervention (Ariga K et al., 2016). This kind of approach employs two mechanisms, namely extrinsic healing in which the healing agents are embedded in the polymer matrix that helps in recovering the damage and the second is intrinsic healing where the chemical bonds of the matrix itself possess the capability to restore its original structure. In the extrinsic mechanism, the self-healing approach is through the presence of an intentionally added external healing agent, which can be encapsulated in such a manner that when the damage occurs the component along with the capsule tends to break down and releases the healing agents in the damaged regions (Zhu D Y et al., 2015). One of the best extrinsic agents belongs to the category of epoxy nanocomposite nanofillers which include carbon nanotubes (CNTs), nanoclay, and alumina powder as a good additive for improving the mechanical properties of composite materials. Among these nanofillers, CNTs are the primary choice for making polymer composites due to their high strength-to-weight ratio and they can be surface functionalized to encapsulate healing agents within the epoxy matrix. When the damage occurs, the component along with the CNT capsule break down and release the healing agents into the damaged regions and subsequently repair it by undergoing cross-linking and polymerization by forming new bonds to restore its original structure (Gorrasi G et al., 2015). In the case of intrinsic self-healing, the polymers are made to undergo several covalent changes like Diels–Alder reaction, dynamic urea bond, imine bonds, radical exchange, and noncovalent interactions like H-bonding and ligand–metal bonding. The mechanism is often triggered by external stimuli like temperature, pH, light, and pressure, and the healing can be done for multiple cycles without the need for any catalyst. The only disadvantage of intrinsic self-healing is that it is restricted to thermoplastic materials like acrylic, polyester, etc., which lowers its strength and makes the bond weaker. For instance, a multifunctional nanocomposite based on oxidized multiwalled carbon nanotubes (MWCNTs) and polyethylene polyamine (PPA) has been used as an intrinsic factor. In this process, the hierarchical hydrogen bonds act as an internal driving force in the hydrogel preparation and this leads to the formation of oxidized MWCNTs/PPA hydrogels with thermal responsiveness, pH responsiveness, and self-repair properties in the material (Du R et al., 2015). In another approach, the blend of pyrene-functionalized polyamide (π-electron donor), polyimide (π-electron acceptor), and pyrene-functionalized gold NPs can be used that produce thermally induced π–π stacking interactions between functionalized gold NPs and the polymer matrix causing self-healing properties (Tung T T et al., 2017). In this way, these increased self-healing abilities by the advent of NT not only enhance the stability and lifespan of plastic but also reduce the need for frequent replacements, thereby limiting the use of plastic and contributing to a green planet. However, the main disadvantages associated with this process include non-biodegradability of plastic leading to improper disposal and management. Even though there are major environmental issues linked with conventional plastics, no one can deny that plastics are an indispensable part of our lives due to their versatile nature, however there is an urgent need for their replacement. In this context, NT provides enormous opportunities to develop biodegradable plastics with improved characteristics. They are designed in such a way that they can be degraded easily into their natural constituents like CO_2, H_2O, and organic matter. A number of biological materials such as sugar cane, starch, and cellulose can be used quite effectively for the synthesis of bioplastics. One of the most common approaches utilizes the fermentation of sugars by bacteria (*Vibrio harveyi*, *Paracoccus*, *Micrococcus*, *Erythrobacter aquimaris*, *Halomonas elongata*, *Bacillus*, *Alcaligenes*) and fungi. During the process of fermentation, microbes convert sugar like glucose into biopolymers like polyhydroxyalkanoates (PHAs) which are used for the production of bioplastics (Koller M et al., 2018) However, there are certain disadvantages associated with bioplastics such as low mechanical strength and poor barrier properties, and these can be overcome by incorporating nanomaterials during the production of bioplastics (Ciesielski S et al., 2015). In one approach, montmorillonite (MMT), a nanoclay, is reinforced into a polylactic acid biopolymer matrix to create NCs and the combination increases mechanical properties like tensile strength, tensile modulus, and rigidity, resulting in enhanced shelf life and overall biodegradability of the material. Another method is the application of silica NPs in bioplastics causing low toxicity and improved mechanical properties (Tu H et al., 2021; Chandran R R, 2023). Similarly, the use of titanium dioxide (TiO_2) NPs enhances the mechanical properties, thermal properties, barrier properties (modulus, elongation at break and tensile strength), high surface area, and photocatalytic activity of bioplastics by increasing the melting temperature and reducing the permeability of gases (Altaf A et al., 2022).

Another approach to overcoming plastic pollution is by converting plastic waste into carbon dots (CDs) using NT. These CDs, which are spherical particles smaller than 10 nm, offer a range of benefits such as safety, affordability, and diverse applications in fields like bioimaging and electronics. The accumulation of plastic waste, particularly from COVID-19 protective gear such as masks and shields, poses a significant environmental threat due to its inability to biodegrade. However, the recycling of this plastic waste into CDs has emerged as a promising solution, providing both environmental and economic advantages (Arpita et al., 2023). By incorporating heteroatoms and employing hydrothermal treatment, solvothermal methods, or pyrolysis, the optical properties of the CDs are enhanced, resulting in increased sensitivity to heavy metals. Through the breakdown of plastic materials under high temperature and pressure conditions, smaller carbon-based NPs, namely CDs, are formed. This transformation involves the chemical degradation and restructuring of plastic polymers, ultimately leading to the formation of CDs. Once synthesized, these CDs exhibit unique optical, electrical, and catalytic properties, making them highly suitable for various applications in fields such as sensing, catalysis, and LED fabrication. This innovative approach harnesses the potential of nanoscale materials derived from plastic waste to create valuable and versatile carbon-based NPs (Behera et al., 2022).

17.4 MANAGEMENT OF MICROPLASTIC POLLUTANTS

Microplastics (MPs) are tiny plastic particles having a size of less than 5 mm (0.2 inches) in diameter and can be detected in almost all the regions of Earth including oceans, rivers, soil, and even polar regions (Ashrafy A et al., 2022). They are the class of emerging contaminants of concern and are categorized into two types based upon their origin: primary and secondary MPs (Lambert S et al., 2018). The primary MPs are manufactured intentionally as feedstocks for their applications in paints, fertilizers, cosmetics, detergents, and cleaning products including microbeads (found in personal care products), plastic pellets or nurdles (used in industrial manufacturing), and plastic fibers (used in synthetic textiles, such as nylon), whereas the secondary MPs are generated from physical, chemical, or biological degradation of large plastic debris (Arpia et al., 2021). The scale of microplastic waste generation is alarming, with more than 6 billion metric tons of MPs being generated up to 2015 and approximately 4.9 billion metric tons of plastic waste being located in landfills and the environment. A large quantity of different kinds of microplastics have been observed across different global ecosystems (Table 17.2). It has been estimated that by the year 2050, the amount of microplastic waste in landfills and the environment will reach more than 12 billion metric tons. The incorporation of these micro-sized particles in the food chain can be observed through aquatic systems. There are a number of sources that cause the accumulation of MPs in the oceans across the world (Figure 17.5). The presence of these MPs in the marine ecosystem results in their accumulation in the bodies of aquatic animals because they are ingested mistakenly as a portion

TABLE 17.2
Global Aquatic Ecosystems and Microplastic Contents in Them

Global Lakes/Rivers	Size of Filtration Mesh (μm)	MPs (Abundant Polymer Types)	MP Values Recorded (in area m²/volume m³)
River Rhine	300	PE, PS, and PMMA	4960/1000 m³
San Gabriel & Los Angeles Rivers	333	All polymeric types	<1–153 m⁻³
Danube	500	All polymeric types	317/1000 m³
North Shore Channel, Chicago	333	All polymeric types	1.94–17.93 m⁻³
Swiss rivers	300	PE, PP, and PS	7 m⁻³
U.S. Great Lake tributaries	333	All polymeric types	0.05–32 m⁻³
Taihu Lake, China	5	All polymeric types	3.4–25.8 L⁻¹
Saigon River, Vietnam	2.7	PE & PP	172,000–519,000 m⁻³
Three Gorges Reservoir, China	48	PS, PP, and PE	1597–12,611 m⁻³
Teltow Canal, Berlin, Germany	63	PP, PE, and PS	0.01–95.8 L⁻¹
Dongting & Hong Lakes, China	50	PE and PP	900–4650 m⁻³
Lake Geneva	300	PS pellets	48 particles/1000 m²
Raritan River, New Jersey	153	All polymeric types	4–108 m⁻³
Lakes Bolsena & Chiusi, Italy	300	All polymeric types	0.82–4.41 m⁻³
Lake Michigan	333	All polymeric types	0–100 particles/1000 m²
Seine River, Paris	330	All polymeric types	0.28–0.47 m⁻³
Guanabara Bay, Rio de Janeiro, Brazil	300	PE and PP	Abundance of MPs 1.4–21.3 (p/m³)
Lake Superior	333	PS, PVC, PET, PE, and PP	0–110 particles/1000 m²

Source: Based on data from Rai P K et al. (2021).

Abbreviations: PE, polyethylene; PVC, polyvinyl chloride; PET, polyethylene terephthalate; PP, polypropylene; PS, polystyrene; PMMA, polymethylmethacrylate.

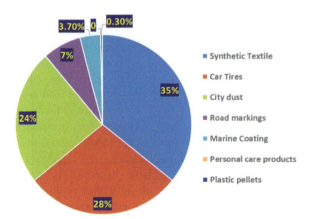

FIGURE 17.5 Estimation of various sources causing microplastic pollution in the world's ocean. (Based on data from Boucher J et al. 2017.)

of food or along with food, and once they enter the food chain they get transferred to higher tropic levels and even to humans (Issac M N et al., 2021). The impact of microplastic on marine biota is an issue of concern as it leads to entanglement and ingestion, which can cause various problems such as false satiation, reproductive issues, impaired enzyme production, reduced growth rate, and oxidative stress, which can be fatal (Zolotova N et al., 2022).

Role of Nanotechnology in Plastic and MP Management 311

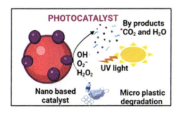
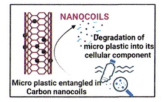
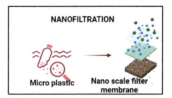
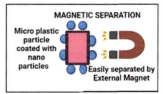

FIGURE 17.6 NT-based approaches and their mechanism employed for mitigating microplastic pollution.

They are not limited to marine creatures but have also been found in the stomachs of several bird species such as *Phalacrocorax bougainvillii, Pelecanoides garnotii, Pelecanoides urinatrix, Pelecanus thagus, Spheniscus humboldti,* and *Larus dominicanus,* and they are associated with alterations in feeding behavior, reproduction, and mortality (Thiel M et al., 2018). In recent years, MPs have also been documented in human blood, stool, tissues, organs, maternal and fetal placental tissues (Ragusa A et al., 2021), as well as in human breast milk (Ragusa A et al., 2022), and they are also associated with various harmful impacts on metabolic processes (Gerstenbacher C M et al., 2022; da Silva Brito W A et al, 2022). It has now become a global problem and there is an urgent need for their remediation through effective strategies and strict environmental regulation. A number of methodologies have been applied to overcome this issue but they are associated with certain disadvantages such as difficulty in capturing and removing MPs from waste due to their small size (Sharma S et al., 2021). Some approaches like source reduction are specifically concerned about only one aspect of microplastic while they persist in the environment from different sources and hence an integrated approach is required to properly assess and address the issue. In recent years of technological innovations, several new materials and techniques have been explored for the advancement in this field. NT offers a great deal with improved monitoring, characterization, and identification, as well as enhanced removal efficiencies with respect to microplastics. The nanotechnology-based approaches have proved to be very effective remediation and management techniques for microplastic pollution (Goh P S et al., 2022). There are a number of ways by which NPs can be used to address the issue of microplastic management (Figure 17.6).

17.4.1 Nanocoils

Nanotubes (NTs) with their significant characteristics manipulate matter by breaking down material into individual atoms and molecules and play a crucial role in reinforcing, improving, and repurposing the polymers of MPs (Singh D V et al., 2020). Nanocoils have been used quite effectively in this aspect and they are reusable nanosized reactors that are made up of carbon nanotubes and metal, which help in the breakdown of microplastics to their core molecular components like carbon dioxide and water (Liu W W et al., 2014). Efficient and sustainable microplastic management can be done with their distinctive coil-like structures and a number of mechanisms have been used through this technique that contribute to its high efficiency and selectivity. Insights are provided by

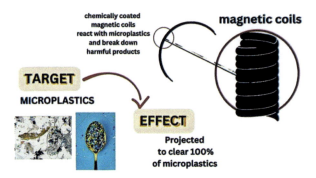

FIGURE 17.7 Microplastic degradation by magnetic coil.

exploring this mechanism in microplastic capture and removal by the addition of functional groups like hydrophobic moieties which can be attached to the nanotool surface, promoting interactions with different types of MPs based on their surface properties, and this phenomenon is known as surface functionalization (Saleem H et al., 2020). van der Waals and electrostatic interactions are also exploited to attract and capture microplastic particles. These forces arise from fluctuations in electron distributions within the molecule, resulting in temporary negative and positive charges, which leads to the adsorption of MPs, while electrostatic attraction facilitates the adhesion of microplastic particles onto the nanotool surface (Jin T et al., 2022). Also, the opportunities for mechanical entanglement are being utilized because of the coil-like structure as whenever a microplastic comes into contact they become entangled with the coil due to its elongated shape and flexibility, which enhance its capturing efficiency (Goh P S, et al., 2022).

Unlike the other technologies that cannot physically reach all MPs, nanotechnologies being work on a small scale which enables them to replicate microplastic distribution patterns and target MPs more efficiently (Enfrin M et al., 2021). One of the most profound examples in this area is helical carbon nanotubes (CNTs) which were engineered by doping with high-level nitrogen and encapsulating them with manganese carbide NPs. This unique innovation has led to the creation of magnetic nanohybrids that can break down MPs derived from cosmetic products (Figure 17.7). The high graphite degree of the nanocarbon in combination with the additive effects of the helical CNTs and encapsulated manganese carbide nanoparticles play a crucial role in ensuring the firmness of the nanocomposite during the highly developed oxidation process. The encapsulation of manganese carbide NPs assists in the delocalization of electrons, leading to the formation of reactive oxygen radicals, which causes mineralization of the microplastic into water and carbon dioxide (Kang J et al., 2019).

17.4.2 Nanomaterials as Photocatalyst

Nanomaterials working as photocatalysts have been found to play a crucial role in the remediation of microplastic (Goh P S et al., 2022). The large surface area in proportion to volume makes NPs very suitable for facilitating more proficient oxidation of MPs (Saleem H et al., 2020). The process utilizes two types of reactions, namely photo-oxidation and photo-reduction, in which photo-oxidation involves the delocalization of electrons from the photocatalyst to the microplastic resulting in oxidation reactions, and subsequent generation of reactive oxygen radicals facilitates the breakdown of microplastic pollutants. Photo-reduction involves the loss of an electron from the microplastic to the photocatalyst, leading to conversion of toxic pollutants to less harmful compounds (Pavel M et al., 2023). One of the most commonly used photocatalysts is green nanoscale semiconductors, which possess two energy bands, namely the valence band (filled with electrons) and the conduction band (empty), so when the light (photons) falls on these nanomaterials they tend to absorb it and if energy is equivalent or greater than the energy difference between the valence and conduction bands, it causes excitation of electrons from the valence to conduction band, creating a positive charge valence known as a hole. The holes along with excited electrons then interact with the microplastic and initiate the process of photo-oxidation and photo-reduction, thereby decomposing the microplastic into carbon dioxide and water (Chellasamy G et al., 2022). This technique has been widely utilized for the degradation of low-density MPs like polypropylene using glass substrates coated with zinc oxide (ZnO) nanorods, as a catalyst, absorbing UV light energy and subsequently leading to the generation of electron–hole pairs. These pairs then react with oxygen and water, forming short-lived reactive radicals like hydroxyl and superoxide radicals, leading to the mineralization of the polymers into water and carbon dioxide (Tofa T S et al., 2019). Likewise, the breakdown of MPs such as polystyrene and polyethylene has been accomplished using TiO_2 NP-based films as a photocatalyst, which converts this microplastic into byproducts like carbon dioxide with 99.9% efficiency in 24 hours. This strategy offers a physical method for removing microplastic particles from water systems using glass fiber substrates by capturing and concentrating them and with the utilization of light energy (Nabi I et al., 2020). The addition of a photocatalytic coating accelerates the degradation process and hence increases its effectiveness. However, this approach encounters a major challenge in accomplishing adequate contact between catalyst and microplastic, which in turn obstructs the whole effect, while problems such as low solubility, precipitation, and agglomeration further cause difficulties in its implementation on a large scale (Goh P S et al., 2022). In order to combat such hurdles a unique strategy has been utilized which involves the development of a novel photocatalyst called nanorobots/microrobots (www.sciencedaily.com/releases/2021/06/210610135744.htm). These NPs exploit solar energy to drive themselves toward MPs, thereby enhancing the contact and attachment between photocatalyst and microparticles, facilitating their degradation (Kumar R, 2023). An example of one such light-driven photocatalytic microrobot/nanorobot is based on bismuth vanadate ($BiVO_4$) and has remarkable capabilities that involve proficient swimming in liquid media under UV visible light, followed by the attachment to different floating MPs with unique polymeric structures like PLA, PET, and polypropylene, causing a subsequent breakdown of these MPs into small-sized organic molecules and oligomers. The incorporation of magnetic constituents allows the easy collection and useful recovery of nanorobots from the natural surroundings. This low-energy method eliminates the need for bulky mechanical stirrers and costly pretreatments utilized in traditional TiO_2 flow bed reactors used for photocatalyst deposition on plastic surfaces (Salaheldeen Elnashaie S et al., 2015). The efficiency of ZnO as a photocatalyst is low in comparison to TiO_2 but they can easily degrade microplastics like low-density polyethylene film and polypropylene (PP). However, both these NPs possess many challenges in terms of their practical applications such as wide bandgap, formation of hydroxyl radicals, and low adsorption capacity (Xie A et al., 2023).

17.4.3 BIONANOMATERIALS

Bionanomaterials are another unique technology that utilizes natural methods for the bioremediation of MPs. These nanomaterials are of either biological origin or made by a combination of inorganic nanomaterials and organic constituents (Chellasamy G et al., 2022). They offer a natural and promising realm for microplastic degradation, and their unique properties and composition make them a proficient catalyst for decomposing MPs by providing a sustainable substitute for microplastic management (Jalvo B et al., 2021). One of the major utilizations of this technique is in the development of the water filtration membrane which is based upon non-woven cellulosic fabric having various polysaccharide nanocrystals such as cellulosic nanocrystals, 2,2,6,6-tetramethylpiperidine-1-oxyl radical (TEMPO)-oxidized cellulose nanofibers (T-CNF), as well as nanocrystals, made up of chitin infused with the fiber via cast coatings technology which have been used effectively. This improves various properties such as the hydrophilicity of the fabric, tensile strength, charges on the surface, as well as the permeability of the membrane, allowing them to remove microplastics as small as 500 nanometers in size (Jalvo B et al., 2021). Another remarkable bionanomaterial is lignin–zeolite composite nanofiber membranes fabricated through a process called electrospinning, exhibiting superior flow and diffusion rates as compared to conventional membranes (Bahi A et al., 2017). The addition of zeolite NPs in a uniform manner along with post-heat treatment causes tremendous improvement in the mechanical properties such as tensile strength and elasticity; thereby making an efficient prefiltration filter material to be used for removing MPs (Chellasamy G et al., 2022). Lysozyme amyloid fibers are another newly introduced natural nanobioflocculant used for the efficient removal of dispersed polystyrene. This flocculent possesses a positively charged surface and interacts with negatively charged microplastic which creates attractive forces between them, leading to their aggregation and the formation of large flocs. These large flocs can further be separated by conventional methods such as filtration or magnetic separation, with a removal efficiency of 98.2% (Peydayesh M et al., 2021).

17.4.4 Magnetic Separation

Magnetic separation based on nanomaterials is one of the most simplified and cost-effective strategies for the efficient removal of microplastics. The basic principle behind this method is that the nanomaterials having dual nature are being used to separate MPs. These nanomaterials can adhere to the surface of the microplastic so that the smaller MPs can be magnetized and are easily removed from water using magnetic recovery techniques (Li X et al., 2021). One of the most commonly used NPs for the effective removal of MPs made of various materials, including polyethylene, polypropylene, polystyrenes and PET, is magnetite (Fe_3O_4) NPs. All magnetized MPs can be extracted using external forces, and with the addition of positively charged Fe_3O_4 NPs, the mechanisms of adsorption and magnetism are enhanced (Shi X et al., 2022). However, PET MPs had a lower removal rate than PE, PP, and PS due to their inferior hydrophobic qualities. In order to overcome this issue, immobilization of enzymes like PETase on Fe_3O_4 nanoparticles has been done, but due to the limited enzyme activity and tightly packed polymer chains in MPs this approach still struggles to endure the harsh conditions required for the degradation of MPs (Schwaminger S P et al., 2021). At lower temperatures, the machine learning-guided engineered PETase has made some progress, but it is still far from completely degrading the PET MPs. On the other hand, the peroxidase-like nanozyme (Zandieh M et al., 2022) activity of bare Fe_3O_4 NPs makes them appealing substitutes for natural enzymes with improved stability for the complete degradation of microplastic pollutants (Zha, J. et al., 2022). Iron NPs can also be used for removing MPs based upon the same principle. However, iron NPs do not adhere to MPs effectively and need to be treated with certain compounds in order to use them, such as adding hexadecyltrimethoxysilane (HDTMS) to iron NPs improves their hydrophobicity and binding capacity toward MPs. The iron NPs so formed can effectively remove PE and PS beads of 10–20 μm size and the efficiency of removal toward other MPs of size larger than 1 mm is around 92–93% (Martin L M et al., 2022). Furthermore, carbon nanotubes (CNTs) surface modified with magnetite possess a strong affinity toward hydrophobic microplastic, thereby improving their removal process through external magnetic forces. These carbon nanotubes can also maintain their supermagnetic features even after complete removal of microplastic, thus allowing easy recycling and effective desorption. The latest innovations and technology have allowed hydrophobic magnetite NPs to be made using the aerial parts of plant species like *Anthemis pseudocotula* but they have limited applications in terms of microplastic extraction (Challesamy G. et al., 2022). An easy one-pot microwave synthesis method can also be used to create a highly porous and evenly distributed carbon–Fe_3O_4 composite that can effectively absorb MPs from water which can then easily be removed with an external magnet. These novel methods offer promising substitutes for eco-friendly microplastics removal by utilizing plant extracts and creating composite materials to address the problem of microplastic pollution (Chen X et al., 2023). Apart from these, biohybrid microrobots with magnetic properties have also been created for the efficient removal of microplastics from the aquatic environment. In these microrobots, Fe_3O_4 nanoparticles are integrated on the surface of algae cells and thereafter the movement of these cells can be controlled precisely for dealing with microplastics (Peng X et al., 2023).

17.4.5 NANOFILTRATION

Nanofiltration (NF) is a pressure-driven cross-flow separation technology used to purify solutions by utilizing nanoporous membrane having a pore size ranging from 0.1 to 10 nm (Tian J et al., 2021). This technology works on the principle of size exclusion and a charge-based interaction retaining molecules like microplastics, microbeads, and microfibers, while allowing other solutes to pass through it, which makes it ideal for removing MPs. It is advantageous over ultrafiltration and microfiltration due to its small pore size that allows retaining even small molecules having sizes of 200–1000 g mol^{-1}. It works at a low transmembrane pressure, usually 50–225 psi, requires less energy consumption, and provides higher rejection rates for multivalent ions, allowing the separation of monovalent ions and thereby facilitating the removal of divalent ions and retaining the specific molecules. Due to all such ideal properties this strategy is being widely used to efficiently tackle the global pollution caused by MPs. It is a modification of membrane separation methods and has already proven to be a vital tool in a variety of applications because of its ability to separate molecules according to their size and charge. There are numerous alternative techniques that can be applied, such as size exclusion, adsorption, and charge-based interactions. Size exclusion is the fundamental strategy used in NFs to eliminate minute plastic particles. It is crucial to take extra precautions to guarantee that the membrane's pores are no larger than the smallest microplastic particles that need to be eliminated (Chang E E et al., 2012). Charge-based interactions can be utilized to remove MPs from water by adjusting the surroundings and the charge of the membrane (Yadav D et al., 2022). Adsorption is a different technique for eliminating MPs and is used in conjunction with size exclusion or interactions based on charge to remove MPs. The removal of 70–99% of MPs from water can be achieved using this technique, which uses a graphene oxide membrane and reduced graphene oxide membranes (Mehmood T et al., 2022). These are generally made to work through a combination of size exclusion and charge-based interactions. In another methodology, the incorporation of a special type of iron nanoparticle, named MIL-100, into the cellulose acetate NF membrane causes hydrophilicity and pore size improvement for the efficient extraction of MPs from the textile waste water (Cevallos-Mendoza J et al., 2022). This technique is also widely used in many industrial projects, such as the Netherland, where the "NEO WATER" [new energy optimized water treatment and reuse] project greatly contributes to microplastic management and environmental sustainability. The main working principle utilized in this project is the use of NF technology for managing wastewater discharge from the sewage system, where the small pore size of nanofilter membranes addresses the pollution caused by MPs and other contaminants by removing and preventing their release into the environment, which therefore promotes clear water resources.

17.5 REGULATIONS FOR NANOPARTICLE MANAGEMENT

NT=based approaches may be considered a true paradigm shift in the field of science and can be used for a wide range of applications which also allows for addressing the problem of plastic and microplastic management. However, sometimes the results are indescribable and are associated with unexpected risks (Murashov V et al., 2021). Despite the advantages, there are unforeseen dangers associated with the use of nanomaterials in the remediation process. When nanomaterials are released into the environment, they have the chance to interact with many biological and physicochemical elements, which will have a negative impact on the environment at the level of the ecological unit (Saleem H et al., 2020). The increasing influence of exposure to various types of materials at the nanoscale level needs to be explored using novel legal methods to evaluate liability, with the fact that exposures will occur in a combined form that cannot be defined in contexts and cannot be regulated. Prior to introduction, novel methods or technologies must go through extensive testing for adverse health and environmental effects. The following are the key codes for ecology decisions:

(1) Taking the safest actions in the face of doubtfulness
(2) Changing the burden of proofs to the props of an action
(3) Finding a wide number of substitutes for potentially dangerous actions
(4) Enhancing public participation in decision-making

However, these regulations are not being followed when NPs are developed, creating uncertainty about the risk versus the advantages of nanomaterials. Even though there is disagreement in the research community over the clear safety assessment of nanomaterial for regulation and governance, the regulatory authorities are working to adopt significant limits around NPs. Despite differences of opinion regarding the proper range of hazards among regulatory organizations, scientists, and employers, there are currently no standardized methods for identifying and characterizing NP, which makes it very difficult to provide any appropriate evaluation. In terms of exposure to nanomaterials, there are no specific laws, with a few exceptions, and the European Commission has acknowledged that NPs are "difficult to regulate" because of their complexity and knowledge gaps (Kabir E et al., 2018).

Thus, a standard operating procedure should be established in order to enable the sustainable use of nanomaterials as well as a workable future extension from industrial use to health care and environmental approaches. Various awareness campaigns should be organized to encourage proper management practices, increasing public participation to reduce the generation of plastic and microplastic wastes contributing toward a clean and green environment (Breggin L K et al., 2016).

17.6 CONCLUSION

NT has revolutionized the ways of addressing global issues of plastic and microplastic pollution. Due to its unique properties such as high surface area-to-volume ratio, ease of surface modification, selectivity, sensitivity, real-time monitoring, and integration of nanomaterials with traditional methods such as filtration, magnetic separation, coils, photocatalytic degradation, and so on, it results in precise, effective, and efficient removal of plastic and microplastic pollutants. However, several challenges are associated with it for its applicability on a large scale such as harmful environmental impact, unforeseen dangers that cannot be detected and tackled, as well as skilled personnel being required. Additionally, adequate study in this area is still lacking. In the coming years, by reducing the vast gaps in our understanding of the nature of nanomaterial interactions, we will have acceptable strategies for regulating, handling, and processing plastic pollution, contributing toward a green and sustainable environment.

REFERENCES

Altaf, A., Dar, A.H., Khan, S.A. and Singh, A. 2022. Nanocomposite and food packaging. *NT in Intelligent Food Packaging* 1–23.

American Chemical Society. Bacteria-sized robots take on microplastics and win by breaking them down. ScienceDaily. www.sciencedaily.com/releases/2021/06/210610135744.htm (accessed March 9, 2024).

Ariga, K., Li, J., Fei, J., Ji, Q. and Hill, J.P. 2016. Nanoarchitectonics for dynamic functional materials from atomic-/molecular-level manipulation to macroscopic action. *Advanced Materials* 28:1251–1286.

Arpia, A.A., Chen, W.H., Ubando, A.T., Naqvi, S.R. and Culaba, A.B. 2021. Microplastic degradation as a sustainable concurrent approach for producing biofuel and obliterating hazardous environmental effects: A state-of-the-art review. *Journal of Hazardous Materials* 418:126381.

Arpita, K., P., Kataria, N., Narwal, N., Kumar, S., Kumar, R., Khoo, K.S. and Show, P.L. 2023. Plastic waste-derived carbon dots: Insights of recycling valuable materials towards environmental sustainability. *Current Pollution Reports* 1–21.

Ashrafy, A., Liza, A.A., Islam, M.N., Billah, M.M., Arafat, S.T., Rahman, M.M. and Rahman, S.M. 2022. MP pollution: A brief review of its source and abundance in different aquatic ecosystems. *Journal of Hazardous Materials Advances* 100215.

Bahi, A., Shao, J., Mohseni, M. and Ko, F.K. 2017. Membranes based on electrospun lignin-zeolite composite nanofibers. *Separation and Purification Technology* 187:207–213.

Bayda, S., Adeel, M., Tuccinardi, T., Cordani, M. and Rizzolio, F., 2019. The history of nanoscience and NT: from chemical–physical applications to nanomedicine. *Molecules* 25:112.

Behera, A., Sahini, D. and Pardhi, D. 2022. Procedures for recycling of nanomaterials: a sustainable approach. *Nanomaterials Recycling* 175–207.

Boucher, J. and Friot D. 2017. *Primary Microplastics in the Oceans: A Global Evaluation of Sources.* Gland, Switzerland: IUCN. 43.

Breggin, L.K., Falkner, R., Pendergrass, J., Porter, R. and Jaspers, N. 2016. Addressing the risks of nanomaterials under united states and european union regulatory frameworks for chemicals. *Assessing Nanoparticle Risks to Human Health* 179–254.

Campanale, C., Massarelli, C., Savino, I., Locaputo, V. and Uricchio, V.F. 2020. A detailed review study on potential effects of MP and additives of concern on human health. *International Journal of Environmental Research and Public Health* 17:1212.

Carr, C.M., Clarke, D.J. and Dobson, A.D. 2020. Microbial polyethylene terephthalate hydrolases: current and future perspectives. *Frontiers in Microbiology* 11:571265.

Cevallos-Mendoza, J., Amorim, C.G., Rodríguez-Díaz, J.M. and Montenegro, M.D.C.B. 2022. Removal of contaminants from water by membrane filtration: a review. *Membranes* 12:570.

Chandran, R.R., Thomson, B.I., Natishah, A.J., Mary, J. and Nachiyar, V. 2023. NT in plastic degradation. *Biosciences Biotechnology Research Asia* 20:53–68.

Chang, E.E., Liang, C.H., Huang, C.P. and Chiang, P.C. 2012. A simplified method for elucidating the effect of size exclusion on NF membranes. *Separation and Purification Technology* 85:1–7.

Chellasamy, G., Kiriyanthan, R.M., Maharajan, T., Radha, A. and Yun, K. 2022. Remediation of MP using bionanomaterials: A review. *Environmental Research* 208:112724.

Chen, X., Du, S., Hong, R. and Chen, H. 2023. Preparation of RGO/Fe$_3$O$_4$ NC as a microwave absorbing material. *Inorganics* 11: 143.

Ciesielski, S., Możejko, J. and Pisutpaisal, N. 2015. Plant oils as promising substrates for polyhydroxyalkanoates production. *Journal of Cleaner Production* 106:408–421.

da Silva Brito, W.A., Mutter, F., Wende, K., Cecchini, A.L., Schmidt, A. and Bekeschus, S. 2022. Consequences of nano and microplastic exposure in rodent models: the known and unknown. *Particle and Fibre Toxicology* 19:1–24.

Du, R., Zhao, Q., Zhang, N. and Zhang, J. 2015. Macroscopic carbon nanotube-based 3D monoliths. *Small* 11:3263–3289.

Duis, K. and Coors, A. 2016. MP in the aquatic and terrestrial environment: sources (with a specific focus on personal care products), fate and effects. *Environmental Sciences Europe* 28:1–25.

Ellis, G.A., Dean, S.N., Walper, S.A. and Medintz, I.L. 2020. Quantum dots and gold NP as scaffolds for enzymatic enhancement: recent advances and the influence of nanoparticle size. *Catalysts* 10:83.

Enfrin, M., Hachemi, C., Hodgson, P.D., Jegatheesan, V., Vrouwenvelder, J., Callahan, D.L., Lee, J. and Dumée, L.F. 2021. Nano/micro plastics–challenges on quantification and remediation: a review. *Journal of Water Process Engineering*, 42:102128.

Gerstenbacher, C.M., Finzi, A.C., Rotjan, R.D. and Novak, A.B. 2022. A review of microplastic impacts on seagrasses, epiphytes, and associated sediment communities. *Environmental Pollution* 303:119108.

Goh, P.S., Kang, H.S., Ismail, A.F., Khor, W.H., Quen, L.K. and Higgins, D. 2022. Nanomaterials for microplastic remediation from aquatic environment: Why nano matters? *Chemosphere* 299:134418.

Gondal, A.H., Bhat, R.A., Gómez, R.L., Areche, F.O. and Huaman, J.T. 2022. Advances in plastic pollution prevention and their fragile effects on soil, water, and air continuums. *International Journal of Environmental Science and Technology* 20: 1–16.

Gorrasi, G. and Sorrentino, A. 2015. Mechanical milling as a technology to produce structural and functional bio-NC. *Green Chemistry* 17:2610–2625.

Hale, R.C., Seeley, M.E., La Guardia, M.J., Mai, L. and Zeng, E.Y. 2020. A global perspective on MP. *Journal of Geophysical Research: Oceans* 12:2018JC014719.

Hofstede, L.T., Vasse, G.F. and Melgert, B.N. 2023. Microplastics: A threat for developing and repairing organs? *Cambridge Prisms: Plastics* 1:19. 48,58.

Iannone, F., Casiello, M., Monopoli, A., Cotugno, P., Sportelli, M.C., Picca, R.A., Cioffi, N., Dell'Anna, M.M. and Nacci, A. 2017. Ionic liquids/ZnO NP as recyclable catalyst for polycarbonate depolymerization. *Journal of Molecular Catalysis A: Chemical* 426:107–116.

Issac, M.N. and Kandasubramanian, B., 2021. Effect of MP in water and aquatic systems. *Environmental Science and Pollution Research* 28:19544–19562.

Jalvo, B., Aguilar-Sanchez, A., Ruiz-Caldas, M.X. and Mathew, A.P. 2021. Water filtration membranes based on non-woven cellulose fabrics: effect of nanopolysaccharide coatings on selective particle rejection, antifouling, and antibacterial properties. *Nanomaterials* 11:1752.

Jia, Y., Samak, N.A., Hao, X., Chen, Z., Yang, G., Zhao, X., Mu, T., Yang, M. and Xing, J. 2021. Nano-immobilization of PETase enzyme for enhanced polyethylene terephthalate biodegradation. *Biochemical Engineering Journal* 176:108205.

Jin, T., Tang, J., Lyu, H., Wang, L., Gillmore, A.B. and Schaeffer, S.M. 2022. Activities of MP (MPs) in agricultural soil: a review of MPs pollution from the perspective of agricultural ecosystems. *Journal of Agricultural and Food Chemistry* 70:4182–4201.

Kabir, E., Kumar, V., Kim, K.H., Yip, A.C. and Sohn, J.R. 2018. Environmental impacts of nanomaterials. *Journal of Environmental Management* 225:261–271.

Kang, J., Zhou, L., Duan, X., Sun, H., Ao, Z. and Wang, S. 2019. Degradation of cosmetic MP via functionalized carbon nanosprings. *Matter* 1:745–758.

Krakor, E., Gessner, I., Wilhelm, M., Brune, V., Hohnsen, J., Frenzen, L. and Mathur, S. 2021. Selective degradation of synthetic polymers through enzymes immobilized on nanocarriers. *MRS Communications* 11:363–371.

Khan, W.S., Asmatulu, R., Davuluri, S. and Dandin, V.K. 2014. Improving the economic values of the recycled plastics using NT associated studies. *Journal of Materials Science & Technology* 30:854–859.

Koller, M. and Braunegg, G. 2018. Advanced approaches to produce polyhydroxyalkanoate (PHA) biopolyesters in a sustainable and economic fashion. *EuroBiotech J* 2:89–103.

Kumar, R. 2023. Metal oxides-based nano/microstructures for photodegradation of MP. *Advanced Sustainable Systems:* 2300033.

Lambert, S. and Wagner, M. 2018. Microplastics are contaminants of emerging concern in freshwater environments: an overview. In *Fresh Water Microplastics- Emerging Environmental Contaminants,* 1–23. Springer International Publishing.

Lee, S.H. and Song, W.S. 2011. Enzymatic hydrolysis of polylactic acid fiber. *Applied Biochemistry and Biotechnology* 164:89–102.

Li, J.P.H. 2014. Characterisation of Heterogeneous Acid/Base Catalysts and their Application in the Synthesis of Fine and Intermediate Chemicals. PhD diss., University of Newcastle, Australia.

Li, X., Liu, B., Lao, Y., Wan, P., Mao, X. and Chen, F. 2021. Efficient magnetic harvesting of microalgae enabled by surface-initiated formation of iron NP. *Chemical Engineering Journal* 408:127252.

Lin, Y.D., Huang, P.H., Chen, Y.W., Hsieh, C.W., Tain, Y.L., Lee, B.H., Hou, C.Y. and Shih, M.K. 2023. Sources, degradation, ingestion and effects of microplastics on humans: a review. *Toxics* 11:747

Liu, W.W., Chai, S.P., Mohamed, A.R. and Hashim, U. 2014. Synthesis and characterization of graphene and carbon nanotubes: a review on the past and recent developments. *Journal of Industrial and Engineering Chemistry* 20:1171–1185.

Martin, L.M., Sheng, J., Zimba, P.V., Zhu, L., Fadare, O.O., Haley, C., Wang, M., Phillips, T.D., Conkle, J. and Xu, W. 2022. Testing an iron oxide nanoparticle-based method for magnetic separation of nanoplastics and MP from water. *Nanomaterials* 12:2348.

Mehmood, T., Mustafa, B., Mackenzie, K., Ali, W., Sabir, R.I., Anum, W., Gaurav, G.K., Riaz, U., Xinghui, L. and Peng, L. 2022. Recent developments in microplastic contaminated water treatment: progress and prospects of carbon-based two-dimensional materials for membranes separation. *Chemosphere* 137704.

Murashov, V., Geraci, C.L., Schulte, P.A. and Howard, J. 2021. Nano-and MP in the workplace. *Journal of Occupational and Environmental Hygiene* 18:489–494

Nabi, I., Li, K., Cheng, H., Wang, T., Liu, Y., Ajmal, S., Yang, Y., Feng, Y. and Zhang, L., 2020. Complete photocatalytic mineralization of microplastic on TiO_2 nanoparticle film. *Iscience* 23:101326.

Neupane, S. 2019. Understanding the Interaction between Enzymes and Nanomaterials. PhD diss., North Dakota State University of Agriculture and Applied Science.

Palm, G.J., Reisky, L., Böttcher, D., Müller, H., Michels, E.A., Walczak, M.C., Berndt, L., Weiss, M.S., Bornscheuer, U.T. and Weber, G. 2019. Structure of the plastic-degrading Ideonella sakaiensis MHETase bound to a substrate. *Nature Communications* 10:1717.

Pavel, M., Anastasescu, C., State, R.N., Vasile, A., Papa, F. and Balint, I. 2023. Photocatalytic degradation of organic and inorganic pollutants to harmless products: assessment of practical application potential for water and air cleaning. *Catalysts* 13:380.

Peng, X., Urso, M., Kolackova, M., Huska, D. and Pumera, M. 2023. Biohybrid magnetically driven microrobots for sustainable removal of micro/nanoplastics from the aquatic environment. *Advanced Functional Materials* 2307477.

Peydayesh, M., Suta, T., Usuelli, M., Handschin, S., Canelli, G., Bagnani, M. and Mezzenga, R. 2021. Sustainable removal of MP and natural organic matter from water by coagulation–flocculation with protein amyloid fibrils. *Environmental Science & Technology* 55:8848–8858.

Ragusa, A., Notarstefano, V., Svelato, A., Belloni, A., Gioacchini, G., Blondeel, C., Zucchelli, E., De Luca, C., D'Avino, S., Gulotta, A. and Carnevali, O. 2022. Raman microspectroscopy detection and characterisation of MP in human breastmilk. *Polymers* 14:2700.

Ragusa, A., Svelato, A., Santacroce, C., Catalano, P., Notarstefano, V., Carnevali, O., Papa, F., Rongioletti, M.C.A., Baiocco, F., Draghi, S. and D'Amore, E. 2021. Plasticenta: first evidence of MP in human placenta. *Environment International* 146:106274.

Rorrer, J.E., Beckham, G.T. and Román-Leshkov, Y. 2020. Conversion of polyolefin waste to liquid alkanes with Ru-based catalysts under mild conditions. *Jacs Au* 1:8–12.

Rose, P.K., Yadav, S., Kataria, N. and Khoo, K.S. 2023. MP and nanoplastics in the terrestrial food chain: uptake, translocation, trophic transfer, ecotoxicology, and human health risk. *TrAC Trends in Analytical Chemistry* 117249.

Salaheldeen Elnashaie, S., Danafar, F., Hashemipour Rafsanjani, H., Salaheldeen Elnashaie, S., Danafar, F. and Hashemipour Rafsanjani, H. 2015. From NT to nanoengineering. *NT for Chemical Engineers* 79–178.

Saleem, H. and Zaidi, S.J. 2020. Developments in the application of nanomaterials for water treatment and their impact on the environment. *Nanomaterials* 10:1764.

Sato, S., Saika, A., Shinozaki, Y., Watanabe, T., Suzuki, K., Sameshima-Yamashita, Y., Fukuoka, T., Habe, H., Morita, T. and Kitamoto, H. 2017. Degradation profiles of biodegradable plastic films by biodegradable plastic-degrading enzymes from the yeast Pseudozyma antarctica and the fungus Paraphoma sp. B47–9. *Polymer Degradation and Stability* 141:26–32.

Schwaminger, S.P., Fehn, S., Steegmüller, T., Rauwolf, S., Löwe, H., Pflüger-Grau, K. and Berensmeier, S. 2021. Immobilization of PETase enzymes on magnetic iron oxide NP for the decomposition of microplastic PET. *Nanoscale Advances* 3:4395–4399.

Sharma, S., Basu, S., Shetti, N.P., Nadagouda, M.N. and Aminabhavi, T.M. 2021. MP in the environment: occurrence, perils, and eradication. *Chemical Engineering Journal* 408:127317.

Shi, X., Zhang, X., Gao, W., Zhang, Y. and He, D. 2022. Removal of MP from water by magnetic nano-Fe_3O_4. *Science of The Total Environment* 802:149838.

Singh, D.V., Bhat, R.A., Dervash, M.A., Qadri, H., Mehmood, M.A., Dar, G.H., Hameed, M. and Rashid, N. 2020. Wonders of NT for remediation of polluted aquatic environs. *Fresh Water Pollution Dynamics and Remediation:* 319–339.

Thiel, M., Luna-Jorquera, G., Álvarez-Varas, R., Gallardo, C., Hinojosa, I.A., Luna, N., Miranda-Urbina, D., Morales, N., Ory, N., Pacheco, A.S. and Portflitt-Toro, M. 2018. Impacts of marine plastic pollution from continental coasts to subtropical gyres—fish, seabirds, and other vertebrates in the SE Pacific. *Frontiers in Marine Science* 238.

Thiyagarajan, S., Maaskant-Reilink, E., Ewing, T.A., Julsing, M.K. and Van Haveren, J. 2022. Back-to-monomer recycling of polycondensation polymers: opportunities for chemicals and enzymes. *RSC Advances* 12:947–970.

Thompson, R.C., Moore, C.J., Vom Saal, F.S. and Swan, S.H. 2009. Plastics, the environment and human health: current consensus and future trends. *Philosophical Transactions of the Royal Society B: Biological Sciences* 364:2153–2166.

Tian, J., Zhao, X., Gao, S., Wang, X. and Zhang, R. 2021. Progress in research and application of NF (nf) technology for brackish water treatment. *Membranes* 11:662.

Tofa, T.S., Ye, F., Kunjali, K.L. and Dutta, J. 2019. Enhanced visible light photodegradation of microplastic fragments with plasmonic platinum/zinc oxide nanorod photocatalysts. *Catalysts* 9:819.

Tu, H., Zhu, M., Duan, B. and Zhang, L. 2021. Recent progress in high-strength and robust regenerated cellulose materials. *Advanced Materials* 33:2000682.

Tung, T.T., Nine, M.J., Krebsz, M., Pasinszki, T., Coghlan, C.J., Tran, D.N. and Losic, D. 2017. Recent advances in sensing applications of graphene assemblies and their composites. *Advanced Functional Materials* 27:1702891.

Urbanek, A.K., Kosiorowska, K.E. and Mirończuk, A.M. 2021. Current knowledge on polyethylene terephthalate degradation by genetically modified microorganisms. *Frontiers in Bioengineering and Biotechnology* 9:771133.

Xie, A., Jin, M., Zhu, J., Zhou, Q., Fu, L. and Wu, W. 2023. Photocatalytic technologies for transformation and degradation of MP in the environment: current achievements and future prospects. *Catalysts* 13:846.

Yadav, D., Karki, S. and Ingole, P.G. 2022. Current advances and opportunities in the development of NF (NF) membranes in the area of wastewater treatment, water desalination, biotechnological and pharmaceutical applications. *Journal of Environmental Chemical Engineering* 10:108109.

Yaqoob, A.A., Parveen, T., Umar, K. and Mohamad Ibrahim, M.N. 2020. Role of nanomaterials in the treatment of wastewater: a review. *Water* 12:495.

Yin, Q., You, S., Zhang, J., Qi, W. and Su, R. 2022. Enhancement of the polyethylene terephthalate and mono-(2-hydroxyethyl) terephthalate degradation activity of Ideonella sakaiensis PETase by an electrostatic interaction-based strategy. *Bioresource Technology* 364:128026.

Zandieh, M. and Liu, J. 2022. Removal and degradation of MP using the magnetic and nanozyme activities of bare iron oxide nanoaggregates. *Angewandte Chemie International Edition* 61:202212013.

Zdarta, J., Meyer, A.S., Jesionowski, T. and Pinelo, M. 2018. A general overview of support materials for enzyme immobilization: characteristics, properties, practical utility. *Catalysts* 8:92.

Zha, J., Wu, W., Xie, P., Han, H., Fang, Z., Chen, Y. and Jia, Z. 2022. Polymeric nanocapsule enhances the peroxidase-like activity of Fe3O4 nanozyme for removing organic dyes. *Catalysts* 12:614.

Zhu, D.Y., Rong, M.Z. and Zhang, M.Q. 2015. Self-healing polymeric materials based on microencapsulated healing agents: from design to preparation. *Progress in Polymer Science* 49:175–220.

Ziani, K., Ioniță-Mîndrican, C.B., Mititelu, M., Neacşu, S.M., Negrei, C., Moroşan, E., Drăgănescu, D. and Preda, O.T. 2023. MP: a real global threat for environment and food safety: a state-of-the-art review. *Nutrients* 15:617.

Zichittella, G., Ebrahim, A.M., Zhu, J., Brenner, A.E., Drake, G., Beckham, G.T., Bare, S.R., Rorrer, J.E. and Román-Leshkov, Y. 2022. Hydrogenolysis of polyethylene and polypropylene into propane over cobalt-based catalysts. *JACS Au* 2:2259–2268.

Zolotova, N., Kosyreva, A., Dzhalilova, D., Fokichev, N. and Makarova, O. 2022. Harmful effects of the microplastic pollution on animal health: a literature review. *PeerJ* 10:13503.

Zurier, H.S. and Goddard, J.M. 2022. Directed immobilization of PETase on mesoporous silica enables sustained depolymerase activity in synthetic wastewater conditions. *ACS Applied Bio Materials* 5:4981–4992.

18 Life-cycle Assessment of Microplastics in the Environment

Kassian T.T. Amesho, Sumarlin Shangdiar, E.I. Edoun, Abner Kukeyinge Shopati, Timoteus Kadhila, Sioni Iikela, Nastassia Thandiwe Sithole, Chandra Mohan, and Manoj Chandra Garg

18.1 INTRODUCTION

Microplastics, defined as plastic particles with a size range typically less than 5 millimeters, have garnered increasing attention in recent years due to their ubiquitous presence in various environmental compartments and their potential adverse impacts on ecosystems and human health (Boulay et al., 2021; Borrelle et al., 2020). These minute particles originate from a myriad of sources, including the fragmentation of larger plastic debris, industrial abrasion, and the breakdown of synthetic textiles (Saling et al., 2020; Sonnemann and Valdivia, 2017). Their small size and buoyancy enable them to disperse widely in the environment, with studies documenting their occurrence in oceans, rivers, lakes, sediments, soils, and even the atmosphere (Stefanini et al., 2020; Woods et al., 2019).

The proliferation of microplastics poses multifaceted environmental challenges, ranging from physical and chemical hazards to ecological disruptions. In aquatic environments, microplastics can be ingested by a plethora of marine organisms, leading to adverse effects such as intestinal blockage, reduced feeding efficiency, and bioaccumulation of toxic contaminants (Yang et al., 2021; Zhou et al., 2020; Besseling et al., 2018). Moreover, the sorption of persistent organic pollutants (POPs) onto microplastic surfaces raises concerns about the potential transfer of these contaminants through marine food webs, thereby posing risks to higher trophic levels, including humans (Akhbarizadeh et al., 2021b; Gao et al., 2021).

To comprehensively assess and mitigate the environmental impacts of microplastics, there is a growing recognition of the importance of employing life cycle assessment (LCA) methodologies. LCA is a holistic approach that evaluates the environmental burdens associated with the entire life cycle of a product or system, from raw material extraction to end-of-life disposal (Rodríguez et al., 2021). By quantifying resource consumption, energy use, emissions, and potential ecological damage at each stage of the life cycle, LCA facilitates informed decision-making and the identification of opportunities for pollution prevention and resource conservation (Gandhi et al., 2021). Figure 18.1 shows the keyword network analysis of "Microplastics, Life cycle assessment"-related articles from 2018 to 2024 in the Web of Science (WoS) database.

In this chapter, we delve into the application of LCA to assess the environmental implications of microplastics across their life cycle. We begin by elucidating the definition and characteristics of microplastics, followed by an exploration of the environmental issues stemming from their proliferation. Subsequently, we underscore the significance of employing LCA as a tool for

DOI: 10.1201/9781032706573-18

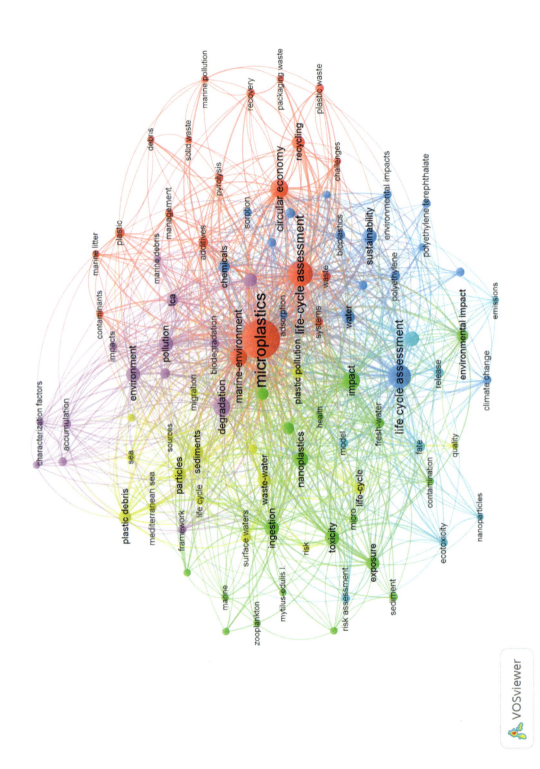

FIGURE 18.1 Keyword network analysis on "Microplastics, Life cycle assessment"-related articles from 2018 to 2024 found in Web of Science (WoS) database.

Life-cycle Assessment of Microplastics in the Environment

comprehensively understanding the impacts of microplastics on ecosystems and human well-being. Through a synthesis of existing literature and case studies, we aim to provide insights into the potential benefits and challenges of applying LCA to address the complex environmental challenges posed by microplastics.

18.2 SOURCES, FATE, AND TRANSPORT OF MICROPLASTICS

Microplastics, ubiquitous in various environmental compartments, originate from a multitude of sources, categorized as primary and secondary. Primary sources include the deliberate addition of microplastics in personal care products, industrial abrasion, and plastic resin pellets used in manufacturing processes (Beiras, R., Schonemann, 2020; Filho et al., 2021). Secondary sources involve the fragmentation of larger plastic debris due to weathering, UV radiation, and mechanical action, such as wave action and abrasion against coastal rocks (Gnaim et al., 2021; Jeong et al., 2023).

The pathways through which microplastics enter the environment are diverse and interconnected. In aquatic environments, wastewater effluents from domestic, industrial, and agricultural activities serve as significant conduits for microplastic discharge into rivers, lakes, and oceans (Li et al., 2020; Ncube et al., 2021). Atmospheric deposition is another pathway, with microplastics being transported through air currents and deposited onto terrestrial and aquatic surfaces (Li et al., 2023). Additionally, the use of plastic mulches in agriculture and the application of plastic-based products in landscaping contribute to soil contamination (Oliveira et al., 2020).

Once released into the environment, microplastics undergo complex fate and transport processes influenced by their physical and chemical properties. In aquatic systems, buoyant microplastics can be transported long distances by ocean currents, while denser particles may settle and accumulate in sediments (Pete et al., 2022). Surface waters act as reservoirs, with microplastics becoming entrapped in floating debris patches or dispersed across vast oceanic regions (Rubin and Zucker, 2022). In terrestrial environments, microplastics can migrate through soil profiles via infiltration and runoff, potentially reaching groundwater reservoirs (Shamskhany et al., 2021).

Understanding the fate and transport of microplastics in various environmental compartments is crucial for comprehensively assessing their environmental impacts and developing targeted mitigation measures. Microplastics exhibit diverse behaviors depending on factors such as size, shape, density, and surface properties, which influence their movement and distribution across different environmental matrices. For instance, buoyant microplastics tend to remain suspended in water bodies, where they can be transported over long distances by ocean currents, posing risks to marine life and ecosystems (Alvarado et al., 2020). Conversely, denser microplastics may settle in sediments, where they can persist for extended periods and potentially serve as long-term sources of contamination (Ziajahromi et al., 2018).

Moreover, the fate and transport of microplastics are influenced by complex interactions with biotic and abiotic factors. Biological processes, such as biofilm formation and ingestion by organisms, can affect the behavior and fate of microplastics in aquatic and terrestrial ecosystems (Sullivan et al., 2021). Additionally, physical processes such as advection, diffusion, and sedimentation play critical roles in determining the spatial distribution and temporal dynamics of microplastics in the environment (van Raamsdonk et al., 2020).

To unravel the complexities of microplastic dynamics, researchers employ a multidisciplinary approach that integrates data from field observations, laboratory experiments, and modeling studies. Field observations provide valuable insights into the real-world distribution and abundance of microplastics across different environmental compartments, while laboratory experiments enable controlled investigations into specific processes such as degradation, aggregation, and bioaccumulation (Gasperi et al., 2018). Furthermore, mathematical models, including numerical simulations and statistical analyses, help elucidate the underlying mechanisms driving microplastic fate and transport, allowing researchers to predict future trends and assess the efficacy of mitigation strategies (Zheng and Suh, 2019).

By synthesizing information from diverse sources and disciplines, researchers can generate a comprehensive understanding of microplastic dynamics and their environmental implications. This knowledge forms the basis for evidence-based decision-making and the development of sustainable management practices aimed at mitigating the impacts of microplastics on ecosystems and human health.

18.3 ECOLOGICAL AND HUMAN HEALTH IMPACTS OF MICROPLASTICS

Microplastics pose significant ecological and human health risks due to their widespread distribution and persistence in the environment. This section provides a comprehensive review of the potential impacts of microplastics on ecosystems and human health, focusing on ingestion, bioaccumulation, and associated effects.

18.3.1 ECOLOGICAL IMPACTS

Microplastics are ingested by a wide range of organisms across various trophic levels, including zooplankton, fish, and marine mammals (Parashar and Hait, 2020). Upon ingestion, microplastics can cause physical harm, such as blockage of the digestive tract, leading to reduced feeding efficiency and nutrient absorption (Sparks et al., 2021). Furthermore, microplastics can act as vectors for the transport of chemical contaminants and pathogens, potentially exacerbating their toxic effects on marine life (Sommer et al., 2018). Additionally, microplastics may serve as substrates for the colonization of harmful microorganisms, altering microbial communities and ecosystem functioning (Ebner and Iacovidou, 2021).

Bioaccumulation of microplastics occurs when organisms ingest and retain these particles in their tissues over time (Cox et al., 2020). This bioaccumulation can lead to biomagnification of microplastics and associated contaminants up the food chain, posing risks to higher trophic levels, including humans (Cox et al., 2019). Moreover, the presence of microplastics in sediments can alter sediment properties and benthic habitats, impacting the biodiversity and ecological balance of marine and freshwater ecosystems (Chen et al., 2018).

18.3.2 HUMAN HEALTH IMPACTS

The potential human health impacts of microplastics have garnered increasing attention in recent years. Humans may be exposed to microplastics through ingestion of contaminated food and water, as well as inhalation of airborne particles (Accinelli et al., 2019). Microplastics can adsorb and concentrate chemical pollutants, including persistent organic pollutants (POPs) and heavy metals, which may leach into tissues upon ingestion, posing risks to human health. Furthermore, the small size of microplastics allows them to translocate across biological barriers and accumulate in organs and tissues, potentially causing inflammatory responses and cellular damage (Woods et al., 2016; 2019).

In summary, microplastics pose significant ecological and human health risks due to their widespread distribution and persistence in the environment. Understanding the impacts of microplastics on ecosystems and human health is essential for informing evidence-based management strategies and promoting sustainable practices. Continued research and monitoring efforts are needed to elucidate the full extent of these impacts and develop effective mitigation measures to protect both the environment and human well-being.

18.4 LIFE CYCLE ASSESSMENT (LCA) METHODOLOGIES

Life cycle assessment (LCA) is a comprehensive methodology used to evaluate the environmental impacts of products or systems throughout their entire life cycle. This section provides an in-depth exploration of LCA principles, methodologies, and applications, focusing on its relevance to the assessment of microplastics in the environment.

Life-cycle Assessment of Microplastics in the Environment

18.4.1 Principles and Methodologies of LCA

LCA follows a systematic approach to quantify the environmental impacts associated with all stages of a product's life cycle, including raw material extraction, manufacturing, use, and disposal (UNEP, 2018). The methodology is guided by four main stages: goal and scope definition, life cycle inventory (LCI) analysis, life cycle impact assessment (LCIA), and interpretation (UNEP, 2018). Goal and scope definition involve defining the purpose of the study, specifying the boundaries, and selecting relevant impact categories for assessment. LCI analysis entails compiling data on energy, material inputs, and emissions associated with each life cycle stage. LCIA involves assessing the potential environmental impacts of these inputs using impact assessment methods and characterization models. Finally, interpretation involves synthesizing the results, identifying hotspots, and drawing conclusions to support decision-making.

18.4.2 Types of LCA and Their Applications

Several types of LCA exist, each tailored to address specific questions or objectives. The most common types include attributional LCA, consequential LCA, and hybrid LCA (Lavoie et al., 2021). Attributional LCA focuses on quantifying the environmental impacts directly attributed to a product or system under study, providing insights into its current environmental performance. Consequential LCA, on the other hand, considers the broader system-wide impacts of changes in product demand or production, allowing for the assessment of downstream effects and potential trade-offs. Hybrid LCA integrates elements of both attributional and consequential approaches, combining detailed process-based analysis with broader economic and environmental modeling. Table 18.1 outlines the different types of LCA and their various applications.

TABLE 18.1
Types of LCA and Their Applications

Type of LCA	Description	Applications
Attributional LCA	• Focuses on quantifying the direct environmental impacts of a product or system, typically based on process-based modeling and inventory data • Assesses the environmental burdens associated with specific processes and inputs throughout the life cycle	• Assessing the environmental performance of specific products • Comparing different manufacturing processes to identify hotspots • Informing eco-design strategies for improving product sustainability
Consequential LCA	• Considers the broader system-wide impacts of changes in product demand or production, incorporating dynamic modeling of market interactions and feedback effects • Evaluates the indirect or downstream environmental consequences of decisions and actions, including market effects, technology substitution, and rebound effects	• Evaluating the environmental implications of policy interventions • Assessing the environmental consequences of changes in consumption patterns • Analyzing the environmental trade-offs associated with different scenarios or strategies
Hybrid LCA	• Integrates elements of both attributional and consequential approaches, combining detailed process-based analysis with broader economic and environmental modeling • Provides a comprehensive assessment of product life cycles, capturing both direct and indirect impacts while considering system-wide effects	• Providing a holistic view of environmental impacts, considering both upstream and downstream effects • Supporting decision-making by considering the full range of environmental consequences • Incorporating dynamic modeling to assess the long-term sustainability of products and systems

LCA methodologies play a crucial role in evaluating the environmental impacts of microplastics throughout their life cycle. By applying rigorous principles and methodologies, researchers can quantify the environmental footprint of microplastics and identify opportunities for mitigation and sustainable management. Continued advancements in LCA techniques and data availability will further enhance our understanding of microplastic impacts and support evidence-based decision-making for environmental protection.

18.5 APPLICATION OF LCA TO MICROPLASTICS

LCA is a robust methodology employed to assess the environmental impacts of products or systems, offering insights into their entire life cycle, from raw material extraction to end-of-life disposal. This comprehensive approach allows researchers to quantify various environmental burdens, including energy consumption, resource depletion, emissions, and waste generation. When applied to microplastics, LCA provides a systematic framework for evaluating their environmental footprint across different stages of their life cycle, from production to disposal.

In this section, we delve into the intricacies of applying LCA to microplastics, shedding light on the data sources, assumptions, and methodologies used in such assessments. LCA studies on microplastics draw upon a wide array of data, including information on the production processes of microplastic-containing products, their usage patterns, and the pathways through which microplastics enter the environment. These data are essential for accurately characterizing the environmental impacts associated with microplastics and guiding informed decision-making. Additionally, LCA analyses often rely on assumptions regarding the fate and behavior of microplastics in various environmental matrices, such as air, water, and soil. Assumptions may encompass factors such as the rate of microplastic degradation, their propensity for bioaccumulation, and their interactions with biotic and abiotic components of ecosystems (Civancik-Uslu et al., 2019). By integrating these data and assumptions, LCA studies provide valuable insights into the environmental implications of microplastic pollution, informing strategies for mitigation and management.

18.5.1 CHALLENGES ASSOCIATED WITH LCA STUDIES ON MICROPLASTICS

Despite the potential benefits of LCA in understanding the environmental impacts of microplastics, several challenges exist:

1. *Data availability*: One of the primary challenges is the limited availability of data on microplastics, particularly regarding their production volumes, distribution, and environmental concentrations. Data gaps hinder the accurate quantification of microplastic emissions and their subsequent environmental impacts.
2. *Uncertainty*: LCA studies on microplastics often face uncertainties due to the complex nature of microplastic behavior in the environment. Uncertainties may arise from factors such as variability in microplastic properties, environmental conditions, and the lack of standardized methodologies for measuring microplastic concentrations and impacts.
3. *Modeling complexity*: The dynamic and interconnected nature of microplastic fate and transport in different environmental compartments adds complexity to LCA modeling. Developing accurate models that account for the interactions between microplastics and environmental factors requires interdisciplinary collaboration and advanced modeling techniques.
4. *Temporal and spatial variability*: Microplastic distribution and impacts vary temporally and spatially, posing challenges for LCA studies. Accounting for temporal and spatial variability

Life-cycle Assessment of Microplastics in the Environment 327

in microplastic concentrations and impacts requires robust sampling strategies and modeling approaches.

5. *Toxicity assessment*: Assessing the toxicity of microplastics and their effects on ecosystems and human health remains challenging. Integrating toxicity data into LCA studies requires standardized toxicity testing protocols and comprehensive risk assessment frameworks.

18.6 IMPLICATIONS OF LCA STUDIES FOR MICROPLASTICS MANAGEMENT AND POLICY

LCA studies play a crucial role in informing decision-making processes related to microplastics management and policy development. Table 18.2 outlines the challenges associated with LCA studies on microplastics. This section examines the potential benefits and limitations of LCA studies in this context, along with an overview of current policies and regulations governing microplastics and the role of LCA in promoting sustainable practices and products.

18.6.1 POTENTIAL BENEFITS OF LCA STUDIES

LCA provides a comprehensive framework for assessing the environmental impacts of microplastics across their entire life cycle, aiding in the identification of hotspots and informing targeted interventions. By quantifying the environmental footprint of microplastics, LCA studies facilitate evidence-based decision-making for policymakers, enabling the prioritization of mitigation

TABLE 18.2
Challenges Associated With LCA Studies on Microplastics

Challenge	Description	Implications	Potential Solutions
Data availability	Limited availability of data on microplastics production, distribution, and environmental concentrations	Difficulty in accurately quantifying microplastic emissions and their environmental impacts	Collaborate with stakeholders to improve data collection efforts Standardize data reporting protocols
Uncertainty	Uncertainties in LCA studies due to the complex behavior of microplastics in the environment	Reduced confidence in LCA results and recommendations	Conduct sensitivity analyses to assess the impact of uncertainty on study outcomes
Modeling complexity	Challenges in developing accurate models that account for the interactions between microplastics and environmental factors	Increased complexity and computational demands of LCA modeling	Integrate advanced modeling techniques, such as machine learning, to improve model accuracy
Temporal and spatial variability	Variability in microplastic distribution and impacts over time and space, requiring robust sampling strategies and modeling approaches	Difficulty in capturing temporal and spatial dynamics of microplastics in LCA studies	Implement longitudinal monitoring programs to track microplastic distribution and impacts over time
Toxicity assessment	Difficulty in assessing the toxicity of microplastics and their effects on ecosystems and human health	Incomplete understanding of the ecological and human health risks associated with microplastics	Develop standardized toxicity testing protocols and comprehensive risk assessment frameworks

measures and resource allocation. Additionally, LCA can help identify opportunities for improving the sustainability of microplastic-containing products and processes, guiding innovation and product design toward more eco-friendly alternatives.

18.6.2 Limitations of LCA Studies

Despite its utility, LCA studies on microplastics encounter several challenges and limitations. These include data gaps and uncertainties regarding the fate and behavior of microplastics in the environment, as well as limitations in modeling complex environmental interactions. Furthermore, the dynamic nature of microplastic pollution presents challenges in accurately capturing its full life cycle impacts, necessitating ongoing refinement and validation of LCA methodologies.

18.6.3 Overview of Current Policies and Regulations

The regulatory landscape surrounding microplastics varies globally, with some regions implementing specific measures to address their environmental impacts. Policies range from bans on certain types of microplastics in consumer products to regulations governing their discharge into the environment. However, gaps and inconsistencies exist in the regulatory framework, highlighting the need for harmonization and strengthening of regulations across jurisdictions (Boulay et al., 2021).

18.6.4 Role of LCA in Promoting Sustainable Practices

LCA can inform the development and implementation of policies aimed at reducing microplastic pollution, providing valuable insights into the environmental consequences of different management strategies. By quantifying the environmental benefits of alternative materials and waste management practices, LCA contributes to the design of more sustainable solutions. Additionally, LCA studies can guide the establishment of eco-labeling schemes and certification programs, empowering consumers to make informed choices and incentivizing the adoption of environmentally friendly products and practices. Table 18.3 provides an overview of current policies and regulations on microplastics.

18.7 CONCLUSION

In conclusion, this chapter has provided a comprehensive overview of the LCA of microplastics in the environment. We began by defining microplastics and highlighting their ubiquitous presence in various environmental compartments, emphasizing the importance of understanding their sources, fate, and transport. Through the application of LCA methodologies, researchers can assess the environmental impacts of microplastics throughout their entire life cycle, from production to disposal. We explored the ecological and human health impacts of microplastics, including their ingestion by organisms, bioaccumulation, and potential transfer through food webs. The discussion underscored the need for robust data and assumptions to accurately quantify these impacts and inform evidence-based decision-making. Furthermore, we examined the challenges associated with LCA studies on microplastics, such as data availability and uncertainty, emphasizing the importance of addressing these challenges to improve the reliability and accuracy of LCA results.

Moving forward, there is a clear call to action for continued research and application of LCA methodologies to better understand the impacts of microplastics on the environment and human health. By addressing knowledge gaps and refining LCA methodologies, we can enhance our ability to assess and mitigate the environmental footprint of microplastics, ultimately contributing to more sustainable management practices and policies. In summary, the integration of LCA into microplastics research offers valuable insights into their environmental impacts and supports

Life-cycle Assessment of Microplastics in the Environment

TABLE 18.3

Overview of Current Policies and Regulations on Microplastics

Policy/Regulation	Description	Role of LCA
Ban on microbeads	Prohibition of microplastic beads in personal care products	Assessing alternative materials and evaluating the environmental impacts
Restrictions on plastic bags	Limitations on the use of single-use plastic bags	Comparing the environmental performance of different bag materials and waste management options
Water quality standards	Standards for microplastic levels in water bodies	Supporting risk assessment and management efforts by quantifying environmental impacts
Microplastic pollution control	Measures to mitigate microplastic pollution in aquatic environments	Evaluating the effectiveness of pollution control measures and identifying opportunities for improvement
Product labeling requirements	Mandates for labeling products containing microplastics	Facilitating consumer awareness and informed decision-making by providing information on microplastic content and environmental impact
Extended producer responsibility (EPR)	Policies requiring producers to take responsibility for the end-of-life management of their products, including microplastics	Assessing the environmental performance of different product designs and waste management strategies
Marine debris action plans	Strategies to address marine debris, including microplastics	Informing the development of targeted interventions and resource allocation for marine conservation efforts
Coastal cleanup initiatives	Programs aimed at removing litter, including microplastics, from coastal areas	Assessing the efficacy of cleanup efforts and identifying sources and pathways of microplastic pollution
Wastewater treatment regulations	Standards for the treatment of wastewater to reduce microplastic discharge	Evaluating the environmental benefits of different treatment technologies and guiding investment decisions

informed decision-making for the protection of ecosystems and human health. It is imperative that we continue to advance our understanding of microplastics through interdisciplinary research and collaborative efforts, with a focus on promoting environmental sustainability and safeguarding future generations from the adverse effects of plastic pollution.

REFERENCES

Accinelli, C., Abbas, H.K., Shier, W.T., Vicari, A., Little, N.S., Aloise, M.R., Giacomini, S., 2019. Degradation of microplastic seed film-coating fragments in soil. *Chemosphere* 226, 645–650. https://doi.org/10.1016/j.chemosphere.2019.03.161

Akhbarizadeh, R., Dobaradaran, S., Torkmahalleh, M.A., Saeedi, R., Aibaghi, R., Ghasemi, F.F., 2021b. Suspended fine particulate matter (PM2.5), microplastics (MPs), and polycyclic aromatic hydrocarbons (PAHs) in air: their possible relationships and health implications. *Environ. Res.* 192, 110339 https://doi.org/10.1016/j.envres.2020.110339.

Alvarado Chacon, F., Brouwer, M.T., Thoden van Velzen, E.U., 2020. Effect of recycled content and rPET quality on the properties of PET bottles, part I: optical and mechanical properties. *Packag. Technol. Sci.* 33, 347–357. https://doi.org/10.1002/pts.2490.

Beiras, R., Schonemann, A.M., 2020. Currently monitored microplastics pose negligible ecological risk to the global ocean. *Sci. Rep.* 10, 1–9. https://doi.org/10.1038/s41598-020-79304-

Besseling, E., Redondo-Hasselerharm, P., Foekema, E.M., Koelmans, A.A., 2018. Quantifying ecological risks of aquatic micro- and nanoplastic. *Crit. Rev. Environ. Sci. Technol.* 49, 32–80. https://doi.org/10.1080/10643389.2018.1531688.

Borrelle, S.B., Ringma, J., Law, K.L., Monnahan, C.C., Lebreton, L., McGivern, A., Murphy, E. et al., 2020. Predicted growth in plastic waste exceeds efforts to mitigate plastic pollution. *Science* 369 (6510), 1515–1518.

Boulay, A.M., Verones, F., Vázquez-Rowe, I., 2021. Marine plastics in LCA: current status and MarILCA's contributions. *Int J Life Cycle Assess* 26, 2105–2108. https://doi.org/10.1007/s11367-021-01975-1

Chen, M., Jin, M., Tao, P., Wang, Z., Xie, W., Yu, X., Wang, K., 2018. Assessment of microplastics derived from mariculture in Xiangshan Bay, China. *Environ. Pollut.* 242, 1146–1156. https://doi.org/10.1016/j.envpol.2018.07.133.

Civancik-Uslu, D., Puig, R., Hauschild, M., Fullana-i-Palmer, P., 2019. Life cycle assessment of carrier bags and development of a littering indicator. *Sci Total Environ* 685, 621–630. https://doi.org/10.1016/j.scitotenv.2019.05.372

Cox, K.D., Covernton, G.A., Davies, H.L., Dower, J.F., Juanes, F., Dudas, S.E., 2019. Human consumption of microplastics. *Environ. Sci. Technol.* 53 (12), 7068–7074. https://doi.org/10.1021/acs.est.9b01517.

Cox, K.D., Covernton, G.A., Davies, H.L., Dower, J.F., Juanes, F., Dudas, S.E., 2020. Correction to human consumption of microplastics. *Environ. Sci. Technol.* 54 (17), 10974. https://doi.org/10.1021/acs.est.0c04032.

Ebner, N., Iacovidou, E., 2021. The challenges of Covid-19 pandemic on improving plastic waste recycling rates. *Sustain. Prod. Consum.* 28, 726–735. https://doi.org/10.1016/j.spc.2021.07.001.

Filho, W.L., Hunt, J., Kovaleva, M., 2021. Garbage patches and their environmental implications in a plastisphere. *J. Mar. Sci. Eng.* https://doi.org/10.3390/jmse9111289.

Gandhi, N., Farfaras, N., Wang, N.H.L., Chen, W.T., 2021. Life cycle assessment of recycling high-density polyethylene plastic waste. *J. Renew. Mater.* 9, 1463–1483. https://doi.org/10.32604/jrm.2021.015529

Gao, F., Shen, Y., Sallach, J.B., Li, H., Liu, C., Li, Y., 2021. Direct prediction of bioaccumulation of organic contaminants in plant roots from soils with machine learning models based on molecular structures. *Environ. Sci. Technol.* 55 (24), 16358–16368. https://doi.org/10.1021/acs.est.1c02376

Gasperi, J., Wright, S.L., Dris, R., Collard, F., Mandin, C., Guerrouache, M., Langlois, V., Kelly, F.J., Tassin, B., 2018. Microplastics in air: are we breathing it in? *Curr. Opinion Environ. Sci. Health* 1, 1–5. https://doi.org/10.1016/j.coesh.2017.10.002.

Gnaim, R., Polikovsky, M., Unis, R., Sheviryov, J., Gozin, M., Golberg, A., 2021. Marine bacteria associated with the green seaweed Ulva sp. for the production of polyhydroxyalkanoates. *Bioresour. Technol.* 328, 124815 https://doi.org/10.1016/j.biortech.2021.124815.

Jeong, Y., Gong, G., Lee, H.-J., Seong, J., Hong, S.W., Lee, C., 2023. Transformation of microplastics by oxidative water and wastewater treatment processes: a critical review. *J. Hazard. Mater.* 443, 130313 https://doi.org/10.1016/j.jhazmat.2022.130313

Lavoie, J., Boulay, A.M., Bulle, C., 2021. Aquatic micro- and nano-plastics in life cycle assessment: development of an effect factor for the quantification of their physical impact on biota. *J Ind Ecol.* 26, 2123–2135.

Li, R., Tao, J., Huang, D., Zhou, W., Gao, L., Wang, X., Chen, H., Huang, H., 2023. Investigating the effects of biodegradable microplastics and copper ions on probiotic (Bacillus amyloliquefaciens): toxicity and application. *J. Hazard. Mater.* 443, 130081 https://doi.org/10.1016/j.jhazmat.2022.130081.

Li, S., Wang, P., Zhang, C., Zhou, X., Yin, Z., Hu, T., Hu, D., Liu, C., Zhu, L., 2020. Influence of polystyrene microplastics on the growth, photosynthetic efficiency and aggregation of freshwater microalgae Chlamydomonas reinhardtii. *Sci. Total Environ.* 714, 136767 https://doi.org/10.1016/j.scitotenv.2020.136767.

Ncube, L.K., Ude, A.U., Ogunmuyiwa, E.N., Zulkifli, R., Beas, I.N., 2021. An overview of plastic waste generation and management in food packaging industries. *Recycling* 6, 1–25. https://doi.org/10.3390/recycling6010012

Oliveira, J., Belchior, A., da Silva, V.D., Rotter, A., Petrovski, Z., ˇAlmeida, P.L., Lourenço, N.D., Gaudˆencio, S.P., 2020. Marine environmental plastic pollution: mitigation by microorganism degradation and recycling valorization. *Front. Mar. Sci.* 7. https://doi.org/10.3389/fmars.2020.567126.

Parashar, N., Hait, S., 2020. Plastics in the time of COVID-19 pandemic: protector or polluter? *Sci. Total Environ.* 144274. https://doi.org/10.1016/j.scitotenv.2020.144274.

Pete, A.J., Brahana, P.J., Bello, M., Benton, M.G., Bharti, B., 2022. Biofilm formation influences the wettability and settling of microplastics. *Environ. Sci. Technol. Lett.* https://doi.org/10.1021/acs.estlett.2c00728.

Rodríguez, N.B., Formentini, G., Favi, C., Marconi, M., 2021. Environmental implication of personal protection equipment in the pandemic era: LCA comparison of face masks typologies. *Procedia CIRP* 98, 306–311. https://doi.org/10.1016/j.procir.2021.01.108.

Rubin, A.E., Zucker, I., 2022. Interactions of microplastics and organic compounds in aquatic environments: a case study of augmented joint toxicity. *Chemosphere* 289, 133212. https://doi.org/10.1016/j.chemosphere.2021.133212.

Saling, P., Gyuzeleva, L., Wittstock, K., Wessolowski, V., Griesshammer, R., 2020. Life cycle impact assessment of microplastics as one component of marine plastic debris. *Int J Life Cycle Assess* 25 (10), 2008–2026.

Shamskhany, A., Li, Z., Patel, P., Karimpour, S., 2021. Evidence of microplastic size impact on mobility and transport in the marine environment: a review and synthesis of recent research. *Front. Mar. Sci.* 8 https://doi.org/10.3389/fmars.2021.760649.

Sommer, F., Dietze, V., Baum, A., Sauer, J., Gilge, S., Maschowski, C., Gieré, R., 2018. Tire abrasion as a major source of microplastics in the environment. *Aerosol Air Qual. Res.* 18 (8), 2014–2028. https://doi.org/10.4209/aaqr.2018.03.0099.

Sonnemann, G., Valdivia, S., 2017. Medellin declaration on marine litter in life cycle assessment and management. *Int J Life Cycle Assess* 22 (10), 1637–1639.

Sparks, C., Awe, A., Maneveld, J., 2021. Abundance and characteristics of microplastics in retail mussels from Cape Town, South Africa. *Mar. Pollut. Bull.* 166, 112186. https://doi.org/10.1016/j.marpolbul.2021.112186

Stefanini, R., Borghesi, G., Ronzano, A., Vignali, G., 2020. Plastic or glass: a new environmental assessment with a marine litter indicator for the comparison of pasteurized milk bottles. *Int J Life Cycle Assess* 26 (4), 767–784. https://doi.org/10.1007/s11367-020-01804-x.

Sullivan, G.L., Delgado-Gallardo, J., Watson, T.M., Sarp, S., 2021. An investigation into the leaching of micro and nano particles and chemical pollutants from disposable face masks – linked to the COVID-19 pandemic. *Water Res.* 196, 117033 https://doi.org/10.1016/j.watres.2021.117033.

UN Environment., 2018. Addressing marine plastics: a systemic approach – stocktaking report. Notten, P. United Nations Environment Programme. Nairobi, Kenya.

van Raamsdonk, L.W.D., van der Zande, M., Koelmans, A.A., Hoogenboom, R.L.A.P., Peters, R.J.B., Groot, M.J., Peijnenburg, A.A.C.M., Weesepoel, Y.J.A., 2020. Current insights into monitoring, bioaccumulation, and potential health effects of microplastics present in the food chain. *Foods* 9 (1), 72. https://doi.org/10.3390/foods9010072.

Woods, J.S., Rødder, G., Verones, F., 2019. An effect factor approach for quantifying the entanglement impact on marine species of macroplastic debris within life cycle impact assessment. *Ecol Indic* 99, 61–66.

Woods, J.S., Veltman, K., Huijbregts, M.A., Verones, F., Hertwich, E.G., 2016. Towards a meaningful assessment of marine ecological impacts in life cycle assessment (LCA). *Environ Int* 89, 48–61.

Yang, J., Li, R., Zhou, Q., Li, L., Li, Y., Tu, C., Luo, Y., 2021. Abundance and morphology of microplastics in an agricultural soil following long-term repeated application of pig manure. *Environ. Pollut.* 272, 116028. https://doi.org/10.1016/j.envpol.2020.116028.

Zheng, J., Suh, S., 2019. Strategies to reduce the global carbon footprint of plastics. *Nat. Clim. Chang.* 9 (5), 374–378. https://doi.org/10.1038/s41558-019-0459-z.

Zhou, B., Wang, J., Zhang, H., Shi, H., Fei, Y., Huang, S., Barceló, D., 2020. Microplastics in agricultural soils on the coastal plain of Hangzhou Bay, East China: multiple sources other than plastic mulching film. *J. Hazard. Mater.* 388, 121814. https://doi.org/10.1016/j.jhazmat.2019.121814.

Ziajahromi, S., Kumar, A., Neale, P.A., Leusch, F.D., 2018. Environmentally relevant concentrations of polyethylene microplastics negatively impact the survival, growth and emergence of sediment-dwelling invertebrates. *Environ. Pollut.* 236, 425–431. https://doi.org/10.1016/j.envpol.2018.01.094.

Index

A

Absorbed, 19, 38, 44, 75, 88, 89, 105, 153, 195, 235, 238, 239, 249, 250
Absorption, 5, 18, 19, 36, 75, 76, 95, 109, 112, 113, 153, 154, 159, 195, 196, 212, 214, 238, 250, 252, 274, 304, 324
Abyssal, 250
Accumulation, 1, 2, 5, 6, 8, 12, 13, 18–21, 28–30, 35, 39, 43, 73, 75, 89, 95, 124, 167, 174, 175, 180, 182, 189, 195, 209, 211, 221, 229, 230, 236, 238, 246–248, 251, 254, 256, 258, 269, 304, 309, 324
Adhesion, 16, 137, 159, 267, 269, 271, 312
Adhesive, 291
Adsorption, 5, 7, 44, 55, 56, 66, 74, 75, 101, 106, 125, 135, 136, 147, 180, 185, 195, 198, 200, 229, 233, 235, 239, 251, 261, 306, 307, 312, 315
Advance, 290, 297, 329
Agglomeration, 4, 132, 313
Agitation, 70, 128
Air, 1, 5, 15, 28–37, 39–50, 70, 73, 98, 105, 115, 134, 135, 137, 149, 200, 206, 209, 210, 213, 215, 216, 220, 230, 231, 246–248, 250, 252, 255, 256, 303, 323, 326
Airborne microplastic, 28, 29, 34, 37, 43, 49, 73, 74, 149, 192
Alanine, 307
Algae, 28, 29, 34–37, 43, 49, 73, 74, 149, 192
Allochthonous, 234
Amyloid, 314
Anaerobic, 3, 18, 39, 272, 273, 277
Analytical techniques, 28, 29, 34–37, 43, 49, 73, 74, 149, 192
Antarctica, 250, 275, 306
Antiandrogen action, 304
Aquatic organism, 16, 18, 19, 21, 88–90, 98, 189, 192, 195, 200, 211, 228, 247, 251
Arctic, 119, 250, 303
Artificial habitat, 33
Asphyxia, 229
Assessment, 34, 39, 46, 98, 125, 199–202, 293–295, 297–299, 316, 321–327, 329
Atmosphere, 2, 15, 16, 19, 28–35, 37, 39–42, 44, 48–50, 96, 147, 206, 209, 210, 213, 216, 247, 248, 255, 256, 321
Atomic force microscopy, 113, 157, 159, 161, 267
Awareness, 6, 45–48, 69, 80, 98, 182, 201, 216, 219, 221, 257, 294, 299, 316, 329

B

Bacillus species, 18, 135
Bacteria, 15, 18, 21, 90, 135, 137, 138, 148, 199, 231, 233–235, 240, 246, 249, 252, 258, 266, 269, 271, 273, 278–280, 290, 306, 308
Benthic life, 13
Benthic zone, 250
Bioavailability, 5, 19, 21, 75, 229, 233, 240, 247, 248, 254
Biocatalysis, 286, 289, 290, 299
Biodeterioration, 271

Biofouling, 16, 21
Bio fragmentation, 271, 272
Biogas, 14
Biogeochemical, 230
Biological health, 4
Bioplastic, 7, 8, 115, 181, 182, 258, 270, 286–299, 308, 309
Biopolymer, 7, 8, 288–293, 299, 309
Bioreactor, 13, 63, 134–136
Bioremediation, 47, 138, 258, 279, 313
Biota, 16, 74, 75, 98, 109, 125, 129, 149, 167–169, 171, 173, 183, 191, 200, 214, 229, 232, 233, 239, 247, 248, 254, 310
Bioturbation, 15, 16, 174, 230
Buffer, 167, 271
Buoyancy, 21, 40, 73–75, 85, 170, 271, 321

C

Carbon dots, 309
Carcinogenicity, 43, 304
Cellular degeneration, 304
Cellular pathway, 5, 288
Chemical additive, 5, 55, 181, 195
Chemical composition, 76, 94, 96, 106, 113, 114, 116, 117, 129–132, 147, 148, 154–158, 160, 161, 174, 208, 248, 252, 268, 279
Chitinase, 94, 127
Circular plastic economy, 6
Circulation, 15, 18
Clean up, 6, 46–48, 170, 182, 197, 329
Climate change, 2, 8, 39, 42, 239, 286
Coastal area, 17, 28, 32, 88, 90, 192, 193, 329
Cognitive development, 304
Collaboration, 46, 49, 98, 201, 220, 299, 326
Consortia, 138, 229, 266, 268
Construction and demolition, 31
Consumers, 5, 17, 47, 48, 86, 167, 175, 220, 247, 255, 257, 294, 328
Contamination, 3, 6, 8, 14, 30, 34, 36, 44, 46, 48–50, 58, 60, 64, 68–79, 80, 87, 88, 92, 97, 105, 115, 118, 125–127, 131, 137, 149, 156, 161, 167–183, 193, 210, 214, 216–220, 233, 235, 239, 252, 253, 255, 256, 266, 294, 303, 323
Contributor, 65, 69, 87, 170
Cosmic radiation, 32
Cost-effective, 151
Cradle, to, gate, 295, 296
Cradle, to, grave, 295, 296
Cryogenic, 96, 175
Cytotoxicity, 5, 251

D

Database, 110, 113, 155, 157, 189, 190, 286, 287, 321, 322
Degradation, 2, 3, 6–8, 12–15, 17–19, 21, 28, 31–33, 45, 48, 56, 60, 64, 68, 71, 73, 74, 80, 85, 86, 92, 113, 124, 133–135, 137, 138, 147, 152, 154, 157, 167, 170, 172,

333

177, 180, 189, 191, 199, 228, 232, 233, 238, 245, 246, 248, 258, 266–280, 288, 293, 294, 304–307, 309, 312–314, 316, 323, 326
Dehydrogenase, 18, 272, 274
Depolymerization 112, 198, 199, 305–307
Desorption, 36, 75, 97, 106, 133, 147, 314
Detection probability, 95
Diffusion, 39, 47, 148, 314, 323
Discharged, 2, 8, 13, 16, 54, 55, 57, 87, 105, 147, 208, 209, 217
Discussion, 69, 133, 328
Disinfection, 57, 64
Disintegration, 15, 31–33, 73, 132, 248, 273, 279
Dispersion, 12, 13, 30, 32, 37, 40, 49, 73, 80, 151, 191, 229, 248, 250, 269
Disposal practices, 2, 13, 245, 294
Distance, 1, 15, 16, 28, 31, 32, 37, 39–42, 70, 73, 79, 149, 158, 200, 211, 220, 256, 323
DNA, 44, 138, 160, 161, 195, 196, 213, 214, 217, 221
Documented, 1, 16, 88, 195, 201, 267, 269, 270, 274, 275, 278, 311
Domestic, 9, 16, 30, 55, 56, 72, 87, 167, 174, 179, 210, 228, 247, 248, 323
Durability, 1, 4, 12, 13, 33, 39, 105, 206, 247, 266
Dyes, 19, 95, 107, 114, 115, 137, 158, 159, 274
Dynamic membrane, 134

E

Eco-friendly, 8, 48, 280
Ecological impact, 75, 89, 98, 189, 193, 194, 200, 201, 324
Ecological niche, 33, 234
Ecotoxicological, 4, 8, 88, 229, 247
Effluent, 13, 14, 16, 19, 30, 54–65, 87, 134, 137, 174, 192, 246, 256, 323
Effluent treatment plants (ETPs), 87
Electrostatic force, 16
Elemental composition, 116, 117, 154–157, 160, 237, 246
Endocrine disruption, 4, 195, 213, 214
End of life, 258, 294, 298, 321, 326, 329
Energy dispersive X-ray spectroscopy (EDX), 96, 106
Engineered enzyme, 138, 273, 276, 277, 307
Entanglement, 189, 195, 196, 200, 229, 303, 310, 312
Erosion, 2, 12, 15, 73, 75, 168, 182, 192, 210, 249
Estuaries, 1, 17, 90, 169, 170, 172, 188, 192, 218, 250
Extraction techniques, 90, 92, 93, 99
Extrusion, 13, 257

F

Family, 272, 274
Farming film, 246
Fibrosis, 44, 304
Fibrous microplastic, 167, 180
Field study, 97, 176
Film mulching, 180, 246
Floatation, 135, 137
Flow cytometry, 36, 108, 109, 154, 156, 162
Fluorescence microscopy, 114, 115, 128, 157–159, 161
Follicles, 304
Food web, 3, 18, 89, 90, 175, 189, 195, 196, 235, 245, 247, 250, 252, 255

Fourier transform infrared spectroscopy (FTIR), 36–38, 58, 95, 99, 106, 109, 132, 149, 153
Fragmentation, 8, 12, 13, 29, 31, 32, 41, 68, 71–73, 86, 124, 147, 148, 155, 189, 191, 192, 199, 229, 278, 321, 323
Froth flotation, 137
Fungi, 18, 135, 137, 138, 199, 266–268, 271, 273, 308

G

Gas chromatography, 36–8, 96, 97, 110, 111, 148, 154, 157, 179
Genome, 292
Genotoxicity, 5, 43, 236
Governmental policies, 29
Government bodies, 215, 221
Greenhouse gases (GHGs), 2, 44, 232, 233
Groundwater reservoirs, 323
Gyres, 170, 192

H

Health impacts, 12, 13, 46, 99, 152, 195, 199, 201, 202, 211, 212, 214, 293, 324, 328
Heterogenous, 13, 42, 69, 124, 126, 132, 133, 137, 233
High-density polyethylene (HDPE), 132, 256
High-temperature gel, permeation chromatography (HT-GPC), 97
Human exposure, 5, 42, 43, 49, 50, 90, 149, 195
Humidity, 40, 73, 77, 235, 277, 278
Hydrodynamic, 16, 99, 171, 173, 183, 191
Hydrolase, 138, 233, 275, 306
Hydrophobicity, 13, 18, 21, 74, 137, 155, 248, 267, 268, 271, 278, 279, 307, 314
Hyporheic zone, 17

I

Immunological damage, 4
Impairment, 19, 195, 196, 229, 247, 304
Indicator, 35, 111, 174, 232, 233, 254, 297, 298
Indoor, 3, 8, 34, 40, 149, 247, 255, 256
Inductively coupled plasma spectroscopy (ICP-MS), 110, 111
Infiltrate, 31, 55, 68, 76, 87, 124, 228, 230
Inflammation, 5, 19, 42–44, 76, 88, 195, 196, 212–214, 304
Innovation, 49, 113, 199, 219, 295, 297, 299, 311, 312, 314, 328
Instrumentation, 8, 105
Insulation, 31, 147
Interdisciplinary, 199, 201, 326, 329
Intestinal absorption, 5
Invertebrates, 18, 127, 148, 173, 175, 192, 195, 196, 200, 229, 249, 266, 270, 273
Isotopes, 110, 160

L

Lactase, 268
Lagoons, 54, 57, 59, 60, 62, 63, 65, 95, 250
Landfill, 12, 16, 69, 85, 138, 147, 266, 303, 309
Leachate, 2, 14, 16
Leukocytes, 304

Index

Life cycle assessment, 294, 295, 298, 321, 322, 324
Lightweight, 1, 105, 303
Low-density polyethylene (LDPE), 115, 135, 138, 174, 177, 182, 231, 237, 256
Luminescence, 238

M

Magnetic separation, 135, 136, 303, 314, 316
Management strategies, 45, 65, 86, 215, 220, 266, 324, 328, 329
Mangrove forest, 167, 175, 177, 183
Marine debris, 3, 48, 90, 180, 329
Marine organism, 4, 5, 12, 18, 19, 40, 88, 91, 173, 192, 195, 208, 216, 249, 321
Mechanical abrasion, 15, 32, 85, 191, 247
Membrane treatment, 6
Mercury cadmium telluride (MCT), 132
Meteorological factors, 16, 31, 33, 34, 78
Microbeads, 6, 12, 13, 28, 29, 41, 42, 47, 48, 86, 87, 97, 107, 135, 137, 181, 191, 199, 202, 207, 213, 215, 232, 233, 257, 269, 309, 315, 329
Microbial community, 39, 230, 234, 235
Microbial degradation, 6, 18, 33, 191, 199, 266, 267, 278
Microfibers, 14, 15, 69, 136, 167, 179, 181, 207, 208, 210, 211, 213, 246, 315
Microfiltration, 46, 63, 133, 135, 196, 315
Microfluidics, 160
Microscopy, 4, 34, 36, 38, 49, 94, 96, 99, 106–110, 113–115, 124, 127–130, 148, 149, 157–162, 178, 267, 274
Mineralisation, 231, 271, 273
Mitigation method, 8, 48, 133, 134
Moldable, 147
Mudskipper, 173–175, 178, 180, 181
Mulch film, 3, 31, 71, 72, 291, 306
Municipal waste, 2, 8, 54, 56, 57, 61, 65

N

Nano coils, 311
Nanocomposite, 7, 288, 306, 308, 312
Nano filter, 303, 315
Nanorods, 313
Nanotechnology, 7, 8, 46, 118, 198, 200, 286, 288, 293, 299, 303, 304, 306, 311
Nano thermal analysis, 114
Nanotubes, 136, 305, 308, 311, 312, 314
Nanozymes, 124, 314
National institute of standard technology (NIST), 48
Nuclear magnetic resonance, 267

O

Oceanographic sea storms, 16
Outdoors, 3, 34, 39, 40, 149, 255, 256
Oxidative digestion, 94
Oxidative stress, 4, 5, 19, 42, 44, 195, 196, 214, 229, 236, 247, 251–254, 256, 258, 304, 310

P

Peroxidase, 138, 237, 251, 266, 268, 269, 271, 274, 314
Persistence, 13, 43, 46, 48, 73, 75, 200, 248, 324

Pervasive impacts, 68, 85
PET, 84, 105, 113, 135, 138, 155, 157, 161, 170, 172, 173, 200, 207, 216, 248, 249, 254, 255, 267, 269, 272, 275–277, 279, 306, 307, 310
Photocatalyst, 198, 200, 312, 313
Photodetector, 152
Physiology, 1, 68, 89, 195, 291, 292
Placenta, 304, 311
Plankton, 30, 88, 91, 93, 94, 125, 126, 149, 177, 229, 247, 248, 271, 324
Plastic pellets, 2, 3, 13, 29, 69, 71, 86, 246, 309
Plastic polymer matrix, 5
Plastisphere, 233, 234, 270, 271
Pneumatophores, 170, 171
Policymakers, 124
Polychlorinated biphenyl (PCB), 4, 5, 19, 239, 256
Polychlorinated phenols, 4, 19, 239, 256
Polycyclic aromatic hydrocarbons (PAH), 4, 43, 217, 218, 239, 256
Polymerization, 117, 286, 289, 305, 308
Polyvinyl chloride, 55, 62, 105, 135, 147, 170, 213, 216, 231, 234, 248, 249, 255, 266, 269–271, 279, 310
Precipitation, 15, 16, 39, 42, 73, 180, 248, 313
Pre-treatment, 58, 65, 93, 96, 113, 114, 116, 117, 126, 128, 131, 134, 136, 138, 150, 156, 159, 161, 269, 313
Producer, 5, 18, 45, 124, 167, 170, 201, 202, 210, 235, 252, 258, 275, 329
Protease, 43, 94, 198, 199, 275
Pyrolysis, 36–38, 96, 111, 132, 154, 157, 179, 307, 309

Q

Quality assurance, 34, 97
Quantification techniques, 2, 99
Quantitative assessment, 19

R

Raman spectroscopy, 4, 36–38, 58, 96, 97, 99, 106, 109, 110, 113, 116, 117, 127, 128, 130, 132, 149, 154–158, 178, 179
Reactive oxygen species (ROS), 43, 55, 137, 198, 214, 239, 251
Reductase, 251
Regulatory measures, 45
Regulatory tool, 257
Remediation, 6, 189, 196, 198–202, 266, 304, 311, 312, 315
Reproductive toxicity, 43
Residence time, 15, 16, 230
Resilience, 40, 228
Resistance, 1, 12, 97, 134, 239
Respiratory roots, 170
Riparian, 17

S

Sampling techniques, 34–36, 49, 90, 91, 99, 149
Satellite image, 46
Scanning electron microscopy (SEM), 4, 37–39, 96, 106, 108, 127, 129, 149, 157, 161, 178, 267, 274
Scavenging, 37, 42
Sedimentation, 35, 42, 56, 57, 64, 98, 136, 137, 191, 192, 196, 198, 200, 323

Index

Shredding, 70
Siphoning, 198
Solar flux, 31
Solvothermal methods, 309
Sonication, 200
Source reduction, 45, 46, 98, 219, 311
Spatiotemporal, 17
Supernatant, 127, 178
Suspended particles, 33
Sustainable environment, 8, 313
Synergistic, 180, 198, 200, 268, 296
Synergistic effects, 180, 198, 200
Synthetic polymer, 7, 94, 115, 136, 206, 279, 280
Synthetic textiles, 2, 14, 15, 29, 189, 199, 209, 250, 309, 321

T

Terrestrial, 1, 8, 13–16, 21, 28, 39, 40, 42, 44, 68–80, 98, 106, 147, 167, 168, 170, 173, 179, 181, 182, 189, 192, 206, 207, 210, 211, 217, 219, 220, 228, 229, 233, 235–239, 245, 248–251, 268, 269, 286, 303, 323
Textiles, 2, 14, 15, 29, 86, 179, 189, 191, 199, 209, 210, 220, 246, 250, 257, 291, 309, 321
Thermogravimetric techniques, 97
Three-D printing, 47
Threshold, 197, 278
Topography, 21, 39, 40, 159, 161, 233
Transboundary, 257
Transmission electron microscopy (TEM), 108, 129, 157, 160
Transpiration, 76, 238
Transportation, 2, 3, 8, 13, 17, 21, 28, 31, 33, 34, 39, 41, 43, 49, 175, 180, 182, 238, 298
Trophic structure, 217
Tropic level, 5, 6, 16, 18, 30, 89, 105, 125, 167, 175, 189, 196, 200, 218, 245, 247, 252, 258, 321, 324
Turbulence, 15, 18, 32, 37, 39, 42, 73, 74, 172

U

Ubiquitous, 3, 19, 55, 85, 170, 173, 193, 245, 247, 321, 323, 328
Ultimate fate, 12, 13, 193
Ultrasonication, 94
Uncertainties, 61, 245, 294, 326–328
Unequivocal, 13

Unsaturated structure, 17
Urbanization, 37, 55
Urban population, 18
Urban runoff, 13, 86, 87, 192
Urban settings, 32
Urban topography, 40
UV radiation, 17, 32, 40, 78, 85, 111, 138, 180, 191, 208, 323

V

Vehicular traffic, 15
Ventilation, 88, 177
Versatility, 12, 289
Vertebrates, 15, 18, 148, 173, 174, 249, 270
Vertical mixing, 15, 17
Visual identification, 58, 77, 95, 97, 99, 107, 108, 114, 124
Vulnerable, 18, 33, 168, 174, 175, 195, 201, 240, 253

W

Waste management, 2, 6, 14, 45, 69, 70, 75, 98, 139, 167, 179, 182, 201, 209, 211, 213–216, 219, 220, 257, 258, 279, 294, 303, 328, 329
Wastewater treatment plants (WWTPs), 13, 14, 16, 54–62, 64, 65, 134–137
Water holding capacity, 4, 229, 230, 232
Wavelength, 36, 109, 114, 115, 153, 159
Weathering, 3, 12, 13, 15, 17, 18, 21, 28, 29, 32, 39, 41, 130, 147, 171, 209, 246, 323
Wetland, 16, 17, 95, 170, 174, 249
Wildlife, 44, 47, 168, 169, 206, 212, 216
Worldwide, 1, 3, 8, 9, 19, 28, 39, 50, 85, 109, 189, 206, 220, 230, 233, 245, 257, 266, 303

X

XPS, 154–156
X-ray photoelectron spectroscopy, 155
XRF, 36
Xylem, 238

Z

Zeolite, 314
Zero plastic, 6
Zooplanktonic, 18

9781032706535